SKILLS LINK

Everyday Mathematics®

Mc
Graw
Hill
Education

www.everydaymath.com

Send all inquiries to:
McGraw-Hill Education
8787 Orion Place
Columbus, OH 43240

ISBN: 978-0-07-672762-9
MHID: 0-07-672762-9

Printed in the United States of America.

4 5 6 7 QVS 21 20 19 18

Contents

Review

Grade 3 Review 1

Practice Sets

Unit 1

Practice Sets 1–13 11

Unit 2

Practice Sets 14–26 25

Unit 3

Practice Sets 27–39 38

Unit 4

Practice Sets 40–52 51

Unit 5

Practice Sets 53–65 65

Unit 6

Practice Sets 66–78 78

Unit 7

Practice Sets 79–91 91

Unit 8

Practice Sets 92–104 104

Test Practice

Test Practice 117

Grade 3 Review: Operations and Algebraic Thinking

Complete the frames-and-arrows diagram.

①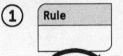

_____

② Describe the number pattern below. Write a number sentence that shows how to find the next number in the pattern.

0, 1, 1, 2, 3, 5, 8, 13

_____

_____

③ Bria earns $5.00 each week for helping around the house. Each week she buys a magazine that costs $3.00. She saves the rest of her money. Write a number sentence that shows how much money Bria will have saved after 4 weeks. Use the letter *m* for the unknown. Then solve.

Number sentence: _____   Answer: _____

_____

Tell whether each number sentence is true or false.

④ $(6 \times 3) \times 5 = (5 \times 6) \times 3$ _____   ⑤ $9 \times 7 = 7 \times 8$ _____

_____

⑥ Five children each bring 3 cans of food for the food drive. Draw a picture and write a number sentence to show how many cans of food the children brought all together.

Number sentence: _____

Grade 3 Review: Operations and Algebraic Thinking
(continued)

Write the fact families.

⑦ 54, 6, 9

_____ × _____ = _____

_____ × _____ = _____

_____ ÷ _____ = _____

_____ ÷ _____ = _____

⑧ 7, 4, 28

_____ × _____ = _____

_____ × _____ = _____

_____ ÷ _____ = _____

_____ ÷ _____ = _____

Write the number that completes both number sentences.

⑨ 42 ÷ 7 = _____

_____ × 7 = 42

⑩ 18 ÷ _____ = 6

6 × _____ = 18

⑪ 64 ÷ 8 = _____

_____ × 8 = 64

⑫ Mrs. Bourne has 32 square pattern blocks she would like to divide evenly
between 4 groups of students. How many pattern blocks should she
give each group? Draw a picture and write a number sentence to show
how many square pattern blocks should be given to each group.

Number sentence: _____

⑬ Oliver has 4 sports trading cards. He uses his allowance to buy 5 packs of trading
cards. Each pack contains 8 cards. How many trading cards does Oliver have now?
Circle the number sentence below that can be used to solve the problem. Then solve.

$5 \times (4 + 8) = t$ $4 + (5 \times 8) = t$

$t =$ _____

Draw equal groups to solve each problem. Write the answer.

⑭ $2 \times 7 =$ _____

⑮ $12 \div 3 =$ _____

Grade 3 Review: Number and Operations in Base Ten

Add or subtract.

(1) $\begin{array}{r} 86 \\ -\ 40 \\ \hline \end{array}$

(2) $\begin{array}{r} 37 \\ +\ 44 \\ \hline \end{array}$

(3) $\begin{array}{r} 90 \\ +\ 21 \\ \hline \end{array}$

(4) $\begin{array}{r} 52 \\ -\ 19 \\ \hline \end{array}$

(5) $\begin{array}{r} 500 \\ -\ 32 \\ \hline \end{array}$

(6) $\begin{array}{r} 796 \\ -\ 387 \\ \hline \end{array}$

(7) $\begin{array}{r} 403 \\ +\ 159 \\ \hline \end{array}$

(8) $\begin{array}{r} 208 \\ +\ 96 \\ \hline \end{array}$

Round each number to the nearest ten and the nearest hundred.

	Ten	**Hundred**			**Ten**	**Hundred**
(9) 285	_____	_____		(10) 7,694	_____	_____
(11) 63	_____	_____		(12) 35,014	_____	_____
(13) 15	_____	_____		(14) 4,003	_____	_____

Multiply.

(15) $10 \times 9 =$ ____

(16) $70 \times 2 =$ ____

(17) $5 \times 40 =$ ____

(18) $8 \times 20 =$ ____

(19) $60 \times 4 =$ ____

(20) $7 \times 50 =$ ____

(21) $80 \times 3 =$ ____

(22) $6 \times 30 =$ ____

---

(23) Mr. Thomas rounded 178 to 200. Did he round to the nearest
ten or hundred? _____

(24) Merei rounded 13,660 to 13,700. Did she round to the nearest
ten or hundred? _____

(25) Asher rounded 2,445 to 2,450. Did he round to the nearest
ten or hundred? _____

(26) Nsenga rounded 51 to 100. Did he round to the nearest
ten or hundred? _____

(27) Madeline rounded 99,311 to 99,310. Did she round to the
nearest ten or hundred? _____

Grade 3 Review: Number and Operations in Base Ten
(continued)

Round each 2-digit number to the nearest ten to estimate the answer to each problem below. Then multiply to find the estimate.

EXAMPLE 62×8

estimate ↓

__60__ $\times 8 =$ __480__

(28) 15×6 (29) 39×5 (30) 74×3

estimate ↓   estimate ↓   estimate ↓

____ $\times 6 =$ ____   ____ $\times 5 =$ ____   ____ $\times 3 =$ ____

(31) 92×2 (32) 61×8 (33) 28×7

estimate ↓   estimate ↓   estimate ↓

____ $\times 2 =$ ____   ____ $\times 8 =$ ____   ____ $\times 7 =$ ____

Solve each problem below. Then write and solve a subtraction problem to check your work.

EXAMPLE

$$\begin{array}{r} 350 \\ + 211 \\ \hline 561 \end{array}$$

$$\begin{array}{r} 561 \\ - 211 \\ \hline 350 \end{array}$$

(34)
$$\begin{array}{r} 89 \\ + 26 \\ \hline \end{array}$$
$-$ _____

(35)
$$\begin{array}{r} 118 \\ + 47 \\ \hline \end{array}$$
$-$ _____

(36)
$$\begin{array}{r} 303 \\ + 572 \\ \hline \end{array}$$
$-$ _____

Solve each problem below. Then write and solve an addition problem to check your work.

EXAMPLE

$$\begin{array}{r} 607 \\ - 210 \\ \hline 397 \end{array}$$

$$\begin{array}{r} 397 \\ + 210 \\ \hline 607 \end{array}$$

(37)
$$\begin{array}{r} 90 \\ - 34 \\ \hline \end{array}$$
$+$ _____

(38)
$$\begin{array}{r} 881 \\ - 52 \\ \hline \end{array}$$
$+$ _____

(39)
$$\begin{array}{r} 400 \\ - 104 \\ \hline \end{array}$$
$+$ _____

Grade 3 Review: Number and Operations–Fractions

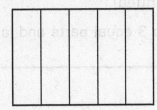

① Color $\frac{1}{5}$ of the equal parts blue.

② Color $\frac{3}{5}$ of the equal parts red.

③ Color the rest of the equal parts green.

④ How many of the equal parts are green? _____

⑤ What fraction of the equal parts is green? _____

Write the following numbers as fractions.

EXAMPLE $8 = \dfrac{8}{1}$

⑥ 9 = _____   ⑦ 2 = _____   ⑧ 10 = _____   ⑨ 4 = _____

⑩ 12 = _____   ⑪ 1 = _____   ⑫ 18 = _____   ⑬ 5 = _____

⑭ Shade $\frac{1}{4}$ of each figure. Write the equivalent fraction.

          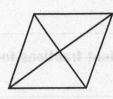

_____   _____   _____

⑮ Circle the fractions that are equivalent to 1.

$\frac{4}{5}$ $\frac{7}{7}$ $\frac{2}{2}$ $\frac{1}{6}$ $\frac{3}{3}$

Grade 3 Review: Number and Operations–Fractions
(continued)

16 Partition the number line into 3 equal parts and label each partition.

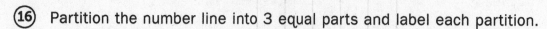

17 Partition the number line into 6 equal parts and label each partition.

18 Partition the number line into 5 equal parts and label each partition.

19 Examine the two number lines below. Which fraction is equivalent to $\frac{2}{2}$? _____

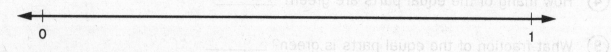

20 Examine the two number lines below. Which fraction is equivalent to $\frac{6}{8}$? _____

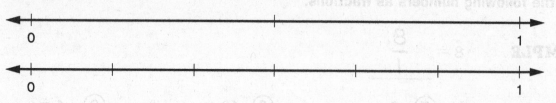

___

Write 2 equivalent fractions for each given fraction.

21 $\frac{1}{3}$ _____

22 $\frac{2}{4}$ _____

23 $\frac{3}{4}$ _____

24 $\frac{1}{5}$ _____

___

Compare the fractions. Write >, <, or =. Draw fraction models to explain your answer.

25 $\frac{1}{5}$ ◯ $\frac{1}{2}$

26 $\frac{6}{8}$ ◯ $\frac{6}{10}$

27 $\frac{3}{7}$ ◯ $\frac{6}{7}$

Grade 3 Review: Measurement and Data

① Use the tally chart to complete the bar graph.

Number of Legs on Animals

Number of Legs	Number of Animals
0	‖‖‖ I
2	‖‖‖
4	‖‖
6	‖‖‖ I
8	‖‖‖

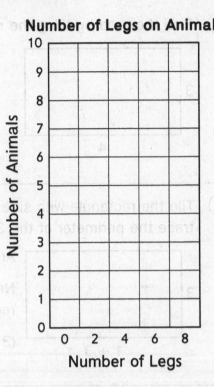

Number of Legs on Animals

Use the graph to answer Problem 2.

② How many animals have more than 2 legs, but less than 8 legs? _____

③ A blue jay has a mass of 85 grams, and a cardinal has a mass of 49 grams. How much more mass does the blue jay have? _____ grams

④ Write the time in the afternoon shown on the clock. Circle A.M. or P.M.

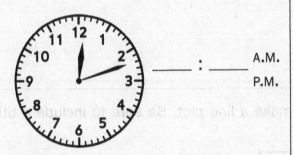

_____ : _____   A.M.
P.M.

What time was it 57 minutes earlier?

Circle A.M. or P.M. _____ : _____   A.M.
P.M.

⑤ Draw a figure with an area of 28 square units.

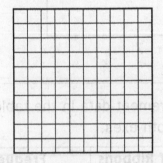

What is the **perimeter** of your figure?

_____ units

⑥ Explain how you knew how to draw the figure in Problem 7. Use the phrases *square units* and *no gaps or overlaps* in your explanation.

_____

_____

Grade 3 Review: Measurement and Data (continued)

⑦ Find the area by tiling the rectangle.

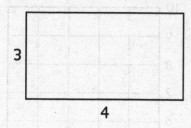

Area = _____ square units

Now, multiply the side lengths to find the area of the rectangle above.

_____ × _____ = _____ square units

---

⑧ Tile the rectangle with side lengths 3 and 1 + 3. Use a color pencil or crayon to trace the perimeter of the 3-by-1 rectangle and the 3-by-3 rectangle.

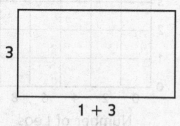

Area = _____ square units

Now, use the Distributive Property to find the area of the rectangle.

(3 × _____) + (3 × _____) = _____ square units

---

⑨ Find the total area of the figure.

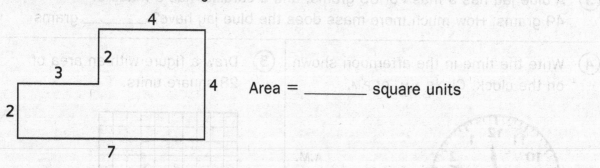

Area = _____ square units

---

⑩ Use measurement data in the table to make a line plot. Be sure to include a title and label both axes.

Lengths of Ribbons	Frequency
8	////
$8\frac{1}{2}$	//
9	////
$9\frac{1}{2}$	///

8   $8\frac{1}{2}$ 9 $9\frac{1}{2}$ 10

_____

Grade 3 Review: Geometry

Circle <u>all</u> the correct answers.

① A rectangle has all the properties of a _____.

  trapezoid   rhombus   quadrilateral   parallelogram

② A parallelogram has all the properties of a _____.

  rectangle   square   rhombus   quadrilateral

③ A square has all the properties of a _____.

  rectangle   rhombus   quadrilateral   parallelogram

④ A rhombus has all the properties of a _____.

  rectangle   trapezoid   quadrilateral   parallelogram

---

⑤ Circle all the polygons.

   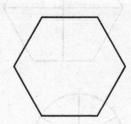

---

⑥ Circle all the triangles.

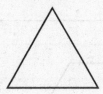  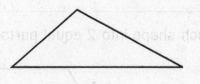 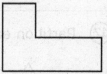

---

⑦ Circle all the parallelograms.

 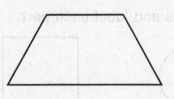  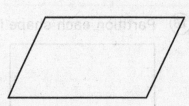

Grade 3 Review: Geometry (continued)

Tell whether each statement is true or false. If the statement is false, rewrite it to make it true.

(8) A rhombus is always a square. _____

_____

(9) A triangle is never a parallelogram. _____

_____

(10) A rectangle is always a square. _____

_____

Is each shape partitioned into equal parts? Circle yes or no.

(11) 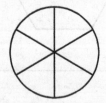   yes
                             no

(12) 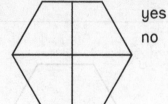   yes
                             no

(13) 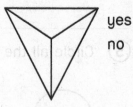   yes
                             no

(14) 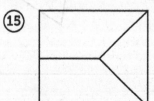   yes
                             no

(15) 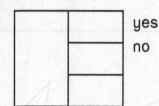   yes
                             no

(16) 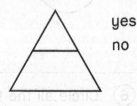   yes
                             no

(17) Partition each shape into 2 equal parts and label each part.

    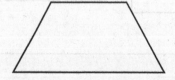   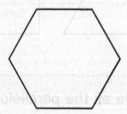

(18) Partition each shape into 3 equal parts and label each part.

Practice Set 1

(1) In 16,075, what is the value of 5? _____ 1? _____ 7? _____

(2) In 239,046, what is the value of 9? _____ 0? _____ 2? _____

(3) In 8,942, what is the value of 2? _____ 8? _____ 4? _____

(4) In 721,385, what is the value of 2? _____ 3? _____ 7? _____

Write the number words for the following.

(5) 564,290 _____

(6) 48,128 _____

(7) 773,963 _____

(8) 102,756 _____

(9) Write a number in which the digit 8 is worth 10 times as much as it is in 81. _____

(10) Write a number in which the digit 3 is worth 10 times as much as it is in 302. _____

(11) Write a number in which the digit 1 is worth 10 times as much as it is in 62,901. _____

(12) Write a number in which the digit 9 is worth 10 times as much as it is in 4,915. _____

Use digits to write the following numbers.

(13) seven hundred eighteen thousand, nine hundred twenty

_____

(14) fourteen thousand, seventy-eight

_____

(15) thirty-eight thousand, five

_____

(16) five hundred sixty-nine thousand, two hundred thirty-one

_____

Practice Set 2

Write the following numbers in expanded form.

① 6,037 _____

② 22,910 _____

③ 304,008 _____

④ 155,972 _____

⑤ 461,205 _____

⑥ Use the clues to complete the place-value puzzle.

1,000s	100s	10s	1s

- Write the result of 21 ÷ 7 in the ones place.

- Multiply 8 * 9. Subtract 65. Write the result in the tens place.

- Double the number in the ones place. Write the result in the thousands place.

- Divide 18 by 6. Add 5 and write the result in the hundreds place.

Fill in the name-collection boxes. Use as many different numbers and operations as you can.

EXAMPLE

19
(6 * 3) + 1
(20 ÷ 2) + 9
(40 − 25) + 4

⑦
38

⑧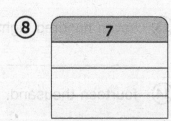
7

⑨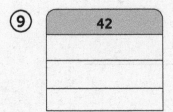
42

⑩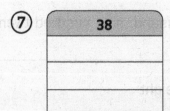
111

⑪
218

Practice Set 3

Round each number to the nearest thousand, ten-thousand, and hundred-thousand.

		Thousand	Ten-Thousand	Hundred-Thousand
①	605,318			
②	89,447			
③	542,366			
④	49,092			

Write >, < or =.

⑤ 6,985 _____ 6,367   ⑥ 459 _____ 489

⑦ 1,640 _____ 1,643   ⑧ 10,387 _____ 10,340

Add.

⑨ 5 + 5 = _____   5 + 50 = _____   50 + 50 = _____

⑩ 2 + 7 = _____   20 + 7 = _____   20 + 70 = _____

⑪ 6 + 4 = _____   60 + 4 = _____   60 + 40 = _____

⑫ 8 + 9 = _____   8 + 90 = _____   80 + 90 = _____

Solve.

⑬ The Coffee-to-Go Café uses about 10 gallons of milk per day.

Ⓐ About how many gallons of milk does it use in a week (7 days)? _____ gallons

Ⓑ How many gallons in 5 weeks? _____ gallons

Ⓒ How many gallons in 10 weeks? _____ gallons

Practice Set 4

Lila and three of her classmates each rolled a number cube seven times to make a 7-digit number. Use the information in the table to solve the problems.

Lila	Simbule	Henry	Anika
6,914,223	8,365,713	2,719,643	5,913,365

(1) Round the number rolled by Simbule to the nearest million. _____

(2) Which number, when rounded to the nearest million, is 6,000,000? _____

(3) Write a number that is more than Anika's number and less than Lila's number.

_____

(4) Use >, <, or = to compare the numbers rolled by Henry and Lila. _____

Subtract.

(5) $12 - 4 =$ _____ $120 - 40 =$ _____

(6) $8 - 6 =$ _____ $80 - 60 =$ _____

(7) $18 - 9 =$ _____ $180 - 90 =$ _____

(8) $11 - 7 =$ _____ $110 - 70 =$ _____

---

(9) What is the value of 2 in 52,017? _____

Now write a number where the 2 is worth 10 times as much. _____

(10) What is the value of 9 in 986,343? _____

Now write a number where the 9 is worth 10 times as much. _____

(11) What is the value of 5 in 475? _____

Now write a number where the 5 is worth 10 times as much. _____

(12) What is the value of 7 in 2,879? _____

Now write a number where the 7 is worth 10 times as much. _____

Use with or after Lesson 1-4.

Practice Set 5

① **Writing/Reasoning** The fourth-grade classes are raising money to donate to an animal shelter. Mr. Cahill's class raised $253, Mrs. Parker's class raised $178, and Mrs. Kline's class raised $440. Anna, a student in Mrs. Parker's class, estimates the three classes raised $300 + $200 + $500 = $1,000. Explain the mistake Anna made.

_____

_____

_____

_____

Now, make a correct estimate for the amount raised by the three

classes: _____

_____

Solve.

②   118
   −117

③   68
   − 34

④   74
   + 27

⑤   510
   − 54

⑥   500
   −290

⑦   402
   +293

⑧   416
   +583

⑨   49
   − 6

⑩   120
   − 30

⑪   81
   + 40

⑫   35
   − 22

⑬   350
   − 150

Fill in the missing numbers. State the rule.

⑭ 10, 22, 34, _____, _____, _____

Rule: _____

⑮ 68, 58, 48, _____, _____, _____

Rule: _____

⑯ 123, 121, 119, _____, _____, _____

Rule: _____

Practice Set 6

① Jaycee and her family are planning a day trip on a bike trail. Jaycee would like to bike 18 miles before lunch and 9 miles afterward. Her brother Alec would like to bike 7 miles before lunch and 13 miles afterward. What is the difference in the total lengths of the two proposed bike trips? First, estimate the answer. Then, write a number model with a letter for the unknown. Then solve.

Estimate: _____ miles

Number model: _____

_____ miles

② **Writing/Reasoning** Explain how you know your answer makes sense.

_____

_____

Write >, <, or =.

③ 265,182 _____ 257,908   ④ 324,077 _____ 2,198,631

⑤ 608,036 _____ 608,036   ⑥ 19,954 _____ 19,904

⑦ 4,501,387 _____ 4,501,837   ⑧ 772,008 _____ 727,115

Write the digit in the hundreds place for each number.

⑨ 16,945 _____   ⑩ 1,439 _____   ⑪ 6,123 _____   ⑫ 753,313 _____

Write the digit in the ten-thousands place for each number.

⑬ 67,204 _____   ⑭ 807,152 _____   ⑮ 729,114 _____   ⑯ 91,360 _____

Order each set of numbers from greatest to least.

⑰ 584,263; 608,422; 59,014; 511,360; 60,338

_____

⑱ 8,100,636; 905,470; 99,468; 8,013,645; 915,391

_____

Practice Set 7

Solve each problem using U.S. traditional addition. Then solve the same problem using either column addition or partial-sums addition.

① 294 + 51	294 + 51	② 87 +66	87 +66	
③ 304 +918	304 +918	④ 1,264 + 550	1,264 + 550	

⑤ Use the clues to complete the place-value puzzle.

10,000s	1,000s	100s	10s	1s

- Divide 72 by 6. Subtract 4 and write the result in the ones place.

- Double the number in the ones place and divide by 8. Write the result in the tens place.

- Multiply 9 * 10. Subtract 83. Write the result in the hundreds place.

- Halve the number in the tens place. Multiply by 3 and write the result in the thousands place.

- Divide 27 by the number in the thousands place. Write the result in the ten-thousands place.

Practice Set 8

Write a number model and solve.

(1) Erin has a collection of 127 seashells. Kip has a collection of 203 seashells. How many more seashells does Kip have?

Number model: _____

_____ seashells

(2) Kip found 57 more seashells. Now how many shells does he have in his collection?

Number model: _____

_____ seashells

Fill in the name-collection box for each number.

(3) 42

(4) 27

(5) 125

Complete the number patterns.

(6) 950, 951, _____, _____, 954, _____, _____, _____, 958

(7) _____; 5,060; _____; _____; 5,180; 5,220; _____; _____

(8) 10,400; _____; _____; 10,325; _____; _____; 10,250

(9) 2,335; _____; 2,535; 2,635; _____; _____; 2,935

Practice Set 9

Solve each problem using U.S. traditional subtraction.

(1)
```
   84
 - 37
```

(2)
```
   602
 -  51
```

(3)
```
   549
 - 190
```

(4)
```
   2,238
 -   409
```

Use rounding and front-end estimation to estimate the answer to each problem below. Then solve the problem. Circle the estimate that is closer to the actual answer.

EXAMPLE

Rounding $\quad \boxed{500 + 400 = 900}$

Front-End $\quad 500 + 300 = 800$

```
   527
 + 369
 -----
   896
```

(5) Rounding _____

Front-End _____
```
   94
 + 18
```

(6) Rounding _____

Front-End _____
```
   203
 +  55
```

(7) Rounding _____

Front-End _____
```
   776
 + 983
```

(8) Rounding _____

Front-End _____
```
   850
 + 259
```

(9) **Writing/Reasoning** In the problems you solved, was the rounding estimate or the front-end estimate you found usually closer to the actual answer? Why do you think that is? _____

_____

(10) Which method do you prefer, rounding or front-end estimation? Why?

_____

_____

Practice Set 10

Convert.

Yards	Feet
①	6
② 4	
③ 7	
④	27

Use the conversion chart to solve the problem below.

⑤ Beau is 3 feet tall. His dad is 2 yards tall. What is the total height of both Beau and his dad, in feet?

Answer: _____ feet

Estimate. Then solve.

⑥ Delaney is exercising on a circular jogging path. Each lap is 625 feet. She jogs 3 laps, and then walks 1 lap. How many total feet did Delaney travel?

Estimate: _____

Number model: _____

Answer: _____ feet

Does your answer make sense? _____ How do you know? _____

_____

_____

⑦ Next Delaney decides to ride her bike to a friend's house. She can follow one route by riding 142 yards, taking a right, and then riding 220 yards. The second route, which goes through the park, is 370 yards long. How much longer is the second route?

Estimate: _____

Number model: _____

Answer: _____ feet

Does your answer make sense? _____ How do you know? _____

_____

_____

Practice Set 11

(1) Draw and label point G.

(2) Draw and label ray YZ.

(3) Draw and label line RS.

(4) Draw and label line segment VW.

(5) Draw and label line AB that is parallel to line CD.

(6) Draw a line. Place four collinear points, J, K, L, and M, on the line.

(7) **Writing/Reasoning** How are a ray and a line similar? How are they different?

_____

_____

_____

Solve.

(8)   377
          + 905

(9)   640
          + 298

(10)   793
          − 555

(11)   400
          − 262

(12)   671
          + 454

(13)   270
          − 234

(14)   6,017
          + 872

(15)   1,829
          −   199

(16)   5,182
          − 3,093

Practice Set 12

① Write S next to each line segment. Write R next to each ray.
Write L next to each line. Circle each right angle.

② Draw $\overrightarrow{AB}$ and $\overleftrightarrow{YZ}$.

A ⋅   Y ⋅

B ⋅   Z ⋅

Are the lines intersecting or parallel?

_____

③ Draw $\overrightarrow{AZ}$ and $\overleftrightarrow{YB}$.

A ⋅   Y ⋅

B ⋅   Z ⋅

Are the lines intersecting or parallel?

_____

Name the shape and fill in the information about each shape.

④ This shape is a _____.

It has _____ sides.

It has _____ vertices.

Are all sides the same length? _____

Are there any parallel sides? _____

Are there any perpendicular

sides? _____

⑤ This shape is a _____.

It has _____ sides.

It has _____ vertices.

Are all sides the same length? _____

Are there any parallel sides? _____

Are there any perpendicular

sides? _____

⑥ This shape is a _____.

It has _____ sides.

It has _____ vertices.

Are all sides the same length? _____

Are there any parallel sides? _____

Are there any perpendicular

sides? _____

⑦ This shape is a _____.

It has _____ sides.

It has _____ vertices.

Are all sides the same length? _____

Are there any parallel sides? _____

Are there any perpendicular

sides? _____

Practice Set 12 (continued)

Convert.

(8) Which figures are polygons? _____

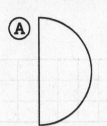 (A)

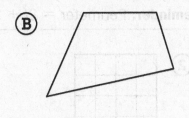 (B)

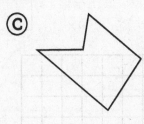 (C)

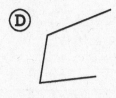 (D)

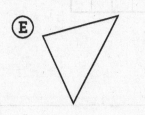 (E)

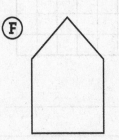 (F)

(9) **Writing/Reasoning** Choose a figure you did not include as your answer to Problem 8. Explain why it is not a polygon.

_____

_____

Write 2 addition and 2 subtraction facts for each group of numbers.

(10) 20, 90, 110

(11) 80, 90, 170

(12) 10, 60, 70

_____ _____ _____

_____ _____ _____

_____ _____ _____

_____ _____ _____

(13) How many inches are in 5 feet? _____ inches

(14) How many feet are in 10 yards? _____ feet

(15) The capybara is the world's largest rodent. It can grow to a length of 4 feet, 4 inches. What is the length of the capybara in inches?

_____ inches

Practice Set 13

Find the perimeter of each rectangle.
Write the number model.

> **Reminder:** Perimeter = 2*l* + 2*w*

①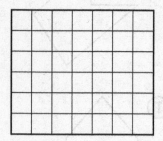

②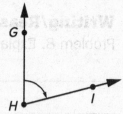

③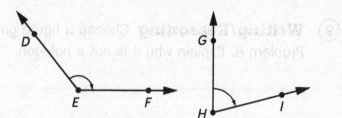

_____   _____   _____

④ Name an angle that has a smaller measure than ∠GHI. _____

⑤ What is another name for ∠ABC? _____

⑥ What is the vertex of ∠DEF? _____

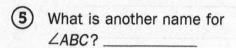

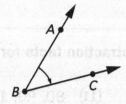

Write the number words for the following numbers.

⑦ 12,743 _____

⑧ 8,054 _____

⑨ 69,231 _____

⑩ 4,782 _____

Practice Set 14

Guess the mystery number. Then draw an array to show the number.

① Clue 1: I am a square number.

Clue 2: I am greater than 20.

Clue 3: The sum of my two digits is 7.

I am _____.

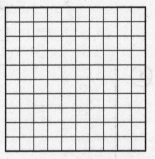

② Clue 1: I am a square number.

Clue 2: I am an even number.

Clue 3: I am a 1-digit number.

I am _____.

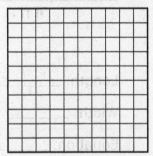

③ Clue 1: I am a square number.

Clue 2: I am greater than 20.

Clue 3: My ones digit is 5 more than my tens digit.

I am _____.

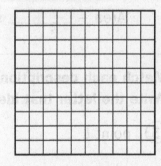

④ What line segments

make up ∠ABC? _____

⑤ What is the vertex of ∠GFH? _____

⑥ What is another name

for ∠DEG? _____

⑦ What type of angle is ∠EFH? _____

⑧ What type of angle is ∠FED? _____

⑨ **Writing/Reasoning** Compare ∠ABC and ∠GFH. Which angle has the greater

measure? _____ How do you know? _____

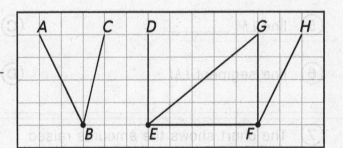

Practice Set 15

Find the area of each rectangle.

Reminder: Area = length (*l*) * width (*w*)

① 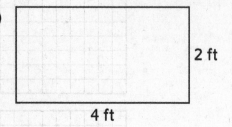 2 ft

4 ft

Length: _____ ft

Width: _____ ft

Equation: _____

Area = _____ square feet

② 3 in.

3 in.

Length: _____ in.

Width: _____ in.

Equation: _____

Area = _____ square inches

Match each description with the correct example.
Write the letter that identifies the example.

③ point L _____

④ ray LM _____

⑤ line LM _____

⑥ line segment LM _____

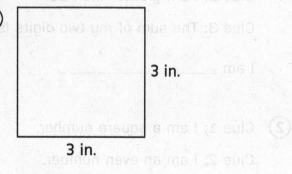

Ⓐ L —————→ M

Ⓑ ←————→ L M

Ⓒ L ————— M

Ⓓ • L

⑦ The chart shows the amounts raised by children in each grade.

Use >, <, or = to write a number sentence that compares the amounts of money raised by Grade 2 and Grade 4.

_____

Amounts Raised	
Grade	Amount (in dollars)
2	$1,248
3	$1,264
4	$1,285
5	$1,259

Use with or after Lesson 2-2.

Practice Set 16

Write all the factor pairs for each number.

① 25 _____

② 8 _____

③ 40 _____

④ 16 _____

⑤ 20 _____

⑥ 24 _____

Add. Use the partial-sums method.

⑦
100s	10s	1s
2	0	4
+1	4	9

⑧
100s	10s	1s
5	5	1
+2	6	7

⑨
1,000s	100s	10s	1s
	8	5	9
+1,	5	9	6

Add. Use the column-addition method.

⑩
100s	10s	1s
7	3	4
+4	7	8

⑪
100s	10s	1s
5	9	2
+8	7	9

⑫
1,000s	100s	10s	1s
2,	7	3	5
+1,	3	0	5

Practice Set 17

List the next ten multiples of each number.

① 7, _____, _____, _____, _____, _____, _____, _____, _____, _____, _____

② 3, _____, _____, _____, _____, _____, _____, _____, _____, _____, _____

③ 9, _____, _____, _____, _____, _____, _____, _____, _____, _____, _____

④ 4, _____, _____, _____, _____, _____, _____, _____, _____, _____, _____

⑤ 8, _____, _____, _____, _____, _____, _____, _____, _____, _____, _____

Add. Use any method you choose.

⑥ 795 + 616 = _____

⑦ 8,838 + 8,956 = _____

⑧ 8,214 + 5,488 = _____

⑨ 50,694 + 39,518 = _____

⑩ Use the clues to complete the place-value puzzle.

10,000s	1,000s	100s	10s	1s

- Divide 88 by 11. Add 1 and write the result in the thousands place.

- Double the number in the thousands place and divide by 3. Write the result in the tens place.

- Multiply 4 * 12. Subtract 42. Write the result in the hundreds place.

- Divide 63 by the number in the thousands place. Write the result in the ones place.

- Halve the number in the tens place. Add 1 and write the result in the ten-thousands place.

Use digits to write the following numbers.

⑪ twenty-four thousand, nine hundred sixty-eight _____

⑫ seventy-six thousand, six hundred fourteen _____

⑬ six thousand, nine hundred two _____

Use with or after Lesson 2-4.

Practice Set 18

List all the factors of each number. Then check whether it is a prime number or a composite number.

		Factors	Prime	Composite
EXAMPLE	18	1, 2, 3, 6, 8, 16		√
①	45			
②	31			
③	26			
④	40			
⑤	7			
⑥	23			

The National Baseball Hall of Fame has over 35,000 bats, balls, uniforms, and gloves. There are over 130,000 baseball cards in the museum collection. The museum also has nearly 500,000 photographs and about 12,000 hours of audio and video recordings. There are 286 Hall of Fame players.

Circle the best answer.

⑦ About how many bats, balls, uniforms, and gloves are displayed in the Baseball Hall of Fame?

Ⓐ between 20,000 and 25,000   Ⓑ between 25,000 and 30,000

Ⓒ between 30,000 and 35,000   Ⓓ between 35,000 and 40,000

⑧ Which of the following is not an estimate?

Ⓐ 500,000 photographs   Ⓑ 130,000 baseball cards

Ⓒ 286 Hall of Fame players   Ⓓ 12,000 hours of audio and video

⑨ **Writing/Reasoning** Why do you think a museum might estimate the number of items in its collection?

_____

_____

Practice Set 19

Complete the "What's My Rule?" tables and rule boxes.

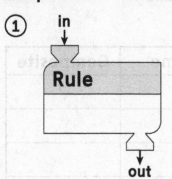

①

in	out
3	9
4	12
7	
9	
10	

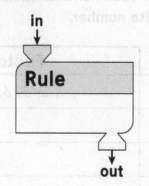

②

in	out
8	4
14	7
	9
20	
22	

Subtract. Use the trade-first method.

③
100s	10s	1s
	9	2
−	3	7

④
100s	10s	1s
6	2	4
− 2	8	6

⑤
100s	10s	1s
3	4	8
− 1	5	9

Subtract. Use U.S. traditional subtraction.

⑥
100s	10s	1s
	7	8
−	5	7

⑦
100s	10s	1s
6	5	8
− 2	7	0

⑧
100s	10s	1s
7	3	4
− 3	8	6

Subtract. Use any method you choose.

⑨ 79 − 21 = _____ ⑩ 33 − 17 = _____ ⑪ 636 − 498 = _____

Solve mentally. Use the counting-up strategy.

⑫ 961 − 185 = _____ ⑬ 70 − 51 = _____ ⑭ 130 − 97 = _____

Practice Set 20

① How many seconds are in 2 minutes? _____ seconds

② How many seconds are in 10 minutes? _____ seconds

③ How many seconds are in 6 minutes? _____ seconds

④ How many minutes are in 3 hours? _____ minutes

⑤ How many minutes are in 8 hours? _____ minutes

⑥ How many minutes are in 5 hours? _____ minutes

⑦ Sofia listened to three songs while cleaning her room. The first
song was 4 minutes, 30 seconds long, the second song was
3 minutes, 10 seconds long, and the third song was 4 minutes,
57 seconds long. What is the total length of all three songs,

in seconds? _____ seconds

Complete the "What's My Rule?" tables and rule boxes.

⑧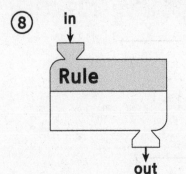

in	out
9	3
18	6
27	
36	

⑨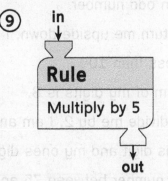

Rule
Multiply by 5

in	out
6	
8	
10	
12	

⑩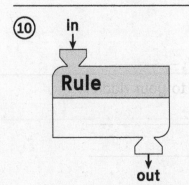

in	out
10	40
20	80
30	
	160

⑪

Rule
Multiply by 8

in	out
	32
	48
	64
	80

Practice Set 21

Write an equation with a letter for the unknown. Then solve.

(1) What number is 7 times as much as 3?

Equation with unknown:

_____

Answer: _____

(2) What number is 10 times as much as 6?

Equation with unknown:

_____

Answer: _____

(3) 49 is 7 times as much as what number?

Equation with unknown:

_____

Answer: _____

(4) 24 is 3 times as much as what number?

Equation with unknown:

_____

Answer: _____

Solve the "Who am I?" riddles.

(5) Clue 1: I am less than 10.

Clue 2: I am an odd number.

Clue 3: If you turn me upside down, I am an even number. _____

(6) Clue 1: I am less than 100

Clue 2: The sum of my digits is 8.

Clue 3: If you divide me by 2, I am an even number.

Clue 4: My tens digit and my ones digit are the same. _____

(7) Clue 1: I am a number between 75 and 150.

Clue 2: The sum of my digits is 5.

Clue 3: My hundreds digit and my ones digit are the same. _____

(8) Make up your own "Who am I?" riddle. Include the answer to your riddle.

_____

_____

_____

_____

Use with or after Lesson 2-8.

Practice Set 22

Write an equation with an unknown to represent and solve each number story.

① Joy has a package of 6 balloons. She needs 4 times as much to decorate for a party. How many balloons does Joy need?

Equation with unknown: _____ Answer: _____ balloons

② There are 4 friends playing a bean bag toss game at Joy's party. There are 5 times as many people at the party. How many people are at Joy's party?

Equation with unknown: _____ Answer: _____ people

③ 16 party guests had lemonade, and 4 party guests had water to drink. The number of guests who had lemonade is how many times more than the number of guests who had water?

Equation with unknown: _____ Answer: _____ times more than

Complete the "What's My Rule?" tables and rule box.

④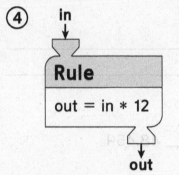

in	out
2	
3	36
	60
	72
8	

⑤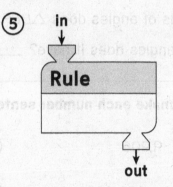

in	out
8	24
14	30
	44
35	
43	59

Solve.

⑥
```
  1,800
-   927
```

⑦
```
  3,684
-   485
```

⑧
```
  3,164
+ 5,791
```

⑨
```
  8,261
- 3,540
```

⑩
```
  600
-  31
```

⑪
```
  475
+ 250
```

⑫
```
  1,834
+ 8,365
```

⑬
```
  469
-  70
```

⑭
```
  1,200
-    30
```

Practice Set 23

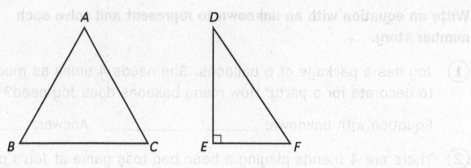

① What type of triangle is △ABC? _____

② How do you know? _____

③ How many kinds of angles does △ABC have? _____

④ What kinds of angles does it have? _____

_____

⑤ What type of triangle is △DEF? _____

⑥ How do you know? _____

⑦ How many kinds of angles does △DEF have? _____

⑧ What kinds of angles does it have? _____

_____

Write >, <, or = to make each number sentence true.

⑨ 9,608 _____ 9,906   ⑩ 48,459 _____ 48,459

⑪ 113,102 _____ 131,102   ⑫ 278,300 _____ 79,309

⑬ 3,780,576 _____ 3,780,576   ⑭ 5,701,318 _____ 5,710,381

⑮ Write the number that has
    4 in the tens place
    7 in the hundred-thousands place
    5 in the ones place
    0 in the thousands place
    6 in the hundreds place
    8 in the ten-thousands place

_____ _____ _____ , _____ _____ _____

Practice Set 24

Write the letter of the description that matches the polygon.

(1) kite _____

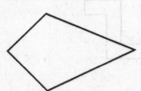

(A) 4 sides of equal length, 4 right angles

(2) parallelogram _____

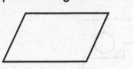

(B) 2 separate pairs of equal sides that are next to each other

(3) square _____

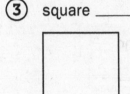

(C) only 1 pair of parallel sides

(4) trapezoid _____

(D) 2 pairs of parallel sides, no right angles

Fill in the missing factors.

(5) 70 * _____ = 210   (6) _____ * 4 = 360   (7) _____ * 80 = 640

(8) 12 * _____ = 960   (9) 5 * _____ = 250   (10) 70 * _____ = 350

Add.

(11) 4,975
    + 1,265

(12) 2,390
    + 1,783

(13) 1,640
    + 9,870

(14)   765
    + 1,832

(15) 5,539
    + 801

(16) 2,892
    + 3,009

(17) 4,774
    + 1,638

(18) 7,057
    + 1,967

Practice Set 25

Circle *yes* or *no* to tell whether the figure is symmetrical.
If you circled *yes*, write how many lines of symmetry it has.

①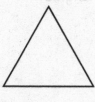

yes   no

_____

②

yes   no

_____

③

yes   no

_____

④

yes   no

_____

⑤

yes   no

_____

⑥

yes   no

_____

⑦ List all the prime numbers less than 10. _____

⑧ List all the prime numbers between 20 and 30. _____

⑨ List all the composite numbers between 10 and 20. _____

⑩ **Writing/Reasoning** Tell how you know whether a number is prime or composite.

_____

_____

⑪ Draw a triangle with an obtuse angle.

⑫ Draw a triangle with 2 perpendicular sides.

⑬ Draw a triangle with at least 2 acute angles.

⑭ Can you draw a triangle with 2 obtuse angles? Why or why not?

_____

_____

Practice Set 26

Complete the "What's My Rule?" tables.
Add your own input and output values in the last row.

① in

Rule
* 5, then + 2
out

in	out
2	
7	
10	
30	

② in
Rule
60 minutes in
an hour
out

in	out
	60
2	
4	
	300

Draw all the lines of symmetry for the figures below.

③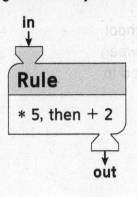

④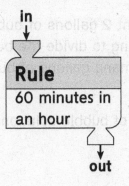

Look at the picture at the right. Write *true* or *false* for each statement.

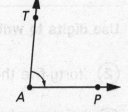

⑤ The angle is formed by 2 line segments. _____

⑥ The vertex of this angle is *A*. _____

⑦ The angle can be named ∠*P*. _____

⑧ Rays form the sides of the angle. _____

⑨ **Writing/Reasoning** If you connected points *T* and *P*, what type of triangle would you make? How do you know?

_____

_____

_____

Practice Set 27

Solve the problem below in two ways. Use drawings to help you solve the problem.

① Mr. Turner bought 2 gallons of bubble solution for the preschool picnic. He is going to divide the bubble solution equally between 8 bottles. How many gallons of bubble solution will be placed in each bottle?

_____ gallon of bubble solution

One way:

Another way:

Use digits to write the following numbers.

② forty-five thousand, three hundred ninety-two _____

③ four hundred fifty-nine thousand, seven hundred three _____

④ six million, four thousand, six hundred nine _____

⑤ one hundred eighteen thousand, seventy _____

⑥ two million, nine hundred fifteen _____

Use with or after Lesson 3-1.

Practice Set 28

1 What fraction of the square below is shaded? _____

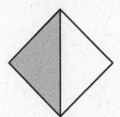

2 Shade a fraction of the square below that is equivalent to the fraction shaded in the first square.

What fraction of this square is shaded?

_____

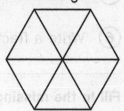

3 What fraction of the hexagon below is shaded? _____

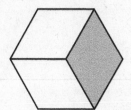

4 Shade a fraction of the hexagon below that is equivalent to the fraction shaded in the first hexagon.

What fraction of this hexagon is shaded?

_____

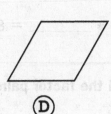

**Match each name with the correct figure.
Write the letter that identifies the figure.**

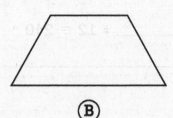
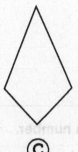

A **B** **C** **D**

5 rhombus _____

6 trapezoid _____

7 square _____

8 kite _____

9 **Writing/Reasoning** Which of these shapes has two names? Explain your answer.

_____

_____

_____

Practice Set 29

Fill in the missing fractions on the number lines below.

①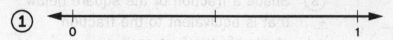

②

③

Use the fraction number lines to answer the questions below.

④ Write a fraction that is equivalent to $\frac{1}{2}$. _____

⑤ Write a fraction that is equivalent to $\frac{4}{6}$. _____

⑥ Write a fraction that is equivalent to $\frac{1}{3}$. _____

Fill in the missing factors.

⑦ 8 * _____ = 24

⑧ _____ * 90 = 360

⑨ _____ * 7 = 49

⑩ 2 * _____ = 160

⑪ 60 * _____ = 360

⑫ _____ * 7 = 630

⑬ 9 * _____ = 810

⑭ _____ * 12 = 240

List all the factor pairs for each number.

⑮ 15 _____

⑯ 35 _____

⑰ 28 _____

⑱ 50 _____

⑲ 12 _____

⑳ 42 _____

Use with or after Lesson 3-3.

Practice Set 30

Use multiplication or division to find an equivalent fraction for the following numbers.
Show your work.

EXAMPLE $\frac{1}{6}$ $\dfrac{* \; 4}{* \; 4} = \dfrac{4}{24}$

① $\frac{3}{7}$ _____

② $\frac{3}{9}$ _____

③ $\frac{6}{8}$ _____

④ $\frac{1}{4}$ _____

⑤ $\frac{3}{5}$ _____

⑥ $\frac{4}{4}$ _____

⑦ $\frac{2}{10}$ _____

⑧ Use the clues to complete the place-value puzzle.

- Divide 18 by 6. Write the result in the ones place.
- Double the number in the ones place. Divide by 3. Write the result in the tens place.
- Write the result of 8 * 5 divided by 10 in the thousands place.
- Multiply 7 by 2. Subtract 7. Write the result in the hundreds place.

1,000s	100s	10s	1s

Make a name-collection box for each number listed below.
Use as many different numbers and operations as you can.

⑨
24

⑩
15

⑪
60

⑫
100

⑬
54

⑭
73

Use with or after Lesson 3-4.

41

Practice Set 31

The fourth graders had a pizza party. They ordered pizzas and
divided each pizza into 6 equal slices. Twenty-one students,
1 teacher, and 4 parents were invited to the party. The students
assumed each person would get one slice of pizza.

(1) How many people were invited to the party? _____ people

(2) How many slices of pizza did the class need? _____ slices

(3) How many pizzas did the class order? _____ pizzas

(4) If everyone ate just one slice, how many slices were left over?

_____ slices

(5) What fraction of a whole pizza is that? _____

(6) If everyone ate two slices of pizza, how many slices did they need?

_____ slices

(7) How many whole pizzas did the class need then? _____ pizzas

(8) What fraction of a whole pizza was left over then? _____

Subtract.

(9) 619
   − 307

(10) 3,510
   − 782

(11) 4,627
   − 4,550

(12) 5,011
   − 2,306

(13) 1,637
   − 540

(14) 800
   − 607

(15) 2,842
   − 2,469

(16) 8,007
   − 492

Use with or after Lesson 3-5.

Practice Set 32

Write > or <.

① $\frac{4}{5}$ _____ $\frac{4}{8}$ ② $\frac{1}{10}$ _____ $\frac{1}{6}$ ③ $\frac{4}{6}$ _____ $\frac{3}{6}$

④ $\frac{2}{3}$ _____ $\frac{2}{4}$ ⑤ $\frac{2}{5}$ _____ $\frac{4}{5}$ ⑥ $\frac{3}{4}$ _____ $\frac{2}{4}$

⑦ $\frac{6}{10}$ _____ $\frac{8}{10}$ ⑧ $\frac{3}{7}$ _____ $\frac{3}{5}$ ⑨ $\frac{5}{6}$ _____ $\frac{5}{10}$

Determine each pattern. Then fill in the missing numbers on the number lines.

⑩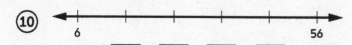

⑪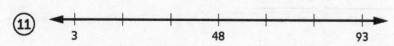

⑫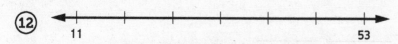

⑬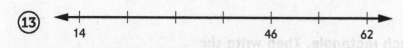

Round each number to the nearest thousand and ten-thousand.

		Thousand	**Ten-Thousand**
⑭	53,046	_____	_____
⑮	605,961	_____	_____
⑯	8,707	_____	_____
⑰	4,451,882	_____	_____
⑱	323,087	_____	_____
⑲	192,480	_____	_____
⑳	4,017,095	_____	_____
㉑	921,500	_____	_____

Practice Set 33

Write the fractions in order from least to greatest.

① $\frac{15}{16}, \frac{5}{16}, \frac{3}{16}, \frac{1}{16}, \frac{7}{16}$

_____, _____, _____, _____, _____

② $\frac{1}{4}, \frac{1}{12}, \frac{1}{9}, \frac{1}{2}, \frac{1}{3}$

_____, _____, _____, _____, _____

③ $\frac{4}{6}, \frac{4}{9}, \frac{4}{8}, \frac{4}{15}, \frac{4}{100}$

_____, _____, _____, _____, _____

④ $\frac{30}{100}, \frac{5}{10}, \frac{6}{10}, \frac{25}{100}, \frac{82}{100}$

_____, _____, _____, _____, _____

Find the area in square units for each rectangle. Then write the number model.

Area = length (*l*) * width (*w*)

⑤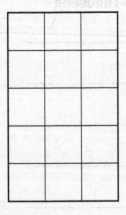

⑥

Length: _____ units   Length: _____ units

Width: _____ units   Width: _____ units

Area: _____ square units   Area: _____ square units

Number model: _____   Number model: _____

Practice Set 34

Match the fraction with the correct decimal. The decimal may be in word or number form.

(1) $\frac{6}{10}$ eight-tenths

(2) $\frac{1}{10}$ 0.3

(3) $\frac{4}{10}$ one-tenth

(4) $\frac{8}{10}$ 0.6

(5) $\frac{3}{10}$ 0.4

For each number model, write _T_ if it is true, _F_ if it is false, or _?_ if you can't tell.

(6) $8 * 9 = 76$ _____

(7) $5 + 9 < 20$ _____

(8) $4 = 64 \div 8$ _____

(9) $26 + 19 = 47$ _____

(10) $450 - 119$ _____

(11) $18 < 15 + 6$ _____

(12) $70 - 21 = 49$ _____

(13) $9 * 4 > 36$ _____

What would the underlined number be worth if it were moved one place to the left?

EXAMPLE   6<u>1</u>2   _100_

(14) 4,<u>7</u>62 _____

(15) 202,<u>3</u>36 _____

(16) <u>8</u> _____

(17) 3,<u>9</u>75,109 _____

(18) <u>1</u>,557 _____

(19) <u>2</u>3,963 _____

(20) 48,<u>2</u>91,005 _____

Practice Set 35

Write the value of the shaded part as a decimal and a fraction.

①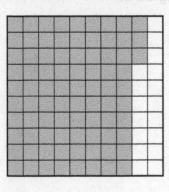

②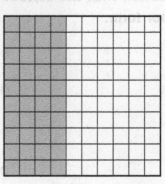

_____ , _____

0.4 _____ , _____

Identify each polygon. Then, write a fact about the polygon's sides.

EXAMPLE

③

rectangle
2 sets of
parallel sides

_____

_____

④

_____

_____

⑤

_____

_____

Write the value of the digit 8 in the numbers.

⑥ 589 _____

⑦ 482,391 _____

⑧ 87,402 _____

⑨ 8,946,326 _____

⑩ 719,538 _____

⑪ 68,457 _____

Practice Set 36

Complete the table. The ☐ is ONE.

Base-10 Blocks	Fraction Notation	Decimal Notation
① ‖‖‖.......	$\frac{58}{100}$	
② ‖‖	$\frac{3}{10}$	
③ ‖‖‖........		0.49
④  ☐ ‖..		1.12
⑤	$\frac{90}{100}$	0.90
⑥ 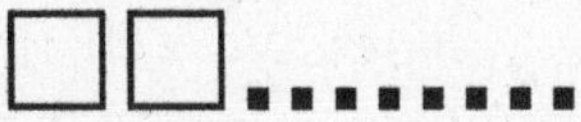 ☐☐........		2.08
⑦  ‖‖‖.	$\frac{41}{100}$	

⑧ Write the following in digits: five-hundredths.

_____

⑨ Write the number words for 0.03.

_____

⑩ Write the number words for 0.76.

Practice Set 37

Measure the line segments to the nearest centimeter. Then record the measurement in meters.

① _____

_____ cm = _____ m

② _____

_____ cm = _____ m

③ _____

_____ cm = _____ m

④ _____

_____ cm = _____ m

Write the fractions in order from greatest to least.

⑤ $\frac{1}{4}$ $\frac{3}{6}$ $\frac{1}{10}$ $\frac{9}{12}$ $\frac{16}{16}$

_____, _____, _____, _____, _____

Write the amounts using a dollar sign and a decimal point.

⑥ ⓆⓆⓆⓆⓆⒹⓃⓃⓃⓅⓅ

_____

⑦ [$1] [$1] ⓆⓆⓆⓆⒹⒹⒹⓃⓅⓅ

_____

⑧ [$5] [$5] [$5] [$1] ⓆⓃⓃⓃ

_____

⑨ [$100] [$20] [$20] [$5] [$1] [$1] [$1] Ⓠ

_____

Use with or after Lesson 3-11.

SRB
136-142
180-183
238

Practice Set 38

**Measure the line segments to the nearest tenth of a centimeter.
Then record the measurement in millimeters.**

① _____

_____ cm = _____ mm

② _____

_____ cm = _____ mm

③ _____

_____ cm = _____ mm

Write two equivalent fractions for each of the following numbers.

④ $\frac{1}{3}$ _____   ⑤ $\frac{3}{4}$ _____   ⑥ $\frac{3}{6}$ _____   ⑦ $\frac{5}{12}$ _____

⑧ $\frac{10}{16}$ _____   ⑨ $\frac{14}{7}$ _____   ⑩ $\frac{5}{5}$ _____   ⑪ $\frac{6}{9}$ _____

⑫ $\frac{3}{1}$ _____   ⑬ 6 _____   ⑭ $\frac{0}{4}$ _____   ⑮ $\frac{2}{2}$ _____

---

⑯ Draw a figure with exactly 1 line of symmetry. Draw the line of symmetry.

⑰ Draw a figure with exactly 4 lines of symmetry. Draw the lines of symmetry.

⑱ Draw a figure with NO lines of symmetry.

⑲ Draw a figure with exactly 2 lines of symmetry. Draw the lines of symmetry.

Use with or after Lesson 3-12.

Practice Set 39

Compare. Write >, <, or =.

① 6.43 _____ 6.69   ② 5.50 _____ 5.5   ③ 2.59 _____ 2.6

④ 9.9 _____ 9.7   ⑤ 3.82 _____ 3.82   ⑥ 7.2 _____ 7.20

⑦ 16.02 _____ 16.2   ⑧ 9.52 _____ 9.33   ⑨ 4 _____ 4.01

⑩ 3.2 _____ 3.03   ⑪ 0.57 _____ 0.75   ⑫ 6.5 _____ 6.32

Use digits to write the following numbers.

⑬ nine hundred sixteen thousand, five hundred seven _____

⑭ eight and two-tenths _____

⑮ seventy-two and thirty-one hundredths _____

⑯ one hundred six and four-hundredths _____

Write the missing numbers.

⑰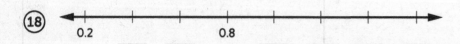

9   17   33   65

⑱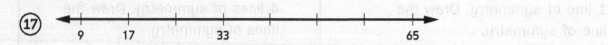

0.2   0.8

⑲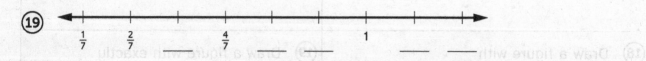

$\frac{1}{7}$ $\frac{2}{7}$ $\frac{4}{7}$ 1

Complete.

⑳ 4.3 cm = _____ mm   ㉑ 30 mm = _____ cm

㉒ 6 mm = _____ cm   ㉓ 5.0 cm = _____ mm

Practice Set 40

Write a multiplication fact for each Fact Triangle.
Then extend this fact by changing both factors to multiples of 10.

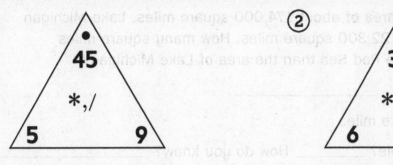

①

45
*,/
5   **9**

Original fact: _____

Extended fact: _____

②

36
*,/
6   **6**

Original fact: _____

Extended fact: _____

③

42
*,/
6   **7**

Original fact: M_____

Extended fact: _____

④

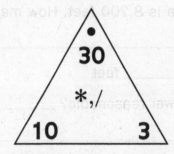

30
*,/
10   **3**

Original fact: _____

Extended fact: _____

Write the next three numbers in each pattern.

⑤ 36, 33, 30, _____, _____, _____

⑥ 10, 25, 40, _____, _____, _____

⑦ 48, 42, 36, _____, _____, _____

⑧ 140, 125, 110, _____, _____, _____

Write >, <, or = to make a true number sentence.

⑨ 6,029 + 33 _____ 1,045 + 5,017

⑩ 20 + 30 + 40 _____ 35 + 55

⑪ 335 + 213 _____ 432 + 118

⑫ 130 − 38 _____ 50 * 2

⑬ 0 + 12 _____ 24 ÷ 2

⑭ 118 + 220 _____ 1,000 − 692

Use with or after Lesson 4-1.

51

Practice Set 41

Write an estimate and show your thinking. Use a calculator to solve the problem. Check to see that your answer is reasonable based on your estimate.

① The Red Sea covers an area of about 174,000 square miles. Lake Michigan covers an area of about 22,300 square miles. How many square miles greater is the area of the Red Sea than the area of Lake Michigan?

Estimate: _____

Answer: _____ square miles

Is your answer reasonable? _____ How do you know? _____

_____

② The deepest part of Lake Michigan is 923 feet. The deepest part of the Red Sea is 8,200 feet. How many feet deeper is the Red Sea?

Estimate: _____

Answer: _____ feet

Is your answer reasonable? _____ How do you know? _____

_____

③ About how many more times deeper is the Red Sea than Lake Michigan?

Estimate: _____

Answer: about _____ times

Is your answer reasonable? _____ How do you know? _____

_____

④ The Red Sea contains about 1,100 types of fish. Lake Michigan has about 100 types of fish. How many more types of fish are in the Red Sea?

Estimate: _____

Answer: _____ types of fish

Is your answer reasonable? _____ How do you know? _____

_____

⑤ **Writing/Reasoning** Given the information in Problem 4, can you tell if there are more fish in the Red Sea than in Lake Michigan? Explain your answer.

_____

_____

Use with or after Lesson 4-2.

Practice Set 41 (continued)

Determine each pattern. Then fill in the missing numbers on the number lines.

⑥

8   16   24   32   48

⑦

0   54

⑧

0   200

⑨

6   24

Write a number model to estimate the sum.

⑩ 695 + 205

Number model: _____

⑪ 316 + 890

Number model: _____

⑫ 574 + 98

Number model: _____

⑬ 412 + 1,878

Number model: _____

Measure the line segments to the nearest centimeter. Then record the measurement in meters.

⑭ _____

_____ cm = _____ m

⑮ _____

_____ cm = _____ m

⑯ _____

_____ cm = _____ m

⑰ _____

_____ cm = _____ m

Practice Set 42

Solve the multiplication problems by partitioning a rectangle. Then write an equation to show how you added each part of the rectangle to get the product.

---

① 8 * 45

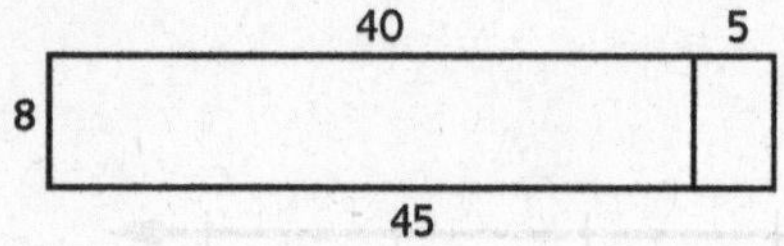

```
        40              5
  8 [              |    ]
        45
```

+ _____

---

② 4 * 32

---

③ 3 * 63

---

④ 6 * 89

Write the next three numbers in the pattern.

⑤ 23,610; 23,615; 23,620; _____; _____; _____

⑥ 39.55, 39.50, 39.45, _____, _____, _____

⑦ 151, 148, 145, _____, _____, _____

⑧ 1,780; 1,455; 1,130; _____; _____; _____

Use with or after Lesson 4-3.

Practice Set 43

Solve.

1. A restaurant serves 4 liters of soup on Thursday, 12 liters of soup on Friday, and 15 liters of soup on Saturday. How many total milliliters of soup did the restaurant serve during those three days? _____ milliliters

2. A punch recipe calls for 1,000 milliliters of orange juice, 500 milliliters of pineapple juice, and 500 milliliters of ginger ale. If the recipe is doubled, how many liters of punch will be made? _____ liters

Write the amounts using a dollar sign and a decimal point.

3. Ⓠ Ⓠ Ⓠ Ⓓ Ⓓ Ⓓ Ⓝ Ⓟ Ⓟ Ⓟ Ⓟ Ⓟ

   _____

4. $1 Ⓠ Ⓠ Ⓠ Ⓓ Ⓓ Ⓓ Ⓝ Ⓝ Ⓝ

   _____

5. $10 $5 Ⓠ Ⓠ Ⓝ Ⓝ Ⓟ Ⓟ Ⓟ Ⓟ Ⓟ Ⓟ

   _____

Compare. Write >, <, or =.

6. $\frac{1}{4}$ _____ $\frac{2}{3}$ 7. $\frac{4}{16}$ _____ $\frac{1}{4}$

8. $\frac{5}{11}$ _____ $\frac{5}{10}$ 9. 43,201 _____ 43,201

10. 17,300 _____ 13,700   11. 765,983 _____ 725,983

12. In 3,905, what is the value of the . . .   13. In 87,641, what is the value of the . . .

    9? _____   7? _____

    0? _____   4? _____

    3? _____   6? _____

    5? _____   8? _____

Practice Set 44

Patrick's garden has vegetables, herbs, and flowers.

	16 ft	6 ft
8 ft	vegetables	herbs
6 ft	flowers	

(1) What is the total length of the garden? _____ feet

The total width of the garden? _____ feet

(2) What is the perimeter of the garden? _____ feet

(3) What is the area of the vegetable section of the garden? _____ square feet

(4) What is the area of the herb section of the garden? _____ square feet

(5) What is the area of the flower section of the garden? _____ square feet

(6) What is the total area of Patrick's garden? _____ square feet

(7) **Writing/Reasoning** Explain how you found the answer to Problem 5.

_____

_____

In the numeral 1,527, the 2 stands for 20.

(8) What does the 5 stand for? _____

(9) What does the 7 stand for? _____

(10) What does the 1 stand for? _____

In the numeral 28,490, the 8 stands for 8,000.

(11) What does the 4 stand for? _____

(12) What does the 2 stand for? _____

(13) What does the 9 stand for? _____

(14) What does the 0 stand for? _____

Use with or after Lesson 4-5.

NAME DATE SRB 106

Practice Set 45

Use partial-products multiplication to solve the problems.

① 82 * 9	② 135 * 6
③ 545 * 3	④ 417 * 2

Tell whether each fraction is less than $\frac{1}{2}$, equal to $\frac{1}{2}$, or greater than $\frac{1}{2}$.

⑤ $\frac{3}{6}$ ______

⑥ $\frac{10}{25}$ ______

⑦ $\frac{12}{15}$ ______

⑧ $\frac{10}{20}$ ______

⑨ $\frac{7}{10}$ ______

⑩ $\frac{1}{12}$ ______

⑪ $\frac{9}{16}$ ______

⑫ $\frac{12}{30}$ ______

SRB
65-68
188-189

Practice Set 46

A nickel has a mass of about 5 grams. A liter of soda has a mass of about 1 kilogram.

Match the object with a possible mass. Write the letter of the possible mass.

(1) pair of scissors _____

(A) 200 kg

(2) mug of hot chocolate _____

(B) 1 kg

(3) loaf of bread _____

(C) 1 g

(4) full-grown grizzly bear _____

(D) 50 g

(5) safety pin _____

(E) 350 g

Complete the "What's My Rule?" tables.

(6)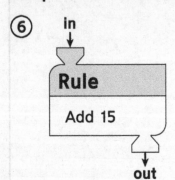

in	out
18	
34	
	56
48	
90	

Rule: Add 15

(7)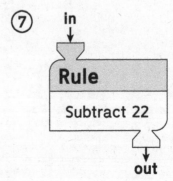

in	out
43	
34	
	9
58	
	77

Rule: Subtract 22

(8)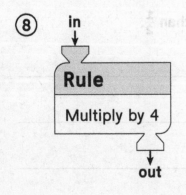

in	out
0	
	8
4	
10	
15	

Rule: Multiply by 4

(9)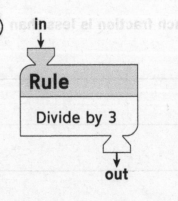

in	out
18	
	12
60	
	30
	40

Rule: Divide by 3

Use with or after Lesson 4-7.

Practice Set 47

Lightbulbs
4-pack $4

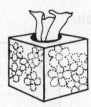
Tissues
$2

Batteries
4-pack $3

Transparent
Tape $1

Ballpoint
Pen $1

DVR disc
$3

① John must buy supplies for his company. He needs 5 lightbulbs, 4 boxes of tissues, and 6 rolls of transparent tape. How much money does he need for these supplies?

_____

② Ms. Larson has $4. How many boxes of tissues can she buy?

_____ boxes

③ **Writing/Reasoning** Jude and Muriel are going to record their school's play. They need 8 batteries and 2 DVR discs for the camera. They have $10.00. Do they have enough money to buy what they need?

_____

Explain. _____

_____

④ The tissue box contains 100 tissues. About how much does each tissue cost?

_____

⑤ About how much is each of the batteries in the 4-pack?

_____

Practice Set 48

Solve using partial-products multiplication.

①   27
     * 13

②   36
     * 46

③   90
     * 19

④   23
     * 62

Use digits to write the following numbers.

⑤   seventy-four million, nine thousand, sixty-four   _____

⑥   nineteen and sixty-eight hundredths   _____

⑦   four hundred nine and seven-tenths   _____

Write the number words for the following.

⑧   0.76

_____

⑨   18.04

_____

Practice Set 49

Solve.

(1) (3 + 4) * (5 + 2) = _____   (2) (2 + 1) * (9 + 6) = _____

(3) (5 + 8) * (7 + 1) = _____   (4) (9 + 2) * (8 + 3) = _____

(5)   26   (6)   965   (7)   541
      * 30   − 86   −   8

(8)   160   (9)   6,045   (10)   2,289
     + 1,400   +   248   + 1,374

(11)   18   (12)   4,371   (13)   890
      * 11   + 3,148   −   15

(14) (70 + 15) * 4 = _____   (15) 67 − (8 * 4) = _____

(16) 430 + 70 + 145 = _____   (17) (72 ÷ 8) * 3 = _____

Write >, <, or = to make each number sentence true.

(18) $\frac{2}{5}$ _____ $\frac{7}{8}$ (19) $\frac{1}{3}$ _____ $\frac{2}{3}$

(20) $\frac{7}{8}$ _____ $\frac{1}{2}$ (21) $\frac{5}{8}$ _____ $\frac{10}{16}$

(22) $\frac{9}{10}$ _____ $\frac{2}{10}$ (23) $\frac{6}{12}$ _____ $\frac{3}{6}$

(24) $\frac{6}{30}$ _____ $\frac{3}{15}$ (25) $\frac{9}{12}$ _____ $\frac{1}{5}$

Practice Set 50

Find the area of each polygon.

①
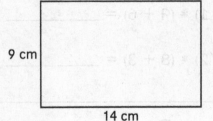
9 cm

14 cm

Equation: _____

_____

Area: _____ square centimeters

②
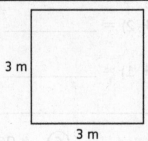
3 m

3 m

Equation: _____

_____

Area: _____ square meters

③
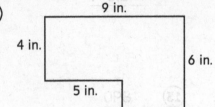
9 in.

4 in.

6 in.

5 in.

Equation(s): _____

_____

_____

Area: _____ square inches

④
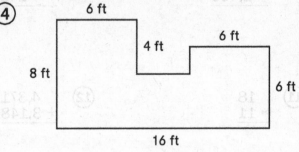
6 ft

4 ft

6 ft

8 ft

6 ft

16 ft

Equation(s): _____

_____

_____

Area: _____ square feet

Complete the missing factors.

⑤ 70 * _____ = 2,100

⑥ _____ * 4 = 360

⑦ _____ * 8 = 6,400

⑧ 12 * _____ = 960

⑨ 40 * _____ = 480

⑩ _____ * 50 = 3,500

⑪ 6 * _____ = 360

⑫ _____ * 7 = 840

Practice Set 51

Write estimates and number models for each problem. Then solve.

① **Writing/Reasoning** Brianna needs enough wrapping paper to cover five presents. She estimates one roll of wrapping paper will cover two presents. Each roll costs $2.00. She has $10.00. Does she have enough money? Explain your answer.

Number models with unknowns: _____

_____

Answer: _____

Explain: _____

_____

② Mr. Baker is shopping with his two children. He bought each of them a pair of sunglasses and a toothbrush. The sunglasses cost $7 each and the toothbrushes cost $3 each. Tax is included in the price. He pays with a $50 bill. How much change should he receive?

Number models with unknowns: _____

_____

Answer: $_____

Determine each pattern. Then fill in the missing numbers on the number lines.

③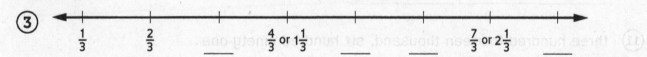

④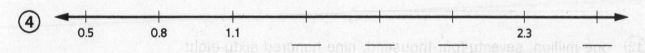

⑤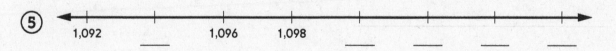

Practice Set 52

Use the lattice method to find the following products.

① 61 * 7 = _____

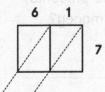

② 58 * 16 = _____

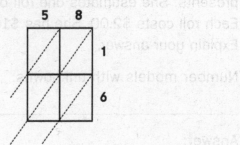

For each number sentence, write *T* if it is true or *F* if it is false.

③ 92 − 40 = 42 _____

④ 80 = 18 + 62 _____

⑤ 9 * 7 = 67 _____

⑥ 6 > 72 ÷ 9 _____

⑦ 500 + 80 < 550 _____

⑧ 448 − 15 > 400 _____

Use digits to write the following numbers.

⑨ two hundred sixty thousand, four hundred fifty-three

_____

⑩ five hundred eighty-six and thirty-eight hundredths

_____

⑪ three hundred fourteen thousand, six hundred ninety-one

_____

⑫ one million, seventy-four thousand, nine hundred sixty-eight

_____

⑬ six million, seven hundred nine thousand, eight hundred forty-five

_____

Use with or after Lesson 4-13.

NAME ____________ DATE ____________ SRB 182-183 233-235

Practice Set 53

Write an equation to show each fraction as the sum of unit fractions.

1. $\frac{3}{7}$ ____________________

2. $\frac{8}{12}$ ____________________

3. $1\frac{2}{10}$ ____________________

4. $\frac{4}{5}$ ____________________

5. $2\frac{5}{9}$ ____________________

Complete.

6. 18 cm = ________ m

7. 3 m = ________ cm

8. ________ mm = 42 cm

9. ________ cm = 450 mm

10. 1.5 m = ________ cm

11. ________ mm = 8 cm

12. 100 mm = ________ cm

13. 200 cm = ________ m

14. ________ m = 75 cm

15. 155 cm = ________ mm

Match each term with the correct figure. Write the letter that identifies the figure.

16. acute triangle ________

17. right triangle ________

18. rhombus ________

19. rectangle ________

A 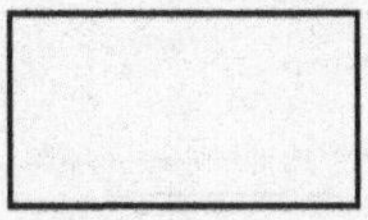

B 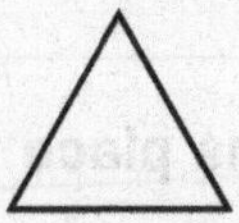

C 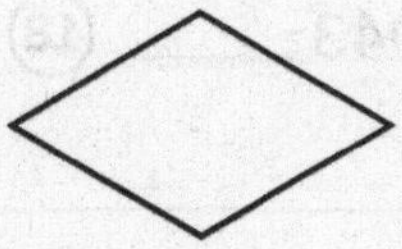

D 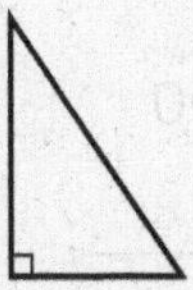

Practice Set 54

The first figure is $\frac{1}{2}$ of the whole. What fraction of the whole is each of the other figures?

 $\frac{1}{2}$ ①   ②   ③

_____   _____   _____

Solve.

④   20
   * 40

⑤   214
   − 67

⑥   8,462
   − 2,700

⑦   3,008
   + 5,002

⑧   350
   − 192

⑨   3,082
   + 8,369

⑩   6,743
   + 5,921

⑪   9
   * 80

⑫ $(50 + 20) * 4 =$ _____

⑬ $227 − (25 \div 5) =$ _____

⑭ $600 − (77 + 115) =$ _____

⑮ $(482 − 315) + 91 =$ _____

Write the digit in the hundredths place for each number.

⑯ 5.92 _____ ⑰ 1.043 _____ ⑱ 8.10 _____ ⑲ 0.280 _____ ⑳ 3.13 _____

㉑ Put these numbers in order from greatest to least.

    14,001   114,000   110.41   41,000

_____ ; _____ ; _____ ; _____

Use with or after Lesson 5-2.

Practice Set 55

Add.

① $\frac{1}{6} + \frac{5}{6} =$ _____

② $\frac{4}{9} + \frac{8}{9} =$ _____

③ $\frac{3}{4} + \frac{1}{4} =$ _____

④ $\frac{5}{12} + \frac{5}{12} =$ _____

⑤ $\frac{7}{8} + \frac{2}{8} =$ _____

⑥ $\frac{6}{10} + \frac{2}{10} =$ _____

⑦ $\frac{3}{7} + \frac{2}{7} =$ _____

⑧ $\frac{2}{5} + \frac{3}{5} =$ _____

⑨ If 9 counters are $\frac{1}{2}$, then _____ counters are ONE.

⑩ If 4 counters are $\frac{1}{3}$, then _____ counters are ONE.

⑪ If 7 counters are $\frac{1}{5}$, then _____ counters are ONE.

⑫ If 10 counters are $\frac{2}{9}$, then _____ counters are ONE.

⑬ If 6 counters are $\frac{2}{3}$, then _____ counters are ONE.

⑭ If 15 counters are $\frac{5}{8}$, then _____ counters are ONE.

Write the amounts using a dollar sign and a decimal point.

⑮ 18 dimes _____

⑯ 13 quarters _____

⑰ 35 nickels _____

⑱ 20 quarters and 6 dimes _____

⑲ 11 dimes and 4 nickels _____

⑳ 14 quarters and 10 nickels _____

㉑ 7 quarters and 35 pennies _____

Practice Set 56

(1) Lexi ate $1\frac{2}{3}$ peaches, and her brother ate $2\frac{1}{3}$ peaches. How many peaches did they eat in all?

Number model: _____

Answer: _____ peaches

(2) Charlotte, TaJuan, and Jax each ran $2\frac{1}{2}$ miles. How many miles did they run altogether?

Number model: _____

Answer: _____ miles

(3) Keigo read for $1\frac{1}{3}$ hours on Sunday, $2\frac{1}{3}$ hours on Tuesday, and 4 hours on Wednesday. How many hours in all did he read on those three days?

Number model: _____

Answer: _____ hours

Round each number to the place value indicated.

(4) Round 61 to the nearest ten.

_____

(5) Round 310 to the nearest hundred.

_____

(6) Round 19 to the nearest hundred.

_____

(7) Round 24,079 to the nearest thousand.

_____

(8) Round 83,553 to the nearest ten-thousand.

_____

(9) Round 14,070 to the nearest ten-thousand.

_____

Write the value of the shaded part as a decimal and a fraction.

(10)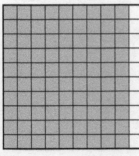

_____ , _____

(11)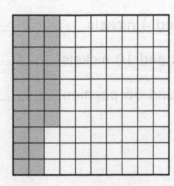

_____ , _____

Use with or after Lesson 5-4.

Practice Set 57

Use what you know about equivalent fractions to add. Write an equation to show your work.

① 8 tenths + 9 hundredths

Equation: _____

② 1 hundredth + 4 tenths + 25 hundredths

Equation: _____

③ 30 hundredths + 35 hundredths + 6 tenths

Equation: _____

④ $\frac{92}{100} + \frac{5}{10}$

Equation: _____

⑤ $8\frac{2}{100} + \frac{3}{10}$

Equation: _____

⑥ $7\frac{5}{10} + \frac{81}{100}$

Equation: _____

**For each pair of fractions, write *yes* if the fractions are equivalent.
Write *no* if the fractions are not equivalent.**

⑦ $\frac{1}{6}, \frac{2}{12}$ _____   ⑧ $\frac{7}{12}, \frac{3}{4}$ _____

⑨ $\frac{1}{2}, \frac{5}{10}$ _____   ⑩ $\frac{1}{3}, \frac{2}{9}$ _____

⑪ $\frac{4}{5}, \frac{12}{20}$ _____   ⑫ $\frac{40}{50}, \frac{8}{10}$ _____

⑬ $\frac{6}{8}, \frac{3}{4}$ _____   ⑭ $\frac{1}{4}, \frac{4}{12}$ _____

⑮ $\frac{9}{10}, \frac{18}{24}$ _____   ⑯ $\frac{1}{10}, \frac{10}{100}$ _____

Practice Set 58

Write *yes* if the fractions add up to one whole. Write *no* if the fractions do not add up to one whole. Use fraction circles if needed.

(1) $\frac{1}{4} + \frac{2}{4} + \frac{1}{4}$ _____

(2) $\frac{6}{7} + \frac{0}{7}$ _____

(3) $\frac{1}{2} + \frac{2}{8} + \frac{1}{4}$ _____

(4) $\frac{2}{10} + \frac{60}{100} + \frac{1}{10}$ _____

(5) $\frac{1}{6} + \frac{4}{6} + \frac{1}{3}$ _____

Complete the "What's My Rule?" tables and rule box.

(6)

Rule

* 200

in	out
7	1,400
9	
12	
14	
35	

(7)

Rule

in	out
7	$14\frac{1}{2}$
10	$17\frac{1}{2}$
$12\frac{1}{2}$	
$13\frac{1}{2}$	
$22\frac{1}{2}$	

Solve.

(8) $180 \div 60 =$ _____

(9) $24 * 6 =$ _____

(10) $810 = 90 *$ _____

(11) $40 * 700 =$ _____

(12) $80 *$ _____ $= 3,200$

(13) _____ $\div 100 = 6$

(14) $8,400 \div 700 =$ _____

(15) $36 *$ _____ $= 72$

(16) _____ $\div 5 = 70$

(17) $9 * 200 =$ _____

(18) $5,400 \div$ _____ $= 9$

(19) $12 * 85 =$ _____

Practice Set 59

① Mrs. Cantrell's class watched $\frac{5}{8}$ of a movie before lunch. They watched the rest of the movie after lunch. What fraction of the movie did they watch after lunch?

Number model: _____

Answer: _____

② Stefon hopped on one foot for $\frac{3}{4}$ minute. Joelle hopped on one foot for $\frac{1}{4}$ minute. How much longer did Stefon hop on one foot?

Number model: _____

Answer (with unit): _____

Find the area of each polygon.

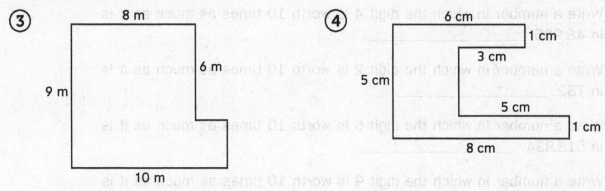

③ 8 m · 6 m · 9 m · 10 m

④ 6 cm · 1 cm · 3 cm · 5 cm · 5 cm · 1 cm · 8 cm

Equation(s): _____

_____

_____

Area: _____ square meters

Equation(s): _____

_____

_____

Area: _____ square centimeters

Write the next three numbers in each pattern.

⑤ 204, 212, 220, _____,

_____, _____

⑥ 6,085; 6,090; 6,095; _____;

_____; _____

⑦ 116, 112, 108, _____,

_____, _____

⑧ 1.5, 2.1, 2.7, _____,

_____, _____

Practice Set 60

① For a class party, Mr. Cole brought in $2\frac{1}{2}$ liters of lemonade and 4 liters of fruit punch. How many more liters of punch are there than lemonade?

Number model: _____

Answer (with unit): _____

② A red panda has a mass of $8\frac{2}{8}$ kilograms. A giant panda has a mass of $130\frac{5}{8}$ kilograms. How much more mass does the giant panda have?

Number model: _____

Answer (with unit): _____

③ Write a number in which the digit 6 is worth 10 times as much as it is in 620. _____

④ Write a number in which the digit 4 is worth 10 times as much as it is in 48,905. _____

⑤ Write a number in which the digit 2 is worth 10 times as much as it is in 732. _____

⑥ Write a number in which the digit 5 is worth 10 times as much as it is in 513,934. _____

⑦ Write a number in which the digit 9 is worth 10 times as much as it is in 89,630. _____

Write an equation to show each fraction as the sum of unit fractions.

⑧ $\frac{3}{4}$ _____

⑨ $\frac{4}{6}$ _____

⑩ $1\frac{5}{15}$ _____

⑪ $\frac{7}{8}$ _____

⑫ $2\frac{3}{7}$ _____

Use with or after Lesson 5-8.

Practice Set 61

Mr. Adema asked his piano students how many hours they practice each week. The tally chart shows the data he collected. Use the table to help you answer the questions below.

Number of Hours	Number of Students
1	//
$1\frac{1}{2}$	~~HHT~~ //
2	////
$2\frac{1}{2}$	///
3	//
$3\frac{1}{2}$	/
4	/

① Construct a line plot for the data.

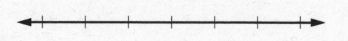

② What is the greatest number of hours spent practicing each week? _____ hours

③ How many students practiced less than 2 hours? _____ students

④ How many more students practiced for $1\frac{1}{2}$ hours than for $3\frac{1}{2}$ hours?

_____ students

⑤ How many students practiced more than 1 hour but less than 3 hours?

_____ students

⑥ **Writing/Reasoning** Explain how you found your answer to Problem 5.

_____

_____

_____

Use the lattice method to find the following products.

⑦ 29 * 55 = _____

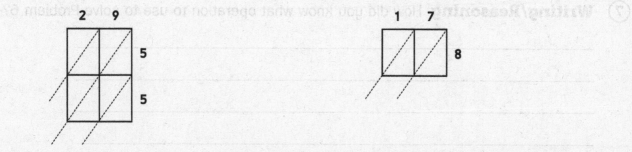

⑧ 17 * 8 = _____

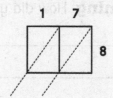

Practice Set 62

Tell whether each angle shows a $\frac{1}{4}$, $\frac{1}{2}$, $\frac{3}{4}$, or full turn.

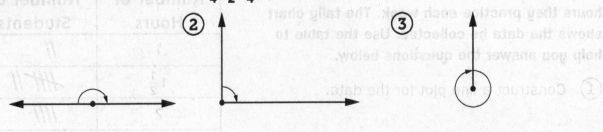

① ② ③

_____   _____   _____

Solve. Write a number model.

④ On Monday, Mr. Tong drove his car $10\frac{3}{8}$ miles to work. Later that day, Mr. Tong drove his car $10\frac{3}{8}$ miles home. How many miles did he drive to and from work?

Number model: _____

_____ miles

⑤ The T-Shirt Mart sells small, medium, and large T-shirts. There are 342 small T-shirts, 496 medium T-shirts, and 683 large T-shirts in stock. The clerk sells 24 large T-shirts to a group of tourists. How many more large T-shirts are now in stock than small T-shirts?

Number model: _____

_____ more large T-shirts

⑥ Ellen has two pieces of string. One is 143 cm in length. The other is 257 cm in length. What is the difference between the lengths of the two pieces?

Number model: _____

_____ cm

⑦ **Writing/Reasoning** How did you know what operation to use to solve Problem 6?

_____

_____

_____

_____

Use with or after Lesson 5-10.

Practice Set 63

Use a clock to help you answer the questions.

(1) In 1 hour, how many degrees does the minute hand rotate? _____

(2) In 1 minute, how many degrees does the minute hand rotate? _____

(3) In 5 minutes, how many degrees does the minute hand rotate? _____

(4) In 15 minutes, how many degrees does the minute hand rotate? _____

(5) In 30 minutes, how many degrees does the minute hand rotate? _____

(6) In 45 minutes, how many degrees does the minute hand rotate? _____

Write the number of degrees the minute hand moves.

(7) from 1:00 to 1:15 _____

(8) from 2:00 to 2:30 _____

(9) from 11:00 to 11:03 _____

(10) from 7:00 to 7:25 _____

(11) from 10:35 to 11:00 _____

(12) from 1:00 to 2:00 _____

Solve.

(13) 207 − _____ = 65

(14) 6,521 + 3,227 = _____

(15) 4,190 + 448 = _____

(16) 1,416 − 948 = _____

(17) 70 = 4,900 ÷ _____

(18) 720 ÷ 9 = _____

(19)   21
     * 49
     _____

(20)   54
     * 38
     _____

(21)   66
     * 87
     _____

Practice Set 64

Draw the other half of the symmetrical shapes below.

①

②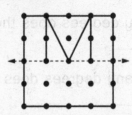

③ **Writing/Reasoning** Explain how you knew what to draw for Problem 2.

_____

_____

_____

Match the fraction with the shape that is the ONE for that fraction.

④ If is $\frac{1}{3}$,

the ONE is _____.

Ⓐ

⑤ If 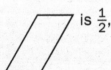 is $\frac{1}{2}$,

the ONE is _____.

Ⓑ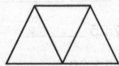

⑥ If 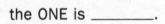 is $\frac{3}{4}$,

the ONE is _____.

Ⓒ

⑦ If 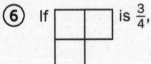 is $\frac{1}{2}$,

the ONE is _____.

Ⓓ

Use with or after Lesson 5-12.

Practice Set 65

**Solve. Write a number model with a letter for the unknown.
Then record the answer with the correct unit.**

(1) Mrs. Urban makes clocks out of wood from old fences and playgrounds. She sells the large clocks for $100 each and the small clocks for $60 each. At an art show, she sells 8 large clocks and 12 small clocks. How much did she make in all?

Number model with unknown: _____

Answer (with unit): _____

(2) The Bisons football team scored 4 touchdowns and 2 field goals. Touchdowns are worth 6 points each and field goals are worth 3 points each. How many points did the Bisons score altogether?

Number model with unknown: _____

Answer (with unit): _____

The students in Ms. Gonzalez's class estimate the time they spent on homework each week. The tally chart shows the data collected.

(3) Construct a line plot for the data.

Number of Hours per Week Spent Studying	Number of Students
3	⊮
$3\frac{1}{2}$	/
4	//
$4\frac{1}{2}$	⊮ /
5	⊮ ⊮
$5\frac{1}{2}$	///
6	//

(4) What is the difference between the greatest number of hours spent studying and the least number of hours spent studying? _____ hours

(5) How many students studied more than 5 hours? _____ students

Practice Set 66

Write a basic division fact and an extended division fact for each Fact Triangle.

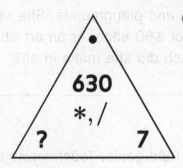

① **630**
*, /
?   7

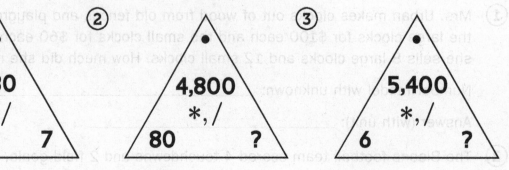

② **4,800**
*, /
80   ?

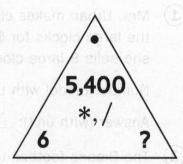

③ **5,400**
*, /
6   ?

Basic fact: _____   Basic fact: _____   Basic fact: _____

Extended fact: Extended fact: Extended fact:

_____   _____   _____

Write the missing numbers below.

④ 30 * _____ = 150   ⑤ _____ * 8 = 4,800   ⑥ _____ / 70 = 6

⑦ 30 * _____ = 300   ⑧ 1,800 / _____ = 90   ⑨ _____ * 700 = 4,200

⑩ _____ * 40 = 320   ⑪ _____ / 20 = 70   ⑫ 200 / _____ = 5

⑬ 5 * _____ = 2,500   ⑭ _____ / 7 = 40   ⑮ 3,000 / _____ = 60

Compare. Write >, < or =.

⑯ 15.26 ____ 16   ⑰ 19.62 ____ 19.62   ⑱ 201.5 ____ 200.9

⑲ 4,023.1 ____ 4,023.10   ⑳ 21.63 ____ 20.36   ㉑ 2.60 ____ 2.61

㉒ 10.78 ____ 17.8   ㉓ 39.09 ____ 39.10   ㉔ 5,213.2 ____ 5,213.15

㉕ 15.80 ____ 15.8   ㉖ 19.6 ____ 19.56   ㉗ 852.01 ____ 852.01

Practice Set 67

Fill in the unknown information for the rectangular and square areas below.

	Length in Units	Width in Units	Area in Square Units
①	8		64
②		2	12
③	5	8	
④	3		18
⑤		7	49

⑥ The bases on a baseball diamond are placed exactly 90 feet apart.

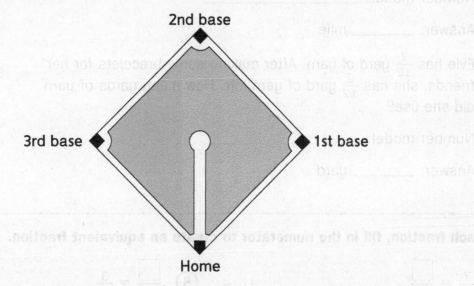

2nd base

3rd base   1st base

Home

If a batter hits a home run, how many feet does she run? _____ feet

⑦ If there are runners on first base and third base when a batter hits a home run, what is the total distance all three players run? _____ feet

Add.

⑧ $5\frac{1}{8} + 2\frac{6}{8} =$ _____   ⑨ $1\frac{4}{5} + 6\frac{2}{5} =$ _____   ⑩ $7\frac{1}{2} + 6 + 2\frac{1}{2} =$ _____

⑪ $3\frac{4}{6} + 3\frac{4}{6} =$ _____   ⑫ $6\frac{3}{10} + 1\frac{4}{10} + 2\frac{3}{10} =$ _____   ⑬ $2\frac{9}{12} + 5\frac{4}{12} =$ _____

NAME _____ DATE _____

Practice Set 68

(1) How many 6s are in 342?

10 [6s] = _____   Number model with unknown: _____

20 [6s] = _____   Answer: _____

30 [6s] = _____   Number model with answer: _____

40 [6s] = _____

50 [6s] = _____

(2) Ashton walked $\frac{1}{10}$ mile to the park, and then $\frac{4}{10}$ mile to his friend's house. How far did he walk in all?

Number model: _____

Answer: _____ mile

(3) Evie has $\frac{7}{12}$ yard of yarn. After making some bracelets for her friends, she has $\frac{2}{12}$ yard of yarn left. How many yards of yarn did she use?

Number model: _____

Answer: _____ yard

For each fraction, fill in the numerator to create an equivalent fraction.

(4) $\frac{7}{10} = \frac{\square}{100}$

(5) $\frac{\square}{100} = \frac{3}{10}$

(6) $\frac{80}{100} = \frac{\square}{10}$

(7) $\frac{20}{100} = \frac{\square}{10}$

(8) $\frac{\square}{10} = \frac{50}{100}$

(9) $\frac{\square}{100} = \frac{1}{10}$

(10) $\frac{40}{100} = \frac{\square}{10}$

(11) $\frac{9}{10} = \frac{\square}{100}$

Use with or after Lesson 6-3.

Practice Set 69

Estimate. Write a number model with an unknown to represent the problem. Then solve using partial-quotients division.

① Mr. Kotter's class paid a total of $184 for a field trip to the history museum. Each student paid $8. How many students are in Mr. Kotter's class?

Estimate: _____

Number model with unknown: _____

Answer: _____ students

Complete the "What's My Rule?" tables.

②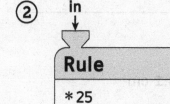

in	out
3	
4	
8	
	250
16	

③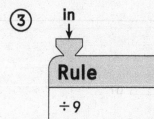

in	out
81	
54	
	12
117	
	7

Draw a picture. Write a number model. Then solve.

④ Juan has 12 apple slices, which he gives to his friends. Each friend gets 3 slices. How many friends get apple slices?

Number model: _____

Answer (with unit): _____

⑤ Nika collected 24 rocks. She shares them equally between 7 of her classmates and herself. How many rocks does each person receive?

Number model: _____

Answer (with unit): _____

Practice Set 70

① Mr. Wyatt is arranging 68 trading cards into an album. Each page in the album holds 9 trading cards. How many pages will Mr. Wyatt need? Show or explain how you know.

Answer (with units): _____

Complete.

② 500 cm = _____ m

③ _____ mm = 8.1 cm

④ _____ mm = 72 cm

⑤ 150 cm = _____ mm

⑥ 0.35 m = _____ cm

⑦ 63 cm = _____ m

⑧ _____ mm = 9.8 cm

⑨ _____ m = 375 cm

⑩ **Writing/Reasoning** Explain how you found the answer to Problem 9.

_____

_____

_____

Use digits to write the numbers.

⑪ one hundred seventy-five thousand, three hundred _____

⑫ twenty and eight-tenths _____

⑬ three million, two thousand, six hundred four _____

Use with or after Lesson 6-5.

Practice Set 71

Use the statements below to help you solve the problems. Be sure to include the unit in the answer.

- The average person throws away about 5 pounds of trash per day.
- One ton is equal to 2,000 pounds.
- There are about 300 million people in the United States.

① How much trash does the average person throw away in one week?

_____

② How much trash does the average person throw away in one year?

_____

③ About how many tons is that? _____

④ About how many tons of trash does the average family of four throw away in one year? _____

⑤ Does the population of the United States produce more or less than 10 million tons of trash per year? _____

⑥ About how many tons of trash does the United States produce in one year? _____

⑦ **Writing/Reasoning** Explain how you found the answer to Problem 6.

_____

_____

_____

Estimate the answer. Write a number model to show how you estimated.

⑧ There are 12 cans in each case. How many cans are in 37 cases?

_____

⑨ A large can of peaches holds 112 ounces. How many ounces are in 6 cans?

_____

⑩ Ellen uses 1 gallon of lemonade to serve 22 people. How many people could she serve with 53 gallons of lemonade?

_____

Practice Set 72

Estimate. Write a number model to represent the unknown. Then use partial-quotients division to solve.

(1) James is reading a book with 245 pages. If he reads 5 pages each day, how long will it take him to read the entire book?

Estimate: _____ days

Number model with unknown:

_____

(2) Maddy and 3 friends make $113 at a lemonade stand. If the profits are split equally, how many whole dollars will each person get?

Estimate: $_____

Number model with unknown:

_____

Answer: _____ days

How many pages are left over?

_____ pages

Answer: $_____

How much money is left over?

$_____

---

(3) 40 is _____ times as many as 8.

(4) 18 is 9 times as many as _____.

(5) 60 is _____ times as many as 12.

(6) 56 is 7 times as many as _____.

(7) 33 is _____ times as many as 3.

(8) 72 is _____ times as many as 9.

(9) 36 is 4 times as many as _____.

(10) 64 is _____ times as many as 8.

Use with or after Lesson 6-7.

Practice Set 73

Write a number model to represent the unknown. Then use partial-quotients division to solve.

① A total of 129 fourth graders are going on a field trip to the botanical garden. Each van will hold 6 fourth graders. How many vans are needed?

Number model with unknown:

_____

② A shipment of 397 vases is being prepared. Each box can hold 8 vases. How many boxes are needed for the shipment?

Number model with unknown:

_____

Answer: _____ vans

What did you do about the remainder?

_____

_____

Answer: _____ boxes

What did you do about the remainder?

_____

_____

Write the numbers in order from least to greatest.

③ 1.2, 0.2, 2.10, 2.2

_____

④ 0.23, 1.2, 0.04, 5.1

_____

⑤ 4.01, 1.4, 2.14, 1.41

_____

⑥ 9.5, 1.95, 19.5, 0.59

_____

⑦ 0.3, 3.3, 0.03, 3.03

_____

⑧ 5.20, 5.12, 5.02, 5.21

_____

Practice Set 74

Use your angle measurer to estimate the angle measures. Label each angle as *right*, *acute*, or *obtuse*.

①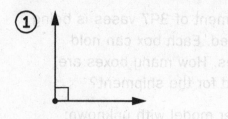

Estimate: _____

Type: _____

②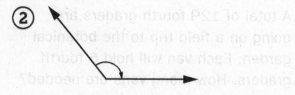

Estimate: _____

Type: _____

③

Estimate: _____

Type: _____

④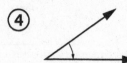

Estimate: _____

Type: _____

Mr. Lopez's class is growing pea plants. The heights of the students' plants are shown in the table below.

⑤ Construct a line plot for the data.

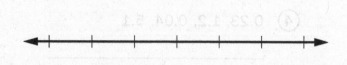

Height in Inches	Number of Plants
2	/
$2\frac{1}{4}$	///
$2\frac{1}{2}$	✝✝✝
$2\frac{3}{4}$	/
3	//
$3\frac{1}{4}$	✝✝✝ /
$3\frac{1}{2}$	///

⑥ Write a question that can be answered using the line plot you created. Answer your question.

_____

_____

_____

NAME DATE

SRB 208-210

Practice Set 75

Use a half-circle protractor and a straightedge to draw angles with the following measures.

① 20°

② 145°

③ 72°

④ 95°

Write number sentences for the following. Then tell whether they are *true* or *false*.

⑤ 6 is 2 times as much as 12.

__________ ______

⑥ 81 is 9 times as much as 8.

__________ ______

⑦ 15 is 5 times as much as 5.

__________ ______

⑧ 110 is 11 times as much as 10.

__________ ______

⑨ 20 is 4 times as much as 5.

__________ ______

⑩ 72 is 12 times as much as 8.

__________ ______

Practice Set 76

Find the unknown angle measures. Do *not* use a protractor.

①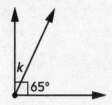

Equation with unknown: _____

Answer: _____°

②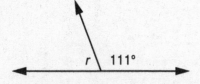

Equation with unknown: _____

Answer: _____°

③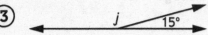

Equation with unknown: _____

Answer: _____°

④ If 10 counters are $\frac{1}{2}$, then _____ counters are ONE.

⑤ If 3 counters are $\frac{1}{4}$, then _____ counters are ONE.

⑥ If 8 counters are $\frac{2}{3}$, then _____ counters are ONE.

⑦ If 70 counters are $\frac{7}{4}$, then _____ counters are ONE.

⑧ If 4 counters are $\frac{1}{5}$, then _____ counters are ONE.

⑨ If 3 counters are $\frac{3}{7}$, then _____ counters are ONE.

⑩ If 15 counters are $\frac{5}{8}$, then _____ counters are ONE.

Practice Set 77

**Write a number model with an unknown to represent each problem.
Then solve.**

① Ms. Helton's fourth-grade class recycled $10\frac{5}{8}$ pounds of paper in
January and $13\frac{6}{8}$ pounds paper in February. How many total
pounds of paper did they recycle in the two months?

Number model with unknown: _____

Answer: _____ pounds

What is the weight, in ounces, of the amount of paper recycled

in January? _____ ounces   in February? _____ ounces

in both months? _____ ounces

② A beluga whale is $6\frac{2}{3}$ yards long. A killer whale is 10 yards long.
How much longer is the killer whale?

Number model with unknown: _____

Answer: _____ yards

What is the length, in feet, of each whale?

beluga whale: _____ feet   killer whale: _____ feet

_____

③ What is the value of 8 in 1,682? _____

Now write a number where the 8 is worth 10 times as much. _____

④ What is the value of 1 in 100,514? _____

Now write a number where the 1 is worth 10 times as much. _____

⑤ What is the value of 5 in 95,710? _____

Now write a number where the 5 is worth 10 times as much. _____

⑥ What is the value of 9 in 6,903? _____

Now write a number where the 9 is worth 10 times as much. _____

⑦ What is the value of 3 in 463,221? _____

Now write a number where the 3 is worth 10 times as much. _____

Practice Set 78

(1) Each day Xavier practices on his violin for $\frac{3}{4}$ hour. How many hours will he have practiced in 5 days?

Drawing:

Words: $\frac{3}{4}$ hour for 5 days = _____ hours

Multiplication equation: $\frac{3}{4} * 5 =$ _____

Hours Practiced per Day	Number of Days	Total Hours Practiced
$\frac{3}{4}$	5	

Answer: _____ hours

Compare. Write >, <, or =.

 (2) 629.12 _____ 629.12

(3) 13.09 _____ 13.9

(4) 8.20 _____ 8.2

(5) 1.96 _____ 2.1

(6) 441.71 _____ 441.17

(7) 78.58 _____ 75.85

Measure each angle using a half-circle protractor.

(8)

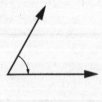

_____ °

(9)

_____ °

(10)

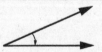

_____ °

(11)

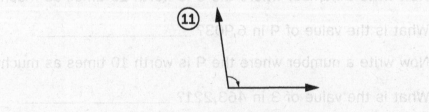

_____ °

Practice Set 79

Complete.

(1) 4 cups = _____ pints

(2) 3 quarts = _____ cups

(3) _____ gallons = 8 quarts

(4) _____ pints = 10 cups

(5) $1\frac{1}{2}$ quarts = _____ cups

(6) 3 gallons = _____ quarts

Write an equation with an unknown to represent and solve each number story.

(7) Each hour 11 canoes and 33 kayaks are rented from the Rivers Rental Shop. How many times more kayaks do people rent than canoes?

Equation with unknown: _____ Answer: _____ times

(8) There are 27 children and 9 adults in line for a ride at an amusement park. How many times more children than adults are in line?

Equation with unknown: _____ Answer: _____ times

(9) A smoothie shop sold 24 banana berry smoothies and 12 blueberry pineapple smoothies. How many times more banana berry smoothies were sold than blueberry pineapple smoothies?

Equation with unknown: _____ Answer: _____ times

---

(10) Draw a figure with exactly 3 lines of symmetry. Draw the lines of symmetry.	(11) Draw a figure with exactly 1 line of symmetry. Draw the line of symmetry.	(12) Draw a figure with exactly 5 lines of symmetry. Draw the lines of symmetry.

Practice Set 80

Solve the number stories. In the space below each problem, use pictures or equations to show what you did to find your answers.

① Tobin and his sister sold 5 pitchers of lemonade at their lemonade stand. Each pitcher holds $\frac{6}{10}$ gallon of lemonade. How many gallons of lemonade did they sell in all?

_____ gallons

② Yuka runs $\frac{3}{5}$ mile every day for a week. How many miles does she run in all in one week?

_____ miles

Find the unknown angle measures. Do _not_ use a protractor.

③

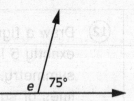

Equation with unknown: _____

Answer: _____ °

④

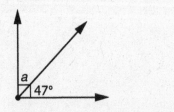

Equation with unknown: _____

Answer: _____ °

Use with or after Lesson 7-2.

NAME _____   DATE _____

Practice Set 81

Complete with a unit fraction.

① $\frac{5}{6}$ is a multiple of the unit fraction _____.

② $\frac{3}{2}$ is a multiple of the unit fraction _____.

③ $\frac{2}{12}$ is a multiple of the unit fraction _____.

④ $\frac{8}{7}$ is a multiple of the unit fraction _____.

At the beginning of soccer practice, each of the players ran a lap around the soccer fields. The time it took for each player to run a lap, in minutes, is shown in the table.

Time (in minutes)	Number of Players
5	/
$5\frac{1}{4}$	///
$5\frac{1}{2}$	////
$5\frac{3}{4}$	/
6	///
$6\frac{1}{4}$	
$6\frac{1}{2}$	/

⑤ Construct a line plot for the data.

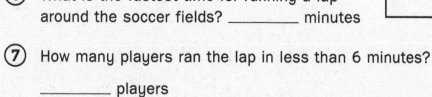

⑥ What is the fastest time for running a lap around the soccer fields? _____ minutes

⑦ How many players ran the lap in less than 6 minutes?

_____ players

⑧ How many more players ran the lap in $5\frac{1}{2}$ minutes than in $6\frac{1}{2}$ minutes?

_____ players

⑨ What is the difference between the fastest and slowest times for running

the lap? _____ minutes

⑩ **Writing/Reasoning** Explain how you found your answer to Problem 9.

_____

_____

Practice Set 82

1. Wenbi is making a tower by stacking blocks that are each $2\frac{1}{10}$ inches high. If she stacks 8 blocks, how tall will her tower be, in inches?

   A Draw a picture.

   B Write a multiplication equation. _____

   C Answer: _____ inches

   D Between what two whole numbers does your answer lie?

   _____ and _____

Use a half-circle protractor and a straightedge to draw angles with the following measures.

2. 110°

3. 30°

4. 86°

5. 163°

Use with or after Lesson 7-4.

Practice Set 83

Multiply.

(1) $4 * 2\frac{2}{5} =$ _____	(2) $6 * 4\frac{1}{9} =$ _____
(3) $3 * 1\frac{3}{4} =$ _____	(4) $5 * 3\frac{4}{7} =$ _____

Complete.

(5) 20 cm = _____ mm   (6) 5,000 mm = _____ cm

(7) 20,000 mm = _____ cm   (8) 2 m = _____ cm

(9) 15 m = _____ cm   (10) 2,000 m = _____ cm

A threadworm is about 30 cm long.

(11) What is its length in millimeters? _____ mm

(12) What is its length in meters? _____ m

Solve.

(13)   88,574
    − 9,695

(14)   5,983
   + 11,389

(15)   96,312
   − 45,160

(16)   73,418
   − 24,972

Practice Set 84

You may choose tools such as fraction circles or the Number-Line Poster to help you solve these problems. Write *yes* or *no* to tell whether each fraction makes an exact number of wholes. Explain or show why or why not for each.

① $\frac{16}{4}$ Exact number of wholes? _____

② $\frac{42}{8}$ Exact number of wholes? _____

③ $\frac{21}{2}$ Exact number of wholes? _____

④ $\frac{30}{6}$ Exact number of wholes? _____

Draw the other half of the symmetrical shapes below.

⑤

⑥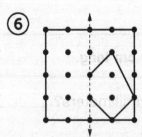

Write the following numbers with digits.

⑦ four million, seventy-nine thousand _____

⑧ eighty-eight thousand, fifteen _____

⑨ six hundred two thousand, three hundred three _____

⑩ nine million, five thousand, two hundred eleven _____

⑪ twenty-nine thousand, forty _____

Use with or after Lesson 7-6.

SRB
65-68
82-84

Practice Set 85

Make an estimate, and write an equation with an unknown. Then solve.

① Mr. Lin paid a total of $162 for 9 plastic water bottles. Ms. Kawai paid a total of $147 for 7 glass water bottles. Which costs more, a plastic water bottle or a glass water bottle? How much more?

Estimate: _____

_____

Equation with letter for unknown: _____

A _____ water bottle is $_____ less than a _____ water bottle.

Complete the "What's My Rule?" tables and rule boxes.

②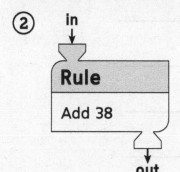

in	out
39	
412	
715	
925	
1,601	

③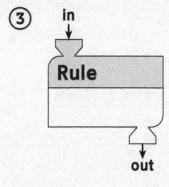

in	out
7	490
10	700
	420
11	
3	210

④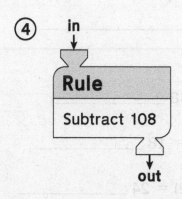

in	out
808	
160	
	90
2,400	
	1,300

⑤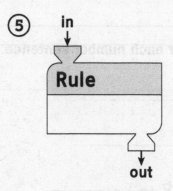

in	out
160	40
440	110
	15
800	
4,440	

Practice Set 86

Write a number model with an unknown. Then solve.

(1) A melon has a mass of 6 kilograms. Mr. Gibbs uses some of the melon in a fruit salad. Then he evenly divides 1,200 grams of melon between his three children. What is the mass of the melon, in grams, that he gave to each of his children?

Number model with unknown: _____

Answer: _____ grams

(2) Green sea turtles are one of the largest sea turtles in the world. End to end, five green sea turtles are 7.5 meters long. If each turtle is the same length, how long is each green sea turtle, in centimeters?

Number model with unknown: _____

Answer: _____ centimeters

Solve.

(3) $5 * \frac{1}{6} =$ _____

(4) $\frac{1}{10} * 8 =$ _____

(5) $\frac{1}{3} * 7 =$ _____

(6) $4 * \frac{1}{9} =$ _____

(7) $\frac{1}{12} * 2 =$ _____

(8) $\frac{1}{7} * 7 =$ _____

(9) $9 * \frac{1}{5} =$ _____

(10) $6 * \frac{1}{3} =$ _____

(11) $\frac{1}{10} * 10 =$ _____

(12) $\frac{1}{4} * 12 =$ _____

Write *true* or *false* for each number sentence.

(13) $14 + 13 = 27$ _____

(14) $6 * 9 = 48$ _____

(15) $4 * 7 < 30$ _____

(16) $41 - 25 = 18$ _____

(17) $18 * 3 = 54 + 6$ _____

(18) $4 * (6 + 2) = 24$ _____

Use with or after Lesson 7-8.

Practice Set 87

① Draw the next two rectangular numbers.

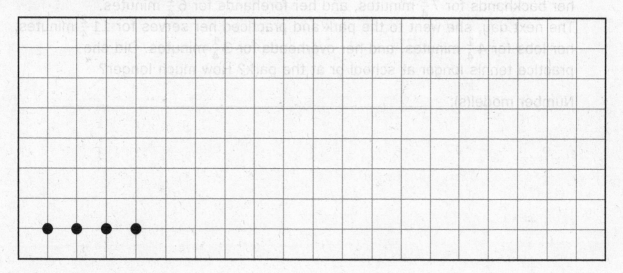

② Describe the pattern. _____

_____

Find the unknown angle measures. Do *not* use a protractor.

③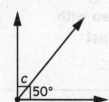

Equation with unknown:

Answer: _____°

④

Equation with unknown:

_____

Answer: _____°

⑤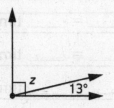

Equation with unknown:

_____

Answer: _____°

⑥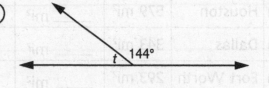

Equation with unknown:

_____

Answer: _____°

Practice Set 88

(1) At her school's tennis courts, Riley practiced her volleys for $7\frac{3}{4}$ minutes, her backhands for $7\frac{1}{4}$ minutes, and her forehands for $6\frac{2}{4}$ minutes. The next day, she went to the park and practiced her serves for $11\frac{2}{4}$ minutes, her lobs for $4\frac{1}{4}$ minutes, and her overheads for $3\frac{1}{4}$ minutes. Did she practice tennis longer at school or at the park? How much longer?

Number model(s): _____

_____

Answer: Riley practiced _____ minutes longer at _____.

How many seconds is this? _____ seconds

Sitka, Alaska, has a land area of 2,870 square miles. Compare this area with the area of several cities in Texas by filling in the table below. In the last column, the area of Sitka has been rounded to the nearest thousand.

Cities of Texas				
City	Area	Area (rounded to the nearest hundred)	Estimate the number of times it would fit in the area of Sitka.	Divide the rounded areas. (Sitka area ÷ city area)
(2) Houston	579 mi²	_____ mi²		3,000 ÷ _____ = _____ times
(3) Dallas	343 mi²	_____ mi²		3,000 ÷ _____ = _____ times
(4) Fort Worth	293 mi²	_____ mi²		3,000 ÷ _____ = _____ times
(5) Lubbock	115 mi²	_____ mi²		3,000 ÷ _____ = _____ times
(6) Waco	84 mi²	_____ mi²		3,000 ÷ _____ = _____ times

Use with or after Lesson 7-10.

SRB
188-192

Practice Set 89

Convert from pounds to ounces.

①

Pounds	Ounces
1	
4	
12	
20	
35	

②

Pounds	Ounces
$\frac{1}{2}$	
$\frac{1}{4}$	
$\frac{1}{8}$	
$\frac{3}{8}$	
$\frac{7}{8}$	

Solve.

③ Two paper clips have a mass of about 1 gram.

About how many paper clips have a mass of 10 grams? _____ paper clips

④ About how many paper clips have a mass of 1 kilogram?

(1 kilogram = 1,000 grams) _____ paper clips

⑤ About how much does a box of 1,000 paper clips weigh
if the empty box has a mass of 15 grams? about _____ grams

⑥ One ounce is about 30 grams. About how many
paper clips weigh 1 ounce? _____ paper clips

⑦ About how many paper clips weigh 1 pound? _____ paper clips

---

Solve.

⑧ $8 * 9 = $ _____

⑨ $96 \div 8 = $ _____

⑩ $60 \div 12 = $ _____

⑪ $12 * 7 = $ _____

⑫ $6 * 11 = $ _____

⑬ $9 * 50 = $ _____

⑭ $60 * 40 = $ _____

⑮ $800 \div 10 = $ _____

⑯ $100 * 11 = $ _____

⑰ $70 * 12 = $ _____

Use with or after Lesson 7-11.

Practice Set 90

Write an equivalent decimal for each fraction.

① $\frac{1}{100}$ = _____   ② $\frac{74}{100}$ = _____   ③ $\frac{3}{10}$ = _____   ④ $\frac{98}{100}$ = _____

⑤ $\frac{4}{10}$ = _____   ⑥ $\frac{14}{100}$ = _____   ⑦ $\frac{39}{100}$ = _____   ⑧ $\frac{6}{100}$ = _____

⑨ $\frac{7}{10}$ = _____   ⑩ $\frac{55}{100}$ = _____   ⑪ $\frac{2}{10}$ = _____   ⑫ $\frac{81}{100}$ = _____

Measure each angle using a half-circle protractor.

⑬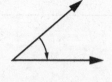

_____°

⑭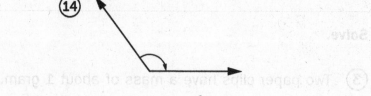

_____°

⑮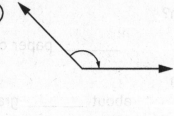

_____°

⑯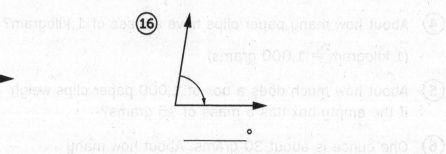

_____°

⑰ Dakota draws a 160° angle, which he divides into two smaller angles. He says one of the angles is a right angle and the other is an acute angle. Is this possible? _____ Show or explain how you know.

Practice Set 91

Stella delivers newspapers every Saturday. She keeps track of the total number of hours her paper route takes each week:

$$2, 1\frac{3}{4}, 2\frac{1}{4}, 2\frac{2}{4}, 1\frac{2}{4}, 2, 1\frac{3}{4}, 3, 1\frac{3}{4}, 2\frac{1}{4}, 1\frac{3}{4}, 2$$

(1) Construct a line plot for the data.

(2) How many times did it take Stella $2\frac{1}{4}$ hours to complete her

paper route? _____ times

(3) What is the amount of time it most often took Stella to complete

her route? _____ hours

(4) What is the shortest time it took Stella to complete her route?

_____ hours

(5) What is the longest time it took Stella to complete her route?

_____ hours

(6) What is the difference between the shortest and longest times for

completing her route? _____ hours

Solve.

(7)	(8)	(9)	(10)
440 115 + 711	7,910 +2,896	784 − 426	23 * 8

(11)	(12)	(13)	(14)
52 * 9	263 1,357 + 195	4,315 − 78	96 * 23

Practice Set 92

① A salesperson bought a case of 8 laptops for $4,800. He sold
3 laptops for $850 each at an electronics fair, and he sold the rest
online for $775 each. How much was the salesperson's profit?

Number model with unknown: _____

Answer: $ _____

② Oliver earns $12 an hour doing yard work, and he also takes care of
neighbors' pets. Last month he did yard work for 24 hours and pet
sitting for 13 hours. He made a total of $366. How much does
Oliver make per hour pet sitting?

Number model with unknown: _____

Answer: $ _____

Divide.

③ 37 ÷ 2 = _____   ④ 2,001 ÷ 2 = _____   ⑤ 183 ÷ 5 = _____

⑥ 2,264 ÷ 8 = _____   ⑦ 680 ÷ 7 = _____   ⑧ 1,773 ÷ 3 = _____

Write the amounts using a dollar sign and a decimal point.

⑨ $1 ⓆⓆⓆⓆⓆⓆⓆⒹⒹⒹⒹⒹⒹⓅⓅⓅ $_____

⑩ $5 $1 $1 ⓆⓆⓆⓃⓃⓃ $_____

⑪ $100 $20 $5 $5 $5 $1 $1 $_____

⑫ ⓆⓆⓆⓆⓆⓆⓃⓃⓃⓅⓅ $_____

⑬ $5 $5 $1 $1 ⓆⒹⓃⓅⓅ $_____

⑭ $20 $20 $10 $5 ⓆⓆⓆⓆⓃⓃ $_____

Use with or after Lesson 8-1.

Practice Set 93

① One of a square's angles is divided into two smaller angles. One angle measures 63°. What is the measure of the other angle? _____°

② One of a rectangle's angles is divided into three smaller angles. One angle measures 16°. A second angle measures 55°. What is the measure of the third angle? _____°

③ A clock's minute hand turns 360° in one hour. If the hand has already turned 315°, how many more degrees must it turn this hour? _____°

④ A clock's minute hand has 11° more to turn until it has turned a full 360°. How many degrees has it already turned? _____°

Write the number model. Then solve.

⑤ Kian has 516 trading cards, which is 4 times as many trading cards as Leo has. How many trading cards does Leo have?

Number model: _____

Answer (with unit): _____

⑥ Alex earned $59 in one day. His dad earned 6 times as much in one day. How much did his dad earn?

Number model: _____

Answer (with unit): _____

⑦ Meryl has a collection of glass animals. She has a total of 40 glass animals, which she stores in boxes. One box holds 8 glass animals. The total number of glass animals she has is how many times more than the number of animals she stores in one box?

Number model: _____

Answer (with unit): _____

⑧ To raise money, the nature club sold boxes of note cards. Sierra sold 6 boxes. Her sister sold 15 times as many boxes. How many boxes of note cards did Sierra's sister sell?

Number model: _____

Answer (with unit): _____

Practice Set 94

① Jace places a square, a triangle, and a narrow rhombus pattern
block on his desk. He knows that each angle of a square is a right
angle and measures 90°. He wants to find the measure of one of
the triangle's angles. He fits both a narrow rhombus angle and
a triangle angle together to form one of the square's angles.
Jace knows the measure of the narrow rhombus angle is 30°.
What is the measure of the triangle angle? _____°

Draw a picture and explain how you know.

First estimate and then use a half-circle protractor to measure each angle.

②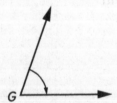

This angle is _____ (>, <) 90°.

∠G: _____°

③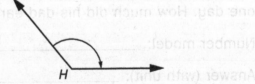

This angle is _____ (>, <) 90°.

∠H: _____°

④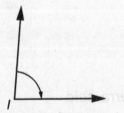

This angle is _____ (>, <) 90°.

∠I: _____°

⑤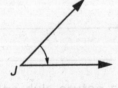

This angle is _____ (>, <) 90°.

∠J: _____°

Use with or after Lesson 8-3.

Practice Set 95

Make your own 9-Patch Patterns on 3-by-3 grids. Use squares, rectangles, and triangles to make your patterns.

① Make a pattern that has 2 lines of symmetry.

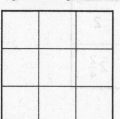

② Make a pattern that has no lines of symmetry.

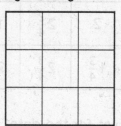

Use what you know about equivalent fractions to add. Write an equation to show your work.

③ 45 hundredths + 3 tenths

Equation: _____

④ 6 tenths + 10 hundredths

Equation: _____

⑤ 45 hundredths + 2 tenths + 3 tenths

Equation: _____

⑥ $\frac{1}{10} + \frac{86}{100}$

Equation: _____

⑦ $8\frac{7}{100} + \frac{4}{10}$

Equation: _____

⑧ $7\frac{5}{10} + \frac{13}{100}$

Equation: _____

Write the next three numbers in each pattern.

⑨ $7\frac{3}{4}$, $7\frac{2}{4}$, $7\frac{1}{4}$, _____, _____, _____

⑩ 0.04, 0.06, 0.08, _____, _____, _____

⑪ 0.44, 0.49, 0.54, _____, _____, _____

Practice Set 96

A scientist is measuring the lengths of red-eyed tree frogs in the rain forest. She recorded her measurements in the table below:

Frog Lengths to the Nearest $\frac{1}{4}$ inch

$1\frac{3}{4}$	$2\frac{1}{4}$	$1\frac{2}{4}$	2	$2\frac{2}{4}$	$1\frac{2}{4}$	$2\frac{3}{4}$	$1\frac{3}{4}$	$1\frac{3}{4}$	2
$2\frac{1}{4}$	2	$1\frac{2}{4}$	$1\frac{3}{4}$	$2\frac{3}{4}$	$2\frac{1}{4}$	$1\frac{3}{4}$	$2\frac{3}{4}$	$1\frac{2}{4}$	$2\frac{2}{4}$

(1) Construct a line plot for the data.

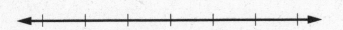

(2) What was the difference between the shortest and longest lengths measured by the scientist? _____ inches

(3) Which frog length was the most common? _____ inches

(4) Write a question that can be answering using the line plot you created. Answer your question.

_____

_____

_____

Write >, <, or = to make each number sentence true.

(5) 150 * 4 _____ 222 + 374

(6) 108 + 892 _____ 5,000 ÷ 5

(7) 85 + 423 _____ 281 + 235

(8) 9 * 12 _____ 404 ÷ 4

(9) 160 ÷ 4 _____ 95 − 56

Practice Set 97

Use a formula to find the perimeter of each rectangle. Show your work.

① $5\frac{1}{8}$ in. $2\frac{3}{8}$ in. $2\frac{3}{8}$ in. $5\frac{1}{8}$ in.

_____

② $3\frac{7}{10}$ m $1\frac{1}{10}$ m $1\frac{1}{10}$ m $3\frac{7}{10}$ m

_____

③ $3\frac{1}{2}$ cm $3\frac{1}{2}$ cm $3\frac{1}{2}$ cm $3\frac{1}{2}$ cm

_____

④ 10 ft $7\frac{2}{3}$ ft $7\frac{2}{3}$ ft   10 ft

_____

Compare. Write >, <, or =.

⑤ 0.52 _____ 0.25

⑥ 6.23 _____ 10

⑦ 179.8 _____ 179.80

⑧ 5.40 _____ 5.04

⑨ 6,170.18 _____ 6,170.3

⑩ 87.9 _____ 87.19

⑪ 561.02 _____ 561.02

⑫ 14.50 _____ 14.5

⑬ 88.08 _____ 88.80

⑭ 34.09 _____ 34.1

⑮ 293.55 _____ 239.55

⑯ 852.01 _____ 852.01

Complete with a unit fraction.

⑰ $\frac{4}{10}$ is a multiple of the unit fraction _____.

⑱ $\frac{2}{3}$ is a multiple of the unit fraction _____.

⑲ $\frac{5}{8}$ is a multiple of the unit fraction _____.

⑳ $\frac{4}{5}$ is a multiple of the unit fraction _____.

Practice Set 98

Solve. (Prices include tax.)

A

$3.99

B

$1.99

C

$1.59

① Ms. Jackson wants to buy enough crayons to give one crayon to each of her 29 students. She has $3.50.

Ⓐ What can she buy? _____

Ⓑ How many crayons will she have left over? _____

② How many boxes of 16 crayons would it take to equal the number in the 64-crayon box? _____

③ How much would this cost? _____

④ **Writing/Reasoning** Estimate whether $18 is enough to buy 5 boxes of 64 crayons. Explain your reasoning.

_____

_____

Solve.

⑤ 29 * 3 = _____   ⑥ 57 * 8 = _____   ⑦ 495 * 6 = _____

⑧ 307 * 4 = _____   ⑨ 86 * 71 = _____   ⑩ 34 * 19 = _____

Practice Set 99

Use a formula to find the area of each rectangle. Show your work.

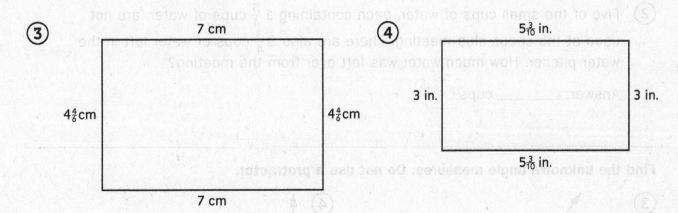

① 2 yd
$1\frac{7}{8}$ yd $1\frac{7}{8}$ yd
2 yd

Area: _____ square yards

② $\frac{2}{5}$ mm   8 mm $\frac{2}{5}$ mm
8 mm

Area: _____ square millimeters

③ 7 cm
$4\frac{4}{6}$ cm $4\frac{4}{6}$ cm
7 cm

Area: _____ square centimeters

④ $5\frac{3}{10}$ in.
3 in.   3 in.
$5\frac{3}{10}$ in.

Area: _____ square inches

Who am I?

⑤ Clue 1: I am less than 20.

Clue 2: I am an odd number.

Clue 3: I am the third multiple of 5.

I am _____.

⑥ Clue 1: I am less than 1,000

Clue 2: The sum of my digits is 26.

Clue 3: My ones digit is 8.

Clue 4: The other two digits are identical to each other.

I am _____.

Practice Set 100

Solve the number stories below. Use equations or drawings to show how you solved each problem.

(1) At a chess club meeting, Mr. Samuel pours $1\frac{3}{4}$ cups of water into each of 20 small cups for the student members and $2\frac{1}{4}$ cups of water into each of 3 large cups for the adult advisors. How many total cups of water does he pour?

Answer: _____ cups

(2) Five of the small cups of water, each containing $1\frac{3}{4}$ cups of water, are not used at the chess club meeting. There are also $3\frac{3}{4}$ cups of water left in the water pitcher. How much water was left over from the meeting?

Answer: _____ cups

Find the unknown angle measures. Do *not* use a protractor.

(3)

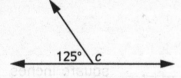

125° c

Equation with unknown: _____

Answer: _____°

(4)

j

31°

Equation with unknown: _____

Answer: _____°

(5) In the number 906, if 9 where moved once place to the left, what would it be worth? _____

(6) In the number 14,305, if 4 where moved once place to the left, what would it be worth? _____

(7) In the number 22.1, if 1 where moved once place to the left, what would it be worth? _____

(8) In the number 7,655,324, if 6 where moved once place to the left, what would it be worth? _____

(9) In the number 8.03, if 3 where moved once place to the left, what would it be worth? _____

Practice Set 101

Complete the tables.
Add your own input and output values in the last row.

① in ↓

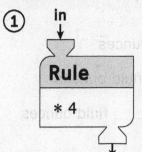

Rule

* 4

out ↓

in	out
	4 quarts
$2\frac{1}{4}$ quarts	
10 quarts	

② in ↓

Rule

$-\frac{1}{8}$ pint

out ↓

in	out
	$\frac{7}{8}$ pint
$2\frac{3}{8}$ pints	
$5\frac{1}{8}$ pints	

③ in ↓

Rule

$+ 1\frac{1}{2}$ cups

out ↓

in	out
3 cups	
	$8\frac{1}{2}$ cups
$5\frac{1}{2}$ cups	

④ in ↓

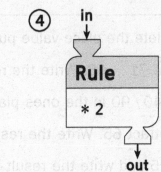

Rule

* 2

out ↓

in	out
20 gallons	
$10\frac{5}{6}$ gallons	
	$12\frac{6}{8}$ gallons

Solve.

⑤ $\frac{1}{7} * 2 =$ _____

⑥ $9 * \frac{1}{3} =$ _____

⑦ $\frac{1}{5} * 5 =$ _____

⑧ $3 * \frac{1}{12} =$ _____

⑨ $\frac{1}{9} * 8 =$ _____

⑩ $4 * \frac{1}{10} =$ _____

⑪ $10 * \frac{1}{6} =$ _____

⑫ $\frac{1}{8} * 6 =$ _____

⑬ $\frac{1}{2} * 12 =$ _____

⑭ $\frac{1}{4} * 7 =$ _____

Write two multiplication and two division facts for each Fact Triangle.

⑮

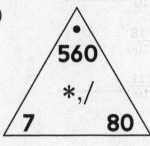

560

*,/

7   80

_____
_____
_____
_____

⑯

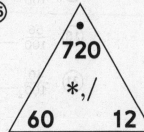

720

*,/

60   12

_____
_____
_____
_____

Practice Set 102

1. How many fluid ounces are in $3\frac{3}{4}$ gallons of orange juice? _____ fluid ounces

2. How many ounces are in $10\frac{1}{2}$ pounds of soil? _____ ounces

3. How many ounces are in $2\frac{1}{4}$ pounds of watermelon? _____ ounces

4. How many fluid ounces are in $5\frac{1}{2}$ pints of dish soap? _____ fluid ounces

5. How many fluid ounces are in $2\frac{3}{8}$ cups of salad dressing? _____ fluid ounces

6. How many ounces are in $25\frac{1}{8}$ pounds of ice? _____ ounces

7. Use the clues to complete the place-value puzzle.

   • Add 3 to the result of $71 - 68$. Write the result in the hundredths place.

   • Write the result of $540 / 90$ in the ones place.

   • Multiply $6 * 12$. Subtract 65. Write the result in the tens place.

   • Divide 24 by 6. Add 5 and write the result in the tenths place.

10s	1s		0.1s	0.01s
		.		

Rename the following fractions as decimals.

8. $\frac{1}{10}$ _____   9. $\frac{2}{100}$ _____   10. $\frac{6}{100}$ _____

11. $\frac{6}{10}$ _____   12. $\frac{50}{100}$ _____   13. $\frac{47}{100}$ _____

14. $\frac{7}{10}$ _____   15. $\frac{3}{100}$ _____   16. $\frac{9}{10}$ _____

17. $\frac{34}{100}$ _____   18. $\frac{56}{100}$ _____   19. $\frac{18}{100}$ _____

20. $\frac{5}{10}$ _____   21. $\frac{40}{100}$ _____   22. $\frac{11}{100}$ _____

Practice Set 103

Use the numbers below to create the 7-digit numbers described in the clues.
Use each number exactly once to create each 7-digit number.

6 1 3 4 8 7 2

① What is the *largest* 7-digit number you can make? _____

② What is the *smallest* 7-digit number you can make? _____

③ What is the *largest* 7-digit number you can make that starts with an odd number? _____

④ What is the *smallest* 7-digit number you can make with 1 in the hundreds place? _____

⑤ What is the *largest* 7-digit number you can make with 8 in the thousands place? _____

⑥ What is the *largest* 7-digit number you can make that ends with an even number? _____

Complete with a unit fraction.

⑦ _____ is the tenth multiple of the unit fraction $\frac{1}{12}$.

⑧ _____ is the fourth multiple of the unit fraction $\frac{1}{3}$.

⑨ _____ is the eighth multiple of the unit fraction $\frac{1}{9}$.

⑩ _____ is the fifth multiple of the unit fraction $\frac{1}{5}$.

⑪ _____ is the ninth multiple of the unit fraction $\frac{1}{7}$.

Write the number words for the following numbers.

⑫ 21,894 _____

⑬ 14.1 _____

⑭ 6,048,560 _____

⑮ 903.75 _____

Practice Set 104

Fill in the name-collection boxes. Use as many different numbers and
operations as you can.

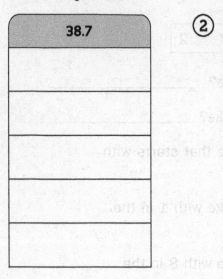

① 38.7

② 7,049

③ $8\frac{12}{100}$

Determine each pattern. Then fill in the missing numbers on the number lines.

④

$\frac{2}{10}$ ___ $\frac{4}{10}$ ___ ___ $\frac{7}{10}$ ___

⑤

3.8   4.1   ___   ___   5.0   ___

⑥

___   2,016   2,018   ___   ___

⑦

20   ___   ___   ___   ___   62

⑧ **Writing/Reasoning** How did you determine the pattern for Problem 7?

_____

_____

_____

Test Practice 1

Fill in the circle next to your answer.

① Tennessee covers an area of 42,126 square miles. What is the place value of the digit 1 in 42,146?

Ⓐ tens

Ⓑ hundreds

Ⓒ thousands

Ⓓ ten-thousands

② Which of the following attributes does the shape below **not** have?

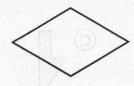

Ⓐ All angles are different sizes.

Ⓑ All sides are equal in length.

Ⓒ Opposite sides are parallel.

Ⓓ Four line segments form its sides.

③ Which number shows 915,328 rounded to the ten-thousands place?

Ⓐ 915,000

Ⓑ 916,000

Ⓒ 910,000

Ⓓ 920,000

④ Which figure below shows ray XY?

Ⓐ

Ⓑ

Ⓒ

Ⓓ

Test Practice 1 (continued)

Fill in the circle next to your answer.

Ⓢ Chen wants to draw the other half of the picture to make it symmetrical. Which figure shows the other half of the figure, after it has been drawn over the dotted line?

Ⓐ Ⓑ Ⓒ Ⓓ

Ⓖ In 1787, Delaware became a state. Florida became a state in 1845. Which expression should you use to find out how many years passed between the time Delaware became a state and Florida became a state?

Ⓐ 1845 + 1787   Ⓑ 1845 − 1787

Ⓒ 1845 * 1787   Ⓓ 1845 ÷ 1787

Ⓖ What are all the factor pairs for 24?

Ⓐ 1 and 24, 3 and 8, 4 and 6

Ⓑ 2 and 12, 3 and 8, 4 and 6

Ⓒ 1 and 24, 2 and 12, 3 and 8, 4 and 6

Ⓓ 1 and 24, 3 and 8, 4 and 6, 5 and 5

Ⓗ Tim has twice as many baseball cards as Monica. Let c stand for the number of baseball cards Monica has. Which of the following represents the number of cards Tim has?

Ⓐ 2 + c   Ⓑ c − 2

Ⓒ c * 2   Ⓓ c ÷ 2

Test Practice 2

Fill in the circle next to your answer.

① The answer to a multiplication problem is 3,120. What is the multiplication problem?

(A) 32 * 97   (B) 82 * 39

(C) 48 * 65   (D) 55 * 55

② In Mr. Johnson's fourth-grade class, $\frac{11}{20}$ of the students chose math as their favorite subject, $\frac{3}{10}$ of the students chose science as their favorite subject, and $\frac{3}{20}$ of the students chose reading as their favorite subject. Which of the statements below is true?

(A) More students like reading than science.

(B) More students like math than science.

(C) The same number of students like science and reading.

(D) The same number of students like math and reading.

③ Mandy uses these blocks to show the decimal 0.32.

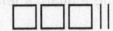

Mandy wants to write an equivalent fraction for this decimal.
Which of the fractions below are equivalent to 0.32?

(A) $2\frac{3}{10}$ (B) $3\frac{2}{10}$

(C) $\frac{32}{10}$ (D) $\frac{32}{100}$

④ What is the next number in the pattern?

0.26, 0.36, 0.46, _____

(A) 0.47   (B) 0.56

(C) 0.72   (D) 0.92

Test Practice 2 (continued)

Fill in the circle next to your answer.

(5) In a game of darts, Conrad hit 30 bull's-eyes in 100 shots. Which of the following has the same value as $\frac{30}{100}$?

(A) $\frac{300}{10}$ (B) $\frac{300}{100}$

(C) $\frac{30}{10}$ (D) $\frac{3}{10}$

(6) Which of the following belongs in the name-collection box?

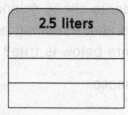

(A) 25 milliliters   (B) 250 milliliters

(C) 2,500 milliliters   (D) 25,000 milliliters

(7) A tree swallow is 5 inches long. A mallard duck is 25 inches long.

The length of the tree swallow is $\frac{5}{25}$ of the mallard's length.

Which fraction has a value equal to $\frac{5}{25}$?

(A) $\frac{1}{5}$ (B) $\frac{1}{4}$

(C) $\frac{1}{2}$ (D) $\frac{25}{2}$

(8) A stadium has 378 seats. Each week, 9 games are played in the stadium. The owner estimates that about 3,600 people can see a game each week. Which of the following correctly tells how the owner made her estimate?

(A) She multiplied 8 * 300.   (B) She multiplied 8 * 400.

(C) She multiplied 9 * 300.   (D) She multiplied 9 * 400.

Test Practice 3

Fill in the circle next to your answer.

① Greg used a protractor to draw a 75° angle. Use a half-circle protractor to determine which of the following shows the angle Greg drew.

(A)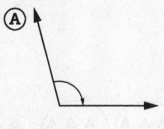

(B)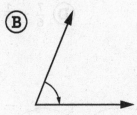

(C)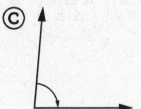

(D)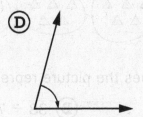

② A black-chinned hummingbird is $8\frac{75}{100}$ centimeters long.

A blue-throated hummingbird is $12\frac{8}{10}$ centimeters long.

How much longer is the blue-throated hummingbird than the black-chinned hummingbird?

(A) $3\frac{5}{100}$ centimeters

(B) $4\frac{5}{100}$ centimeters

(C) $4\frac{15}{100}$ centimeters

(D) $21\frac{55}{100}$ centimeters

③ Look at the division sentences below.

$$70 \div 7 = 10$$
$$700 \div 7 = 100$$
$$7{,}000 \div 7 = \boxed{}$$

What is the missing quotient?

(A) 100   (B) 700   (C) 1,000   (D) 7,000

Test Practice 3 (continued)

Fill in the circle next to your answer.

④ Which of the following equations shows $\frac{5}{6}$ as a sum of unit fractions?

Ⓐ $\frac{1}{6} + \frac{1}{6} + \frac{1}{6} + \frac{1}{6} + \frac{1}{6} = \frac{5}{6}$ Ⓑ $\frac{1}{6} + \frac{1}{6} + \frac{1}{6} + \frac{1}{6} = \frac{5}{6}$

Ⓒ $\frac{3}{6} + \frac{1}{6} + \frac{1}{6} + \frac{1}{6} = \frac{5}{6}$ Ⓓ $\frac{7}{6} - \frac{1}{6} = \frac{5}{6}$

⑤ Caroline drew this picture.

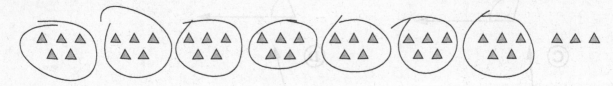

Which of the following does the picture represent?

Ⓐ $38 \div 7 = 5$ Ⓑ $38 \div 7 = 5 \text{ R}3$

Ⓒ $35 \div 7 = 5 \text{ R}3$ Ⓓ $35 \div 7 = 5$

⑥ Which of the following number sentences is **not** true?

Ⓐ $3\frac{9}{10} + 1\frac{1}{10} = 4\frac{9}{10}$

Ⓑ $2\frac{7}{8} + 2\frac{4}{8} = 5\frac{3}{8}$

Ⓒ $8\frac{4}{12} - 6\frac{1}{12} = 2\frac{3}{12}$

Ⓓ $7 - 2\frac{1}{6} = 4\frac{5}{6}$

⑦ Emily found the measure of the unknown angle below without using a protractor. Which equation did she use?

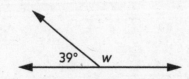

Ⓐ $39° + w = 360°$ Ⓑ $180° - 39° = w$

Ⓒ $90° - 39° = w$ Ⓓ $39° + w = 100°$

Use with or after Unit 6.

Test Practice 4

Fill in the circle next to your answer.

① What is the product of 6 and $2\frac{3}{12}$?

  Ⓐ $13\frac{6}{12}$ Ⓑ $13\frac{3}{12}$

  Ⓒ $12\frac{6}{12}$ Ⓓ $12\frac{3}{12}$

② The largest tree in the world has been named General Sherman and is located in the Sequoia National Park in California. It is 91 yards, 2 feet tall.

What is the height of the General Sherman Tree in feet?

  Ⓐ 93 feet   Ⓑ 275 feet

  Ⓒ 912 feet   Ⓓ 1,094 feet

③ Use the line plot below to answer the question.

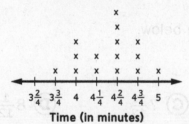

Relay Race Times

Time (in minutes)

How many relay race teams ran the race in less than $4\frac{2}{4}$ minutes?

  Ⓐ 4 teams   Ⓑ 5 teams

  Ⓒ 6 teams   Ⓓ 11 teams

④ Both the bobcat and the lynx can be found in the forests of North America. A bobcat can weigh up to 30.9 pounds, and a lynx can weigh up to 44.1 pounds. What is the difference in the weights of the two animals?

  Ⓐ 13.1 pounds   Ⓑ 13.2 pounds

  Ⓒ 14.2 pounds   Ⓓ 14.8 pounds

Test Practice 4 (continued)

Fill in the circle next to your answer.

(5) Luciana practiced on her flute for $\frac{1}{3}$ hour every day for a week.
How many hours did she practice in one week?

(A) $1\frac{1}{3}$ hours

(B) $1\frac{2}{3}$ hours

(C) 2 hours

(D) $2\frac{1}{3}$ hours

(6) Ms. Alberti is buying carpet for her study.
The picture shows the size and shape of her study.

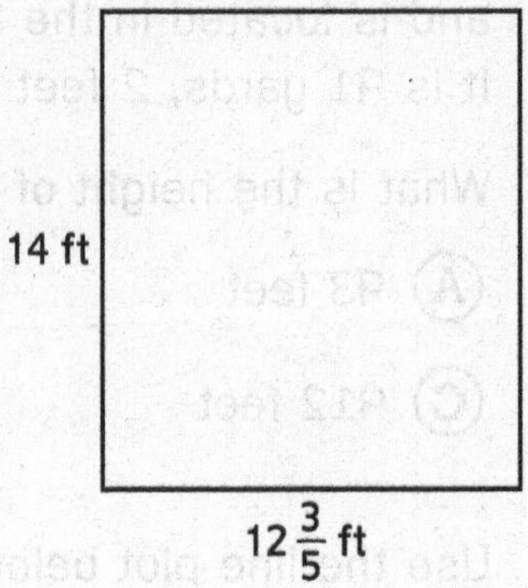

How much carpet does Ms. Alberti need?

| Area of a rectangle = length * width |

14 ft

$12\frac{3}{5}$ ft

(A) $168\frac{3}{5}$ square feet

(B) $174\frac{2}{5}$ square feet

(C) $176\frac{2}{5}$ square feet

(D) $183\frac{3}{5}$ square feet

(7) Find the unknown in the number sentence below.

$$8 * \frac{1}{12} = \square$$

(A) $\frac{8}{12}$

(B) $\frac{9}{12}$

(C) $7\frac{11}{12}$

(D) $8\frac{1}{12}$

(8) Carlos used the numbers 5, 9, 0, 3, and 4 to create a 5-digit number.
He used each number exactly once. The number he made is shown below.

| 54,390 |

Which of the clues did he follow to make the 5-digit number?

(A) Make the smallest 5-digit number possible.

(B) Make the smallest 5-digit number that starts with an odd number.

(C) Make the largest 5-digit number with 9 in the tens place.

(D) Make the largest even 5-digit number possible.

Visit classzone and get connected

Online resources for students and parents

ClassZone resources provide instruction, practice, and learning support.

eEdition Plus ONLINE

This online version of the text features interactive grammar, activities, and video and audio clips.

Online Workbook

Interactive, leveled practice supports skill development.

Webquest

Guided Web activities introduce students to real-world Spanish.

Flashcards

Interactive review of vocabulary and pronunciation includes audio prompts.

Self-Check Quizzes

Self-scoring quizzes help students assess their comprehension.

Writing Center

Unit-level writing workshops invite students to share their creations online.

Now it all clicks!™

CLASSZONE.COM

McDougal Littell

3 tres

California Edition

MᴄDᴏᴜɢᴀʟ Lɪᴛᴛᴇʟʟ

¡En español!

AUTHORS

Estella Gahala

Patricia Hamilton Carlin

Audrey L. Heining-Boynton

Ricardo Otheguy

Barbara J. Rupert

CULTURE CONSULTANT

Jorge A. Capetillo-Ponce

McDougal Littell
A HOUGHTON MIFFLIN COMPANY
Evanston, Illinois • Boston • Dallas

Cover Photography

Foreground: Photo by Martha Granger/EDGE Productions.
Background: Aerial view of Las Ramblas, Barcelona, Spain, AGE Fotostock (also appears on spine).

Back cover, top: Peruvian instrument, Wood River Gallery/PNI; From left to right: El Morro Castle, San Juan, Puerto Rico, Bruce Adams/Corbis; Quito, Ecuador, Joseph F. Viesti/The Viesti Collection; Pyramid of the Sun at Teotihuacán, Mexico City, Michael T. Sedam/Corbis; View of Arenal Volcano from Tabacón Hot Springs, Costa Rica, Kevin Schafer; Aerial view of Las Ramblas, Barcelona, Spain, AGE Fotostock; Machu Picchu, Urubamba Valley, Peru, Robert Fried

Front Matter Photography

vi *top right* R. E. Barber / Visuals Unlimited; *bottom right* Jean-Leo Dugast / Panos Pictures; **vii** *top right* courtesy of Miami Mensual; **viii** *top right* School Division, Houghton Mifflin Company; *bottom right* Frank Siteman/Stock Boston; **xi** *bottom right* Ulrike Welsch; **xii** *top right* Roy Morsch / Corbis; **xiii** *top right* School Division, Houghton Mifflin Company; **xiv** *bottom right* Robert Frerck / Odyssey Productions; *top right* Suzanne Murphy-Larronde; **xvi** *top right* School Division, Houghton Mifflin Company; **xvii** *top* Jo Prater/ Visuals Unlimited; *bottom right* "New Chicago Athletic Club," (1937) Antonio Berni. Oil on canvas, 6′ 3/4″ x 9′ x 101/4″ (184.9 cm x 600.1 cm). The Museum of Modern Art, New York, Inter-American Fund. Photograph © 2003 The Museum of Modern Art, New York; **xviii** "La ventana," Wifredo Lam. Copyright © 2003 Artists Rights Society (ARS), New York / ADAGP, Paris / Christie's Images; **xx** *top right* Justin Kerr; bottom right Liba Taylor / Panos Pictures; **xxi** Emece Editores S.A. Argentina; **xxiii** *top right* PhotoDisc; *bottom right* Courtesy of Maloka Park, Columbia; **xxiv** Editorial Canelas S.A. Bolivia; **xxx** *top* Newberry Library/Stock Montage, *bottom* G. Reynolds/Robertstock; **xxxi** Cleaning Beans, Harvesting Beans, Madera County, 1900, BANC PIC 1983.172 -- PIC, The Bancroft Library, University of California, Berkeley, *center* Michel Boutefeu, *bottom left* Corbis, *bottom right* Courtesy of JPL/NASA.

Illustration

xxxiv-xxxix Gary Antonetti/Ortelius Design.

ISBN-13: 978-0-618-30450-9
ISBN-10: 0-618-30450-9

6 7 8 9 - VJM - 09 08 07 06

Internet: www.mcdougallittell.com

CALIFORNIA LANGUAGE LEARNING CONTINUUM

The following chart indicates pages that provide benchmarks for language learning. Page numbers that appear in bold are core references that appear as annotations on the pages of the text. The remaining page numbers are supporting references that illustrate the full scope of the learning process.

Stage III	Exemplars		
Function	**Introduction**	**Practice**	**Mastery**
Clarify and ask for and comprehend clarification	p. 4	pp. 20, 38, 118, 156, 186, 234, 260, 308, 334, 342, 408, 433	p. 452
Express and understand opinions	p. 63	pp. 112, 120, 160, 190, 194, 212, 216, 238, 282, 290, 304, 360, 386, 412, 436	p. 456
Narrate and understand narration in the present	p. 8	pp. 34, 41, 57, 78, 108, 152, 182, p. 230, 300, 330, 356, 374, 378, 404, 426	p. 448
Narrate and understand narration in the past	p. 16	pp. 44, 78, 134, 272, 278, 288, 294, 374	p. 411
Narrate and understand narration in the future	p. 68	pp. 78, 152, 204, 300, 312	p. 458
Identify, state, and understand feelings and emotions	p. 85	p. 159	p. 172

Context	**Introduction**	**Practice**	**Mastery**
Converse in face-to-face social interactions	p. 24	pp. 42, 46, 60, 82, 116, 138, 162, 188, 208, 268, 286, 338, 380	p. 416
Converse in simple transactions on the phone	p. 264	p. 264	p. 420
Listen during face-to-face social interactions	p. 24	pp. 42, 46, 60, 82, 116, 138, 162, 188, 208, 268, 286, 338, 380	p. 416
Listen to audio texts	p. 36	pp. 58, 80, 110, 154, 184, 206, 228, 258, 280, 302, 332, 354, 376, 406, 428	p. 450
Listen to video texts	p. 2	pp. 6, 10, 14, 18	p. 22
Read short stories	p. 48	p. 166	p. 314
Read poems	p. 196	pp. 237, 240	p. 241
Read essays	p. 144	p. 144	p. 226

Context	Introduction	Practice	Mastery
Read articles	p. 242	pp. 316, **340**, 366, 374, **384**	p. **418**
Write journals	p. **222**	p. **222**	p. **222**
Write letters	p. 9	pp. 41, **246**	p. 323
Write essays	p. 172	p. **348**	p. **396**

Stage III		Exemplars	
Text Type	**Introduction**	**Practice**	**Mastery**
Use strings of related sentences when speaking	p. 66	pp. **88**, **140**, **236**, **266**, **362**	p. **438**
Understand most spoken language when the message is deliberately and carefully conveyed by a speaker accustomed to dealing with learners when listening	p. 12	pp. **41**, **64**, **85**, **90**, **115**, **137**, **142**, **164**, **192**, **210**, **214**, **232**, **263**, **285**, **307**, **310**, **337**, **359**, **364**, **382**, **414**, **434**, **455**	p. **460**
Create paragraphs when writing	p. **100**	pp. **174**, **248**, 302, **322**, **396**	p. **470**
Acquire knowledge and new information from comprehensive, authentic texts when reading	p. **70**	pp. **130**, **218**, **316**, **366**, **430**	p. **440**

Content	Introduction	Practice	Mastery
Understand and convey information about (with an emphasis on significant people and events):			
history	p. **242**	p. **242**	p. 246
art	p. 324	p. **390**	p. 394
literature	p. **270**	p. **344**	p. **462**
music	p. **94**	p. 331	p. 347
current affairs	p. 137	p. **144**	p. 147
civilization	p. **352**	p. **352**	p. 368
Understand and convey information about:			
career choices	p. **256**	p. 261	p. 273
the environment	p. **132**	p. 135	p. 147
social issues	p. 86	p. 144	p. **440**
political issues	p. 108	p. 113	p. 122

Stage III	Exemplars
Accuracy/Assessments	
Tend to become less accurate as the task or message becomes more complex, and some patterns of error may interfere with meaning	pp. **26, 52, 98, 148, 220, 274, 320, 370, 442**
Generally choose appropriate vocabulary for familiar topics, but as the complexity of the message increases, there is evidence of hesitation and groping for words, as well as patterns of mispronunciation and intonation	pp. **72, 96, 124, 170, 200, 296, 368, 392, 422, 468**
Generally use culturally appropriate behavior in social situations	pp. **50, 74, 126, 146, 198, 244, 318, 346, 394, 444, 466**
Are able to understand and retain most key ideas and some supporting detail when reading and listening	pp. **92, 168, 218, 292, 388, 464**

CONTENIDO

Etapa preliminar: ¡Bienvenidos al mundo hispano! xxxiv

ESTADOS UNIDOS Present tense of regular verbs 2

MÉXICO Y CENTROAMÉRICA Present tense verbs with irregular **yo** forms 6

EL CARIBE The preterite tense of regular verbs 10

EL CONO SUR Verbs with spelling changes in the preterite 14

ESPAÑA Verbs with stem changes in the preterite 18

BOLIVIA, COLOMBIA, ECUADOR, PERÚ Y VENEZUELA Irregular preterites 22

EN USO Repaso y más comunicación 26

EN RESUMEN Vocabulary 27

LANGUAGE LEARNING CONTINUUM

- Listen to video texts, *pp. 2, 6, 10, 14, 18, 22*
- Clarify, ask for and comprehend clarification, *p. 4, Activities 2-3; p. 20, Activities 16-17*
- Narrate in the present, *p. 8, Activity 5*
- Understand most spoken language when the message is deliberately and carefully conveyed by a speaker accustomed to dealing with learners when listening, *p. 12, Activity 7*
- Narrate in the past, *p. 16, Activity 13*
- Converse and listen during face-to-face social interaction, *p. 24, Activities 20-21*
- Tend to become less accurate as the task or message becomes more complex, and some patterns of error may interfere with meaning, *p. 26, Activity 2*

ASÍ SOMOS

ETAPA 1

LANGUAGE LEARNING CONTINUUM

- Understand narration in the present, *p. 34*
- Listen to audio texts, *p. 36*

- Clarify, ask for and comprehend clarification, *p. 38, Activity 4*
- Understand most spoken language when the message is deliberately and carefully conveyed by a speaker accustomed to dealing with learners when listening, *p. 40, Activity 7*
- Narrate in the present, *p. 40, Activity 8*
- Converse and listen during face-to-face social interactions, *p. 42, Activity 11; p. 46, Activity 18*
- Narrate in the past, *p. 44, Activity 15*
- Read short stories, *p. 48*
- Generally use appropriate behavior in social situations, *p. 50, Activity 4*
- Tend to become less accurate as the task or message becomes more complex, and some patterns of error may interfere with meaning, *p. 52, Activity 6*

¿Cómo soy?   32

En contexto
VOCABULARIO Características físicas   34

En vivo
SITUACIONES ¡Eres director o directora!   36
LISTENING STRATEGY Recognize descriptions   36

En acción
VOCABULARIO Y GRAMÁTICA   38
   Ser vs. **estar**   40
   Imperfect tense   42
   Preterite vs. imperfect   43
   Present and past perfect tenses   46
SPEAKING STRATEGY Add details to descriptions   39
Refrán   47

En voces
LECTURA *Soñar en cubano* por Cristina García   48
READING STRATEGY Observe how verb tenses reveal time 48

En uso
REPASO Y MÁS COMUNICACIÓN   50
**SPEAKING STRATEGY Describe personal characteristics
   and actions**   52
Interdisciplinary Connection: El arte   52

En resumen
REPASO DE VOCABULARIO   53

¿Cómo me veo? 54

La elegancia está en los detalles

Joyería y Accesorios de Hoy

LANGUAGE LEARNING CONTINUUM

• Understand narration in the present, *p. 56*

• Listen to audio texts, *p. 58*

• Converse and listen during face-to-face social interactions, *p. 60, Activity 4*

• Express and understand opinions, *p. 62, Activities 6-7*

• Understand most spoken language when the message is deliberately and carefully conveyed by a speaker accustomed to dealing with learners when listening, *p. 64, Activity 9*

• Use strings of related sentences when speaking, *p. 66, Activity 14*

• Narrate in the future, *p. 68, Activity 17*

• Acquire knowledge and new information from comprehensive, authentic texts when reading, *p. 70*

• Generally choose appropriate vocabulary for familiar topics, but as the complexity of the message increases, there is evidence of hesitation and groping for words, as well as patterns of mispronunciation and intonation, *p. 72, Activity 1*

• Generally use culturally appropriate behavior in social situations, *p. 74, Activities 5-6*

En contexto
VOCABULARIO Joyas y accesorios — 56

En vivo
SITUACIONES ¡Persigue la moda! — 58
LISTENING STRATEGY Listen and distinguish admiring and critical remarks — 58

En acción
VOCABULARIO Y GRAMÁTICA — 60
   Verbs like **gustar** — 62
   Por and **para** — 63
   The future tense — 65
   Future tense to express probability — 68
SPEAKING STRATEGY Use familiar vocabulary in a new setting — 61
Refrán — 69

En colores
CULTURA Y COMPARACIONES Un gran diseñador — 70
CULTURAL STRATEGY Examine the cultural role of fashion — 70

En uso
REPASO Y MÁS COMUNICACIÓN — 72
SPEAKING STRATEGY Brainstorm to get lots of ideas — 74
Interdisciplinary Connection: Las matemáticas — 74

En resumen
REPASO DE VOCABULARIO — 75

ETAPA
3

**LANGUAGE
LEARNING CONTINUUM**

- Understand narration in the present, past, and future, *p. 78*
- Listen to audio texts, *p. 80*

- Converse and listen during face-to-face social interactions, *p. 82, Activity 4*
- Understand most spoken language when the message is deliberately and carefully conveyed by a speaker accustomed to dealing with learners when listening, *p. 84, Activity 5; p. 90, Activity 13*
- Understand information about social issues, *p. 86, Nota cultural*
- Use strings of related sentences when speaking, *p. 88, Activity 11*
- Are able to understand and retain most key ideas and some supporting detail when reading, *p. 92*
- Understand and convey information about music with an emphasis on significant people and events, *p. 94*
- Generally choose appropriate vocabulary for familiar topics, but as the complexity of the message increases, there is evidence of hesitation and groping for words, as well as patterns of mispronunciation and intonation, *p. 96, Activities 1-2*
- Tend to become less accurate as the task or message becomes more complex, and some patterns of error may interfere with meaning, *p. 98, Activities 5-6*
- Create paragraphs when writing, *p. 100*

¡Hay tanto que hacer!   76

En contexto
VOCABULARIO Los quehaceres   78

En vivo
SITUACIONES ¡Qué desastre!   80
**LISTENING STRATEGY Make an argument for and against
hiring others to maintain a home**   80

En acción
VOCABULARIO Y GRAMÁTICA   82
Reflexive verbs   84
Reflexive verbs used reciprocally   86
Impersonal constructions with **se**   89
Refrán   91

En voces
LECTURA *La casa en Mango Street* por Sandra Cisneros   92
**READING STRATEGY Chart constrasts between dreams
and reality in a personal narrative**   92

En colores
CULTURA Y COMPARACIONES El legendario rey del mambo   94
**CULTURAL STRATEGY Interview, report, and value
musical influences**   94

En uso
REPASO Y MÁS COMUNICACIÓN   96
**SPEAKING STRATEGY Identify feelings important
in a friendship**   98
Tú en la comunidad   98

En resumen
REPASO DE VOCABULARIO   99

En tu propia voz
ESCRITURA Presentaciones personales   100
WRITING STRATEGY Use details to enrich a description   100

¡EL MUNDO ES NUESTRO!

ETAPA 1

LANGUAGE LEARNING CONTINUUM

- Understand narration in the present, *p. 108*
- Listen to audio texts, *p. 110*

- Express and understand opinions, *p. 112, Activity 3; p. 120, Activity 15*
- Converse and listen during face-to-face social interactions, *p. 116, Activity 8*
- Understand most spoken language when the message is deliberately and carefully conveyed by a speaker accustomed to dealing with learners when listening, *p. 114, Activity 5:*
- Clarify, ask for and comprehend clarification, *p. 118, Activity 10*
- Understand and convey information about political issues, *p. 122*
- Generally choose appropriate vocabulary for familiar topics, but as the complexity of the message increases, there is evidence of hesitation and groping for words, as well as patterns of mispronunciation and intonation, *p. 124, Activity 3*
- Generally use culturally appropriate behavior in social situations, *p. 126, Activities 5-6*

Pensemos en los demás   106

En contexto
VOCABULARIO Trabajos voluntarios   108

En vivo
SITUACIONES ¡Los candidatos!   110
LISTENING STRATEGY Anticipate, compare and contrast   110

En acción
VOCABULARIO Y GRAMÁTICA   112
  Command forms   114
  Nosotros commands   116
  Speculating with the conditional   119
SPEAKING STRATEGY Name social problems, then propose solutions   121
Refrán   121

En voces
LECTURA Rigoberta Menchú   122
READING STRATEGY Comprehend complex sentences   122

En uso
REPASO Y MÁS COMUNICACIÓN   124
SPEAKING STRATEGY Identify the general ideas, then delegate responsibilities   126
Tú en la comunidad   126

En resumen
REPASO DE VOCABULARIO   127

LANGUAGE LEARNING CONTINUUM

- Acquire knowledge and new information from comprehensive, authentic texts when reading, *p. 130*
- Understand and convey information about the environment, *p. 132*

- Narrate in the past, *p. 134, Activity 3*
- Understand most spoken language when the message is deliberately and carefully conveyed by a speaker accustomed to dealing with learners when listening, *p. 136, Activity 6; p. 142, Activity 15*
- Converse and listen during face-to-face social interactions, *p. 138, Activity 10*
- Use strings of related sentences when speaking, *p. 140, Activity 13*
- Read essays. Understand and convey information about currents affairs with an emphasis on significant people and events, *p. 144*
- Generally use culturally appropriate behavior in social situations, *p. 146, Activity 1*
- Tend to become less accurate as the task or message becomes more complex, and some patterns of error may interfere with meaning, *p. 148, Activity 6*

Un planeta en peligro   128

En contexto
VOCABULARIO Ecología   130

En vivo
SITUACIONES ¡Viva el medio ambiente!   132
LISTENING STRATEGY Inventory local efforts
   to save the environment   132

En acción
VOCABULARIO Y GRAMÁTICA   134
  The present subjunctive of regular verbs   136
  The present subjunctive of irregular verbs   138
  The present subjunctive of stem-changing verbs 139
  The present perfect subjunctive   141
SPEAKING STRATEGY Consider the effect of words
   and tone of voice   137
Refrán   143

En colores
CULTURA Y COMPARACIONES Unidos podemos hacerlo   144
CULTURAL STRATEGY Gather and analyze information
   about literacy   144

En uso
REPASO Y MÁS COMUNICACIÓN   146
SPEAKING STRATEGY Express support or lack of support   148
Interdisciplinary Connection: Las ciencias   148

En resumen
REPASO DE VOCABULARIO   149

UNIDAD 2

ETAPA 3

LANGUAGE LEARNING CONTINUUM

- Understand narration in the present and in the future, *p. 152*
- Listen to audio texts, *p. 154*

- Clarify, ask for, and comprehend clarification, *p. 156, Activities 2-4*
- Identify, state, and understand feelings and emotions, *p. 158, Activity 7*
- Express and understand opinions, *p. 160, Activity 8*
- Converse and listen during face-to-face social interactions, *p. 162, Activity 10*
- Understand most spoken language when the message is deliberately and carefully conveyed by a speaker accustomed to dealing with learners when listening, *p. 164, Activity 13*
- Read short stories, *p. 166*
- Are able to understand and retain most key ideas and some supporting detail when reading, *p. 168*
- Generally choose appropriate vocabulary for familiar topics, but as the complexity of the message increases, there is evidence of hesitation and groping for words, as well as patterns of mispronunciation and intonation. *p. 170, Activity 2*
- Write essays, *p. 174*
- Create paragraphs when writing, *p. 174*

La riqueza natural   150

En contexto
VOCABULARIO Animales y deportes acuáticos   152

En vivo
SITUACIONES El Parque SalvaNatura   154
LISTENING STRATEGY Determine your purpose for listening   154

En acción
VOCABULARIO Y GRAMÁTICA   156
   The subjunctive with expressions of emotion   158
   The subjunctive to express doubt and uncertainty   160
   The subjunctive with **cuando** and other conjunctions of time   162
SPEAKING STRATEGY Gain thinking time before speaking   159
Refrán   165

En voces
LECTURA *Baby H.P.* por Juan José Arreola   166
READING STRATEGY Recognize uses of satire, parody, and irony   166

En colores
CULTURA Y COMPARACIONES Un país de encanto   168
CULTURAL STRATEGY Analyze the advantages and disadvantages of ecotourism   168

En uso
REPASO Y MÁS COMUNICACIÓN   170
SPEAKING STRATEGY Reassure others   172
Interdisciplinary Connection: Las ciencias   172

En resumen
REPASO DE VOCABULARIO   173

En tu propia voz
ESCRITURA ¡A todos nos toca!   174
WRITING STRATEGY Persuade by presenting solutions to problems   174

ETAPA
1

CELEBRACIÓN DE MI MUNDO

¡Al fin la graduación!   180

En contexto

VOCABULARIO La graduación   182

En vivo

SITUACIONES La graduación de Rosanna   184

LISTENING STRATEGY Listen and recognize major transitions   184

En acción

VOCABULARIO Y GRAMÁTICA   186

The subjunctive for expressing wishes   188

The subjunctive with conjunctions   190

The imperfect subjunctive   192

SPEAKING STRATEGY Accept or reject advice   189

Refrán   195

En voces

LECTURA «Ébano real» por Nicolás Guillén   196

READING STRATEGY Interpret metaphors   196

En uso

REPASO Y MÁS COMUNICACIÓN   198

SPEAKING STRATEGY Give advice and best wishes   200

Tú en la comunidad   200

En resumen

REPASO DE VOCABULARIO   201

Centro de Estudios José Reyes
Diploma
Técnico

LANGUAGE LEARNING CONTINUUM

- Understand narration in the present, *p. 182*
- Listen to audio texts, *p. 184*

- Clarify, ask for and comprehend clarification, *p. 186, Activity 4*
- Converse and listen during face-to-face social interactions, *p. 188, Activity 7*
- Express and understand opinions, *p. 190 Activity 10; p. 194, Activity 15*
- Understand most spoken language when the message is deliberately and carefully conveyed by a speaker accustomed to dealing with learners when listening, *p. 192, Activity 12*
- Read poems, *p. 196*
- Generally uses culturally appropriate behavior in social situations, *p. 198, Activity 1–4*
- Generally choose appropriate vocabulary for familiar topics, but as the complexity of the message increases, there is evidence of hesitation and groping for words, as well as patterns of mispronunciation and intonation, *p. 200, Activity 5*

UNIDAD 3

ETAPA 2

¡Próspero Año Nuevo! 202

En contexto
VOCABULARIO El año nuevo 204

En vivo
SITUACIONES ¡Próspero Año Nuevo! 206
LISTENING STRATEGY Observe interview techniques 206

En acción
VOCABULARIO Y GRAMÁTICA 208
   Subjunctive with nonexistent and indefinite antecedents 210
   The subjunctive for disagreement and denial 212
   Conditional sentences 214
SPEAKING STRATEGY Socialize as host or guest 209
Refrán 217

En colores
CULTURA Y COMPARACIONES
   Una tradición de Puerto Rico 218
CULTURAL STRATEGY Recognize and describe uses of disguise 218

En uso
REPASO Y MÁS COMUNICACIÓN 220
SPEAKING STRATEGY Encourage participation 222
Interdisciplinary Connection: El arte 222

En resumen
REPASO DE VOCABULARIO 223

LANGUAGE LEARNING CONTINUUM

- Understand narration in the future, *p. 204*

- Listen to audio texts, *p. 206*

- Converse and listen during face-to-face social interactions, *p. 208, Activity 4*

- Understand most spoken language when the message is deliberately and carefully conveyed by a speaker accustomed to dealing with learners when listening, *p. 210, Activity 6; p. 214, Activity 13*

- Express and understand opinions, *p. 212, Activity 9; p. 216, Activity 15*

- Acquire knowledge and new information from comprehensive, authentic texts when reading, *p. 218*

- Are able to understand and retain most key ideas and some supporting detail when reading, *p. 218*

- Tend to become less accurate as the task or message becomes more complex, and some patterns of error may interfere with meaning, *p. 220, Activity 3*

- Write journals, *p. 222, Activity 7*

LANGUAGE LEARNING CONTINUUM

- Read essays, *p. 226*
- Listen to audio texts, *p. 228*

- Narrate in the present, *p. 230, Activity 4*
- Understand most spoken language when the message is deliberately and carefully conveyed by a speaker accustomed to dealing with learners when listening, *p. 232, Activity 5*
- Clarify, ask for and comprehend clarification, *p. 234, Activity 8*
- Use strings of related sentences when speaking, *p. 236, Activity 12*
- Express and understand opinions, *p. 238, Activity 16*
- Read poems, *p. 240*
- Understand and convey information about history with an emphasis on significant people and events, *p. 242*
- Generally use culturally appropriate behavior in social situations, *p. 244, Activity 3*
- Write letters, *p. 246, Activity 7*
- Create paragraphs when writing, *p. 248*

Celebraciones de patria — 224

En contexto
VOCABULARIO Celebraciones históricas — 226

En vivo
SITUACIONES Los viajes del Almirante — 228
LISTENING STRATEGY Listen and take notes — 228

En acción
VOCABULARIO Y GRAMÁTICA — 230
    Summary of the subjunctive (Part 1) — 232
    Summary of the subjuntive (Part 2) — 234
    Subjunctive vs. Indicative — 238
SPEAKING STRATEGY Describe celebrations — 231
Refrán — 239

En voces
LECTURA *Versos sencillos* por José Martí — 240
READING STRATEGY Observe what makes poetry — 240

En colores
CULTURA Y COMPARACIONES Una historia única — 242
CULTURAL STRATEGY Analyze national celebrations — 242

En uso
REPASO Y MÁS COMUNICACIÓN — 244
SPEAKING STRATEGY Express yourself — 246
Interdisciplinary Connection: Los estudios sociales — 246

En resumen
REPASO DE VOCABULARIO — 247

En tu propia voz
ESCRITURA ¡Les deseamos mucho éxito! — 248
WRITING STRATEGY Use transitions to make text flow smoothly — 248

UNIDAD 4

ETAPA 1

UN FUTURO BRILLANTE

El próximo paso   254

En contexto
VOCABULARIO Tu futuro   256

En vivo
SITUACIONES ¡Eso sí me interesa!   258
LISTENING STRATEGY Evaluate recommendations   258

En acción
VOCABULARIO Y GRAMÁTICA   260
    Interrogative words   262
    The present progressive   264
    The progressive with **ir, andar,** and **seguir**   266
    The past progressive   268
SPEAKING STRATEGY Establish closer relationships   263
Refrán   269

En voces
LECTURA Jorge Luis Borges   270
READING STRATEGY Analyze the role of identity and fantasy   270

En uso
REPASO Y MÁS COMUNICACIÓN   272
SPEAKING STRATEGY Extend a conversation   274
Tú en la comunidad   274

En resumen
REPASO DE VOCABULARIO   275

LANGUAGE LEARNING CONTINUUM

- Understand and convey information about career choices, *p. 256*
- Listen to audio texts, *p. 258*

- Clarify, ask for and comprehend clarification, *p. 260, Activity 4*
- Understand most spoken language when the message is deliberately and carefully conveyed by a speaker accustomed to dealing with learners when listening, *p. 263, Activity 6*
- Converse in simple transactions on the phone, *p. 264, Activity 8*
- Use strings of related sentences when speaking, *p. 266, Activity 13*
- Converse and listen during face-to-face social interaction, *p. 268, Activity 16*
- Understand and convey information about literature with emphasis on significant people and events, *p. 270*
- Narrate in the past, *p. 272, Activity 4*
- Tend to become less accurate as the task or message becomes more complex, and some patterns of error may interfere with meaning, *p. 274, Activity 5*

ETAPA
2

En contexto
VOCABULARIO Las profesiones 278

En vivo
SITUACIONES Y yo, ¿qué quiero ser? 280
LISTENING STRATEGY Identify key information for careers 280

En acción
VOCABULARIO Y GRAMÁTICA 282
   Affirmative and negative expressions 284
   Past perfect subjunctive 286
   The conditional perfect tense 289
SPEAKING STRATEGY Anticipate what others want to know 283
Refrán 291

En colores
CULTURA Y COMPARACIONES Los jóvenes y el futuro 292
CULTURAL STRATEGY Formulate plans for the future 292

En uso
REPASO Y MÁS COMUNICACIÓN 294
SPEAKING STRATEGY Conduct an interview 296
Interdisciplinary Connection: Los estudios sociales 296

En resumen
REPASO DE VOCABULARIO 297

LANGUAGE LEARNING CONTINUUM

- Understand narration in the past, *p. 278*
- Listen to audio texts, *p. 280*

- Express and understand opinions, *p. 282, Activity 3; p. 290, Activity 14*
- Understand most spoken language when the message is deliberately and carefully conveyed by a speaker accustomed to dealing with learners when listening, *p. 285, Activity 5*
- Converse and listen during face-to-face social interaction, *p. 286, Activity 8*
- Narrate in the past, *p. 288, Activity 10; p. 294, Activity 3*
- Are able to understand and retain most key ideas and some supporting detail when reading, *p. 292*
- Generally choose appropriate vocabulary for familiar topics, but as the complexity of the message increases, there is evidence of hesitation and groping for words, as well as patterns of mispronunciation and intonation. *p. 296, Activity 4*

UNIDAD 4

ETAPA **3**

Un mundo de posibilidades

298

LANGUAGE LEARNING CONTINUUM

- Understand narration in the present and in the future, *p. 300*
- Listen to audio texts, *p. 302*

- Express and understand opinions, *p. 304, Activity 4*
- Understand most spoken language when the message is deliberately and carefully conveyed by a speaker accustomed to dealing with learners when listening, *p. 307, Activity 5; p. 310, Activity 12*
- Clarify, ask for and comprehend clarification, *p. 308, Activity 7*
- Narrate in the future, *p. 312, Activity 14*
- Read short stories, *p. 314*
- Acquire knowledge and new information from comprehensive, authentic texts when reading, *p. 316*
- Generally use appropriate behavior in social situations, *p. 318, Activity 3*
- Tend to become less accurate as the task or message becomes more complex, and some patterns of error may interfere with meaning, *p. 320, Activities 5-6*
- Create paragraphs when writing, *p. 322*

En contexto
VOCABULARIO La economía — 300

En vivo
SITUACIONES ¡Encuéntralo por Internet! — 302
LISTENING STRATEGY Use statistics to evaluate predictions — 302

En acción
VOCABULARIO Y GRAMÁTICA — 304
   Subject and stressed object pronouns — 306
   Possessive pronouns — 308
   The future perfect tense — 310
SPEAKING STRATEGY Guess cognates — 305
Refrán — 313

En voces
LECTURA *Paula* por Isabel Allende — 314
READING STRATEGY Speculate about the author — 314

En colores
CULTURA Y COMPARACIONES
   Se hablan… ¡muchos idiomas! — 316
CULTURAL STRATEGY Observe how language reflects culture — 316

En uso
REPASO Y MÁS COMUNICACIÓN — 318
SPEAKING STRATEGY Speculate about the past — 320
Interdisciplinary Connection: Los estudios sociales — 320

En resumen
REPASO DE VOCABULARIO — 321

En tu propia voz
ESCRITURA Una carrera: ¿Dónde empezar? — 322
WRITING STRATEGY Use cause and effect to demonstrate ability — 322

UNIDAD

ETAPA

1

ARTES EN ESPAÑA Y LAS AMÉRICAS

LANGUAGE LEARNING CONTINUUM

- Understand narration in the present, *p. 330*
- Listen to audio texts, *p. 332*

- Clarify, ask for and comprehend clarification, *p. 334, Activity 3; p. 342, Activity 13*
- Understand most spoken language when the message is deliberately and carefully conveyed by a speaker accustomed to dealing with learners when listening, *p. 336, Activity 4*
- Converse and listen during face-to-face social interaction, *p. 338, Activity 6*
- Read articles, *p. 340, Activity 9*
- Understand and convey information about literature with an emphasis on significant people and events, *p. 344 Activity 4*
- Generally use culturally appropriate behavior in social situations, *p. 346*
- Write essays, *p. 348, Activity 6*

Tradiciones españolas   328

En contexto
VOCABULARIO Arte y bailes típicos   330

En vivo
SITUACIONES Un paseo por El Prado   332
LISTENING STRATEGY Use advance knowledge of the topic   **332**

En acción
VOCABULARIO Y GRAMÁTICA   334
   Demonstrative adjectives and pronouns   336
   ¿Qué? vs. **¿cuál?**   338
   Relative pronouns   341
SPEAKING STRATEGY Discuss a painting   **340**
Refrán   343

En voces
LECTURA Miguel de Unamuno y Ana María Matute   344
READING STRATEGY Compare famous authors   **344**

En uso
REPASO Y MÁS COMUNICACIÓN   346
SPEAKING STRATEGY Organize ideas for research   **348**
Interdisciplinary Connection: Los estudios sociales   348

En resumen
REPASO DE VOCABULARIO   349

ETAPA 2

En contexto
VOCABULARIO Culturas de América   352

En vivo
SITUACIONES Una visita virtual   354
LISTENING STRATEGY Improve your auditory memory 354

En acción
VOCABULARIO Y GRAMÁTICA   356
   Direct object pronouns   358
   Indirect object pronouns   360
   More on relative pronouns   362
   Lo que   364
SPEAKING STRATEGY Mantain a discussion   356
Refrán   365

En colores
CULTURA Y COMPARACIONES Un arquitecto y sus obras 366
CULTURAL STRATEGY Use architecture as a cultural text 366

En uso
REPASO Y MÁS COMUNICACIÓN   368
SPEAKING STRATEGY Discuss Latin American dance 370
Interdisciplinary Connection: Las matemáticas   370

En resumen
REPASO DE VOCABULARIO   371

LANGUAGE LEARNING CONTINUUM

• Understand and convey information about civilization with an emphasis on significant people and events, *p. 352*

• Listen to audio texts, *p. 354*

• Narrate in the present, *p. 356, Activity 3*

• Understand most spoken language when the message is deliberately and carefully conveyed by a speaker accustomed to dealing with learners when listening, *p. 358, Activity 4; p. 364, Activity 13*

• Express and understand opinions, *p. 360, Activity 8*

• Use strings of related sentences when speaking, *p. 362, Activity 12*

• Acquire knowledge and new information from comprehensive, authentic texts when reading, *p. 366*

• Generally choose appropriate vocabulary for familiar topics, but as the complexity of the message increases, there is evidence of hesitation and groping for words, as well as patterns of mispronunciation and intonation, *p. 368, Activity 4*

• Tend to become less accurate as the task or message becomes more complex, and some patterns of error may interfere with meaning, *p. 370, Activity 5*

UNIDAD 5

ETAPA 3

LANGUAGE LEARNING CONTINUUM

- Understand narration in the present and in the past, *p. 374*
- Listen to audio texts, *p. 376*

- Narrate in the present, *p. 378, Activity 4*
- Converse and listen during face-to-face social interactions, *p. 380, Activities 5-6*
- Understand most spoken language when the message is deliberately and carefully conveyed by a speaker accustomed to dealing with learners when listening, *p. 382, Activity 9*
- Read articles, *p. 384, Activity 11*
- Express and understand opinions, *p. 386, Activity 14*
- Are able to understand and retain most key ideas and some supporting detail when reading, *p. 388*
- Understand and convey information about art with an emphasis on significant people and events, *p. 390*
- Generally choose appropriate vocabulary for familiar topics, but as the complexity f the message increases, there is evidence of hesitation and groping for words, as well as patterns of mispronunciation and intonation, *p. 392, Activity 4*
- Generally use culturally appropriate behavior in social situations, *p. 394, Activity 5*
- Write essays, *p. 396*
- Create paragraphs when writing, *p. 396*

Lo mejor de dos mundos — 372

En contexto
VOCABULARIO La literatura — 374

En vivo
SITUACIONES El club de cine — 376
LISTENING STRATEGY Evaluate discussions — 376

En acción
VOCABULARIO Y GRAMÁTICA — 378
    Double object pronouns — 380
    Nominalization — 382
    More on nominalization — 385
SPEAKING STRATEGY Discuss a novel — 379
Refrán — 387

En voces
LECTURA *La casa de Bernarda Alba* por Federico García Lorca — 388
READING STRATEGY Interpret a drama — 388

En colores
CULTURA Y COMPARACIONES Tres directores — 390
CULTURAL STRATEGY Reflect on the international appeal of movies — 390

En uso
REPASO Y MÁS COMUNICACIÓN — 392
SPEAKING STRATEGY Critique a film — 394
Tú en la comunidad — 394

En resumen
REPASO DE VOCABULARIO — 395

En tu propia voz
ESCRITURA Mitos, leyendas, ficciones: ¡la literatura es de todos! — 396
WRITING STRATEGY Support an opinion with facts and examples — 396

UNIDAD

6

ETAPA

1

¡YA LLEGÓ EL FUTURO!

¿Qué quieres ver?   402

En contexto
VOCABULARIO Programas de televisión   404

En vivo
SITUACIONES ¿Qué vamos a ver?   406
**LISTENING STRATEGY Keep up with what
is said and agreed**   406

En acción
VOCABULARIO Y GRAMÁTICA   408
   Preterite vs. imperfect   410
   Indicative vs. subjunctive   412
   Reported speech   414
   Sequence of tenses   415
SPEAKING STRATEGY Negotiate   414
Refrán   417

En voces
LECTURA *Brillo afuera, oscuridad en casa*   418
READING STRATEGY Distinguish facts and interpretation   418

En uso
REPASO Y MÁS COMUNICACIÓN   420
SPEAKING STRATEGY Retell memories   422
Tú en la comunidad   422

En resumen
REPASO DE VOCABULARIO   423

LANGUAGE LEARNING CONTINUUM

- Understand narration in the present, *p. 404*

- Listen to audio texts, *p. 406*

- Clarify, ask for and comprehend clarification, *p. 408, Activities 1-4*

- Narrate in the past, *p. 410, Activity 7*

- Express and understand opinions, *p. 412, Activity 9*

- Understand most spoken language when the message is deliberately and carefully conveyed by a speaker accustomed to dealing with learners when listening, *p. 414, Activity 11*

- Converse and listen during face-to-face social interactions, *p. 416, Activity 17*

- Read articles, *p. 418*

- Converse in simple transactions on the phone, *p. 420, Activity 2*

- Generally choose appropriate vocabulary for familiar topics, but as the complexity of the message increases, there is evidence of hesitation and groping for words, as well as patterns of mispronunciation and intonation, *p. 422, Activity 5*

UNIDAD 6

ETAPA 2

LANGUAGE LEARNING CONTINUUM

- Understand narration in the present, *p. 426*
- Listen to audio texts, *p. 428*

- Acquire knowledge and new information from comprehensive, authentic texts when reading, *p. 430, Activity 3*
- Clarify, ask for and comprehend clarification, *p. 432, Activity 4*
- Understand most spoken language when the message is deliberately and carefully conveyed by a speaker accustomed to dealing with learners when listening, *p. 434, Activity 10*
- Express and understand opinions, *p. 436, Activity 12*
- Use strings of related sentences when reading, *p. 438, Activity 15*
- Acquire knowledge and new information from comprehensive, authentic texts when reading, *p. 440*
- Understand and convey information about social issues, *p. 440*
- Tend to become less accurate as the task or message becomes more complex, and some patterns of error may interfere with meaning, *p. 442, Activity 4*
- Generally use culturally appropriate behavior in social situations, *p. 444, Activity 5*

Aquí tienes mi número...   424

En contexto
VOCABULARIO Artículos electrónicos   426

En vivo
SITUACIONES ¡Grandes rebajas!   428
LISTENING STRATEGY Analyze the appeal in radio ads   428

En acción
VOCABULARIO Y GRAMÁTICA   430
  Conjunctions   432
  Prepositions and adverbs of location   433
  Pero vs. **sino**   435
  Se for unplanned occurrences   437
SPEAKING STRATEGY Make excuses   438
Refrán   439

En colores
CULTURA Y COMPARACIONES ¿Un aparato democrático? 440
CULTURAL STRATEGY Survey technology in daily life   440

En uso
REPASO Y MÁS COMUNICACIÓN   442
SPEAKING STRATEGY Consider the factors for or against an electronic purchase   444
Interdisciplinary Connection: El arte   444

En resumen
REPASO DE VOCABULARIO   445

UNIDAD 6

ETAPA 3

¡Un viaje al ciberespacio! 446

LANGUAGE LEARNING CONTINUUM

- Understand narration in the present, *p. 448*

- Listen to audio texts, *p. 450*

- Clarify, ask for and comprehend clarification, *p. 452, Activity 2*

- Understand most spoken language when the message is deliberately and carefully conveyed by a speaker accustomed to dealing with learners when listening, *p. 454, Activity 5; p. 460, Activity 13*

- Express and understand opinions, *p. 456, Activities 6-7*

- Narrate in the future, *p. 458, Activity 11*

- Understand and convey information about literature with an emphasis on significant people and events, *p. 462*

- Are able to understand and retain most key ideas and some supporting detail when reading, *p. 464*

- Generally use culturally appropriate behavior in social situations, *p. 466, Activity 2*

- Generally choose appropriate vocabulary for familiar topics, but as the complexity of the message increases, there is evidence of hesitation and groping for words, as well as patterns of mispronunciation and intonation, *p. 468, Activity 5*

- Create paragraphs when writing, *p. 470*

En contexto
VOCABULARIO Las computadoras   448

En vivo
SITUACIONES El mejor sistema   450
LISTENING STRATEGY Identify important   450
computer vocabulary

En acción
VOCABULARIO Y GRAMÁTICA   452
   Comparatives and superlatives   454
   Prepositions   457
   Verbs with prepositions   459
SPEAKING STRATEGY Compare and evaluate films   456
Refrán   461

En voces
LECTURA Gabriel García Márquez   462
READING STRATEGY Monitor comprehension   462

En colores
CULTURA Y COMPARACIONES Bolivia en la red   464
CULTURAL STRATEGY Evaluate the Internet as a means of
developing cultural knowledge and understanding   464

En uso
REPASO Y MÁS COMUNICACIÓN   466
SPEAKING STRATEGY Compare and evaluate
computer configurations   468
Interdisciplinary Connection: La tecnología   468

En resumen
REPASO DE VOCABULARIO   469

En tu propia voz
ESCRITURA La tecnología del mundo de hoy   470
WRITING STRATEGY Prioritize information in order
of importance   470

About the Authors

Estella Gahala holds a Ph.D. in Educational Administration and Curriculum from Northwestern University. A career teacher of Spanish and French, she has worked with a wide range of students at the secondary level. She has also served as foreign language department chair and district director of curriculum and instruction. Her workshops at national, regional, and state conferences as well as numerous published articles draw upon the current research in language learning, learning strategies, articulation of foreign language sequences, and implications of the national Standards for Foreign Language Learning upon curriculum, instruction, and assessment. She has coauthored nine basal textbooks.

Patricia Hamilton Carlin completed her M.A. in Spanish at the University of California, Davis, where she also taught as a lecturer. Previously she had earned a Master of Secondary Education with specialization in foreign languages from the University of Arkansas and had taught Spanish and French at levels K–12. Her secondary programs in Arkansas received national recognition. A coauthor of the *¡DIME! UNO* and *¡DIME! DOS* secondary textbooks, Patricia currently teaches Spanish and foreign language/ESL methodology at the University of Central Arkansas, where she coordinates the second language teacher education program. In addition, Patricia is a frequent presenter at local, regional, and national foreign language conferences.

Audrey L. Heining-Boynton received her Ph.D. in Curriculum and Instruction from Michigan State University. She is a Professor of Education and Romance Languages at The University of North Carolina at Chapel Hill, where she is a second language teacher educator and Professor of Spanish. She has also taught Spanish, French, and ESL at the K–12 level. Dr. Heining-Boynton was the president of the National Network for Early Language Learning, has been on the Executive Council of ACTFL, and involved with AATSP, Phi Delta Kappa, and state foreign language associations. She has presented both nationally and internationally, and has published over forty books, articles, and curricula.

Ricardo Otheguy received his Ph.D. in Linguistics from the City University of New York, where he is currently Professor of Linguistics at the Graduate School and University Center. He has written extensively on topics related to Spanish grammar as well as on bilingual education and the Spanish of the United States. He is coauthor of *Tu mundo: Curso para hispanohablantes,* a Spanish high school textbook for Spanish speakers, and of *Prueba de ubicación para hispanohablantes,* a high school Spanish placement test.

Barbara J. Rupert has taught Level 1 through A.P. Spanish and has implemented a FLES program in her district. She completed her M.A. at Pacific Lutheran University. Barbara is the author of CD-ROM activities for the *¡Bravo!* series and has presented at local, regional, and national foreign language conferences. She is the president of the Washington Association for Language Teaching. In 1996, Barbara received the Christa McAuliffe Award for Excellence in Education, and in 1999, she was selected Washington's "Spanish Teacher of the Year" by the Juan de Fuca Chapter of the AATSP.

Culture Consultant

Jorge A. Capetillo-Ponce is currently Assistant Professor of Sociology at University of Massachusetts, Boston, and Researcher at the Mauricio Gastón Institute for Latino Community Development and Public Policy. His graduate studies include an M.A. and a Ph.D. in Sociology from the New School for Social Research in New York City, and an M.A. in Area Studies at El Colegio de México in Mexico City. He is the editor of the book *Images of Mexico in the U.S. News Media* and has published essays on a wide range of subjects such as media, art, politics, religion, international relations, and cultural theory. Dr. Capetillo's geographical areas of expertise are Latin America, the United States, and the Middle East. During the years 2000 and 2001 he was the Executive Director of the Mexican Cultural Institute of New York. He has also worked as an advisor to politicians and public figures, as a researcher and an editor, and as a university professor and television producer in Mexico, the United States, and Central America.

Consulting Authors

Dan Battisti
Dr. Teresa Carrera-Hanley
Bill Lionetti
Patty Murguía Bohannan
Lorena Richins Layser

Senior Reviewers

O. Lynn Bolton
Dr. Jane Govoni
Elías G. Rodríguez
Ann Tollefson

Contributing Writers

Ronni L. Gordon
Christa Harris
Debra Lowry
Sylvia Madrigal Velasco
Sandra Rosenstiel
David M. Stillman
Jill K. Welch

Regional Language Reviewers

Dolores Acosta (Mexico)
Jaime M. Fatás Cabeza (Spain)
Grisel Lozano-Garcini (Puerto Rico)
Isabel Picado (Costa Rica)
Juan Pablo Rovayo (Ecuador)

California Teacher Reviewers

Linda Amour
Highland High School
Bakersfield, CA

Dawne Ashton
Sequoia High School
Redwood City, CA

Gail Block
Daly City, CA

Art Edwards
Canyon High School
Santa Clarita, CA

Rubén D. Elías
Roosevelt High School
Fresno, CA

Paula Hirsch
Windward School
Los Angeles, CA

Ann Hively
Orangevale, CA

Janet King
Long Beach Polytechnic
High School
Long Beach, CA

Maria Leinenweber
Crescenta Valley High School
La Crescenta, CA

Sandra Martín
Palisades Charter High School
Pacific Palisades, CA

Karen McDowell
Aptos, CA

Sue McKee
Tustin, CA

Robert Miller
Woodcreek High School
Roseville, CA

Barbara Mortanian
Tenaya Middle School
Fresno, CA

Leslie Ogden
Nordhoff High School
Ojai, CA

Teri Olsen
Alameda High School
Alameda, CA

Lewis Olvera
Hiram Johnson West Campus
High School
Sacramento, CA

Margery Sotomayor
Ferndale, CA

Carol Sparks
Foothill Middle School
Walnut Creek, CA

Dana Valverde
Arroyo Grande High School
Arroyo Grande, CA

Teacher Reviewers

Susan Arbuckle
Mahomet-Seymour High School
Mahomet, IL

Sheila Bayles
Rogers High School
Rogers, AR

Warren Bender
Duluth East High School
Duluth, MN

Amy Brewer
Stonewall Jackson Middle School
Mechanicsville, VA

William Brill
Hollidaysburg Area Junior High School
Hollidaysburg, PA

Adrienne Chamberlain-Parris
Mariner High School
Everett, WA

Norma Coto
Bishop Moore High School
Orlando, FL

Roberto del Valle
Shorecrest High School
Shoreline, WA

José Esparza
Curie Metropolitan High School
Chicago, IL

Lorraine A. Estrada
Cabarrus County Schools
Concord, NC

Vincent Fazzolari
East Boston High School
East Boston, MA

Alberto Ferreiro
Harrisburg High School
Harrisburg, PA

Judith C. Floyd
Henry Foss High School
Tacoma, WA

Valarie L. Forster
Jefferson Davis High School
Montgomery, AL

Michael Garber
Boston Latin Academy
Boston, MA

Becky Hay de García
James Madison Memorial High School
Madison, WI

Lucy H. García
Pueblo East High School
Pueblo, CO

Marco García
Lincoln Park High School
Chicago, IL

Raquel R. González
Odessa High School
Odessa, TX

Linda Grau
Shorecrest Preparatory School
St. Petersburg, FL

Myriam Gutiérrez
John O'Bryant School
Roxbury, MA

Deborah Hagen
Ionia High School
Ionia, MI

Sandra Hammond
St. Petersburg High School
St. Petersburg, FL

Bill Heller
Perry Junior/Senior High School
Perry, NY

Joan Heller
Lake Braddock Secondary School
Burke, VA

Robert Hughes
Martha Brown Middle School
Fairport, NY

Jody Klopp
Oklahoma State Department
of Education
Edmond, OK

Richard Ladd
Ipswich High School
Ipswich, MA

Carol Leach
Francis Scott Key High School
Union Bridge, MD

Laura McCormick
East Seneca Senior High School
West Seneca, NY

Rafaela McLeod
Southeast Raleigh High School
Raleigh, NC

Kathleen L. Michaels
Palm Harbor University High School
Palm Harbor, FL

Teacher Reviewers (continued)

Vickie A. Mike
Horseheads High School
Horseheads, NY

Patty Murray
Cretin-Derham Hall High School
St. Paul, MN

Linda Nanos
West Roxbury High School
West Roxbury, MA

Terri Nies
Mannford High School
Mannford, OK

María Emma Nunn
John Tyler High School
Tyler, TX

Judith Pasco
Sequim High School
Sequim, WA

Anne-Marie Quihuis
Paradise Valley High School
Phoenix, AZ

Rita Risco
Palm Harbor University High School
Palm Harbor, FL

James J. Rudy, Jr.
Glen Este High School
Cincinnati, OH

Kathleen Solórzano
Homestead High School
Mequon, WI

Sarah Spiesman
Whitmer High School
Toledo, OH

M. Mercedes Stephenson
Hazelwood Central High School
Florissant, MO

Carol Thorp
East Mecklenburg High School
Charlotte, NC

Elizabeth Torosian
Doherty Middle School
Andover, MA

Pamela Urdal Silva
East Lake High School
Tarpon Springs, FL

Wendy Villanueva
Lakeville High School
Lakeville, MN

Helen Webb
Arkadelphia High School
Arkadelphia, AR

Jena Williams
Jonesboro High School
Jonesboro, AR

Janet Wohlers
Weston Middle School
Weston, MA

Teacher Panel

Jeanne Aréchiga
Northbrook High School
Houston, TX

Dena Bachman
Lafayette Senior High School
St. Joseph, MO

Sharon Barnes
J. C. Harmon High School
Kansas City, KS

Paula Biggar
Sumner Academy of Arts & Science
Kansas City, KS

Hercilia Breton
Highlands High School
San Antonio, TX

Edda Cárdenas
Blue Valley North High School
Leawood, KS

Laura Cook
Evans Junior High School
Lubbock, TX

Mike Cooperider
Truman High School
Independence, MO

Judy Dozier
Shawnee Mission South High School
Shawnee Mission, KS

Terri Frésquez
Del Valle High School
El Paso, TX

Rose Jenkins
Clements High School
Sugarland, TX

Susanne Kissane
Shawnee Mission Northwest
    High School
Shawnee Mission, KS

Rudy Molina
McAllen Memorial High School
McAllen, TX

Rob Ramos
J. T. Hutchinson Junior High School
Lubbock, TX

Montserrat Rey
Hightower High School
Fort Bend, TX

Sandra Rivera
Mary Carroll High School
Corpus Christi, TX

Terrie Rynard
Olathe South High School
Olathe, KS

Beth Slinkard
Lee's Summit High School
Lee's Summit, MO

Rosa Stein
Park Hill High School
Kansas City, MO

Shannon Zerby
North Garland High School
Garland, TX

California Teacher Panel

Linda Amour
Highland High School
Bakersfield, CA

Ben Barrientos
Calvin Simmons Junior High School
Oakland, CA

Gwen Cannell
Cajon High School
San Bernardino, CA

Joyce Chow
Crespi Junior High School
Richmond, CA

Maggie Elliott
Bell Junior High School
San Diego, CA

Dana Galloway-Grey
Ontario High School
Ontario, CA

Nieves Gerber
Chatsworth Senior High School
Chatsworth, CA

April Hansen
Livermore High School
Livermore, CA

Janet King
Long Beach Polytechnic High School
Long Beach, CA

Ann López
Pala Middle School
San Jose, CA

Beatrice Marino
Palos Verdes Peninsula High School
Rolling Hills, CA

Anna Marxson
Laguna Creek High School
Elk Grove, CA

Barbara Mortanian
Tenaya Middle School
Fresno, CA

Vickie Musni
Pioneer High School
San Jose, CA

Teri Olsen
Alameda High School
Alameda, CA

Rodolfo Orihuela
C. K. McClatchy High School
Sacramento, CA

Marianne Villalobos
Modesto High School
Modesto, CA

El español en California

Un resumen
Desde la llegada de los exploradores europeos hasta nuestros días, los pueblos hispanohablantes han constituido una parte integral de la historia y cultura de California. Los hablantes de español que se encontraban bajo el mando de España, México o Estados Unidos se han destacado como exploradores, escritores, reformistas y políticos. Éstas son sólo algunas de las importantes maneras en que los hispanohablantes han influenciado a California y contribuido con la identidad de Estados Unidos.

1530–1542
Las primeras expediciones a *Baja California* estuvieron al mando del español Hernán Cortés. Al principio, los exploradores pensaron que se trataba sólo de una isla, como se ve en el mapa antiguo, pero luego notaron que el territorio se extendía hacia el norte. A esa parte la llamaron *Alta California,* la cual más tarde pasó a ser el actual estado de California.

1500   **1600**   **1700**   **1800**

Estatua de Junípero Serra delante de la misión de San Antonio en Jolon

1769
Los exploradores españoles que llegaban a *Alta California* lo hacían acompañados de frailes franciscanos que querían establecer misiones para enseñar la religión católica en ese lugar. En total se establecieron veintiuna misiones. El fraile Junípero Serra fundó la primera de ellas en lo que hoy es San Diego. Actualmente, muchas de esas misones son un lugar de atracción para los turistas.

1885

Desde que la región de California era parte de México, sus habitantes, conocidos como *californios*, se dedicaban en su mayoría a trabajar en las cosechas de uva, una de las principales fuentes de trabajo de esa zona. Durante la transición y el reclamo de California por parte de Estados Unidos, muchos *californios* fueron obligados a abandonar sus tierras, perdiendo su trabajo y teniendo que migrar a otros lugares. La novelista mexicoamericana María Amparo Ruiz de Burton se dedicó a escribir los testimonios de esta época.

Californios trabajando en los campos de Madera

1965

El escritor dramático Luis Valdez crea el Teatro Campesino, una organización artística formada por trabajadores del campo que representaban los problemas a los que se

enfrentaban a diario. Más tarde, este teatro se convertiría en el reflejo de la cultura chicana. Desde su creación ha recibido muchos premios por su forma de presentar problemas sociales y políticos. Actualmente, el Teatro Campesino sigue reuniendo a grandes actores y directores.

1900

2000

1822–1850

California ha pertenecido a diferentes países. España fue la primera en tomar posesión. En 1822 México se independiza de España y California pasa a ser territorio mexicano. Más tarde, los californianos se rebelan contra el gobierno mexicano. Después de dicha revuelta, Estados Unidos reclama el territorio de California. En 1848, por medio del Tratado de Guadalupe Hidalgo, México le cede oficialmente a Estados Unidos el territorio de California junto con otros territorios.

Tratado de Guadalupe Hidalgo

1998

La científica Adriana Ocampo es elegida para formar parte del equipo de la NASA (National Aeronautics and Space Administration). Esta científica se crió en Argentina y realizó sus estudios universitarios en California,

obteniendo una maestría en geología. Desde muy joven ha participado en proyectos de investigación relacionados con las naves espaciales *Viking* y *Voyager*, las cuales realizaron misiones a otros planetas. Ha organizado muchas conferencias relacionadas con las ciencias planetarias y actualmente está estudiando para convertirse en piloto espacial.

Cómo estudiar el español

Puedes usar lo que ya sabes y aprender cosas nuevas con estas partes de tu libro.

Estrategias

Tu libro te da la oportunidad de practicar estas estrategias:

Para escuchar: te preparan para escuchar y entender.

Para conversar: te ayudan a expresarte en español.

Para leer: te ayudan a leer los pasajes del libro.

Para escribir: son para mejorar tu habilidad de escribir.

Para comparar: te ayudan a comparar culturas.

STRATEGY: SPEAKING

Gain thinking time before speaking Sometimes ideas do notcome to us as quickly as we would like. One way to gain time is to restate what was just said which may in turn trigger a fresh idea. Example: **Sí, es una lástima. Espero que se proteja la selva también.**

El Apoyo para estudiar

Esta sección te sugiere ideas para estudiar el español más efectiva y eficientemente.

Apoyo para estudiar

Negative command

Remember that in a negative command, object pronouns precede the verb. So you can advise against an action (**¡No se la compre!**), but you should give a reason why (**porque ella prefiere la poesía**).

Siente más seguridad

Recuerda y aprende información nueva mediante **Repaso** y **Gramática**. **Repaso** presenta lo que aprendiste en los Niveles 1 y 2 y **Gramática** es información nueva. La combinación de palabras y gráficas facilitan el aprendizaje de todos los estudiantes.

REPASO — Preterite vs. Imperfect

 ¿RECUERDAS? *p. 42* You already know two tenses that refer to past time, the **preterite** and the **imperfect**. You use each of these tenses to talk about **past** actions in a different way.

- Use the **preterite** tense to describe an action or series of actions **completed** in the past.

  Aquel día, Pedro **salió** del colegio y **caminó** hasta el café.
  *That day, Pedro **left** school and **walked** to the café.*

Practice:
Actividades
12 13 14 15

Más práctica
cuaderno pp. 19
Para hispanohablantes
cuaderno pp. 15–16

Online Workbook
CLASSZONE.COM

GRAMÁTICA — The Imperfect Subjunctive

You already know the **present** and **present perfect subjunctive**. There are also **past** forms of the subjunctive. Use the **imperfect subjunctive** instead of the present subjunctive when the context of the sentence is in the **past.** Compare the following pairs of sentences.

Present context
Los padrinos **quieren que** felicitemos al graduando.
*The godparents **want** us to **congratulate** the graduate.*

Past context
Los padrinos **querían que** felicitáramos al graduando.
*The godparents **wanted** us to **congratulate** the graduate.*

Practice:
Actividades
11 12 13 14

Más práctica
cuaderno pp. 71–72
Para hispanohablantes
cuaderno pp. 69–70

Online Workbook
CLASSZONE.COM

Leer y escuchar el español

Lee para aprender palabras nuevas en español.

La sección **En contexto** presenta vocabulario nuevo en un contexto real e interesante.

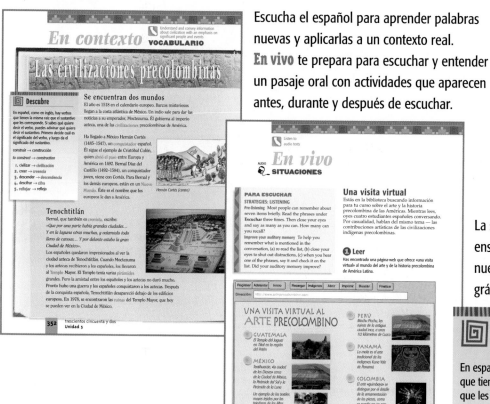

Escucha el español para aprender palabras nuevas y aplicarlas a un contexto real.

En vivo te prepara para escuchar y entender un pasaje oral con actividades que aparecen antes, durante y después de escuchar.

La sección **Descubre** te enseña palabras y expresiones nuevas a través de claves gráficas y textuales.

Descubre

En español, como en inglés, hay verbos que tienen la misma raíz que el sustantivo que les corresponde. Si sabes qué quiere decir el verbo, puedes adivinar qué quiere decir el sustantivo. Primero decide cuál es el significado del verbo, y luego da el significado del sustantivo.

construir → construcción

to construct → construction

1. civilizar → **civilización**
2. creer → **creencia**
3. descender → **descendencia**
4. descifrar → **cifra**
5. **reflejar** → reflejo

¡Diviértete!

Aprender otro idioma puede ser divertido e interesante. Como ya sabes mucho español, puedes expresarte mejor y comunicarte con tus compañeros de clase a través de las actividades de tu libro. Tienes todas las herramientas que necesitas; ¡aprende y disfruta!

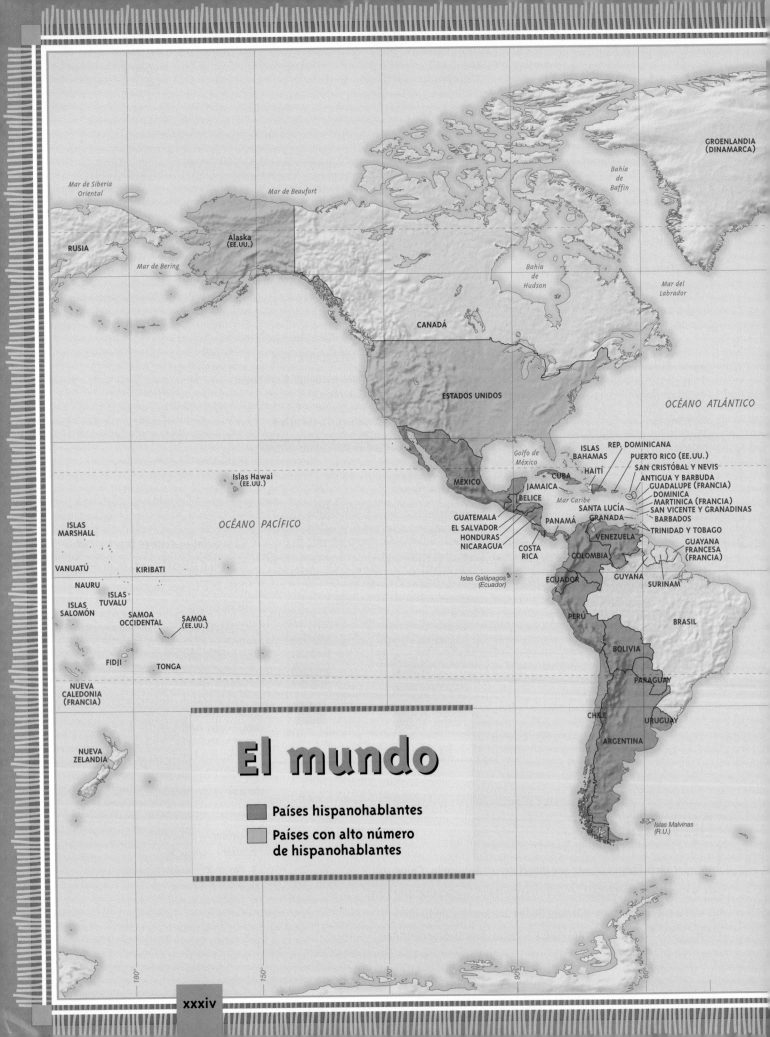

El mundo

■ Países hispanohablantes
■ Países con alto número
de hispanohablantes

OCÉANO ÁRTICO

Mar de Laptev

Mar de Kara

Mar de Barents

Mar de Noruega

ISLANDIA

SUECIA FINLANDIA

NORUEGA

ESTONIA

LETONIA

LITUANIA

RUSIA

REINO UNIDO

Mar del Norte

IRLANDA

POLONIA

BIELORRUSIA

ALEMANIA

UCRANIA

FRANCIA

AUSTRIA

MOLDAVIA

ANDORRA

RUMANIA

ESPAÑA

ITALIA

PORTUGAL

GRECIA

TURQUÍA

GEORGIA

ARMENIA

AZERBAIYÁN

Mar Mediterráneo

CHIPRE

LÍBANO

SIRIA

IRAQ

IRÁN

Mar Negro

Mar Caspio

KAZAKSTÁN

Mar de Aral

UZBEKISTÁN

KIRGUISTÁN

TURKMENISTÁN

TADJIKISTÁN

AFGANISTÁN

Lago Baikal

MONGOLIA

Mar de Ojotsk

COREA DEL NORTE

Mar de Japón

COREA DEL SUR

JAPÓN

CHINA

1	DINAMARCA	9	ESLOVENIA
2	HOLANDA	10	CROACIA
3	BÉLGICA	11	BOSNIA Y HERZEGOVINA
4	LUXEMBURGO	12	YUGOSLAVIA
5	SUIZA	13	ALBANIA
6	REPÚBLICA CHECA	14	MACEDONIA
7	ESLOVAQUIA	15	BULGARIA
8	HUNGRÍA	16	MALTA

GIBRALTAR (R.U.)

Islas Canarias (Esp.)

MARRUECOS

ISRAEL

JORDANIA

EGIPTO

BAHREIN

QATAR

E.A.U.

ARABIA SAUDITA

OMÁN

KUWAIT

PAQUISTÁN

BHUTÁN

NEPAL

INDIA

MYANMAR

Trópico de Cáncer

TAIWÁN

GUAM (EE.UU.)

MAURITANIA

MALÍ

NÍGER

CHAD

SUDÁN

ERITREA

YEMEN

JIBUTI

ETIOPÍA

Mar Rojo

SENEGAL

BURKINA FASO

BENIN

NIGERIA

GUINEA

COSTA DE MARFIL

TOGO

REP. CENTRO-AFRICANA

SOMALIA

GAMBIA

GUINEA BISSAU

LIBERIA

GHANA

SIERRA LEONA

GUINEA ECUATORIAL

CAMERÚN

GABÓN

CONGO

REP. DEL CONGO

UGANDA

KENIA

RUANDA

BURUNDI

TANZANÍA

Mar Arábigo

Golfo de Bengala

BANGLADESH

LAOS

TAILANDIA

VIETNAM

CAMBOYA

Mar de China

BRUNEI

FILIPINAS

MALAYSIA

PALAU

MICRONESIA

SINGAPUR

Ecuador

ISLAS MALDIVAS

SRI LANKA

INDONESIA

PAPUASIA NUEVA GUINEA

CABINDA (ANGOLA)

ANGOLA

ZAMBIA

MALAWI

SEYCHELLES

COMORES

MOZAMBIQUE

TIMOR ORIENTAL

NAMIBIA

ZIMBABWE

BOTSWANA

MADAGASCAR

MAURICIO

OCÉANO ÍNDICO

Trópico de Capricornio

AUSTRALIA

SUDÁFRICA

LESOTHO

SUAZILANDIA

N

| 0 | 1000 | 2000 kilómetros |
| 0 | 1000 | 2000 millas |

ANTÁRTIDA

30° 60° 90° 120°

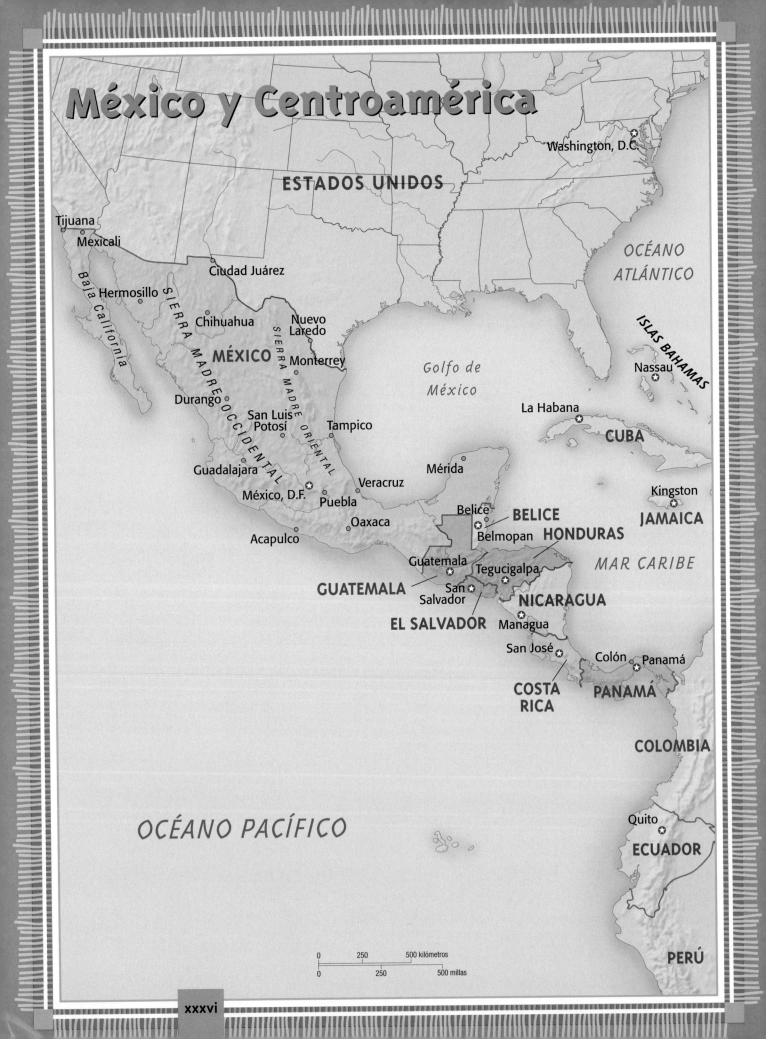

México y Centroamérica

ESTADOS UNIDOS

Washington, D.C.

OCÉANO ATLÁNTICO

Tijuana

Mexicali

Ciudad Juárez

ISLAS BAHAMAS

Hermosillo

Baja California

Chihuahua

Nuevo Laredo

SIERRA MADRE OCCIDENTAL

MÉXICO

Monterrey

Durango

SIERRA MADRE ORIENTAL

Golfo de México

Nassau

San Luis Potosí

La Habana

CUBA

Tampico

Guadalajara

Mérida

Veracruz

Kingston

México, D.F.

Puebla

Belice

BELICE

JAMAICA

Oaxaca

Belmopan

HONDURAS

Acapulco

Guatemala

Tegucigalpa

MAR CARIBE

GUATEMALA

San Salvador

NICARAGUA

EL SALVADOR

Managua

San José

Colón

Panamá

COSTA RICA

PANAMÁ

COLOMBIA

OCÉANO PACÍFICO

Quito

ECUADOR

| 0 | 250 | 500 kilómetros |
| 0 | 250 | 500 millas |

PERÚ

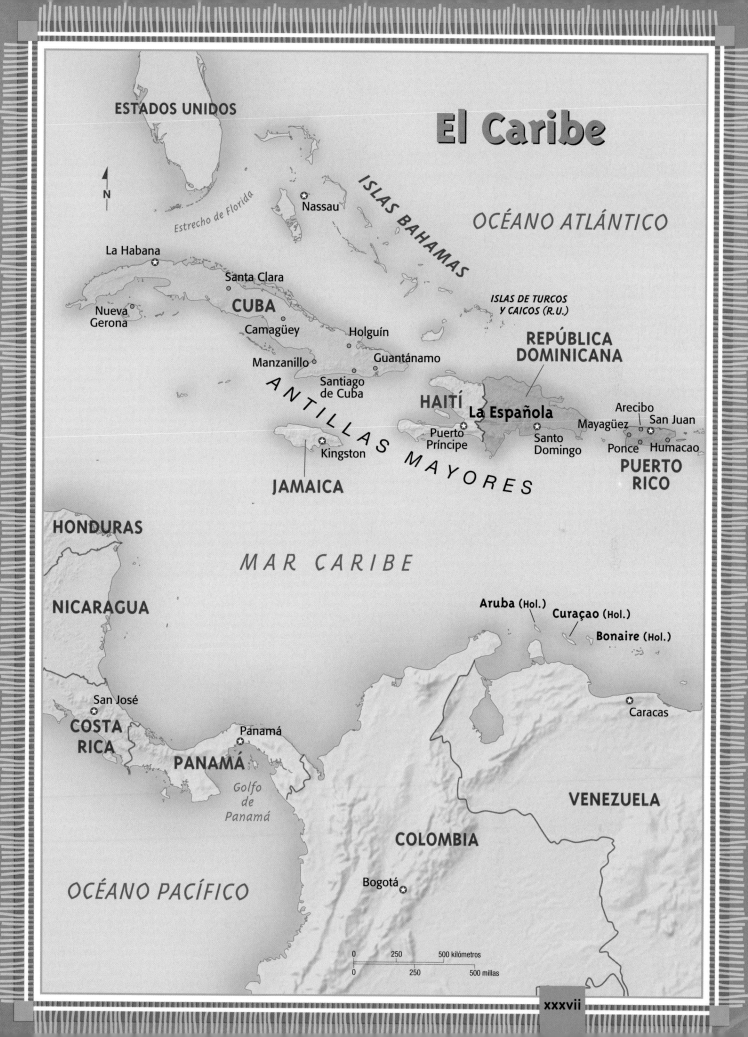

El Caribe

ESTADOS UNIDOS

N

Estrecho de Florida

Nassau

ISLAS BAHAMAS

OCÉANO ATLÁNTICO

La Habana

Santa Clara

CUBA

Nueva
Gerona

Camagüey   Holguín

Manzanillo   Guantánamo

Santiago
de Cuba

ISLAS DE TURCOS
Y CAICOS (R.U.)

REPÚBLICA
DOMINICANA

HAITÍ   La Española

Arecibo
Mayagüez   San Juan

Puerto
Príncipe

Santo
Domingo

Ponce   Humacao

PUERTO
RICO

A N T I L L A S

Kingston

M A Y O R E S

JAMAICA

HONDURAS

MAR CARIBE

NICARAGUA

Aruba (Hol.)   Curaçao (Hol.)

Bonaire (Hol.)

San José

Caracas

COSTA
RICA

Panamá

PANAMÁ

Golfo
de
Panamá

VENEZUELA

COLOMBIA

OCÉANO PACÍFICO

Bogotá

| 0 | 250 | 500 kilómetros |
| 0 | 250 | 500 millas |

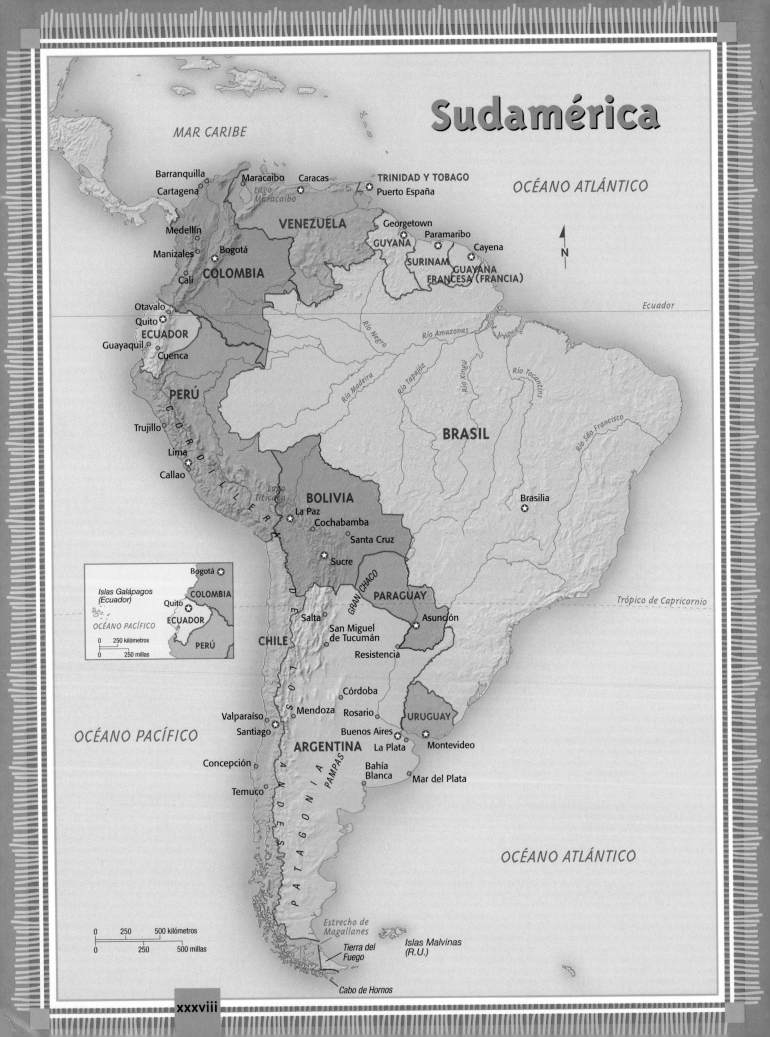

Sudamérica

MAR CARIBE

OCÉANO ATLÁNTICO

Barranquilla
Cartagena
Maracaibo
Caracas
TRINIDAD Y TOBAGO
Puerto España
Lago Maracaibo
Medellín
VENEZUELA
Georgetown
Paramaribo
GUYANA
Cayena
Manizales
Bogotá
SURINAM
GUAYANA FRANCESA (FRANCIA)
Cali
COLOMBIA
N
Otavalo
Ecuador
Quito
ECUADOR
Guayaquil
Cuenca
Río Negro
Río Amazonas
Río Tapajóz
Río Xingú
Río Tocantins
PERÚ
Río Madeira
Río São Francisco
Trujillo
BRASIL
Lima
Callao
Lago Titicaca
BOLIVIA
Brasilia
La Paz
Cochabamba
Santa Cruz
Sucre
GRAN CHACO
PARAGUAY
Trópico de Capricornio
Salta
Asunción
San Miguel de Tucumán
CHILE
Resistencia
Córdoba
Valparaíso
Mendoza
Rosario
URUGUAY
Santiago
Buenos Aires
Montevideo
OCÉANO PACÍFICO
ARGENTINA
La Plata
Concepción
Bahía Blanca
Mar del Plata
Temuco
PAMPAS
PATAGONIA
ANDES
CORDILLERA DE LOS

OCÉANO ATLÁNTICO

Estrecho de Magallanes
Tierra del Fuego
Islas Malvinas (R.U.)
Cabo de Hornos

Islas Galápagos (Ecuador)
Bogotá
COLOMBIA
Quito
ECUADOR
OCÉANO PACÍFICO
PERÚ
0   250 kilómetros
0   250 millas

0   250   500 kilómetros
0   250   500 millas

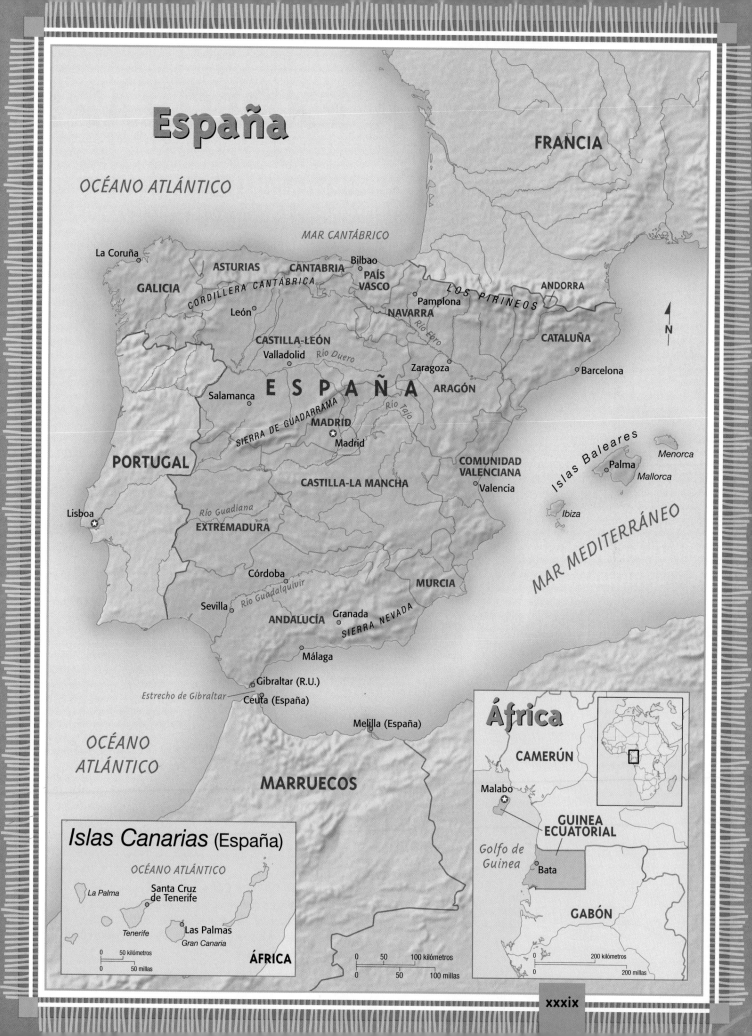

España

OCÉANO ATLÁNTICO

FRANCIA

MAR CANTÁBRICO

La Coruña

GALICIA

ASTURIAS

CANTABRIA

Bilbao

PAÍS VASCO

CORDILLERA CANTÁBRICA

LOS PIRINEOS

ANDORRA

León

Pamplona

NAVARRA

CATALUÑA

CASTILLA-LEÓN

Valladolid

Río Duero

Zaragoza

ARAGÓN

Barcelona

E S P A Ñ A

Salamanca

Río Tajo

SIERRA DE GUADARRAMA

MADRID

Madrid

PORTUGAL

COMUNIDAD VALENCIANA

Valencia

Islas Baleares

Menorca

Palma

Mallorca

CASTILLA-LA MANCHA

Ibiza

Río Guadiana

MAR MEDITERRÁNEO

Lisboa

EXTREMADURA

Córdoba

MURCIA

Sevilla

Río Guadalquivir

ANDALUCÍA

Granada

SIERRA NEVADA

Málaga

Gibraltar (R.U.)

Estrecho de Gibraltar

Ceuta (España)

Melilla (España)

OCÉANO ATLÁNTICO

MARRUECOS

África

CAMERÚN

Malabo

GUINEA ECUATORIAL

Golfo de Guinea

Bata

GABÓN

Islas Canarias (España)

OCÉANO ATLÁNTICO

La Palma

Santa Cruz de Tenerife

Tenerife

Las Palmas

Gran Canaria

| 0 | 50 kilómetros |
| 0 | 50 millas |

ÁFRICA

N

| 0 | 50 | 100 kilómetros |
| 0 | 50 | 100 millas |

| 0 | 200 kilómetros |
| 0 | 200 millas |

ETAPA PRELIMINAR

¡Bienvenidos al mundo hispano!

OBJECTIVES

- Talk about present activities

- Talk about past activities

¡A EXPLORAR!

¡**L**a diversidad cultural y geográfica de los países hispanohablantes es impresionante! En esta etapa preliminar, vas a conocer las seis regiones del mundo hispano que corresponden a las seis unidades de tu libro. Además, vas a practicar el español que ya sabes para prepararte a aprender más... ¡y a explorar nuestro mundo de posibilidades!

UNIDAD 1

ESTADOS UNIDOS
EL INSTITUTO DE CULTURAS TEJANAS Este museo de San Antonio, Texas, ofrece exhibiciones sobre las diferentes culturas que forman la población del estado de Texas. En la Unidad 1, vas a aprender más sobre la identidad y el estilo personal de los hispanohablantes de EE.UU.

UNIDAD 2

La selva de Darién

PANAMÁ
LA SELVA DE DARIÉN Esta selva tropical es muy famosa. Los viajeros tienen que pasar por la selva a pie o tomar un barco desde la costa para seguir su viaje. En la Unidad 2, vas a aprender más sobre la naturaleza y la ecología y cómo podemos preservarlas.

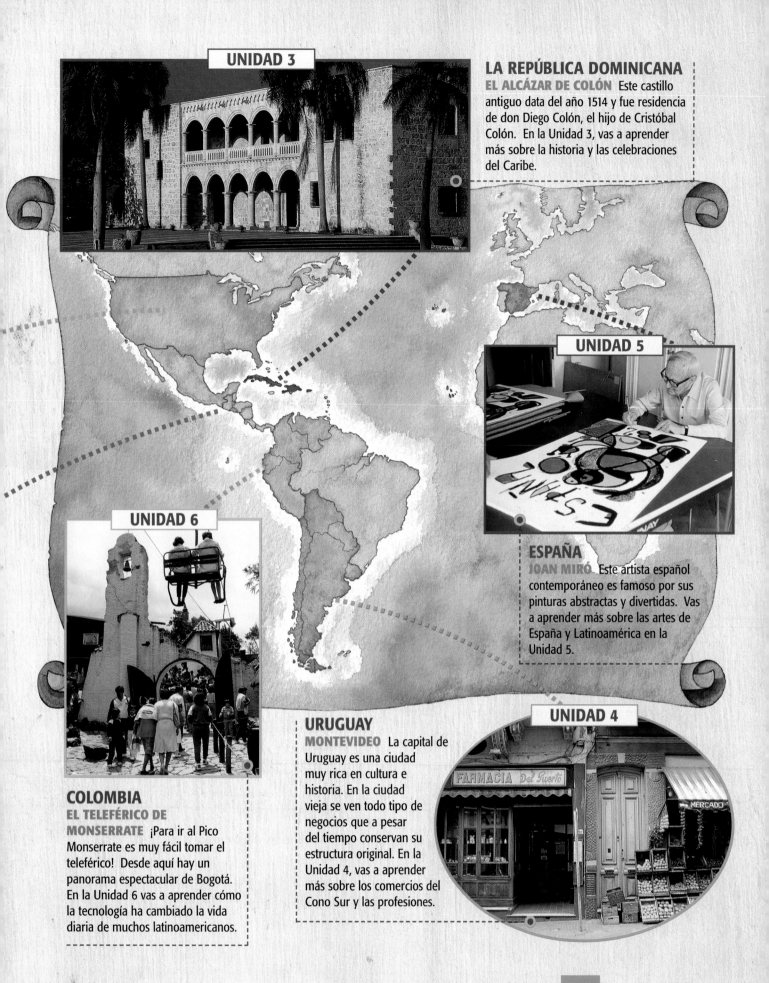

UNIDAD 3

LA REPÚBLICA DOMINICANA
EL ALCÁZAR DE COLÓN Este castillo antiguo data del año 1514 y fue residencia de don Diego Colón, el hijo de Cristóbal Colón. En la Unidad 3, vas a aprender más sobre la historia y las celebraciones del Caribe.

UNIDAD 5

ESPAÑA
JOAN MIRÓ Este artista español contemporáneo es famoso por sus pinturas abstractas y divertidas. Vas a aprender más sobre las artes de España y Latinoamérica en la Unidad 5.

UNIDAD 6

COLOMBIA
EL TELEFÉRICO DE MONSERRATE ¡Para ir al Pico Monserrate es muy fácil tomar el teleférico! Desde aquí hay un panorama espectacular de Bogotá. En la Unidad 6 vas a aprender cómo la tecnología ha cambiado la vida diaria de muchos latinoamericanos.

URUGUAY
MONTEVIDEO La capital de Uruguay es una ciudad muy rica en cultura e historia. En la ciudad vieja se ven todo tipo de negocios que a pesar del tiempo conservan su estructura original. En la Unidad 4, vas a aprender más sobre los comercios del Cono Sur y las profesiones.

UNIDAD 4

I

Estados Unidos

VIDEO   DVD

NUESTRA
cultura e historia

Los hispanohablantes de Estados Unidos somos un grupo diverso con una larga historia. ¿Sabías que ya existían pueblos españoles en el sur y oeste de EE.UU. cuando los ingleses llegaron en el Mayflower para establecer sus propias colonias?

Llegamos al este y noreste del país a fines de los 1800. Como resultado de conflictos entre España y Estados Unidos, Cuba se separó de España y Puerto Rico pasó a ser parte de EE.UU. Después, muchos cubanos y puertorriqueños emigraron a Estados Unidos.

Al empezar el año 1900, la mayoría de los hispanohablantes estadounidenses eran de descendencia española-mexicana y caribeña. Desde entonces han llegado otros grupos étnicos, incluso salvadoreños, nicaragüenses, guatemaltecos, mexicanos, colombianos y dominicanos.

Nos unen la lengua y las tradiciones hispanoamericanas. Pero cada grupo representa una cultura distinta y una historia particular. ¡Representamos una gran variedad de culturas y experiencias!

Robert Castro

¡Hola! Yo soy Robert Castro. Soy actor y hago el papel de Francisco García Flores en el video de los niveles 1 y 2 de ¡En español! En la unidad 1 del video para el nivel 3, hablo sobre mi vida y experiencias: la identidad y el estilo personal, mi familia y los latinos en Estados Unidos.

¡Pero eso viene más tarde! Ahora vamos a conocer un poco de la historia y cultura de los hispanohablantes que viven en Estados Unidos.

¡Nos vemos!

Robert

DESFILE Cada año se reúnen miles de personas en Nueva York para celebrar la comunidad puertorriqueña con el «Puerto Rican Pride Parade». ¡No te lo pierdas!

INDEPENDENCIA Para recordar el día de su independencia, los dominicanos de la Ciudad de Nueva York celebran el 27 de febrero con fiestas, música y baile.

MURAL El barrio conocido como el «Mission District», en la ciudad de San Francisco, es un oasis de la cultura mexicano-americana, famoso por sus taquerías, tiendas latinas, cultura y arte.

GUAYABERA La guayabera es una camisa tradicional de las islas del Caribe. En Miami, La Casa de las Guayaberas es el lugar preferido para comprarse una.

LA CALLE OCHO La cultura cubana es importante en la ciudad de Miami, que celebra un carnaval cada año. Los mejores artistas hispanohablantes participan en este carnaval, llamado el Carnaval de la Calle Ocho.

En acción
VOCABULARIO Y GRAMÁTICA

REPASO **Present Tense of Regular verbs**

You use the **present tense** to talk about what you are doing now and what you plan to do in the immediate future.

Veo la tele.
I'm watching T.V.

Veo una película por semana.
I see one movie a week.

Veo a Carmen esta noche.
I'm seeing (I'll see) Carmen this evening.

Regular verbs

	-ar hablar	-er comer	-ir vivir
yo	hablo	como	vivo
tú	hablas	comes	vives
él, ella, usted	habla	come	vive
nosotros(as)	hablamos	comemos	vivimos
vosotros(as)	habláis	coméis	vivís
ellos, ellas, ustedes	hablan	comen	viven

Remember that in **stem-changing verbs** you change the vowel of the stem in all the forms of the singular and in the third-person plural of the present tense.

Stem-changing verbs

	e → ie pensar	o → ue dormir	e → i pedir
yo	pienso	duermo	pido
tú	piensas	duermes	pides
él, ella, usted	piensa	duerme	pide
nosotros(as)	pensamos	dormimos	pedimos
vosotros(as)	pensáis	dormís	pedís
ellos, ellas, ustedes	piensan	duermen	piden

Practice:
Actividades
1 2 3

Más práctica
cuaderno p. 1
Para hispanohablantes
cuaderno p. 1

 Online Workbook
CLASSZONE.COM

---

1 Tu rutina

Hablar/Escribir Un(a) nuevo(a) estudiante hispanohablante quiere saber más sobre tu rutina. Primero, tu compañero(a) hace el papel del (de la) estudiante. Luego, cambien de papel.

modelo

estudiar (todas las tardes en casa)

Compañero(a): *¿Cuándo estudias?*
o *¿Dónde estudias?*

Tú: *Estudio todas las tardes en casa.*

1. correr en el parque (tres veces por semana)
2. visitar a tus amigos (los fines de semana)
3. escribir correo electrónico (antes de acostarme)
4. leer el periódico estudiantil (en el colegio)
5. almorzar (a las doce)
6. jugar al tenis (después de clases)
7. trabajar (en la tienda de deportes)
8. regresar a casa (a las seis para la cena)
9. dormir (en mi habitación)
10. estudias (en la biblioteca)

---

2 La playa

Hablar/*Escribir* Imagina que tú y tu compañero(a) viven en Los Ángeles y van a la playa a menudo. Tú quieres saber qué hacen tu compañero(a) y los miembros de su familia allí. Él (Ella) quiere saber lo mismo de tu familia. Busquen ideas en el dibujo.

modelo

Tú: *¿Qué haces cuando vas a la playa?*

Compañero(a): *¿Yo? Generalmente, tomo el sol o nado. ¿Y tú?*

Tú: *Pues yo llevo mis patines y patino todo el día. ¿Y tu hermano?*

Compañero(a): *A mi hermano no le gusta nadar. Así que generalmente escucha la radio o juega al voleibol.*

3 Los fines de semana

Hablar/*Escribir* Entrevista a cuatro compañeros(as) de clase. Quieres saber qué hacen los fines de semana. Haz una gráfica *(chart)* y escribe cómo responden.

modelo

Tú: *¿Qué haces los fines de semana?*

Compañero(a): *Generalmente, los sábados por la mañana me levanto temprano y desayuno. Luego, alquilo un video o tomo el sol en la playa.*

Tú: *¿Y los domingos?*

Compañero(a): …

Nombre	Sábado	Domingo
_____	_____	_____
_____	_____	_____
_____	_____	_____
_____	_____	_____

Listen to video texts

México y Centroamérica

VIDEO DVD

TRADICIONES
del pasado

Guadalupe González

¿Sabes quién soy?

Tal vez me conoces como Isabel Palacios del video de los niveles 1 y 2. Pero soy actriz y me llamo Guadalupe González. En el video de la unidad 2 de este libro, vamos a hablar de nuestras comunidades, la naturaleza y los problemas que las pueden afectar.

Ahora quiero mostrarte los países en la unidad 2: México, Guatemala, Honduras, Nicaragua, El Salvador, Costa Rica y Panamá. ¡Es una región interesante y bellísima a la vez!

¡Hasta luego! Guadalupe

En realidad, ¡el «Nuevo Mundo» de los conquistadores era tan viejo como Europa! Grandes civilizaciones, como la maya, la azteca y la tolteca, ya existían cuando los conquistadores llegaron aquí por primera vez. Nuestra cultura tiene estas dos historias.

A principios de los 1800 estos países se separaron de España. Luego, formaron un imperio desde Costa Rica hasta el suroeste de los Estados Unidos. Pero gradualmente, diferencias étnicas, culturales y políticas formaron los países que conocemos hoy.

La influencia de los españoles ha sido grande, pero hoy encontramos millones de personas que todavía hablan lenguas antiguas como el maya y el náhuatl. También mantienen vivas las costumbres y tradiciones que vienen de la época antes de la llegada de los europeos. ¡Estas tradiciones son parte de nuestra identidad contemporánea!

HONDURAS El huracán Mitch ha sido el peor desastre natural del hemisferio occidental en los últimos doscientos años. Más de un millón de personas perdieron sus casas, familiares y amigos. El huracán destruyó muchos pueblos y vecindades. Más de 20 años pasarán antes de que Honduras pueda reconstruir su infraestructura.

EL SALVADOR Aunque es el país más pequeño de Centroamérica, El Salvador tiene una gran variedad geográfica. Cerca de la capital, San Salvador, se encuentra este volcán de dos picos. El más grande se llama Picacho y el más pequeño Boquerón.

BAJA CALIFORNIA Personas de todo el mundo visitan el estado mexicano de Baja California para ver todo tipo de animales acuáticos que pasan por aquí durante su migración anual.

COSTA RICA ¿Un jardín de mariposas? Lo puedes encontrar en San José, Costa Rica. Este país es conocido por la naturaleza que se encuentra allí.

 **Activity 5:** Narrate in the present

En acción
VOCABULARIO Y GRAMÁTICA

REPASO **Irregular yo Forms**

Remember that some verbs are irregular in the present tense only in the first person singular (yo) form. Compare the yo and tú forms of these verbs.

	yo	tú

- Verbs like **hacer**

caer	caigo	caes
hacer	hago	haces
poner	pongo	pones
salir	salgo	sales
traer	traigo	traes
valer	valgo	vales

- Verbs with a spelling change: c → zc

conocer	conozco	conoces

- Other verbs irregular in the yo form

dar	doy	das
saber	sé	sabes
ver	veo	ves

Other irregular verbs that you have already learned (**estar, ir, ser, tener, venir**) are conjugated for you on pp. R33–R35.

Practice:
Actividades ❹ ❺ ❻

 **Online Workbook**
CLASSZONE.COM

Más práctica *cuaderno pp. 2, 3*
Para hispanohablantes *cuaderno pp. 2, 3*

---

4 **La encuesta**

Hablar/*Escribir* Estás de vacaciones en Costa Rica y ves esta encuesta (*survey*) en una revista para jóvenes. La revista quiere saber más de los hábitos de los jóvenes. Contesta las preguntas.

modelo

¿Cuándo haces la tarea?

☑ *después de clases* ☐ *después de la cena* ☐ *¿...?*

Hago la tarea después de clases.

Encuesta

1. ¿Con qué frecuencia sales con tus amigos? ☐ todos los días ☐ los fines de semana ☐ ¿...?
2. ¿Cuándo ves televisión? ☐ después de clases ☐ después de la cena ☐ ¿...?
3. ¿Conoces la música de Maná? ☐ sí ☐ no
4. ¿Quién pone la mesa en tu casa? ☐ yo ☐ hermano o hermana ☐ ¿...?
5. ¿Conduces al colegio o tomas el autobús? ☐ conducir ☐ tomar el autobús ☐ ¿...?
6. ¿Haces tu cama todos los días? ☐ sí ☐ no
7. ¿Sabes usar Internet? ☐ sí ☐ no

5 Correo electrónico

Escribir Escribe una carta por correo electrónico a un(a) amigo(a) nicaragüense. Usa por lo menos cinco palabras de cada columna para describir tu vida.

modelo

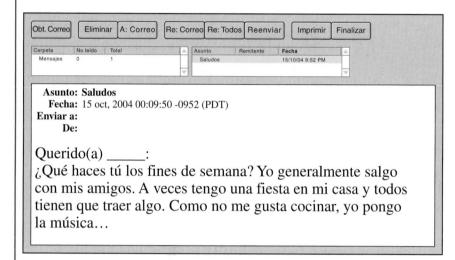

| Obt. Correo | Eliminar | A: Correo | Re: Correo | Re: Todos | Reenviar | Imprimir | Finalizar |

| Carpeta | No leído | Total | | Asunto | Remitente | Fecha |
| Mensajes | 0 | 1 | | Saludos | | 15/10/04 9:52 PM |

Asunto: Saludos
Fecha: 15 oct, 2004 00:09:50 -0952 (PDT)
Enviar a:
De:

Querido(a) _____:
¿Qué haces tú los fines de semana? Yo generalmente salgo con mis amigos. A veces tengo una fiesta en mi casa y todos tienen que traer algo. Como no me gusta cocinar, yo pongo la música...

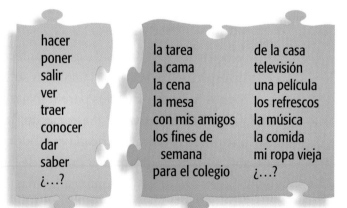

hacer	la tarea	de la casa
poner	la cama	televisión
salir	la cena	una película
ver	la mesa	los refrescos
traer	con mis amigos	la música
conocer	los fines de	la comida
dar	semana	mi ropa vieja
saber	para el colegio	¿...?
¿...?		

6 ¿Lo conoces?

Hablar Acabas de conocer a tu compañero(a) mexicano(a). Quieres saber si él (ella) conoce varias cosas y personas. Hazle varias preguntas. Luego, cambien de papel.

modelo

Tú: *¿Conoces el museo de Frida Kahlo en Coyoacán?*

Compañero(a): *No, no lo conozco, pero sí conozco sus pinturas.*

Tú: *¿De veras? ¿Qué piensas de ellas?*

Compañero(a): *...*

libros   pinturas

ciudad   lugar

música   persona

¿...?

Listen to video texts

El Caribe

VIDEO   DVD

UNA HISTORIA
dramática

Nilka Desirée

¡Bienvenidos al Caribe!

Me llamo Nilka Desirée y soy actriz.
En el video del nivel 1 hago el papel de
Diana Ortiz Avilés. Vas a verme otra
vez en el video de la unidad 3 de este
libro, donde hablo un poco sobre las
celebraciones y los festivales del Caribe.

Antes de tratar estos temas, quiero
darte un poco de la historia de los países
que forman el Caribe hispanohablante:
Puerto Rico, la República Dominicana
y Cuba.

¡Hasta pronto!

Cuando Cristóbal Colón llegó al Caribe
en 1492, la vida de sus habitantes cambió
dramáticamente. La mayoría de éstos eran de
origen taíno, una tribu de indios que vivían en
Puerto Rico y el este de Cuba.

Los europeos conquistaron a los taínos y
usaron las islas para la conquista de otros
imperios, como el azteca y el inca. El Caribe fue
un teatro de conflicto entre las naciones europeas
y los piratas de todo el mundo. Todos querían
controlar estas islas.

Uno de los resultados de estos conflictos
fue la destrucción de la cultura taína. En su lugar,
los europeos trajeron africanos para trabajar la
tierra y producir el azúcar. Por este hecho, la
contribución de la cultura africana es evidente en
el Caribe, aunque aquí hay una mezcla de culturas
como en otras regiones de América Latina.

Gradualmente, Cuba y la República
Dominicana se separaron de España y Puerto
Rico pasó a ser parte de EE.UU. Hoy en día, estos
países celebran su historia y diversidad cultural,
mostrando un aprecio por las tres culturas que
son parte de ellos.

PUERTO RICO Las ferias de artesanía en Puerto Rico son una celebración de la cultura e historia popular de las regiones distintas de la isla.

PUERTO RICO En 1508, el explorador Juan Ponce de León conquistó la isla de Borinquen (hoy Puerto Rico). Construyó una casa donde puedes encontrar parte de la historia de los gobernantes de Puerto Rico.

LA HABANA La Habana fue fundada en 1519, en Cuba. Esta sección de la capital cubana se conoce como «La Habana Vieja». Conserva el estilo colonial, incluso calles estrechas y palacios.

REPÚBLICA DOMINICANA Cerca de la capital dominicana de Santo Domingo se encuentra el Parque de los Tres Ojos. Los «ojos» se refieren a lagunas pequeñas de agua azul que «miran» desde las cuevas del parque.

Activity 7: Understand most spoken language when the message is deliberately and carefully conveyed by a speaker accustomed to dealing with learners when listening

En acción
VOCABULARIO Y GRAMÁTICA

REPASO **The Preterite Tense of Regular Verbs**

> You use the **preterite** to talk about actions that you or others completed in the past.

Escribí cartas por una hora.
I wrote letters for an hour.

Bailamos toda la noche.
We danced all night.

Remember that **-er** and **-ir** verbs have the **same endings** in the preterite.

Regular Preterite Verbs

	-ar hablar	-er comer	-ir vivir
yo	hablé	comí	viví
tú	hablaste	comiste	viviste
usted, él, ella	habló	comió	vivió
nosotros(as)	hablamos	comimos	vivimos
vosotros(as)	hablasteis	comisteis	vivisteis
ustedes, ellos, ellas	hablaron	comieron	vivieron

Practice:
Actividades
7 **8** **9** **10**

Más práctica
cuaderno p. 4
Para hispanohablantes
cuaderno p. 4

Online Workbook
CLASSZONE.COM

7 **¡Pobre Adriana!**

Escuchar/*Escribir* Adriana está contando lo que le pasó ayer. Escúchala y escribe oraciones para describir su día.

modelo

despertarse
Adriana se despertó muy tarde.

1. planchar
2. salir
3. trabajar
4. perder
5. regresar
6. cenar
7. revisar
8. comprar

8 **La isla de Puerto Rico**

Hablar/*Escribir* Marcela fue a Puerto Rico durante el verano. ¿Qué hicieron ella y su familia?

modelo

(yo) mandar muchas tarjetas postales

Mandé muchas tarjetas postales.

1. (yo) visitar el Centro Ceremonial Indígena de Tibes
2. (mi familia y yo) comer comida puertorriqueña muy sabrosa
3. (mi hermano) comprar unos discos compactos de salsa
4. (mi familia y yo) caminar por el Viejo San Juan
5. (mi hermana y yo) tomar el sol en la playa de Luquillo
6. (yo) escribir un poema sobre la belleza de la isla
7. (mi hermana) recibir un regalo de su amigo puertorriqueño
8. (yo) comprender por qué llaman a Puerto Rico «la Isla del Encanto»

9 La familia

Hablar/Escribir El sábado pasado todos los miembros de tu familia hicieron cosas diferentes. Tu amigo(a) te pregunta qué pasó. Mira los dibujos y dile qué hizo cada persona (por lo menos dos cosas). ¡Usa tu imaginación!

modelo

Compañero(a): *¿Qué hizo tu hermano el sábado pasado?*

Tú: *Primero limpió su cuarto y luego alquiló un video.*

Compañero(a): *¡Qué divertido! Y tu mamá, ¿qué hizo?*

Tú: *Pues,…*

1. yo

2. mi abuelo

3. mis primos

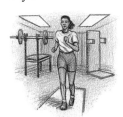

4. mi prima

5. mi mamá

6. mi tía

7. mi hermanito

8. mi hermana

9. mi tío

10 El fin de semana

Hablar Es lunes por la mañana y no has hablado con tu mejor amigo(a) en todo el fin de semana. Los dos quieren saber cómo les fue. Conversen sobre sus actividades.

modelo

Tú: *¿Cómo pasaste el sábado?*

Compañero(a): *¿Yo? Pues el sábado por la tarde llamé a Juan. Alquilamos un video y compramos una pizza para comer en casa. Después…*

Tú: *Y el domingo, ¿qué pasó?*

Compañero(a): *…*

More Practice: **Más comunicación** *p. R2*

El Cono Sur

VIDEO   DVD

VECINOS hispanohablantes

Marcelo Abramo

Ahora estamos en el Cono Sur: los países de Argentina, Chile, Uruguay y Paraguay. ¿Sabes que también se dice «el Uruguay», «la Argentina» y «el Paraguay»? Me llamo Marcelo Abramo y soy de Argentina. ¿Tal vez me reconoces del video del nivel 1 de ¡En español? Hago el papel del arquitecto González. En el video para la unidad 4 de este libro, hablo un poco sobre mis experiencias en el colegio y el mundo del trabajo. Pero ya es hora de aprender algo sobre este grupo de países sudamericanos. ¡Hablaremos después!

Marcelo

El Cono Sur es la región que ocupan Argentina, Uruguay, Paraguay y Chile. Si miras el mapa, verás que estos forman una especie de cono en el extremo sur del continente americano. Tienen una geografía muy variada. La Cordillera de los Andes pasa por Chile y Argentina. Chile está en la costa del Pacífico y Argentina y Uruguay están en la costa del Atlántico. Paraguay está al noreste de Argentina.

El Cono Sur comparte historia, idioma, y geografía. Santiago, una de las primeras ciudades españolas, se convirtió en la capital de Chile. Con el tiempo, crecieron Buenos Aires, Montevideo y Asunción, las capitales de Argentina, Uruguay y Paraguay. Los indígenas huyeron con la llegada de los europeos, menos en Paraguay, donde hoy la cultura guaraní es tan importante como la europea.

España dominó esta región, mientras que Portugal controló el Brasil. En los años 1800, estos países se separaron de España, bajo líderes como José de San Martín y Bernardo O'Higgins. Luego, inmigrantes de Italia, Francia y Alemania vinieron en la década de 1850. Hoy estos países forman parte del comercio internacional.

CHILE Chile tiene una de las industrias mineras más grandes del mundo. Además de otros minerales, el país tiene minerales preciosos como el lapislázuli, del cual se hacen bellas artesanías.

URUGUAY Si la nieve y el frío te molestan, ¿por qué no visitas el famoso destino turístico de Punta del Este en Uruguay? ¡Aquí puedes pasar el Año Nuevo tomando el sol en la playa!

LOS ANDES Si estás cansado(a) del sol en el verano del hemisferio norte, ¿por qué no vienes al hemisferio sur para esquiar en la nieve durante su invierno? Aquí hay excelentes instalaciones para este deporte, como Portillo en Chile y Bariloche en Argentina.

ARGENTINA El famoso e impresionante Aconcagua, el volcán más alto del continente americano, tiene una altura de 6.959 metros. Es un paraíso para los deportistas de todo el mundo.

ARGENTINA Buenos Aires es una de las ciudades más importantes de las Américas. Tiene un importante centro financiero. Se ha convertido en un centro para muchas compañías internacionales.

En acción
VOCABULARIO Y GRAMÁTICA

REPASO
Verbs with Spelling Changes in the Preterite

▶ Certain verbs change the spelling of their **yo forms** in the **preterite**. The rest of the preterite forms are regular.

Verbs with Spelling Changes in the Preterite

c → qu	g → gu	z → c
buscar	llegar	almorzar
busqué	llegué	almorcé
buscaste	llegaste	almorzaste
buscó	llegó	almorzó
buscamos	llegamos	almorzamos
buscasteis	llegasteis	almorzasteis
buscaron	llegaron	almorzaron

Practice:
Actividades
11 12 13 14

Más práctica
cuaderno pp. 5, 6
Para hispanohablantes
cuaderno pp. 5, 6

Online Workbook
CLASSZONE.COM

---

11 ¡Tantas preguntas!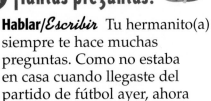

Hablar/*Escribir* Tu hermanito(a) siempre te hace muchas preguntas. Como no estaba en casa cuando llegaste del partido de fútbol ayer, ahora quiere saber más sobre el partido. Primero, tu compañero(a) hará el papel del (de la) hermanito(a). Luego, cambien de papel.

modelo

Compañero(a): *¿Jugaste en el partido de fútbol?*

Tú: *Sí, (No, no) jugué en el partido de fútbol.*

1. ¿Practicaste antes del partido?
2. ¿Sacaste fotos del equipo?
3. ¿Almorzaste con el equipo después del partido?
4. ¿Pagaste la cuenta para todo el equipo?
5. ¿Llegaste tarde a casa?
6. ¿Le explicaste a papá por qué llegaste tarde?
7. ¿Buscaste los libros de historia?
8. ¿Visitaste a tus abuelos?
9. ¿Hablaste con el profesor de español?
10. ¿Comiste en tu casa ayer?

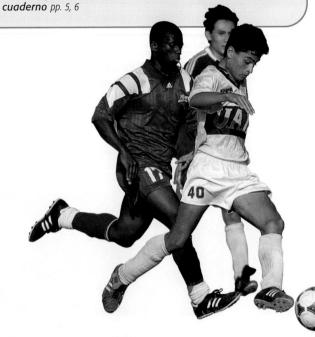

12 Buenos Aires

Escribir Caminas por la Avenida 9 de Julio en Buenos Aires y escuchas las siguientes conversaciones. Usa el pretérito.

«_____ (llegar) al mercado con mucho dinero. Después de que _____ (pagar) la comida, ¡no me quedó nada!»

«_____ (sacar) la cámara. _____ (empezar) a sacar fotos. Un turista se enojó.»

«_____ (almorzar) en un restaurante. _____ (salir) a la calle pero no vi que estaba lloviendo. Cuando lo vi, _____ (comenzar) a correr.»

13 Ayer

Escribir Escribe una descripción de todo lo que hiciste ayer. Usa los verbos de la lista.

modelo

Ayer jugué al fútbol. Luego fui a mi clase de piano. Después…

buscar	llegar	jugar
cruzar	almorzar	empezar
sacar	pagar	tocar

14 De vacaciones

Hablar Fuiste de vacaciones a la República Dominicana. Tu compañero(a) quiere saber más de tu viaje. Luego cambien de papel.

modelo

Tú: *Fui de vacaciones a la República Dominicana.*

Compañero(a): *¿De veras? ¿Sacaste fotos?*

Tú: *Sí, claro. Saqué muchas fotos.*

Compañero(a): *¿Qué más hiciste?*

Tú: *…*

Listen to video texts

España

VIDEO   DVD

UN PAÍS DE
múltiples culturas

Cuando estudiamos las historias de otros países hispanohablantes, aprendimos cómo España los conquistó y trajo su propia cultura. ¿Pero sabías que hace muchos años la misma cosa ocurrió en España?

La región que hoy es España fue conquistada por varios grupos durante su larga historia — los romanos, los visigodos y los árabes, entre otros. Su nombre viene de los romanos, quienes la llamaron «Hispania». Todavía existen ruinas en España que datan de esta época.

Los árabes llegaron en el año 711 d.C. y estuvieron hasta 1492, cuando las fuerzas de los monarcas Fernando e Isabel los expulsaron de España. Estos reyes unificaron las regiones y culturas diversas del país y establecieron una lengua común, el castellano, como también se llama al español.

La España de hoy es un país muy diverso. Las diferentes influencias culturales se ven en regiones donde se hablan una multitud de lenguas además del español. Entre ellas están el catalán, el vasco y el gallego. Vivimos en un período de libertad cultural. Somos parte de la comunidad europea, ¡pero también mantenemos nuestro estilo propio!

Javier Morcillo

Ahora llegamos

a mi patria: España. Permítanme presentarme... En el video del nivel 1 hago el papel de Luis Paz Villarreal. En realidad soy Javier Morcillo y soy actor. En la unidad 5 de este libro vamos a ver cómo las artes y la arquitectura de España influyeron en los países del Nuevo Mundo ¡y viceversa! También voy a dar mis opiniones sobres las bellas artes.

Antes de empezar, vamos a aprender un poco de la historia de España.

¡Adiós!

Javier

GALICIA En Galicia existen costumbres celtas muy antiguas. Su instrumento musical, la gaita, se parece al *bagpipe* celta. Y su baile folklórico, la muñeira, es muy similar a las danzas típicas de Irlanda y Escocia.

TOLEDO Un ejemplo excelente de la fusión de varias culturas es la ciudad de Toledo. Cada parte de esta ciudad evoca su pasado formidable y sus influencias cristianas, árabes y judías.

BARCELONA

CATALUÑA El Palau de la Música Catalana es una obra creativa difícil de olvidar. Su arquitecto, Domènech i Montaner, la construyó entre 1905 y 1908. Es muy famoso por su bella y espectacular cúpula.

BILBAO Uno de los edificios más interesantes es el nuevo Museo Guggenheim, en Bilbao. ¡Esta construcción del arquitecto Frank Gehry ya es una de las más famosas del planeta!

En acción
VOCABULARIO Y GRAMÁTICA

REPASO
Verbs with Stem Changes in the Preterite

▶ Remember that **-ir** verbs that have a change in the stem in the present tense also have a stem change (**e → i** or **o → u**) in the **preterite**.

▶ Other verbs like **sentir**:

  despedirse, divertirse, pedir, preferir, repetir, sugerir, vestirse.

-ir Verbs with Stem Changes in the Preterite

	se**ntir**	**d**o**rmir**
yo	**sent**í	**dorm**í
tú	**sent**iste	**dorm**iste
usted, él, ella,	**s**i**nt**ió	**d**u**rm**ió
nosotros(as)	**sent**imos	**dorm**imos
vosotros(as)	**sent**isteis	**dorm**isteis
ustedes, ellos, ellas	**s**i**nt**ieron	**d**u**rm**ieron

Practice: **Actividades** 15 16 17 18

Más práctica *cuaderno pp. 7, 8*
Para hispanohablantes *cuaderno pp. 7, 8*

Online Workbook
CLASSZONE.COM

15 Al día siguiente

Escribir Al día siguiente, Mariana decidió escribirle a su mejor amiga para describir la fiesta de su prima Ángela. Completa su carta con el pretérito de los verbos entre paréntesis.

Nota: Gramática

Remember to use the pronouns **me, te, se, nos** and **os** with reflexive verbs like **vestirse, dormirse** and **despedirse**.

Querida Ileana,

 Anoche fui a la fiesta de Ángela. Ella ___1___ (sugerir) que llegáramos antes de las seis. Mi hermano Ricardo ___2___ (preferir) no ir porque no se sentía bien. Yo decidí ir con Gustavo. Él ___3___ (vestirse) con un traje muy elegante.

 Gustavo no sabía llegar, así que le ___4___ (pedir) direcciones a un policía que estaba en la esquina. El policía le ___5___ (repetir) las direcciones varias veces pero como quiera nos perdimos.

 Por fin llegamos. Ángela ___6___ (servir) comida muy sabrosa: tortilla española y otras tapas. Todos los invitados ___7___ (divertirse) mucho en la fiesta. Gustavo estaba tan cansado que ¡___8___ (dormirse) en el sofá! Los invitados ___9___ (despedirse) muy tarde. ¡Qué noche más divertida!

Abrazos,

Mariana

16 ¡Qué colores!

Hablar/Escribir Tu amigo español, Juan Felipe, invitó a todos sus amigos a una fiesta con una condición: tenían que vestirse de colores brillantes. ¿Cómo se vistieron todos?

Mario

modelo

Tú: ¿Cómo se vistió Mario en la fiesta?

Compañero(a): No sé, ¡pero dicen que se vistió de jeans morados y camiseta roja!

1. Marta y Mariana   **2.** Álvaro

3. Anita   **4.** Daniel y Donaldo

17 El desayuno

Hablar Tú y tu familia desayunaron en el Hotel Prisma. Tu amigo(a) quiere saber qué pidieron todos. ¿Qué te pregunta y cómo le contestas?

modelo

Compañero(a): ¿Qué pediste para el desayuno?

Tú: Yo pedí huevos revueltos con jamón.

Compañero(a): Y papá, ¿qué pidió?...

Hotel Prisma

Desayuno

Café con leche
Bollos de pan
Chocolate y churros
Zumo de naranja
Cereales

18 ¿Cómo estuvo?

Hablar Tú y tu compañero(a) hablan sobre fiestas distintas con los verbos de la lista.

modelo

Tú: ¿Se divirtieron todos en la fiesta de Elena?

Compañero(a): ¡Bailamos y comimos! ¿Y en tu fiesta?

Tú: También. Bailamos música del grupo...

despedirse	seguir	preferir	repetir
divertirse	servir	sugerir	vestirse

Bolivia, Colombia, Ecuador, Perú y Venezuela

VIDEO DVD

Xmena Barros

VISTAS de los Andes

¡Hola! ¡Volvemos a Sudamérica otra vez! Yo me llamo Ximena Barros, pero en el video del nivel 1, me conoces como Patricia López Carrera. También estoy en el video de este libro. En la unidad 6 hablo de la tecnología personal y cómo nos ha cambiado la vida.

Pero vamos a dejar ese tema hasta más tarde y hablar sobre los países que vas a conocer al empezar la unidad 6: Perú (o el Perú), Colombia, Bolivia, Venezuela y Ecuador (o el Ecuador).

¡Hasta pronto! *Ximena*

La Cordillera de los Andes pasa por todos estos países. Con la excepción de Bolivia, que no tiene costa, los divide en tres regiones: la montaña, la costa y la selva amazónica. Como puedes ver, ¡es una geografía muy diversa!

También son muy diversas las culturas que se encuentran en estas regiones. Cuando los españoles llegaron aquí en los años 1500, descubrieron una multitud de civilizaciones indígenas, incluso el famoso imperio inca. Por lo general, en la región montañosa la población es europea, mestiza e indígena. En la costa se ve mucho la influencia africana. Y en la selva tropical todavía existen sociedades indígenas.

Simón Bolívar, el «Libertador de América», dirigió los movimientos de independencia de España en esta región. Hoy estos países son un grupo de naciones unidas por diversos elementos naturales y culturales. Sin embargo cada país mantiene su identidad individual y su propia historia económica, social y política.

COLOMBIA Artefactos y objetos de arte de todo tipo forman las exhibiciones en el Museo de Oro en Bogotá. Aquí se puede ver una impresionante colección de joyería de oro que data desde antes de la conquista española.

VENEZUELA El parque nacional Canaima es uno de los más grandes del mundo. Aquí puedes explorar más de 11.500 millas cuadradas de naturaleza protegida.

BOLIVIA/PERÚ El lago Titicaca es el lago navegable más elevado del mundo, a una altura de 3.810 metros. Aquí se pueden ver pueblos y costumbres que nos recuerdan cómo era el mundo americano antes de la llegada de los españoles.

PERÚ Cuzco fue la capital del imperio inca, que dominaba una extensa región de Sudamérica antes de la llegada de los españoles.

ECUADOR En Ecuador, como en los demás países andinos, las antiguas costumbres coexisten con la tecnología. Un ejemplo son las hermosas granjas de flores que exportan plantas al mundo entero.

En acción
VOCABULARIO Y GRAMÁTICA

REPASO **Irregular Preterites**

A number of verbs have **irregular preterite** forms.
These verbs are grouped by similar stem changes or other like forms.

	ser and ir
yo	fui
tú	fuiste
usted, él, ella	fue
nosotros(as)	fuimos
vosotros(as)	fuisteis
ustedes, ellos, ellas	fueron

tener	estar	andar
tuve	estuve	anduve
tuviste	estuviste	anduviste
tuvo	estuvo	anduvo
tuvimos	estuvimos	anduvimos
tuvisteis	estuvisteis	anduvisteis
tuvieron	estuvieron	anduvieron

poder	poner	saber
pude	puse	supe
pudiste	pusiste	supiste
pudo	puso	supo
pudimos	pusimos	supimos
pudisteis	pusisteis	supisteis
pudieron	pusieron	supieron

hacer	venir	querer
hice	vine	quise
hiciste	viniste	quisiste
hizo	vino	quiso
hicimos	vinimos	quisimos
hicisteis	vinisteis	quisisteis
hicieron	vinieron	quisieron

decir	traer	producir
dije	traje	produje
dijiste	trajiste	produjiste
dijo	trajo	produjo
dijimos	trajimos	produjimos
dijisteis	trajisteis	produjisteis
dijeron	trajeron	produjeron

dar	ver
di	vi
diste	viste
dio	vio
dimos	vimos
disteis	visteis
dieron	vieron

Practice: **Actividades** 19 20 21

Más práctica *cuaderno pp. 9, 10*
Para hispanohablantes *cuaderno pp. 9, 10*

Online Workbook
CLASSZONE.COM

19 Irma y Javier

Escuchar/*Escribir* Irma y Javier fueron al centro comercial. Escucha su conversación y luego completa las siguientes oraciones con el pretérito del verbo correcto.

modelo

Irma **anduvo** por todo el centro comercial buscando a Javier.

1. Javier _____ en la tienda de música un rato.
2. Entonces _____ a la tienda de deportes.
3. Irma y Javier _____ de acuerdo en dónde se iban a encontrar.
4. Irma _____ las compras que quería hacer.
5. Irma _____ encontrar todo lo que estaba en su lista.
6. Javier _____ que ir al banco.
7. Javier no _____ su tarjeta de crédito.
8. Marín y Lupita _____ que iban a comprar una computadora nueva.

20 ¿Adónde fueron?

Hablar/*Escribir* Tú y tu compañero(a) conversan sobre sus vacaciones. Imagínense que fueron a unos de los países andinos.

modelo

Tú: *¿Adónde fueron de vacaciones este verano?*

Compañero(a): *Fuimos a Ecuador.*

Tú: *¿Ah, sí? ¿Cuánto tiempo estuvieron en Ecuador?*

Compañero(a): *Estuvimos allá dos semanas.*

adónde   cuánto tiempo   regalos

cambiar dinero   la excursión de...

monumentos   por el centro   ¿ ?

21 ¿Quién dijo qué?

Hablar En grupos de cinco o seis, jueguen a «¿Quién dijo qué»? Inventen un personaje como la María del modelo. Cada persona añade algo más sobre María a la oración. ¡Sean originales!

modelo

Tú: *¿Qué dijo Miguel?*

Compañero(a) 1: *Miguel dijo que María no hizo la tarea. ¿Qué dijo Arturo?*

Compañero(a) 2: *Arturo dijo que María no hizo la tarea y llegó tarde a clase.*

More Practice: **Más comunicación** *p. R2*

ETAPA PRELIMINAR

Activity 2: Tend to become less accurate as the task or message becomes more complex, and some patterns of error may interfere with meaning

Self-Check Quiz
CLASSZONE.COM

En uso
REPASO Y MÁS COMUNICACIÓN

OBJECTIVES

- Talk about present activities
- Talk about past activities

Now you can...

- talk about present activities.

To review

- the present tense see pp. 4, 8.

Now you can...

- talk about past activities.

To review

- the preterite see pp. 12, 16, 20, 24.

1 ¿Con qué frecuencia?

Tu compañero(a) quiere saber con qué frecuencia haces varias cosas. ¿Qué te pregunta y cómo le respondes?

modelo

¿alquilar un video?

Compañero(a): *¿Con qué frecuencia alquilas un video?*

Tú: *Alquilo un video una vez por semana.*

I. ¿patinar en el parque?
2. ¿escribir en la computadora?
3. ¿correr?
4. ¿leer una novela?
5. ¿salir con tus amigos?
6. ¿hacer la tarea?

2 Mis últimas vacaciones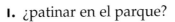

En grupos de dos o tres, conversen sobre sus últimas vacaciones. Digan adónde fueron, qué hicieron, qué compraron, qué vieron y si se divirtieron.

modelo

Tú: *¿Adónde fuiste para tus vacaciones?*

Compañero(a) 1: *Fui con mi familia a Buenos Aires, Argentina.*

Compañero(a) 2: *¿Ah, sí? ¿Qué hicieron? ...*

3 *En tu propia voz*

ESCRITURA Piensa en un evento cómico de tu pasado, o inventa uno si prefieres. Escribe una descripción del evento. (Si prefieres, escríbelo como un diálogo.) Luego, lee tu diálogo a la clase.

modelo

Un día, decidí cortarme el pelo yo mismo(a). Saqué las tijeras y fui al baño para usar el espejo. Empecé a cortarme el pelo de atrás...

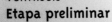

En resumen

♻ YA SABES

TALK ABOUT PRESENT ACTIVITIES

Common -ar verbs

acampar	to camp
alquilar	to rent
ayudar	to help
bailar	to dance
cambiar	to change
caminar	to walk
cantar	to sing
cenar	to eat dinner
cocinar	to cook
comprar	to buy
desayunar	to have breakfast
descansar	to rest
desear	to desire
enseñar	to teach, to show
escuchar	to listen
esperar	to wait for, to hope
estudiar	to study
ganar	to win
hablar	to talk, to speak
lavar	to wash
limpiar	to clean
llamar	to call
llevar	to take along, to wear, to carry
mandar	to send
mirar	to look at, to watch
nadar	to swim
pasar	to pass, to happen
patinar	to skate
planchar	to iron
quedar	to be (in a specific place), to agree (on)
terminar	to finish
tomar	to take, to eat or drink
trabajar	to work

Common -er verbs

aprender	to learn
beber	to drink
comer	to eat
comprender	to understand
correr	to run
deber	should, ought to
leer	to read
vender	to sell

Common -ir verbs

abrir	to open
compartir	to share
escribir	to write
insistir	to insist
recibir	to receive
vivir	to live

Common stem-changing verbs

cerrar (e→ie)	to close
contar (o→ue)	to tell, to count
encontrar (o→ue)	to find, to meet
entender (e→ie)	to understand
llover (o→ue)	to rain
pensar (e→ie)	to think
perder (e→ie)	to lose
recordar (o→ue)	to remember
sentarse (e→ie)	to sit down
volver (o→ue)	to return, to come back

Common reflexive verbs

acostarse (o→ue)	to go to bed, to lie down
afeitarse	to shave
bañarse	to bathe
despertarse	to wake up
ducharse	to take a shower
lavarse	to wash oneself
levantarse	to get up
maquillarse	to put on makeup
peinarse	to comb one's hair
ponerse	to put on

Common verbs with irregular yo form

caer	to fall
conocer	to know
oír	to hear
salir	to leave
valer	to be worth

TALK ABOUT PAST ACTIVITIES

Common preterite stem-changing verbs

competir	to compete
despedirse	to say goodbye
divertirse	to have fun
dormir	to sleep
pedir	to ask for, to order
preferir	to prefer
repetir	to repeat
sentir	to feel
sugerir	to suggest
vestirse	to dress oneself

Common spelling-change preterite verbs

almorzar	to have lunch
apagar	to turn off
buscar	to look for
caer	to fall
comenzar	to start
cruzar	to cross
empezar	to begin
explicar	to explain
jugar	to play (a game)
llegar	to arrive
pagar	to pay
practicar	to practice
sacar	to take
tocar	to play (an instrument)

Common irregular preterite verbs

andar	to walk
conducir	to drive
dar	to give
decir	to say, to tell
estar	to be
hacer	to do, to make
ir	to go
poder	to be able, can
poner	to put
querer	to want
saber	to know
ser	to be
tener	to have
traer	to bring
venir	to come
ver	to see

UNIDAD 1

ASÍ SOMOS

STANDARDS

Communication
- Describing people
- Talking about life experiences and accomplishments
- Describing fashions
- Talking about pastimes
- Predicting future actions
- Talking about household chores
- Expressing feelings

Cultures
- The influence of Spanish speakers in the United States
- The cultural role of fashion
- Musical influences

Connections
- Art: Creating a self-portrait
- Math: Preparing an annual clothing budget

Comparisons
- Childhood experiences
- Geography, climate, and customs and how they influence choice of clothing
- Musical instruments and influences

Communities
- Using Spanish in the workplace
- Using Spanish in Spanish-speaking communities for personal enjoyment

SAN ANTONIO
LA PRENSA Éste es el periódico bilingüe que se ha publicado por más tiempo. Lo fundó el Sr. Durán en 1914. ¿Hay un periódico bilingüe en tu comunidad? ¿Cómo se llama?

INTERNET Preview
CLASSZONE.COM

- More About Latinos
- Webquest
- Self-Check Quizzes
- Flashcards
- Writing Center
- Online Workbook
- eEdition Plus Online

28

ALMANAQUE CULTURAL

POBLACIÓN: Porcentaje latino de la población de Estados Unidos: 12,5%

GENTE FAMOSA: Tito Puente (músico), Oscar de la Renta (diseñador), Cristina García (escritora), Sandra Cisneros (escritora)

LUGARES CON NOMBRES DEL ESPAÑOL: Los Ángeles, San Antonio, El Paso, Colorado, Florida, Calle Ocho

VIDEO   DVD   Mira el video para más información.

CLASSZONE.COM
More About Latinos

LOS ÁNGELES
OSCAR DE LA HOYA Este campeón del boxeo ayuda a su comunidad construyendo lugares como el Resurrection Gym donde los atletas jóvenes pueden entrenarse. ¿Piensas que los deportes son buenos para la juventud? ¿Por qué?

CHICAGO
COMIDA MEXICANA El burrito, el taco y la tostada son comidas muy populares en la comunidad mexicana de Chicago. ¿Por qué crees que la comida mexicana también es popular con otros grupos?

NUEVA YORK
REPERTORIO ESPAÑOL Desde 1968, este grupo de teatro presenta obras en español. ¿Alguna vez viste una obra o función en español?

Flamenco & Spanish Dance

PILAR RIOJA

A SALUTE TO GARCÍA LORCA ON HIS CENTENNIAL

August 26 thru October 4

REPERTORIO ESPAÑOL
138 East 27th Street, NYC
(between Lexington & Third Ave.)
(212) 889-2850

MIAMI
ELLEN OCHOA La Dra. Ellen Ochoa es una astronauta muy conocida y recibió varios premios por su trabajo. Ella investiga los efectos del sol en el clima de la tierra. ¿Qué supones que la Dra. Ochoa estudió para prepararse?

- Comunicación

- Culturas

- Conexiones

- Comparaciones

- Comunidades

ASÍ SOMOS

Al seguir estudiando con ¡En español!, vas a
desarrollar nuevas habilidades en cinco áreas importantes:

Comunicación

Comunícate en otra lengua.

¿Cómo nos comunicamos los unos con los
otros? Describimos, preguntamos, explicamos;
en fin, nos expresamos. También escuchamos.
En la foto, ¿quién habla y quiénes escuchan?
¿Qué tipo de conversación crees que tienen?

Comparaciones

**Compara el español con el inglés:
las culturas y las lenguas.**

Al comparar el español con el
inglés aprendes a identificar y a
apreciar las semejanzas y las
diferencias entre las culturas y las
lenguas. ¿Qué semejanzas y
diferencias ya conoces? ¿Cuáles se
encuentran en este anuncio?

¡Disfruta de la naturaleza! Ven al

Campamento Monteverde

Aquí podrás...

esquiar en el agua

escalar montañas

acampar bajo la luna

hacer alpinismo

y también podrás...
¡navegar por Internet!

volar en planeador

remar en una canoa

Ya verás cuánto te divertirás aquí en
Campamento Monteverde
Una semana aquí y nunca querrás regresar a casa.

Conexiones

Haz conexiones con otras materias.

El español te ayuda a aprender más de otras materias, como el arte, las ciencias y las matemáticas. ¿Qué materias crees que están representadas en la gráfica y en la foto?

¿De dónde son los latinos de Estados Unidos?

- Cuba 3.5%
- Centro y Sudamérica 8.6%
- República Dominicana 2.2%
- Puerto Rico 9.6%
- Otros 17.6%
- México 58.5%

Culturas

Aprende de otras culturas.

Cuando estudiamos las costumbres y los productos de otra cultura, descubrimos las perspectivas que caracterizan a la gente de esa cultura. ¿Qué piensas descubrir al conocer la música de Tito Puente?

Comunidades

Usa el español en tu comunidad.

El español se habla en muchas comunidades del mundo. ¿Se habla español en tu comunidad? Esta alumna usa el español en un hospital. ¿Qué piensas que les pregunta a los pacientes?

Fíjate

¿A cuál de las cinco áreas corresponde cada una de las siguientes situaciones?

1. Estudias y escuchas música hispana para aprender más de la cultura.
2. Buscas en Internet información en español sobre los muralistas hispanos para una presentación en la clase de arte.
3. Trabajas en un lugar donde puedes hablar español.
4. Comparas los anuncios de un periódico hispano con los anuncios del periódico que lees normalmente.
5. En un mensaje electrónico le describes a un(a) amigo(a) tu ropa nueva y le pides su opinión.

UNIDAD 1

ETAPA 1

¿Cómo soy?

OBJECTIVES

- Describe people

- Talk about experiences

- List accomplishments

¿Qué ves?

Mira la foto y contesta las preguntas.

1. ¿Dónde están estas personas?

2. ¿Qué cosas te dicen dónde están?

3. ¿Crees que son amigos? ¿Por qué?

4. ¿Qué te dice la revista sobre la cultura de Miami?

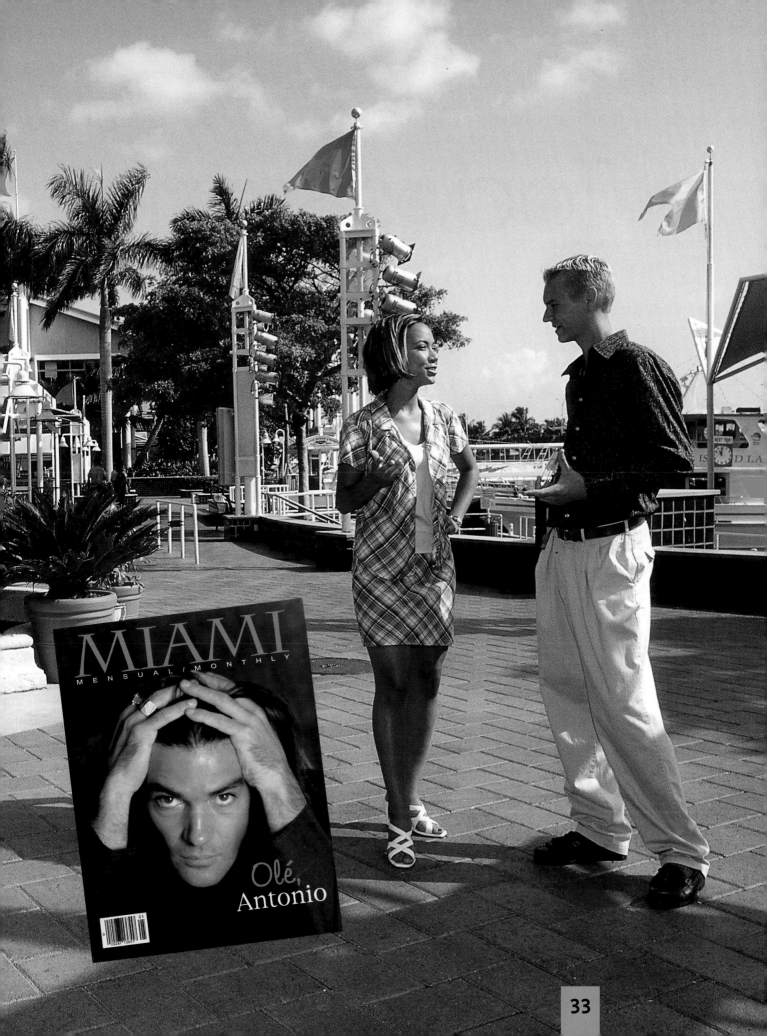

Todos somos diferentes

Descubre

A. Palabras dentro de las palabras

A veces hay palabras dentro de las palabras que ya conoces. Estudia las palabras a continuación y descubre su significado.

La forma de la cara

1. ovalado = óvalo ⭕

2. cuadrado = cuadro ⬜

3. triangular = triángulo 🔻

B. Sinónimos y antónimos

A veces ya sabes el sinónimo (una palabra que tiene el mismo significado) o el antónimo (una palabra que tiene el significado opuesto) de una palabra.

La forma del cuerpo

1. grueso	sinónimo	=	gordo
	antónimo	=	delgado
2. esbelto	sinónimo	=	delgado
	antónimo	=	gordo
3. redondo	sinónimo	=	circular
	antónimo	=	cuadrado

Hay tipos de personas que son fáciles de reconocer... ¿Conoces algunos?

Cola de caballo porque no quiere el pelo en la cara cuando juega al fútbol.

Usa lentes de contacto en vez de gafas, ¡claro!, todo para el deporte.

Siempre tiene el balón a la mano.

La deportista

Edad: 15
Talla: de estatura mediana
Cuerpo: ni gorda ni delgada
Ojos: cafés (y usa lentes de contacto)
Pelo: cabello largo en cola de caballo, color castaño
Cara: ovalada
Característica: lunar bajo el ojo
Uniforme: shorts, camiseta «polo» y balón
Frase favorita: «¡Te gano!»

Tiene el pelo teñido de ¡naranja!, para distinguirse del resto del mundo.

Sus anteojos redondos le dan ese «look» individual.

Camiseta de los sesenta porque le gusta ser diferente — hace lo opuesto de lo que hacen los otros.

El rebelde

Edad: 16
Talla: alto
Cuerpo: delgado
Ojos: azules
Pelo: ¡teñido de naranja!
Cara: triangular
Característica: anteojos redondos
Uniforme: camiseta de los sesenta
Frase favorita: «¡Al contrario!»

El caballero

Edad: 20
Talla: de estatura mediana
Cuerpo: grueso
Ojos: negros
Pelo: calvo (¡no tiene pelo!)
Cara: redonda
Característica: bigote, barba
Uniforme: toda su ropa es de color negro
Frase favorita: «Tú primero.»

¡Es calvo! Se afeita la cabeza porque cree que es un «look» más sofisticado.

Cree que el bigote y la barba le dan un aire misterioso.

Siempre elegante, ¡de pies a cabeza! Y ¿su color preferido? Negro, ¡por supuesto!

Tiene el pelo rojizo y ondulado y lo lleva con flequillo para verse más atractiva.

Tiene un espejo siempre a la mano para admirarse frecuentemente.

La vanidosa

Edad: 16
Talla: baja
Cuerpo: esbelta
Ojos: verdes
Pelo: rojizo y ondulado, con flequillo
Cara: cuadrada
Característica: pecas
Uniforme: vestido nada convencional
Frase favorita: «¿Cómo me veo?»

Lleva un vestido nada convencional para llamar la atención.

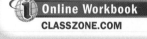
Online Workbook
CLASSZONE.COM

¿Comprendiste?

1. ¿Conoces a personas como éstas?
2. ¿Crees que son tipos verdaderos o estereotipos?
3. ¿Quién te cae bien? ¿Quién te cae mal?
4. ¿Cómo eres tú? Descríbete. Menciona tus ojos, pelo, cara y alguna característica particular.
5. ¿Cuál es tu «uniforme» preferido?
6. ¿Cuál es tu «frase favorita»?

En vivo

AUDIO

SITUACIONES

PARA ESCUCHAR • STRATEGY: LISTENING

Pre-listening How do you recognize someone that you are going to meet but have never seen? List what will help you.

Recognize descriptions Jot down the physical characteristics that best help you identify someone you don't know. Then, after listening to the descriptions, compare that list with yours. How are they the same? How are they different? Would you revise yours?

¡Eres director o directora!

Vas a filmar una película de misterio. Tienes que escoger los actores para los cuatro personajes principales.

1 Leer

Tienes los siguientes dibujos y descripciones de los cuatro personajes principales de la película para ayudarte a seleccionar a los actores.

Dolores MalaGente

Alta, esbelta, siempre lleva gafas negras y ropa elegante. Tiene el pelo largo, negro y liso. Tiene la cara ovalada y los ojos negros. Tiene un lunar bajo el ojo izquierdo. Tiene unos treinta años.

Marcos Deportista

Marcos tiene el cuerpo de deportista: alto, delgado y sin un kilo de más. Lleva el pelo castaño en cola de caballo que va muy bien con su cara triangular. Es atlético de pie a cabeza: va siempre en ropa cómoda. Tiene entre dieciséis y diecisiete años.

Eduardo Vanidoso

De estatura mediana, el señor Vanidoso es grueso y calvo. Tiene bigote y barba porque cree que así puede esconder su cara redonda. También usa lentes de contacto porque cree que se ve mejor sin anteojos. Viste siempre de trajes italianos muy caros. Es un tipo poco agradable. Tiene entre treinta y cuarenta años.

Celia Simpática

Celia es baja.. Tiene el pelo rojizo y ondulado, con flequillo. Su cara es un poco cuadrada y tiene pecas. Es la mejor clase de amiga que uno pueda encontrar. Tiene entre dieciséis y diecisiete años.

SE SOLICITAN ACTORES
para una película de misterio:

2 mujeres y 2 hombres

experiencia
en actuación
necesaria

Por favor llamar al 555-3689 y
dejar un mensaje detallado que
incluya descripción física y edad

2 Leer

Tú pusiste este anuncio en los
periódicos estudiantiles de la
ciudad y también en Internet.

3 Escuchar

Escucha las llamadas que recibiste y decide a quién le vas
a hacer una audición para cada papel. Toma apuntes en una
hoja aparte usando las siguientes categorías. ¡Ojo! No siempre
te darán toda la información que necesites.

Notas del director (de la directora)

Llamada número _____   Ojos: _____

Edad: _____   Pelo: _____

Nombre: _____   Actitud: _____

Talla: _____   Experiencia: _____

Cara: _____   Audición para el papel de _____

4 Hablar

En grupos de tres o cuatro, comparen sus notas y sus selecciones.
¿Están todos de acuerdo? ¿Por qué o por qué no?

Activity 4: Clarify, ask for and comprehend clarification

En acción

Práctica del vocabulario

Objectives for Activities 1–4
• Describe people

1 Mi amigo es...

Escribir Francisco está describiendo a sus amigos y parientes. Completa su descripción con las frases en la segunda columna.

1. Mi amigo Armando es pelirrojo.
2. A mi amiga Susana no le gusta como se ve con anteojos.
3. El color del cabello de mi tío no es natural.
4. ¡Papá se acaba de afeitar!
5. Mi amiga Alma siempre tiene el pelo en la cara.
6. El pelo de Marcos no es lacio.
7. Mi prima no tiene la cara redonda. ¡Es todo lo opuesto!
8. ¡Mi abuelo no tiene pelo!

a. Es ondulado.
b. Ya no tiene ni bigote ni barba.
c. Tiene el pelo rojizo.
d. Tiene la cara muy cuadrada.
e. Está teñido.
f. Debe usar cola de caballo.
g. Es calvo.
h. Prefiere usar lentes de contacto.

También se dice

Hay muchas maneras de decir que alguien es atractivo:

• **majo(a)**, España;
• **galán**, México
• **buen mozo(a)**, varios países;
• **bien parecido(a)**, varios países.

2 ¿Cómo son?

Hablar/*Escribir* Describe a las personas de los dibujos. Nombra por lo menos dos características físicas de cada persona.

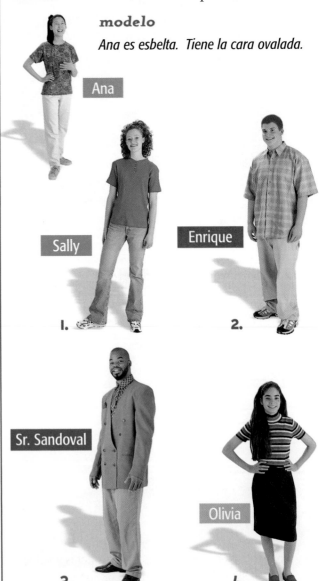

modelo

Ana es esbelta. Tiene la cara ovalada.

Ana

Sally
1.

Enrique
2.

Sr. Sandoval
3.

Olivia
4.

3 **¿Conoces a Arturo?**

Hablar/Escribir En pares, comenten sobre el aspecto de las siguientes personas. Luego, cambien de papel.

> **modelo**
>
> *Arturo*
>
> **Compañero(a):** *¿Conoces a Arturo?*
>
> **Tú:** *Sí, Arturo tiene el pelo ondulado.*

1. el (la) profesor(a) de inglés
2. mi mejor amigo(a)
3. el (la) director(a)
4. el señor [nombre]
5. mi amigo(a) [nombre]
6. el (la) profesor(a) de español
7. mi papá
8. el chofer del autobús
9. mi hermano(a)
10. el (la) profesor(a) de matemáticas

4 **En la cafetería**

Hablar Un(a) alumno(a) nuevo(a) llegó a tu escuela. Ustedes están en la cafetería y él(ella) te hace preguntas sobre los otros alumnos y los profesores. Tú los describes con todos los detalles que puedas.

> **modelo**
>
> *muchacho*
>
> **Compañero(a):** *¿Quién es el muchacho alto con pelo lacio?*
>
> **Tú:** *Es mi amigo Héctor. Es muy buen futbolista, pero es modesto. También es muy cómico.*

1. muchacho	3. muchacha	5. muchacha
2. profesor	4. muchacho	6. profesora

Vocabulario

Características

atrevido(a) *daring*
comprensivo(a) *understanding*
considerado(a) *considerate*
desagradable *unpleasant*
descarado(a) *insolent, shameless*
fiel *faithful*
mimado(a) *spoiled*
modesto(a) *modest*
vanidoso(a) *vain*

 Ya sabes

amable
cómico(a)
impaciente
obediente
paciente
sociable
tímido(a)

▶ ¿Conoces a algunas personas con estas características?

Práctica: gramática y vocabulario

Objectives for Activities 5–15
• Describe people • Talk about experiences • List accomplishments

REPASO   Ser vs. Estar

▶ Remember that in Spanish you can use two different verbs to mean **to be:**
    ser and estar

▶ **Use ser...**

* to identify people and things.

El señor Ortega **es** profesor.
*Mr. Ortega **is** a professor.*

* to express possession.

Esos libros **son** de Pablo.
*Those **are** Pablo's books.*

* with **de** to express origin and to say what something is made of.

Es de oro y **es** de Perú.
*It **is** gold and it **is** from Perú.*

* to express time and date.

Son las doce. Hoy **es** lunes.
*It **is** twelve. Today **is** Monday.*

* to tell where or when an event takes place.

El concierto **es** en el estadio.
*The concert **is** in the stadium.*

* to describe unchanging characteristics, such as color, nationality, size, physical characteristics, personality.

Tus primos **son** muy simpáticos.
*Your cousins **are** very likable.*

▶ **Use estar...**

* to express location.

Los Ángeles **está** en California.
*Los Angeles **is** in California.*

* to express a state or condition, such as health, emotions, and feelings.

Mi hermano **está** enfermo.
*My brother **is** sick.*
Estoy triste y preocupado.
*I **am** sad and worried.*

▶ You also use **estar** to form the **present progressive.**

Estoy **escuchando**.
*I **am** listening.*

▶ You can often use **ser** and **estar** with the same **adjectives,** but with a difference in meaning:

Francisco **es** **nervioso**.
*Francisco **is** nervous. (He is a nervous person.)*

Francisco **está** **nervioso** hoy porque tiene exámenes.
*Francisco **is** nervous today because he has exams. (He is nervous today.)*

Practice:
Actividades
5 6 7 8

Más práctica
cuaderno pp. 15–16
Para hispanohablantes
cuaderno p. 13

Online Workbook
CLASSZONE.COM

5 ¿Soy o estoy?

Escribir Completa las oraciones con la forma correcta de **ser** o **estar**.

modelo

Yo soy de Estados Unidos. Soy estadounidense.

1. Yo _____ estudiante. Ahora _____ en la escuela.

2. Él _____ mi tío. _____ profesor de matemáticas.

3. Ella _____ actriz. _____ en California.

4. Yo _____ del barrio Washington Heights que _____ en Nueva York.

5. ¡_____ las cinco! ¿Dónde _____ los niños?

6. _____ muy preocupada. Tengo un examen.

7. Los niños _____ tristes. Su perrito _____ enfermo.

8. Tú _____ simpático. Siempre _____ de buen humor.

Nota cultural

En algunas ciudades de Estados Unidos, se le da el nombre de **barrio** a zonas urbanas donde hay una gran población latina. La gente allí mantiene vivos las tradiciones y el idioma de su país de origen.

Activity 7: Understand most spoken language when the message is deliberately and carefully conveyed by a speaker accustomed to dealing with learners when listening

Activity 8: Narrate in the present

6 ¡Hola, Futbolista29!

Leer/Escribir Completa la carta electrónica de Atleta37 usando la forma correcta de **ser** o **estar**.

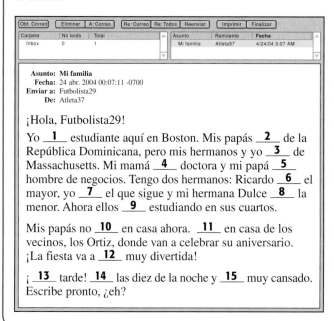

| Obt. Correo | Eliminar | A: Correo | Re: Correo | Re: Todos | Reenviar | Imprimir | Finalizar |

Carpeta	No leído	Total		Asunto	Remitente	**Fecha**
Inbox	0	1		Mi familia	Atleta37	4/24/04 3:07 AM

Asunto: Mi familia
Fecha: 24 abr. 2004 00:07:11 -0700
Enviar a: Futbolista29
De: Atleta37

¡Hola, Futbolista29!

Yo __1__ estudiante aquí en Boston. Mis papás __2__ de la República Dominicana, pero mis hermanos y yo __3__ de Massachusetts. Mi mamá __4__ doctora y mi papá __5__ hombre de negocios. Tengo dos hermanos: Ricardo __6__ el mayor, yo __7__ el que sigue y mi hermana Dulce __8__ la menor. Ahora ellos __9__ estudiando en sus cuartos.

Mis papás no __10__ en casa ahora. __11__ en casa de los vecinos, los Ortiz, donde van a celebrar su aniversario. ¡La fiesta va a __12__ muy divertida!

¡ __13__ tarde! __14__ las diez de la noche y __15__ muy cansado. Escribe pronto, ¿eh?

7 La actriz Anilú Pardo

Escuchar/Escribir Escucha la entrevista con Anilú Pardo. Luego, contesta en oraciones completas.

1. ¿Quién es?
2. ¿De dónde es?
3. ¿Cuál es su nacionalidad?
4. ¿Cuál es su profesión?
5. ¿Quién es el director de la película?
6. ¿Quiénes son los otros actores?
7. ¿Cómo se siente Anilú?

8 ¡Mi página-web!

Leer/Escribir Navegando por Internet, encontraste un sitio donde jóvenes de todo el mundo ponen anuncios para buscar amigos por correspondencia electrónica. Lee el anuncio de una joven puertorriqueña. Luego escribe tu propio anuncio.

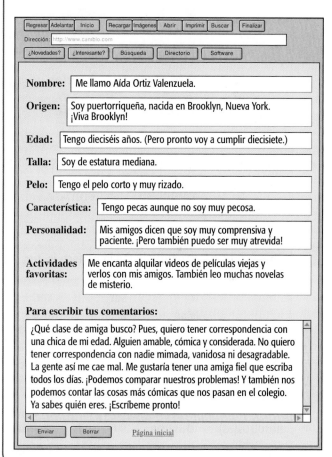

| Regresar | Adelantar | Inicio | Recargar | Imágenes | Abrir | Imprimir | Buscar | Finalizar |

Dirección: http://www.cambio.com

| ¿Novedades? | ¿Interesante? | Búsqueda | Directorio | Software |

Nombre: Me llamo Aída Ortiz Valenzuela.

Origen: Soy puertorriqueña, nacida en Brooklyn, Nueva York. ¡Viva Brooklyn!

Edad: Tengo dieciséis años. (Pero pronto voy a cumplir diecisiete.)

Talla: Soy de estatura mediana.

Pelo: Tengo el pelo corto y muy rizado.

Característica: Tengo pecas aunque no soy muy pecosa.

Personalidad: Mis amigos dicen que soy muy comprensiva y paciente. ¡Pero también puedo ser muy atrevida!

Actividades favoritas: Me encanta alquilar videos de películas viejas y verlos con mis amigos. También leo muchas novelas de misterio.

Para escribir tus comentarios:

¿Qué clase de amiga busco? Pues, quiero tener correspondencia con una chica de mi edad. Alguien amable, cómica y considerada. No quiero tener correspondencia con nadie mimada, vanidosa ni desagradable. La gente así me cae mal. Me gustaría tener una amiga fiel que me escriba todos los días. ¡Podemos comparar nuestros problemas! Y también nos podemos contar las cosas más cómicas que nos pasan en el colegio. Ya sabes quién eres. ¡Escríbeme pronto!

| Enviar | Borrar | Página inicial |

More Practice: **Más comunicación** p. R3

Activity 11: Converse and listen during face-to-face social interactions

The Imperfect Tense

You can use the **imperfect tense** in Spanish to talk about ongoing, habitual, or incomplete actions in the past. You also use the **imperfect** to tell time in the past, and for descriptions in the past.

Use the **endings** shown below to form the **imperfect tense** of regular verbs.

	-ar hablar	-er comer	-ir vivir
yo	habl**aba**	com**ía**	viv**ía**
tú	habl**abas**	com**ías**	viv**ías**
usted, él, ella,	habl**aba**	com**ía**	viv**ía**
nosotros(as)	habl**ábamos**	com**íamos**	viv**íamos**
vosotros(as)	habl**abais**	com**íais**	viv**íais**
ustedes, ellos, ellas	habl**aban**	com**ían**	viv**ían**

Only three verbs are **irregular** in the imperfect: **ir, ser,** and **ver.**

	ir to go	ser to be	ver to see
yo	iba	era	veía
tú	ibas	eras	veías
usted, él, ella,	iba	era	veía
nosotros(as)	íbamos	éramos	veíamos
vosotros(as)	ibais	erais	veíais
ustedes, ellos, ellas	iban	eran	veían

Mi abuela **era** esbelta y baja y **llevaba** anteojos.
*My grandmother **was** slender and short and **wore** eyeglasses.*

Practice: Actividades
 9 10 11

Más práctica
cuaderno pp. 17–18
Para hispanohablantes
cuaderno p. 14

 Online Workbook
CLASSZONE.COM

---

9 **¿Dónde estabas?**

Hablar/*Escribir* Tu amigo(a) te llamó varias veces por teléfono pero no estabas. Contesta su pregunta diciéndole dónde estabas y qué hacías.

modelo

en mi cuarto (hacer la tarea)

Compañero(a): *¿Dónde estabas cuando te llamé anoche?*

Tú: *Estaba en mi cuarto. Hacía la tarea.*

1. en el centro comercial (comprar discos compactos)
2. en la biblioteca (escribir la tarea)
3. en la sala (ver la televisión)
4. en el café (hablar con mis amigos)
5. en el parque (andar en bicicleta)
6. en la tienda de videos (alquilar un video)
7. en el cine (ver una película)
8. en el gimnasio (levantar pesas)
9. la fiesta de Adrián (tocar la guitarra)
10. la casa de mi tía (comer pastel)

10 El verano

Hablar/*Escribir* Hacías cosas durante el verano que ya no puedes hacer durante el año escolar. ¿Qué cosas hacías que ya no puedes hacer?

modelo

navegar por Internet

Yo siempre navegaba por Internet durante el día, pero ahora no puedo.

1. descansar	5. ver
2. levantarse	6. acostarse
3. ir	7. trabajar
4. ponerse	8. viajar

11 La apariencia ayer y hoy

Hablar Habla con un(a) compañero(a) de clase de cómo se veían hace diez años y cómo se ven hoy. ¿Qué diferencias hay?

modelo

Tú: *Cuando tenía seis años, siempre llevaba el pelo en cola de caballo.*

Compañero(a): *Pues, yo usaba anteojos y estaba muy delgado(a). No me vestía muy bien.*

Tú: *Bueno, ¡ahora eres muy elegante!*

 REPASO   **Preterite vs. Imperfect**

> ♻ **¿RECUERDAS?** *p. 42* You already know two tenses that refer to past time, the preterite and the imperfect. You use each of these tenses to talk about **past** actions in a different way.

- Use the preterite tense to describe an action or series of actions completed in the past.

  Aquel día, Pedro **salió** del colegio y **caminó** hasta el café.
  *That day, Pedro **left** school and **walked** to the café.*

- Use the imperfect to describe **ongoing** actions or conditions in the past, without focusing on their beginning or end.

  Yo siempre **salía** del colegio y **caminaba** hasta el café.
  *I always **used to leave** school and **walk** to the café.*

▶ Sometimes you will need to use the imperfect and preterite in the same sentence.

     Use the imperfect to tell what was going on in the background.   **Use the preterite to describe the interrupting action or main event.**

Yo **hacía** la tarea cuando **sonó** el teléfono.
*I **was doing** my homework when the telephone **rang**.*

Practice: **Actividades** 12 13 14 15   **Más práctica** *cuaderno p. 19*   **Para hispanohablantes** *cuaderno pp. 15–16*    **Online Workbook** CLASSZONE.COM

Activity 15:
Narrate in the past

12 Yoli la atrevida

Escribir Cristina escribió una composición sobre una amiga de su niñez. Complétala con las formas correctas de los verbos entre paréntesis para leer sus aventuras.

Cuando yo ___1___ (ser) joven, mi mejor amiga ___2___ (llamarse) Yolanda, pero su apodo ___3___ (ser) Yoli. Yoli y yo siempre ___4___ (andar) juntas.

A nosotras nos ___5___ (gustar) hacer las mismas cosas. Por ejemplo, a ella le ___6___ (gustar) escribir cuentos y a mí me ___7___ (gustar) escribir poemas. Yo ___8___ (compartir) mis ideas con ella. Ella me ___9___ (influir) mucho.

Un día Yoli ___10___ (estar) en mi casa cuando me ___11___ (decir) «¿Por qué no vamos al centro comercial?» Aunque me ___12___ (parecer) buena idea, yo ___13___ (saber) que no ___14___ (poder) ir sin pedir permiso a mis papás. Pero mis papás ___15___ (estar) en el trabajo. «Anda, vamos» me ___16___ (repetir) hasta que me ___17___ (convencer).

Al llegar al centro comercial, la primera persona que ___18___ (ver)... ¡ ___19___ (ser) mi tía! Mi tía inmediatamente ___20___ (llamar) a mis padres para decirles que yo ___21___ (estar) fuera de casa sin permiso. Ese día, ___22___ (resolver) jamás hacerle caso a Yoli.

Vocabulario

Interacciones

compartir *to share*

discutir *to argue*

hacerle caso a *to obey, pay attention to*

influir *to influence*

resolver (o→ue) *to resolve*

respetar *to respect*

tener en común *to have in common*

▶ ¿Tienes estas interacciones con tus amigos?

13 Mi abuelo

Hablar/*Escribir* Hablas con tu abuelo sobre las cosas que él hacía cuando era joven. Hace pocos días que tú hiciste las mismas cosas que él hacía hace años. Sigue el modelo.

modelo

ir al museo / los domingos

Abuelo: *Cuando era joven, yo iba al museo los domingos.*

Tú: *¿De veras? Yo fui al museo el domingo pasado.*

1. ir al cine / los sábados
2. salir a pasear con mis amigos / los viernes
3. jugar al fútbol / los fines de semana
4. ir a conciertos / una vez al mes
5. levantar pesas / los lunes
6. estudiar en la biblioteca / los domingos
7. hacer la limpieza / los fines de semana
8. visitar a mis tíos / todas las semanas
9. cocinar pasta / los viernes
10. nadar en la piscina / los jueves

14 ¡Acción!

Escuchar/*Escribir* Escucha a Luci Pérez. Ella describe una película que vio ayer. Luego, escribe oraciones basadas en su descripción.

modelo

Luci (ver)

Luci vio una película ayer.

1. la película (tener)
2. Lola (ser)
3. los hombres malos (capturar)
4. ella (tener que)
5. los hombres malos (ser)
6. ella (salvar)
7. la actriz (ganar)

Nota cultural

Los apodos *(nicknames)* son comunes en la cultura latina. Muchas veces el apodo viene del nombre de la persona. Algunos ejemplos son:

Chicos

Antonio = Toño, Toni

Guillermo = Memo

Roberto = Beto

Chicas

Graciela = Chela

Isabel = Chabela

Mercedes = Mercha, Meche

FEB California
BETO

15 Cuando era niño(a)...

Escribir Escribe dos párrafos sobre tu niñez, usando expresiones de la lista de vocabulario.

modelo

Cuando era niño(a), yo era muy trabajador(a), a diferencia de mi hermano(a) que era muy perezoso(a). Me gustaba estudiar y jugar con mis amigos. ¡Pero no me gustaba jugar con mi hermano(a)! Por un lado quería...

- ¿Cómo eras?
- ¿Qué te gustaba hacer? ¿Qué no te gustaba?
- ¿Qué querías ser?
- ¿Qué te interesaba?
- ¿Cómo era tu mejor amigo(a)?
- ¿Qué hacían juntos?
- ¿Puedes describir un incidente en particular?
- ¿...?

Vocabulario

Comparaciones

a diferencia de *as contrasted with*

al contrario *on the contrary*

lo bueno/malo *the good thing/bad thing*

lo más/menos *the most/least*

lo mejor/peor *the best/worst*

por otro lado *on the other (hand)*

por un lado *on the one hand*

semejante a *similar to*

▶ ¿Qué dices si quieres comparar dos cosas?

Activity 18: Converse and listen during face-to-face social interactions

GRAMÁTICA Present and Past Perfect Tenses

To express the idea that someone has or had already done something, you use the **present perfect** and **past perfect** tenses.

The **present perfect** tense consists of the **present** tense of the auxiliary verb **haber**, *to have*, plus the **past participle** of the verb. To form regular **past participles**, drop the ending from the **infinitive** and add the following endings.

habl**ar** ➡ habl**ado**   com**er** ➡ com**ido**   viv**ir** ➡ viv**ido**

present perfect of hablar

he **hablado**	hemos **hablado**
has **hablado**	habéis **hablado**
ha **hablado**	han **hablado**

Here are some **irregular past participles:**

abrir → **abierto**, cubrir → **cubierto**, decir → **dicho**, escribir → **escrito**, hacer → **hecho**, morir → **muerto**, poner → **puesto**, resolver → **resuelto**, romper → **roto**, ver → **visto**, volver → **vuelto**.

You use the **present perfect** tense to talk about events or actions that have already occurred.

¿**Has** com**ido**?   No, todavía no **he** com**ido** nada.
Have you eaten?   *No, I still haven't eaten anything.*

The **past perfect** refers to an action that had already occurred when something else happened. Both actions are in the past, one occurring before the other.

past perfect of hablar

había **hablado**	habíamos **hablado**
habías **hablado**	habíais **hablado**
había **hablado**	habían **hablado**

> You form the **past perfect** almost the same way as the **present perfect**, but you use the **imperfect** form of **haber** instead.

Todavía no **había** com**ido** cuando llegó Luis.
I still hadn't eaten when Luis arrived.

In both the **present perfect** and **past perfect**, you place **object pronouns before** the forms of the verb **haber**.

Marta ya tenía el libro.   *before*
Marta already had the book.   ¿Se **lo había** prest**ado** usted?
                                      Had you lent it to her?

Practice: **Actividades** 16 17   **Más práctica** *cuaderno p. 20*
Para hispanohablantes *cuaderno pp. 17–18*

Online Workbook CLASSZONE.COM

Activity 18 brings together all concepts presented.

16 Los quehaceres

Hablar/*Escribir* Haz una lista para tus papás describiendo quién ha hecho cada cosa.

> **modelo**
>
> yo / *pasar la aspiradora*
>
> *Yo he pasado la aspiradora.*

1. yo / poner la mesa
2. tú / lavar los platos
3. mis hermanos / hacer las camas
4. mi hermanita / barrer el piso
5. nosotros / abrir las ventanas

17 No sabía que...

Hablar/*Escribir* Le preguntas a tu compañero(a) qué te perdiste antes de que llegaras a la fiesta.

> **modelo**
>
> *la fiesta empezar*
>
> **Tú:** *Yo no sabía que la fiesta había empezado.*
>
> **Compañero(a):** *Ya había empezado cuando tú llegaste.*

1. los músicos tocar
2. la profesora de español bailar
3. Elisa tocar la guitarra
4. los estudiantes comer el pastel
5. los padres de Elisa hablar
6. el hermano de Elisa entrar
7. los abuelos de Elisa salir
8. la fiesta terminar

18 ¿Lo has hecho?

Hablar Haz una encuesta o añade cosas a la siguiente lista. Luego, pregúntales a varios compañeros si han hecho las cosas de tu lista. Pídeles que cuenten detalles de las cosas que han hecho.

	Sí	No
¿Siempre viviste en la misma casa?		
¿Estuviste en la casa de una persona vanidosa?		
¿Eras mimado(a) cuando eras más joven?		
¿Has dicho mentiras?		
¿Has resuelto un problema difícil?		
¿Habías estudiado español antes?		

More Practice:
Más comunicación *p. R3*

Online Workbook
CLASSZONE.COM

Refrán

Cada cabeza es un mundo.

¿Qué quiere decir el refrán? ¿Crees que cada persona es un individuo? ¿Cómo puedes describir tu «mundo interior»?

Read short stories

AUDIO 🎧

En voces
LECTURA

PARA LEER • **STRATEGY: READING**

Observe how verb tenses reveal time Verb tenses show when different events occur in time. Read Pilar's story of her memories of the past and her plans for the future. Then select at least five major events and place them on a time line. Notice the verb tense of each event. Are the events scattered on the time line or clustered together? Can you say why?

LA NIÑEZ

acariciar	hacer un gesto cariñoso a alguien
agarrarse	hacer que una persona no se mueva
escurrirse a hurtadillas	irse de un lugar sin ser visto
gritar a todo pulmón	hablar a todo volumen
la cría	un(a) niño(a) pequeño(a)
sentar en la falda	sentarse sobre las rodillas de alguien que está sentado

Sobre la autora

Cristina García nació en La Habana, Cuba, en 1958 y se crió en Nueva York. Asistió a Barnard College y a la Escuela de Estudios Internacionales Avanzados de Johns Hopkins University. Ha trabajado como periodista en Miami, San Francisco y Los Ángeles, donde vive actualmente con su esposo. *Soñar en cubano* es su primera novela.

Introducción

Soñar en cubano es una novela que narra la historia de una familia cubana. Celia, la abuela, Lourdes, su hija, y Pilar, su nieta, son los tres personajes principales. Ellas hablan de los sueños y el dolor de la familia que vive en Cuba, y de Lourdes y Pilar. En esta selección habla la nieta, Pilar Puente.

Nota cultural

Según el Servicio de Inmigración y Naturalización, cada año llegan más de 500.000 personas hispanohablantes a Estados Unidos. Estos grupos pueden asimilarse dentro de la nueva cultura, pero sienten mucha nostalgia por su país natal. Es muy común que un(a) inmigrante de un país hispanohablante conserve su identidad, no importa cuántos años viva fuera del país.

Soñar en cubano

Eso es. Ya lo entiendo. Regresaré a Cuba. Estoy harta[1] de todo. Saco todo mi dinero del banco, 120 dólares, el dinero que he ahorrado esclavizada en la pastelería de mi madre, y compro un billete de autocar para irme a Miami. Calculo que una vez allí, podría gestionar[2] mi viaje a Cuba, alquilando un bote, o consiguiendo un pescador que me lleve. Imagino la sorpresa de Abuela Celia cuando me escurriera a hurtadillas por detrás de ella. Estaría sentada en su columpio de mimbre[3] mirando al mar, y olería a[4] sal y a agua de violetas. Habría gaviotas[5] y cangrejos[6] en la orilla del mar. Acariciaría mis mejillas[7] con sus manos frías, y cantaría silenciosamente en mis oídos.

[1] to be fed up with	[5] seagulls
[2] to arrange	[6] crabs
[3] wicker rocking chair	[7] cheeks
[4] she would smell of	

Cuando salí de Cuba tenía sólo dos años, pero recuerdo todo lo que pasó desde que era una cría, cada una de las conversaciones, palabra por palabra. Estaba sentada en la falda de mi abuela jugando con sus pendientes de perlas, cuando mi madre le dijo que nos iríamos de la isla. Abuela Celia le acusó de haber traicionado la revolución. Mamá trató de separarme de la abuela, pero yo me agarré a ella y grité a todo pulmón. Mi abuelo vino corriendo y dijo: «Celia, deja que la niña se vaya. Debe estar con Lourdes.» Ésa fue la última vez que la vi.

Online Workbook
CLASSZONE.COM

¿Comprendiste?

1. ¿Por qué quiere regresar a Cuba Pilar?
2. ¿Cómo piensa llegar?
3. ¿A quién va a ver Pilar?
4. ¿Cuántos años tenía Pilar cuando salió de Cuba?
5. ¿Qué cosas asocia Pilar con su abuela?

¿Qué piensas?

En tu opinión, ¿qué sentimientos tiene Pilar hacia la vida que abandonaron ella y su mamá? ¿Por qué quiere regresar?

Hazlo tú

¿Cómo era tu abuela (abuelo) cuando tenías dos años? Busca una foto de esa época y escribe una descripción de ella (él). También puedes describir a otra persona mayor de tu niñez.

Activity 4: Generally use appropriate behavior in social situations

En uso

REPASO Y MÁS COMUNICACIÓN

OBJECTIVES

• Describe people
• Talk about experiences
• List accomplishments

Now you can...

• describe people.

To review

• **ser** and **estar**, see p. 40.

1 Descripciones

Tú le describes las siguientes personas a tu compañero(a).

Martín

modelo

Martín tiene la cara triangular. Usa lentes de contacto.
¡Es muy vanidoso!

PARADA

Sonia

Daniel

Yolanda

el señor Monsevalles

la profesora Quiñones

Now you can...

• describe people.

To review

• **ser** and **estar**, see p. 40.

2 El equipo de fútbol

Escribe oraciones con la forma correcta de **ser** o **estar**.

modelo

los chicos / estudiantes Los chicos son estudiantes.

1. ellos / futbolistas
2. el equipo / de Chicago
3. sus uniformes / lana

4. ellos / no / listos
5. el partido / las tres de la tarde
6. ellos / descansar / en el hotel

Self-Check Quiz
CLASSZONE.COM

Now you can...

• talk about experiences.

To review

• the preterite and the imperfect see p. 43.

3 ¡Qué desastre!

Ángel describe un día desastroso de su niñez. Para saber qué le pasó, completa el párrafo usando el pretérito o imperfecto de los verbos entre paréntesis.

> Yo ___1___ (tener) diez años. ___2___ (Ser) el día de mi fiesta de cumpleaños. Mamá me ___3___ (decir) que todo ___4___ (estar) listo. La piñata ___5___ (ser) un burro. El pastel ___6___ (ser) de chocolate. Mis amigos y yo ___7___ (estar) muy felices. Nos ___8___ (organizar) para tratar de romper la piñata. Como yo ___9___ (ser) el que ___10___ (cumplir) años, yo ___11___ (ir) a ser el primero. Alguien me ___12___ (poner) un pañuelo en los ojos. Justamente cuando ___13___ (ir) a romper la piñata, ___14___ (empezar) a llover — ¡un aguacero tremendo! Todos ___15___ (correr) hacia la casa. ¡Yo no! ¡Yo ___16___ (querer) seguir con el juego!

Now you can...

• list accomplishments.

To review

• the present and past perfect tenses see p. 46.

4 ¿Qué has hecho?

Tu compañero(a) quiere saber qué han hecho tú y varias personas de tu familia esta semana y la semana pasada.

modelo

tú (trabajar mucho)

Compañero(a): *¿Has trabajado mucho?*

Tú: *Sí, (No, no) he trabajado mucho aunque había trabajado mucho la semana pasada.*

1. tú y tus hermanos (tener mucha tarea)
2. tu padre (viajar mucho)
3. tú (ir al parque)
4. tus primas (comprar muchas cosas)
5. tú y tus hermanos (ver mucha televisión)
6. tú (hablar mucho por teléfono)
7. tú y tus amigos (bailar mucho)
8. tu hermana (visitar a tus abuelos)
9. tú (organizar una fiesta)
10. tú (estudiar español)

5 El (La) amigo(a) ideal

STRATEGY: SPEAKING

Describe personal characteristics and actions In addition to using adjectives to describe ideal friends, think about their actions: **¿Qué dicen o no dicen los amigos ideales? ¿Qué hacen o no hacen?** Is it important for all the lists to be the same? Explain your answer.

Trabajando en grupos, escriban las características del amigo o de la amiga ideal. Comparen sus listas.

modelo

El amigo ideal tiene que ser comprensivo(a).

cómico(a)

amable

sociable

considerado(a)

simpático(a)

6 Tus amigos

¿Qué te gustaría saber de la niñez de tus amigos? Escribe una lista de preguntas y luego entrevista a cuatro o cinco compañeros. Apunta sus respuestas y di a la clase lo más interesante que descubriste.

modelo

¿Qué querías ser cuando eras niño(a)?

¿Cuál era tu programa favorito cuando tenías diez años?

¿A qué le tenías miedo cuando tenías ocho años?

7 *En tu propia voz*

ESCRITURA A veces las películas empiezan con una narración: el protagonista describe una escena de su niñez. Escribe una narración sobre tu niñez para un video. Escoge un incidente impresionante o invéntalo. ¡Sé creativo(a)!

CONEXIONES

El arte Crea un autorretrato. Puedes pintar, dibujar *(to draw)* o hacer un collage de fotografías, artículos o anuncios de revista, etc. Usa tu imaginación. Explica a la clase cómo la obra revela tu personalidad.

En resumen
REPASO DE VOCABULARIO

DESCRIBE PEOPLE

Personality

atrevido(a)	daring
comprensivo(a)	understanding
considerado(a)	considerate
desagradable	unpleasant
descarado(a)	insolent, shameless
fiel	faithful
mimado(a)	spoiled
modesto(a)	modest
vanidoso(a)	vain

Physical appearance

los anteojos	glasses
el balón	soccer ball
la barba	beard
el bigote	mustache
el cabello	hair
calvo(a)	bald
la cola de caballo	ponytail
cuadrado(a)	square
esbelto(a)	slender
el flequillo	bangs
grueso(a)	heavy
los lentes de contacto	contact lenses
el lunar	beauty mark
ondulado(a)	wavy
opuesto(a)	opposite
ovalado(a)	oval
las pecas	freckles
redondo(a)	round
rojizo(a)	reddish
teñido(a)	dyed
triangular	triangular
verse	to look, to appear

♻ **Ya sabes**

amable	nice
cómico(a)	funny
impaciente	impatient
obediente	obedient
paciente	patient
sociable	sociable
tímido(a)	shy

TALK ABOUT EXPERIENCES

Comparisons

a diferencia de	as contrasted with
al contrario	on the contrary
lo bueno/lo malo	the good thing/ bad thing
lo más/menos	the most/least
lo mejor/peor	the best/worst
por un lado	on the one hand
por otro lado	on the other hand
semejante a	similar to

Interactions

compartir	to share
discutir	to discuss, to argue
hacerle caso a	to obey, to pay attention to
influir	to influence
resolver (o→ue)	to resolve
respetar	to respect
tener en común	to have in common

LIST ACCOMPLISHMENTS

Present and Past Perfect Tenses

Ya **hemos resuelto** el problema.
Ellos **habían discutido** mucho.

Juego

El mejor amigo

¿Cuál de estas palabras no se aplica al dibujo?

  a. atrevido

  b. considerado

  c. descarado

  d. fiel

ETAPA

2

¿Cómo me veo?

OBJECTIVES

- Describe fashions

- Talk about pastimes

- Talk about the future

- Predict actions

¿Qué ves?

Mira la foto. Contesta las preguntas.

1. ¿Qué cosas en la foto te dicen dónde están estas personas?

2. ¿Qué relación tienen? ¿Por qué crees esto?

3. ¿Crees que la opinión de los chicos es igual? ¿Por qué?

4. ¿Por qué crees que este lugar se llama La Villita?

En contexto 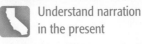 Understand narration in the present

VOCABULARIO

LOS ESTILOS DE LAS ESTRELLAS

Descubre

Busca las palabras dentro de las palabras. ¿Te da alguna idea de qué quieren decir?

llavero: llaves =

monedero: monedas =

billetera: billetes =

prendedor: prender =

sudaderas: sudar =

C: Esa **cadena** de oro tiene inspiración en los años setenta. ¿No sabe que es el siglo 21? ¡Y esa **medalla**!

R: A mí me impresiona el chaleco **estampado**. ¿Dónde lo compró? En 'Nosotros-Somos-Hippies'?

C: ¡Ese **bolso**! ¿Qué llevará adentro? Esos **pendientes** parecen adornos para el árbol de Navidad.

R: ¡Por favor! Ese vestido **suelto** de **lunares** es algo extraterrestre. ¿De dónde es su **diseñador**–del planeta Marte?

C: Pobre chico. Gran actor y no tiene dinero para comprar pantalones. Siempre anda en **sudaderas**.

R: Las sudaderas no me molestan. ¡Pero ese **color brillante**! Necesito mis gafas de sol para mirarlo.

Los críticos

Carlota Jáuregui

Esta venezolana reconocida es la diseñadora que Hollywood más adora. Ha vestido a algunas de las personas más prestigiosas del mundo.

Raúl Montenegro

Después de 15 años de tomarles fotos a las estrellas, se puede decir que este fotógrafo famoso sabe algo de **moda**.

RAÚL   CARLOTA

C: Elegantísimo. El pañuelo de **seda**. El traje de **color oscuro**. **Se destaca** por su elegancia de pie a cabeza.

R: Este chico es **único**. Para él, **poliéster** es una mala palabra.

C: Parece ángel en ese **color claro**. ¿Y ves el **prendedor**? ¡Sus accesorios tienen más estilo que tú y yo juntos!

R: ¿Dónde comprará sus joyas esta mujer? ¿Será amiga de la reina de Inglaterra?

C: ¡Y ese traje!

R: Me encanta su traje **de un solo color**. Dice «elegante», dice «no necesito llamar la atención». Si quieres saber cuál es la moda de **la temporada**, es suficiente ver su **vestuario**.

Online Workbook
CLASSZONE.COM

¿Comprendiste?

1. ¿Lees revistas de moda? ¿Cuáles?
2. ¿Te interesa saber cómo se visten las estrellas de cine? ¿Por qué?
3. Da un ejemplo en el que estás de acuerdo con los críticos y uno en el que no lo estás.
4. ¿Dedicas mucho tiempo a pensar en tu «look» individual o no te importa mucho?
5. ¿Hablas mucho con tus amigos sobre la moda más popular y aceptada en la escuela?
6. ¿Cómo te vistes tú? Describe tu propio estilo.
7. ¿Crees que la moda es importante o que la gente le da demasiada importancia? Explica.

En vivo

AUDIO

SITUACIONES

¡Persigue la moda!

Eres reportero(a) en los Premios de Música en Nueva York y estás ahí para oír todos los detalles que pueden interesar a los lectores de tu revista, *Modas Modernas*. Escucha los últimos chismes de la moda para que tus lectores puedan encontrar lo último del vestir en estas tiendas.

❶ Leer

Estudia los anuncios de las tiendas de moda.

Para los que saben....

EL HOMBRE FINO

La elegancia está en los detalles

Joyería y Accesorios de Hoy

Ropa que juega tan fuerte como usted

Mundo del Atleta

LA VIDA TEJANA

porque la vida es una aventura

El romance del misterio

La Mujer Elegante

EXPRÉSATE

Súper Informal

❷ Escuchar

En los Premios ves a tus viejos amigos, los reporteros Ana Beatriz Castillo y Javier Villanueva. Ellos describen el vestuario de cada persona famosa que pasa y ¡no se les escapa nada! Comentan sobre la ropa, los zapatos, los accesorios, ¡todito! Escucha y haz apuntes en una hoja aparte.

❸ Escribir

Después de escuchar el comentario de los locutores, escribe dónde crees que estos músicos compraron su ropa.

Carson	Los Jaguares
Ana Luisa	Armando Iglesias
Elena	Luis Marcos
Joya	

Activity 4: Converse and listen during face-to-face social interactions

En acción

Práctica del vocabulario

Objectives for Activities 1–4
• Describe fashions

1 ¿Con qué va?

Leer/Escribir Andrea describe el uso de algunos accesorios y otras cosas relacionadas con la moda. Completa sus comentarios.

1. Para no perder mis llaves, las pongo todas en un _____.
   a. llavero
   b. monedero

2. No me gusta tener monedas sueltas en mi bolsillo. Las pongo todas en un _____.
   a. bolso
   b. monedero

3. Papá pone sus billetes en una _____.
   a. billetera
   b. cadena

4. Cuando voy al colegio, uso mi mochila. Pero cuando salgo el fin de semana, prefiero usar mi _____.
   a. bolso
   b. prendedor

5. Me gusta mucho la moda. Sueño con ser _____.
   a. arquitecto(a)
   b. diseñador(a)

2 Estilos diferentes

Hablar/Escribir ¿Cómo se visten las personas siguientes? Con un(a) compañero(a), digan por lo menos dos cosas sobre el vestuario de cada una.

Julia

modelo

Tú: *Julia lleva unas sudaderas de un solo color.*

Compañero(a): *También lleva una camiseta de color brillante.*

2. Gerardo

4. Belisa

1. Mercedes

3. Joaquín

3 La moda de hoy

STRATEGY: SPEAKING

Use familiar vocabulary in a new setting While learning new words about clothing and fashion, use some that you have already learned: **bufanda, collar, anillo, pulsera, aretes, sandalias, a rayas, a cuadros**, as wellas old friends like **blusa, calcetín, camisa, camiseta, falda, jeans, pantalones, suéter, vestido, zapato,** and of course, the colors: **amarillo, anaranjado, azul, blanco, marrón, morado, negro, rojo, verde**.

Hablar/Escribir Tú y tu compañero(a) encontraron una revista de modas. Tu compañero(a) quiere saber si te gustan ciertas cosas. Di lo que piensas de cada artículo.

modelo

chaleco

Compañero(a): *¿Te gusta el chaleco de rayas?*

Tú: *¡Sí! Hace juego con los pantalones de mezclilla.*
o No, prefiero el chaleco de cuero.

1. falda
2. traje
3. blusa
4. sudaderas
5. pantalones
6. suéter
7. vestido
8. chaqueta

4 De compras

Hablar Tú y tu compañero(a) están de compras en una tienda de modas. Tú te pones varias cosas y le preguntas a tu compañero(a) si te quedan bien. Él (Ella) te responde y te hace otra sugerencia. Luego, cambien de papel.

modelo

Tú: *¿Cómo me veo? ¿Te gusta esta chaqueta de lana?*

Compañero(a): *¡Uy, no! ¡Te queda muy floja! Mejor cómprate la chaqueta de seda. Es más elegante.*

Vocabulario

¿De qué es?

el algodón *cotton*
el cuero *leather*
el fleco *fringe*
la lana *wool*
la lentejuela *sequin*
la mezclilla *denim*

Ya sabes

ancho(a)
apretado(a)
estrecho(a)
flojo(a)
hacer juego con...
oscuro(a)
un par de
las rayas
sencillo(a)

▶ ¿Puedes describir tu vestuario con estas palabras?

Objectives for Activities 14–15
• Describe fashions • Talk about pastimes • Talk about the future • Predict actions

REPASO | **Verbs Like gustar**

▶ You most often use verbs like gustar with the **indirect object pronouns** me, te, le, nos, os, and les to express your and others' reactions to things.

Me gusta tu **prendedor.**
I like your pin.

Gracias. ¡Y a mí **me gustan** tus **zapatos!**
Thanks. And I like your shoes!

▶ You already know many verbs that are used like gustar.

encantar	interesar
faltar	quedarle bien
fascinar	quedarle mal
molestar	importar

Me encantan los **deportes.**
I love sports.

¡Qué **bien te queda** ese **sombrero!**
That hat looks good on you!

A Marta **le molestan** los **anteojos** nuevos.
Marta's new glasses are bothering her.

> Notice that the form of these verbs matches **deportes**, not **me**.

Practice: | **Actividades** 5 6 | **Más práctica** *cuaderno pp. 25–27*
Para hispanohablantes *cuaderno pp. 23–25* | **Online Workbook** CLASSZONE.COM

---

 5 Gustos

Hablar/Escribir Todos tenemos gustos diferentes. Di qué le gusta a cada persona.

modelo

a mí: las blusas sencillas
A mí me gustan las blusas sencillas.

1. a ti: las botas de cuero
2. a Hernán: el sombrero negro
3. A Érica y a Laurita: la ropa suelta
4. a nosotros: los zapatos de tenis
5. a usted: el bolso de rayas
6. a mí: los pantalones de lana

6 ¡Me fascina!

Hablar/Escribir Tu compañero te pregunta si te gustan las siguientes cosas. Contéstale usando uno de estos verbos: **gustar, encantar, fascinar** o **interesar.**

modelo

las películas románticas

Compañero(a): *¿Te gustan las películas románticas?*
Tú: *No, no me interesan (o) Sí, me interesan.*

1. la ropa de moda
2. las telenovelas
3. la música clásica
4. la música «rap»
5. las obras de teatro
6. la historia maya
7. los videojuegos
8. los patines en línea

7 ¿De veras?

Hablar ¿Conoces los gustos de tu compañero(a)? Dile cinco cosas que detestas o que te fascinan. ¿Cómo reacciona? Luego, cambien de papel.

modelo

Tú: *No me cae bien la música de los sesenta.*

Compañero(a): *¿De veras? Yo creo que es genial.*

Compañero(a): *Me fascina la moda de los sesenta.*

Tú: *¿De veras? Yo creo que es muy incómoda.*

Vocabulario

¿Cómo es?

caer bien/mal *to like, dislike*

cómodo(a) *comfortable*

detestar *to hate*

formidable *great*

genial *wonderful*

horrible *horrible*

incómodo(a) *uncomfortable*

pesado(a) *boring, heavy*

▶ ¿Puedes describir tus pasatiempos o actividades con estas palabras?

REPASO Por and Para

▶ You already know the prepositions **por** and **para**. Both words can mean *for* in English, but the meaning can change depending on how it is used. Look at these examples.

▶ Use **por** to indicate…

- the idea of passing through.

  Esta carretera pasa **por** Tejas.
  *This highway goes **through** Texas.*

- general rather than specific location.

  No sé si hay una piscina **por** aquí.
  *I don't know if there is a pool **around** here.*

- how long something lasts.

  Vivimos en Puerto Rico **por** muchos años.
  *We lived in Puerto Rico **for** many years.*

- the cause of something.

  No podemos acampar **por** la tormenta.
  *We can't camp **because of** the storm.*

- an exchange.

  Cecilia pagó mucho **por** sus anteojos.
  *Cecilia paid a lot **for** her glasses.*

- doing something in place of or instead of someone else.

  No puedo ir. ¿Puedes ir **por** mí?
  *I can't go. Can you go **in my place**?*

- a means of transportation.

  Viajamos **por** barco.
  *We traveled **by** boat.*

▶ Use **para** to indicate…

- for whom something is done.

  Compraremos un regalo **para** Silvia.
  *We will buy a gift **for** Silvia.*

- destination.

  Francisco tomó el avión **para** San Juan.
  *Francisco took the plane **to** San Juan.*

- the purpose for which something is done.

  Compré anteojos **para** ver mejor.
  *I bought glasses **in order to** see better.*

- to express an opinion.

  Para mí, el montañismo es maravilloso.
  ***To** me, mountaineering is marvelous.*

- to contrast or compare.

  Para programador, no sabe mucho de computadoras.
  ***For** a programmer, he doesn't know much about computers.*

- to express the idea of a deadline.

  Hay que terminar la tarea **para** mañana.
  *The assignment has to be finished **by** tomorrow.*

Practice:
Actividades
8 9 10

Más práctica
cuaderno p. 28
Para hispanohablantes
cuaderno p. 25

Online Workbook
CLASSZONE.COM

Activity 9: Understand most spoken language when the message is deliberately and carefully conveyed by a speaker accustomed to dealing with learners when listening

8 El diario de Maité

Escribir Maité escribió este pasaje en su diario. Complétalo con **para** o **por** para saber más de su viaje a Los Ángeles.

> 3 de julio
>
> Nos vamos de vacaciones ___1___ un mes. Salimos ___2___ Los Ángeles pasado mañana. Primero voy a la agencia de viajes ___3___ los boletos. Luego voy al centro ___4___ comprar una maleta. La maleta que tengo es demasiada pequeña ___5___ toda la ropa que quiero llevar.
>
> He hablado ___6___ teléfono ___7___ dos horas con Carmela, la prima que vamos a visitar. Ella estudia ___8___ ser abogada. Carmela es muy inteligente ___9___ su edad. Siempre me dice «Tienes que ahorrar dinero ___10___ ir a la universidad. ¡Es muy importante!»
>
> Yo le dije que me gustaría viajar ___11___ todo California. No sé si vamos a tener tiempo ___12___ hacer todo lo que quiero. Me prometió que íbamos a dar un paseo ___13___ la playa en cuanto lleguemos. Compré un traje de baño nuevo ___14___ llevar. Pagué treinta dólares ___15___ él.
>
> ¡Qué dicha! ¡ ___16___ fin voy a conocer California!

9 ¿Adónde vas?

Escuchar/*Escribir* Escucha la conversación entre María y Miguel. Luego completa las oraciones con **por** o **para**.

1. María va a la tienda _____ comprar una guayabera _____ su papá.
2. El regalo es _____ su papá.
3. María tiene dinero _____ el regalo.
4. María trabaja en esa oficina _____ tres años.
5. María trabaja _____ la señora Ontiveros.
6. María va a la tienda _____ autobús.
7. María tiene tarea _____ mañana.
8. María estudió _____ tres horas después de clases.

10 La fiesta de cumpleaños

Hablar Tú y tu compañero(a) van a una fiesta de cumpleaños. Usen las palabras **para** y **por** para hablar sobre la persona festejada, el regalo que le darán, cómo van a llegar y cuánto tiempo durará la fiesta.

modelo

Tú: *¿Vas a comprar algo para Miguel para su cumpleaños?*

Compañero(a): *Creo que sí, pero no puedo pagar mucho por el regalo.*

You have already learned two ways to talk about the **future.**

* You can use: ir + a + **infinitive**

    Vamos a estudiar en la biblioteca.
    We'll study (We're going to study) in the library.

* You can use the present tense when the **context** makes it clear that you are talking about the future.

    Mañana alquilamos una película.
    Tomorrow we're renting a film.

You can also use the **future tense.** You form the future tense by adding a special set of **endings** to the **infinitive.**

comer *to eat*

comer**é**	comer**emos**
comer**ás**	comer**éis**
comer**á**	comer**án**

All verbs have the same endings in the future tense.

Nosotros **llegaremos** a las siete.
We will arrive at seven.

With some verbs, you have to change the form of their infinitive slightly before adding the future tense endings.

You still use the same future tense endings.

The future of **hay** is **habrá.**

infinitive	future stem
decir	dir-
hacer	har-
poner	pondr-
salir	saldr-
tener	tendr-
valer	vald-
venir	vendr-
poder	podr-
querer	querr-
saber	sabr-

Apoyo para estudiar

Express future plans or events

Generally you will use **ir + a +** infinitive or the present tense to express the future where intention is strong and the event will happen soon. The future tense is often associated with plans, predictions, or events that are less certain. Which of the following two sentences is more certain? **Voy a México en julio. Algún día iré a Madrid.**

Practice: **Actividades**   **Más práctica** *cuaderno p. 29*   **Online Workbook**
11 12 13 14   **Para hispanohablantes** *cuaderno p. 26*   CLASSZONE.COM

sesenta y cinco
Estados Unidos Etapa 2   **65**

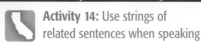
Activity 14: Use strings of related sentences when speaking

11 La cita ideal

Hablar/*Escribir* Ana Bárbara va a salir con Ismael el sábado. Ella imagina cómo va a ser su primera cita con él. ¿Qué dice?

> **modelo**
>
> *yo / salir / con Ismael / sábado por la noche*
>
> *Saldré con Ismael el sábado por la noche.*

1. su papá / prestarle / su carro nuevo
2. Ismael / llegar / a las siete
3. yo / estar / lista
4. yo / ponerse / el nuevo vestido de lunares
5. él / traer [a mí] / flores
6. nosotros / cenar / en un restaurante francés
7. él / poder leer / el menú en francés
8. nosotros / bailar / toda la noche
9. él / hablar / sobre su futuro
10. yo / contar / chistes
11. nuestros amigos / ver [a nosotros]
12. yo / divertirse / mucho
13. nosotros / hacer / otra cita

12 San Antonio

Escuchar/*Escribir* Julia está en San Antonio con su familia por dos días. Al final del primer día, su mamá cuenta lo que hicieron ese día y los planes para el otro día. ¿Qué hicieron ayer y qué van a hacer mañana? Escoge la respuesta correcta.

1. **a.** Fueron al Paseo del Río.
   b. Irán al Paseo del Río.
2. **a.** Fueron a El Álamo.
   b. Irán a El Álamo.
3. **a.** Fueron al Mercado.
   b. Irán al Mercado.
4. **a.** Compraron regalitos para sus amigos.
   b. Comprarán regalitos para sus amigos.
5. **a.** Asistieron a un concierto de música tejana.
   b. Asistirán a un concierto de música tejana.

13 El año que viene

Hablar/*Escribir* Pregunta a un(a) compañero(a) si hará las siguientes actividades el año que viene. Luego, cambien de papel.

> **modelo**
>
> **Tú:** *¿Trabajarás el año que viene?*
>
> **Compañero(a):** *Sí, trabajaré en la tienda de música el año que viene.*

trabajar...   hacer...   dar...   acampar...

viajar a ...   estudiar...   ver...

ir a   volver a...   comprar...   competir...

jugar a...   salir...   celebrar...   ¿...?

14 El Campamento MonteVerde

Hablar Vas a ir al Campamento MonteVerde con dos amigos. Conversen sobre qué les gustaría hacer allí. Hablen de todas las posibilidades y digan por qué les gustan o no las actividades del Campamento.

modelo

Tú: *¿Qué harás tú en el Campamento MonteVerde?*

Amigo 1: *¿Yo? Yo haré alpinismo. Me encanta estar al aire libre y ver las flores y los animales.*

Amigo 2: *Ay, a mí no. No me interesa la naturaleza. Yo navegaré por Internet.*

More Practice: Más comunicación *p. R4*

Vocabulario

Los pasatiempos

acampar *to camp*

coleccionar *to collect*

escalar montañas *to mountain climb*

esquiar en el agua *to waterski*

hacer alpinismo *to go hiking*

hacer montañismo *to go mountaineering*

navegar en tabla de vela *to windsurf*

navegar por Internet *to surf the Internet*

pescar en alta mar *to go deep-sea fishing*

pilotar una avioneta *to fly a single-engine plane*

volar en planeador *to hang-glide*

▶ *¿Has hecho una de estas actividades alguna vez?*

¡Disfruta de la naturaleza! Ven al

Campamento Monteverde

Aquí podrás...

esquiar en el agua

escalar montañas

acampar bajo la luna

hacer alpinismo

y también podrás...
¡navegar por Internet!

volar en planeador

remar en una canoa

Ya verás cuánto te divertirás aquí en
Campamento Monteverde
Una semana aquí y nunca querrás regresar a casa.

Nota cultural

Araceli Segarra es la primera mujer española en escalar el famoso monte Everest. Este pico formidable tiene más de 29.000 pies de altura. Los alpinistas que tratan de conquistar el Everest tienen que entrenarse por mucho tiempo para acostumbrarse a la falta de oxígeno que existe en las altitudes muy elevadas.

Activity 17: Narrate
in the future

GRAMÁTICA **Future Tense to Express Probability**

▶ You can use the **future tense** to speculate about what might occur or what others are doing. When used this way, the **future tense** implies that you are **wondering about an event** or **guessing whether or not it has occurred.**

Marcos ya no quiere jugar al fútbol los sábados.
¿Por qué **será**?
Marcos doesn't want to play soccer on Saturdays any more.
What could be *the reason for that?*

Tendrá novia.
*He **probably has** a girlfriend.*

¿De veras? ¿Quién **será**?
Really? ***I wonder*** *who* ***it might be?***

Practice: Actividades 15 16	**Más práctica** *cuaderno p. 30* **Para hispanohablantes** *cuaderno pp. 27–28*	**Online Workbook** CLASSZONE.COM

Nota cultural

Cuando hablamos de las mascotas, es importante saber algunas diferencias entre el mundo en inglés y el hispanohablante. En inglés nos referimos al **sonido** (sound) que hacen los perros como *bow-wow*, pero en español es guau-guau. Otra diferencia es que los gatos «hispanohablantes» dicen miau en vez de *meow* y los pájaros pío-pío en vez de *tweet-tweet*. Algunos nombres típicos para mascotas son Colita, Mancha y Pelusa para los perros, y Michi o Michifús para los gatos.

15 **¿Dónde estarán?** ♻

Hablar/*Escribir* Nunca sabes dónde están tus cosas, tu familia, tus mascotas (*pets*), tus amigos. Expresa tu frustración.

> **modelo**
>
> *mis libros (sala)*
>
> *¿Dónde estarán mis libros? ¿Estarán en la sala? Tendré que buscarlos.*

1. mi gato (debajo de la cama)
2. mi hermano (afuera)
3. mis zapatos (en el clóset)
4. mi perro (en el jardín)
5. mi mamá (en su habitación)
6. mi raqueta de tenis (en el carro)
7. mis discos compactos (en el cuarto de mi hermano)
8. mi gorro (en mi habitación)
9. mis carpetas (en mi mochila)
10. mi cámara (en casa de mi amiga Delia)
11. Hernán (en la biblioteca)
12. Delia (en el gimnasio)

16 No sé exactamente

Hablar/Escribir Tu compañero(a) te hace muchas preguntas. No estás seguro(a) de la respuesta, pero tratas de contestarle. Sigue el modelo.

modelo

¿Dónde está Ricardo? (en su cuarto)

Compañero(a): *¿Dónde estará Ricardo?*

Tú: *No sé exactamente. Estará en su cuarto.*

1. ¿Cuándo vienen los primos? (la semana que viene)
2. ¿Cuántos años tiene esa señora? (unos ochenta años)
3. ¿Qué hora es? (las diez o las once de la noche)
4. ¿Qué es esa cosa en la calle? (una bolsa de basura)
5. ¿Qué dice la profesora si le dices que no hiciste la tarea? (que no hay problema)
6. ¿Hay suficiente comida para todos? (más en el refrigerador)
7. ¿Dónde están mis sudaderas? (en tu cuarto)
8. ¿Cuándo traen los pasteles? (después del almuerzo)

17 La fiesta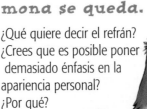

Hablar Tú y tu compañero(a) van a dar una fiesta. Tu compañero(a) siempre se pone nervioso(a) pensando en lo que puede pasar. Te hace muchas preguntas. Contéstale para que se quede tranquilo(a). Luego cambien de papel.

modelo

Compañero(a): *¿Lloverá durante la fiesta?*

Tú: *¡No te preocupes! No va a llover durante la fiesta. Hará mucho calor y vendrá mucha gente.*

Compañero(a): *¿Les gustará la música?*

Tú: *Sí, les encantará.*

llover
venir los invitados
haber suficiente comida
tocar muy mal los músicos
ser aburrido
irse demasiado temprano los invitados
gustar la música

More Practice:
Más comunicación *p. R4*

Online Workbook CLASSZONE.COM

También se dice

Aunque en la mayoría de los países latinoamericanos se dice **sudadera**, en España se usa la palabra **chándal**.

Refrán

Aunque la mona se vista de seda, mona se queda.

¿Qué quiere decir el refrán? ¿Crees que es posible poner demasiado énfasis en la apariencia personal? ¿Por qué?

Acquire knowledge and new information from comprehensive, authentic texts when reading

En colores
CULTURA Y COMPARACIONES

Un gran diseñador

PARA CONOCERNOS
STRATEGY: CONNECTING CULTURES
Examine the cultural role of fashion How do fads, personal style, and high style influence your decisions when you buy clothing? Imagine you are packing for a trip to another country, another culture. How might geography, climate, or local customs influence your choice of clothing? Name an article of clothing you would take and the characteristic that influenced your choice.

Country_____

	Característica	Ropa
Geografía		
Clima		
Costumbres		
Otro		

Oscar de la Renta sabe que para el gusto se hicieron los colores. ¿Cuál es la filosofía de la moda de este gran diseñador? «El vestir[1] es una cosa muy personal; en realidad es un reflejo[2] de la imagen que quieres proyectar», nos dice de la Renta. Esta filosofía y su gran talento artístico lo han hecho uno de los grandes en el mundo de la moda.

[1] dressing
[2] reflection

Oscar de la Renta ha hecho mucho por la niñez de Santo Domingo. Ha ayudado a construir una escuela y un orfelinato[3] para 350 niños, a quienes visita en las Navidades.

«Lo que se quiere es que alguien te vea a ti primero y que te encuentre bien, y que después se fije[4] en tu ropa. Lo contrario, yo creo que es negativo».

[3] orphanage
[4] to look at

De la Renta nació en la República Dominicana y allí estudió pintura en la Escuela de Bellas Artes de Santo Domingo. A la edad de 18 años fue a Madrid, donde conoció al diseñador Cristóbal Balenciaga y se convirtió en su ilustrador. En 1963 fue a Nueva York. Desde entonces, de la Renta ha iniciado muchas tendencias de la moda.

CLASSZONE.COM
More About Latinos

¿Comprendiste?

1. ¿Qué piensa de la Renta del vestir?
2. ¿Cuándo comenzó de la Renta su carrera como diseñador en Estados Unidos?
3. Según el diseñador, ¿cómo debe uno vestirse?
4. ¿Cómo ha ayudado Oscar de la Renta a los dominicanos?

¿Qué piensas?

1. ¿Qué hay que hacer para ser diseñador de alta moda?
2. ¿Estás de acuerdo con la filosofía de la moda de Oscar de la Renta? ¿Por qué?

Hazlo tú

Muchos diseñadores hispanos han influenciado la moda, como Carolina Herrera, Paloma Picasso y Narciso Rodríguez. Busca datos y escribe un párrafo sobre uno de ellos. Si es posible, incluye en tu composición una foto o un dibujo de uno de sus modelos.

 Activity 1: Generally choose appropriate vocabulary for familiar topics, but as the complexity of the message increases, there is evidence of hesitation and groping for words, as well as patterns of mispronunciation and intonation

En uso

REPASO Y MÁS COMUNICACIÓN

Now you can...
- describe fashions.

To review
- verbs like **gustar** see p. 62.

OBJECTIVES

- Describe fashions
- Talk about pastimes
- Talk about the future
- Predict actions

1 ¿Qué lleva?

Hablas por teléfono con un(a) amigo(a). Te pregunta qué llevan varias personas. Contesta y añade tu propio comentario con el verbo indicado.

Martín / quedar

modelo

Martín/quedar

Compañero(a): *¿Qué lleva Martín?*

Tú: *Martín lleva una camisa de un solo color.*

Compañero(a): *¿Le queda bien?*

Tú: *Sí, le queda bien. Hace juego con sus pantalones.*

la profesora García / gustar

Raquel / encantar

Guillermo / gustar

Patricia / quedar

Sabrina / gustar

Gustavo /importar

Now you can...
- talk about pastimes.

To review
- **por** and **para**
  see p. 63.

2 Nueva York

Usa **por** y **para** y completa la nota que te escribió un amigo sobre sus planes.

> *¡Hola! Salgo _____ Nueva York el mes que viene. Voy a viajar _____ tren. Pagué cuarenta dólares _____ el boleto. Voy a visitar a mi hermano que estudia _____ ser doctor. Ha vivido en Nueva York _____ dos años. Me gustaría pasear _____ el Parque Central. También quiero ir al Museo del Barrio _____ ver la exposición de arte taíno. ¡Pero no creas que voy sólo _____ divertirme! Trabajaré _____ mi tío. Ya sabes que necesito ahorrar dinero _____ ir a la universidad. ¡Te mando una postal!*

Now you can...
- talk about the future.

To review
- the future tense
  see p. 65.

3 El club de español

El club de español va a tener su reunión anual. ¿Qué harán todos?

modelo

A Mireya le gusta preparar comida dominicana.

Mireya preparará comida dominicana.

1. A todos les gusta divertirse.
2. A Susana le gusta bailar salsa.
3. A ellos les gusta sacar fotos.
4. A ti te interesa traer los refrescos.
5. A David le gusta hablar.
6. A nosotros nos gusta comer.

Now you can...
- predict actions.

To review
- the future of probability
  see p. 68.

4 ¡Haces demasiadas preguntas!

Tu hermanito(a) quiere saber cosas sobre tus vacaciones con tu amigo(a).

modelo

ir de compras solo(a) o con su amiga

Hermanito(a): *¿Crees que irás de compras solo(a) o con tu amiga?*

Tú: *No sé. Tal vez iré solo(a).*

1. comprar el equipo de acampar o navegar en tabla de vela
2. volar en planeador o hacer alpinismo
3. pescar en alta mar o nadar en la piscina
4. alquilar un video o ir al cine

5 Tu campamento

> **STRATEGY: SPEAKING**
>
> **Brainstorm to get lots of ideas** What kind of camp do you want? What should the emphasis be? Sports, music, art, creative writing, math, or languages… all in the great outdoors? Vote and pick one together. What activities does your camp offer that make it special?

Trabajando en grupos, escriban un anuncio para un campamento donde se pueden hacer muchas actividades. Primero den un nombre a su campamento y escriban un lema (*slogan*) que use el futuro. Hagan dibujos para su anuncio e incluyan una lista de todas las actividades del campamento.

modelo

En el Campamento Mundo Natural usted se sentirá feliz.

6 Desfile de modas

En grupos, escriban una narración para un desfile de modas de la escuela. Describan el vestuario de cuatro modelos detalladamente (*in detail*). Refiéranse a una revista de moda para buscar ideas. Luego, lean su narración a la clase.

modelo

Jim lleva pantalones anchos. Son de mezclilla. Lleva una camisa de rayas que hace juego con los pantalones.

7 En tu propia voz

ESCRITURA ¿Qué piensas del futuro? ¿Cómo será la vida en nuestro planeta, la Tierra? ¿Cómo será tu vida? ¿Qué harás? Escribe diez predicciones para el año 2030.

modelo

Seré periodista para Internet.

No habrá periódicos.

Todos leerán las noticias en Internet.

CONEXIONES

Las matemáticas Prepara un presupuesto (*budget*) anual de lo que vas a gastar en ropa el próximo año. ¿Cuánto vas a gastar en total? ¿Qué porcentaje piensas gastar en zapatos, pantalones, joyas, etc.? Prepara una gráfica con el porcentaje que vas a gastar en cada categoría.

Zapatos 19%
Pantalones 18%
Joyas 15%
Accesorios 10%
Camisas 25%
Vestidos 12%

En resumen
REPASO DE VOCABULARIO

DESCRIBE FASHIONS

Fashion

destacarse	to stand out
el (la) diseñador(a)	designer
la moda	fashion, style
suelto(a)	loose
la temporada	season, period of time
único(a)	unique, only
el vestuario	wardrobe

Items

la billetera	wallet
el bolso	shoulder bag
la cadena	chain
el llavero	keychain
la medalla	medallion
el monedero	change purse
los pendientes	dangling earrings
el prendedor	pin
las sudaderas	sweats

Likes and dislikes

caer bien/mal	to like/dislike
cómodo(a)	comfortable
detestar	to hate
formidable	great
genial	wonderful
horrible	horrible
incómodo(a)	uncomfortable
pesado(a)	boring, heavy

Materials

el algodón	cotton
el color brillante	bright color
el color claro	pastel
el color oscuro	dark color
el cuero	leather
de un solo color	solid color
estampado(a)	print
el fleco	fringe
la lana	wool
la lentejuela	sequin
los lunares	polka-dots
la mezclilla	denim
el poliéster	polyester
la seda	silk

♻ Ya sabes

ancho(a)	wide
apretado(a)	tight
estrecho(a)	narrow
flojo(a)	loose
hacer juego con	to match with
oscuro(a)	dark
un par de	a pair of
las rayas	stripes
sencillo(a)	simple

TALK ABOUT PASTIMES

acampar	to camp
coleccionar	to collect
escalar montañas	to mountain climb
esquiar en el agua	to water-ski
hacer alpinismo	to go hiking
hacer montañismo	to go mountaineering
navegar en tabla de vela	to windsurf
navegar por Internet	to surf the Internet
pescar en alta mar	to go deep-sea fishing
pilotar una avioneta	to fly a single-engine plane
volar en planeador	to hang–glide

TALK ABOUT THE FUTURE

Future tense

Iré a la fiesta.

PREDICT ACTIONS

Future of probability

¿Dónde **estará**?

Juego

Sopa de letras

Pon en orden estas letras para saber qué no cambia pero sí da cambio:

REDONOME

UNIDAD 1

ETAPA **3**

¡Hay tanto que hacer!

OBJECTIVES

- Talk about household chores

- Say what friends do

- Express feelings

¿Qué ves?

Mira la foto. Contesta las preguntas.

1. ¿Qué hacen estas personas?

2. ¿A qué grupo crees que pertenecen?

3. ¿Crees que todos toman decisiones sobre el jardín? ¿Qué más pueden hacer?

4. Mira el póster. ¿Qué cosas hacen los vecinos para cuidar el jardín?

Celebra El día del jardín en el jardín comunitario de

Jamaica Plain

Siembra, cosecha y comparte con familiares, amigos y vecinos.

¡Más de 27 sembrados de flores y verduras!

Descubre

A. El prefijo Cuando una palabra empieza con el prefijo **des-**, esa palabra significa lo opuesto de la misma palabra sin el prefijo **des-**. Trata de descubrir el sentido de las siguientes palabras.

organizado = *organized*

desorganizado = _____

conectar = *to connect*

desconectar = _____

enchufar = *to plug in*

desenchufar = _____

armar = *to assemble*

desarmar = _____

B. La casa Ya sabes los nombres de los cuartos de la casa. Aquí hay dos más. ¿Tiene **sótano** tu casa? ¿**desván**?

el desván

el sótano

Manolo el imposible

Fian Arroyo es un artista de Surfside, Florida. Él dibuja una tira cómica sobre el personaje de Manolo, un niño que siempre causa problemas por todo el mundo. Aquí tienes un episodio de sus aventuras.

Debo cambiar *la bombilla* de la lámpara.

También tengo que *vaciar el basurero.*

Mamá quiere que limpie los *gabinetes.* Están muy *desorganizados.*

¡Pero no tengo ganas! Mejor voy al jardín.

En el jardín

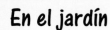

Tengo que regar las plantas.

¡Necesito reparar el cortacésped. ¡No funciona! Para repararlo, lo tengo que desarmar.

Debo desyerbar el jardín. ¡Pero hay tantas malas hierbas! ¡Va a tomarme horas!

¡No tengo ganas! Mejor voy a la oficina de mis padres.

En la oficina

Debo pasar la aspiradora, pero no quiero. ¡Voy a desenchufarla!

Si desconecto el teléfono, puedo ver la televisión sin interrupciones.

¡Enchufé el televisor y voy a encenderlo!

¡Oigo que vienen mis padres! ¡Voy a esconderme en el desván!

También se dice

In Latin America, a potato is **una papa,** but in Spain it is **una patata. Papas fritas** and **patatas fritas** are both french fries and potato chips.

Online Workbook
CLASSZONE.COM

¿Comprendiste?

1. ¿Haces quehaceres en tu casa?
2. ¿Cuántas veces por semana tienes que ayudar en tu casa?
3. ¿Qué quehaceres te tocan a ti?
4. ¿Hay quehaceres que te gustan hacer? ¿que no te gustan hacer? ¿Por qué?
5. ¿Crees que es importante tener responsabilidades en la casa? ¿O tienes suficientes responsabilidades de la escuela?
6. Imagina que eres el jefe o la jefa de tu casa. ¿Cómo compartirías los quehaceres de tu casa?

Listen to
audio texts

En vivo

SITUACIONES

STRATEGY: LISTENING

Pre-listening Most of us like to live in a clean and orderly place, although not all of us like to do the necessary work. Make a list of what is necessary to do both inside and outside your home to maintain it.

Make an argument for and against hiring others to maintain a home How does your list compare with the services of Casa Limpia? Do you think those who live in the home should do the work? Do you think it's o.k. to pay others to do it? What are the reasons for and against either option?

¡Qué desastre!

Acabas de regresar de las vacaciones. Cuando vuelves, ves que tu casa es un desastre. Hay muchos quehaceres. Primero identifica lo que tienes que hacer. Luego llama a un servicio de limpieza para saber si te pueden ayudar.

❶ Mirar

Mira las fotos de los cuartos de esta casa desorganizada. Escribe una lista de todas las cosas que deben hacerse.

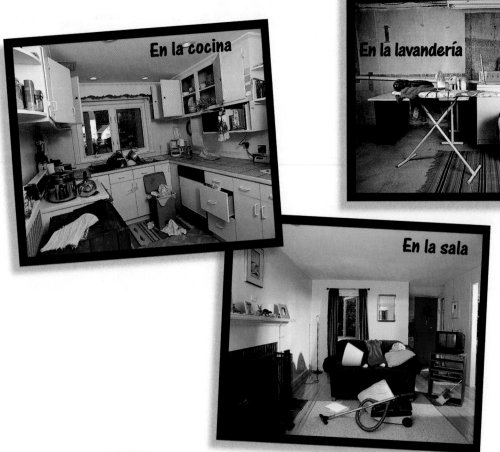

En la cocina

En la lavandería

En la sala

En el jardín

CD-ROM Take-Home Tutor

En el sótano

2 Leer

¡Qué suerte! Ves el anuncio a la derecha en el periódico. Claro, llamas inmediatamente.

3 Escuchar 🎧

Antes de escuchar el mensaje de CasaLimpia, copia este formulario en otro papel. A la izquierda, escribe la lista de quehaceres que ya escribiste. Mientras escuchas el mensaje, marca «sí» en tu formulario si ofrecen el servicio que necesitas y marca «no» si no lo ofrecen.

SERVICIO DE LIMPIEZA

CasaLimpia

¡Hacemos de todo!

¡Relájese!

Limpiamos su casa, su oficina o su apartamento mientras usted descansa.

Llame al 555-3489 para informarse sobre nuestros servicios.
¡Llame hoy! y pronto tendrá...
CasaLimpia

Quehaceres	Servicio de limpieza CasaLimpia		Quehaceres	Servicio de limpieza CasaLimpia	
	Sí	No		Sí	No
En la cocina			En el jardín		
En el sótano			En la sala		
En la lavandería					

4 Escribir

¿Qué tengo que hacer yo? Ahora haz una lista de los quehaceres que tendrás que hacer porque el servicio CasaLimpia no ofrece esos servicios.

En acción

PARTE A

Práctica del vocabulario

Objectives for Activities 1–4
• Talk about households chores • Say what friends do

1 Manolo el bueno

Leer/*Escribir* ¿Qué pasa? ¡Parece que Manolo ha cambiado de personalidad! Ahora quiere hacer todas las cosas que no hizo el otro día. ¿Qué dice? Completa sus oraciones con la respuesta correcta.

1. «No ha llovido en tres semanas. Necesito _____».
   a. regar las plantas
   b. desyerbar el jardín
   c. cortar el césped

2. «No puedo encender el televisor. ¡Ah! Es porque no está _____».
   a. desenchufado
   b. enchufado
   c. desconectado

3. «La lámpara no funciona; no da luz. Probablemente necesito _____».
   a. desarmar la lámpara
   b. reparar el cortacésped
   c. cambiar la bombilla

4. «Tuvimos una fiesta anoche y hoy tenemos mucha basura. Tengo que _____».
   a. organizar los gabinetes
   b. planchar la ropa
   c. vaciar el basurero

5. «Quiero cortar el césped pero no puedo porque _____ no funciona. Necesito repararlo».
   a. el televisor
   b. el cortacésped
   c. el gabinete

2 ¡Hazlo!

Hablar/*Escribir* Hay muchas cosas que hacer. Dile a tu compañero(a) que las haga. Luego, cambien de papel.

vaciar

modelo

Tú: *¡Por favor! Vacía el basurero.*

Compañero(a): *Está bien. Lo haré inmediatamente.*

1. enchufar

2. encender

3. desconectar

4. cambiar

5. desyerbar

6. reparar

③ ¿Sabes?

Hablar/Escribir Hay quehaceres que no sabes hacer. Tú y tu compañero(a) conocen a varias personas que los pueden ayudar. Conversen sobre la situación.

modelo

reparar el cortacésped

Tú: *¿Sabes reparar el cortacésped?*

Compañero(a): *No, no sé hacerlo.*

Tú: *¿Conoces a Arturo?*

Compañero(a): *Sí, sí lo conozco.*

Tú: *Él sabe repararlo. Vamos a pedirle ayuda.*

> **Nota: Gramática**
>
> Remember that you use both **saber** and **conocer** to mean *to know*. Use **saber** when someone knows facts or information. Use **conocer** to express familiarity and acquaintance with people and places.

1. desarmar el televisor
2. reparar la computadora
3. encender las luces en el desván
4. reparar la lámpara
5. apagar la calefacción
6. desarmar el lavaplatos
7. desyerbar el jardín
8. desenchufar la nevera

También se dice

Además de la palabra **césped,** hay muchas otras palabras que significan la misma cosa. Puedes usar también **hierba, pasto, zacate** o **grama** en otras partes del mundo hispanohablante.

④ No puedo

Hablar Quieres invitar a tu mejor amigo(a) a salir. Pero él (ella) te dice que no puede porque tiene algo que hacer en la casa. Conversen sobre la situación. Luego le toca a tu compañero(a) hacerte las preguntas.

modelo

Tú: *¿Quieres ir conmigo al cine? Hay una película nueva en el Cineplex.*

Compañero(a): *Me gustaría, ¡pero no puedo! Tengo que cortar el césped.*

Tú: *¿Por qué no lo haces mañana?*

Compañero(a): *…*

Práctica: gramática y vocabulario

Objectives for Activities 5–13
• Talk about households chores • Say what friends do • Express feelings

REPASO Reflexive Verbs

Remember that you use **reflexive verbs** to describe a person doing something that involves himself or herself. These verbs use **reflexive pronouns** that refer to the person doing the action. Reflexive pronouns are: **me, te, se, nos** and **os**.

> Andrés **se** lastimó.
> *Andrés **hurt himself**.*

> Lucía **se** despertó.
> *Lucía **woke up**.*

You can also use most of these verbs nonreflexively.

nonreflexive

reflexive *matches*

> Desperté a mi hermanito a las siete y media.
> *I woke my little brother up at seven thirty.*

> Me desperté a las siete y media.
> *I woke up at seven thirty.*

Remember that the **reflexive pronoun** and the **verb** always match.

You will often use **reflexive verbs** to refer to:
• emotions • feelings • reactions.

When using a **reflexive verb**, put the **reflexive pronoun** before the **conjugated verb**.

> ¿Cuándo **se levantó** Marcos?
> *When **did Marcos get up**?*

> Todavía no **se ha levantado**.
> *He **hasn't gotten up** yet.*

When you use a reflexive verb in the infinitive, you can put the **reflexive pronouns** either:

• before the **conjugated verb** …

> No **te debes** preocupar.
> *You shouldn't worry.*

• or attach it to the end of the infinitive.

> No **debes** preocupar**te**.

Vocabulario

Las emociones

animarse *to become encouraged, get interested*

dedicarse a *to apply oneself to something*

desanimarse *to get discouraged*

entusiasmarse *to get excited*

oponerse a *to oppose*

ponerse nervioso(a) *to get nervous*

sentirse (e→ie) frustrado(a) *to feel frustrated*

> ¿Cuándo te sientes así?

Practice: **Actividades** 5 6 7 **Más práctica** *cuaderno pp. 35–36*
Para hispanohablantes *cuaderno pp. 33–34*

Online Workbook
CLASSZONE.COM

5 Un día desastroso

Escuchar/*Escribir* ¡Manolo tuvo un día horroroso! ¿Qué le pasó? Escucha y escribe oraciones que describen su día desastroso.

modelo

Manolo / despertarse

Manolo se despertó muy tarde.

1. Manolo / acostarse
2. Manolo / levantarse
3. la hermanita de Manolo / ducharse
4. Manolo / secarse el pelo
5. Manolo / vestirse
6. Manolo / ponerse calcetines

6 Un sábado en Los Ángeles

Hablar/*Escribir* Beatriz describe un sábado en su casa cuando todos tuvieron que ayudar con la limpieza.

modelo

Yo tuve que planchar la ropa. __Me aburrí__ *al planchar la ropa. (aburrirse)*

1. Mamá entró a la casa. _____ al ver la casa tan sucia. (desanimarse)
2. Nuestros primos vinieron a ayudarnos. Nosotros _____ cuando los vimos. (animarse)
3. Mi hermano cortó el césped. _____ mucho. (cansarse)
4. Mi hermana no nos ayudó. Papá _____ con ella. (enojarse)
5. Yo desenchufé la computadora. Mamá _____ cuando vio que la computadora no funcionaba. (ponerse nerviosa)
6. Natalia fue la primera que terminó con sus quehaceres. _____ después de acabarlos. (divertirse)

7 ¿Cómo te sentiste?

STRATEGY: SPEAKING

Express feelings You already know many adjectives like **contento(a), nervioso(a), tímido(a)** to express feelings. Add to your "emotional" vocabulary with verbs or their past participles like **cansarse: me cansé, estaba cansado(a), me sentí cansado(a).** What other verbs express how you feel on different occasions?

Hablar Pregúntale a tu compañero(a) cómo se sintió en estas ocasiones. Luego, cambien de papel.

modelo

Tú: *¿Cómo te sentiste cuando sacaste mala nota en álgebra?*

Compañero(a): *Pues me sentí muy frustrado(a) porque estudié mucho y no pensé que iba a sacar mala nota.*

> sacar mala nota en...
> tener examen final en...
> salir con tus amigos...
> perder el partido de ...
> ¿...?

Nota cultural: Understand information about social issues

GRAMÁTICA Reflexive Verbs Used Reciprocally

▶ You can also use **reflexive verbs** to express the idea of *each other*.

Alicia y **yo nos conocemos** muy bien.
Alicia and I know each other very well.

Mis hermanitos se pelean mucho.
My little brothers fight with each other a lot.

Ustedes deben **ayudarse**.
You ought to help each other.

▶ You can also add the phrase **el uno al otro (la una a la otra)** to emphasize the reciprocal meaning:

Mauricio y **Sara se saludaron.**
Mauricio and Sara said hello.

To each other is implied, but not stated.

Mauricio y **Sara se saludaron el uno al otro.**
Mauricio and Sara said hello to each other.

Vocabulario

Las interacciones

apoyarse *to support each other*

ayudarse *to help each other*

conocerse bien/mal
to know each other well/not very well

contarse (o→ue) secretos/chismes
to tell each other secrets/gossip

llevarse bien/mal (con) *to get along well/badly (with)*

odiarse *to hate each other*

pelearse/no pelearse frecuentemente
to fight/not to fight often

perdonarse *to forgive each other*

quejarse *to complain*

saludarse *to greet, say hello to each other*

telefonearse *to phone each other*

▶ ¿Con quién haces estas cosas?

Practice: **Actividades**   **Más práctica** *cuaderno pp. 37–38*
Para hispanohablantes *cuaderno pp. 35–36*

Online Workbook
CLASSZONE.COM

Apoyo para estudiar

Reflexive verbs used reciprocally

Read these examples of reciprocal use: **Tú y yo nos ayudamos el uno al otro. Tú y Mauricio se entienden bien. Nuestras hermanas no se conocen bien.** Why will reciprocal verbs always have a plural ending and never a singular ending?

8 Se llevan muy bien

Leer/*Escribir* Tienes un(a) amigo(a) que es muy chismoso(a). Lee sus descripciones de varias personas en la primera columna. Luego, escoge la frase de la segunda columna que mejor complete su descripción.

1. Paco y Lola son amigos desde el primer grado.
2. Chela y Juan siempre andan juntos.
3. Mari y Ana hablan todas las noches, no importa dónde están.
4. Pepe y José se mandan cartas a menudo.
5. Pedro y Alicia se hablan cuando se ven en la calle.
6. A Nando y a Berta les gusta hablar de lo que pasa en sus vidas.
7. Tomás y Chepa se pelean, pero luego se piden disculpas.
8. Quique y Rosa siempre están de acuerdo.

a. Se telefonean.
b. Se conocen muy bien.
c. No se pelean.
d. Se escriben.
e. Se saludan.
f. Se llevan muy bien.
g. Se perdonan.
h. Se cuentan todo.

9 Mi padrino ♻

Escribir Tu padrino es el mejor amigo de tu papá. Para saber cómo era su relación cuando eran jóvenes, completa las oraciones de tu papá con el imperfecto del verbo entre paréntesis.

modelo
Nos apoyábamos en momentos difíciles. (apoyarse)

1. _____ con todo. (ayudarse)
2. _____ muy bien. (entenderse)
3. _____ secretos. (contarse)
4. _____ muy bien. (conocerse)
5. _____ todos los días. (hablarse)
6. _____ muy bien. (llevarse)
7. No _____ frecuentemente. (pelearse)
8. Y cuando nos peleábamos, _____. (perdonarse)
9. _____ para ir a pasear. (telefonearse)
10. _____ si no podíamos vernos. (quejarse)

Nota cultural

El compadrazgo Una persona se considera el compadre o la comadre de una familia al compartir la responsabilidad de criar *(to raise)* a un niño o una niña con los padres. El compadre o la comadre es

el padrino o **la madrina** del (de la) niño(a), que se llama **ahijado(a)**. Generalmente, los padrinos van a todas las celebraciones importantes de su ahijado(a), incluso la graduación y la boda.

Activity 11: Use strings of related sentences when speaking

10 Entre nosotros

Hablar/*Escribir* Tú y tu compañero(a) comentan sobre las relaciones entre varios amigos y compañeros de clase. ¿Qué dicen?

modelo

Están en comunicación constante. (telefonearse)

Tú: *¿Sabías que Juan y Olga están en comunicación constante?*

Compañero(a): *Sí, se telefonean todos los días.*

I. A veces se pelean por algo muy tonto. (perdonarse)

2. Nos conocemos desde muy pequeñas. (entenderse)

3. Son amigos por correspondencia de Internet. (escribirse)

4. Casi nunca se hablan. (odiarse)

5. No somos buenos amigos. (llevarse mal)

6. Salen juntos los fines de semana. (verse)

7. No vemos las cosas de la misma manera. (pelearse)

8. Cuando necesito ayuda, yo lo llamo a él. (apoyarse)

11 Las relaciones

Hablar/*Escribir* Las relaciones son muy importantes en la vida. ¿Cómo se sienten algunas personas en tu vida hacia otras? Di cómo se sienten cuatro pares de personas y luego explica por qué se sienten así.

modelo

Mi prima y yo nos entendemos muy bien. Somos de la misma edad.

Mis hermanos se pelean frecuentemente. Siempre quieren usar la computadora al mismo tiempo.

nombre y yo
nombre y nombre

ayudarse
entenderse
escribirse
conocerse bien/mal
pelearse/no pelearse
 frecuentemente
quererse
telefonearse
¿ … ?

More Practice: Más comunicación *p. R5*

Nota cultural

Sammy Sosa es un beisbolista dominicano. Al igual que Mark McGwire, se ha destacado por romper récords y anotar un sinnúmero de jonrones. A pesar de jugar en diferentes equipos, los dos son buenos amigos. Sammy Sosa además creó la Fundación que lleva su nombre para ayudar a los niños necesitados de Chicago y de la República Dominicana. En 1999 fue reconocido como el Hombre del Año durante los Players Choice Awards.

▶ You can use the pronoun **se** in order to avoid specifying the person who is doing the action of the **verb.**

For example, when you say:

Se alquila apartamento.

*Apartment **for rent.***

you are indicating that *someone* is renting an apartment, but that you either don't know who that person is or don't choose to identify him or her.

....

▶ When you use **se**, the verb is always in the third person.

* If the noun that follows the verb is singular, the verb is in the **él/ella** form.

singular

Aquí **se habla** español.

*Spanish **is spoken** here.*

¿Cómo **se apaga** la aspiradora?

*How **do you** (do we, does one) **turn off** the vacuum cleaner?*

* If the noun that follows the verb is plural, you use the **ellos/ellas** form of the verb.

plural

Aquí **se reparan** carros.

*Cars **are repaired** here.*

....

▶ You can use this construction with **se** in all tenses. For example:

Se hizo mucho.

*A lot **was done.***

Se había hecho mucho.

*A lot **had been done.***

Se hará mucho.

*A lot **will be done.***

Practice: **Actividades**   **Más práctica** *cuaderno pp. 39–40*   **Online Workbook**
12 13 14   **Para hispanohablantes** *cuaderno pp. 37–38*   CLASSZONE.COM

Activity 13: Understand most spoken language when the message is deliberately and carefully conveyed by a speaker accustomed to dealing with learners when listening

12 Chicago

Escribir Cuando caminas por el barrio (*neighborhood*) Pilsen de Chicago, ves los siguientes letreros. Completa los letreros.

se venden
CARROS USADOS

modelo

vender

APARTAMENTO DE DOS HABITACIONES

COMPUTADORAS DE TODO TIPO

1. alquilar

2. reparar

PERSONAS BILINGÜES

SECRETARIA EJECUTIVA

3. necesitar

4. buscar

INGLÉS, ESPAÑOL Y FRANCÉS

clases de baile

5. hablar

6. ofrecer

electrodomésticos

comida típica

7. arreglar

8. servir

13 Mundo de Autos

Escuchar/*Escribir* Escucha el anuncio de la radio para Mundo de Autos. Luego, di si las siguientes oraciones son ciertas o falsas. Si la oración es falsa, corrígela.

1. En Mundo de Autos, sólo se venden carros nacionales.

2. Se ofrecen precios baratos.

3. La rebaja se termina en dos meses.

4. En Mundo de Autos sólo se habla inglés.

5. Se dice que los vendedores de Mundo de Autos son los mejores de la ciudad.

6. En Mundo de Autos se reparan carros.

7. No se reparan carros en el mismo día.

8. También se habla ruso y francés.

9. En Mundo de Autos se venden carros usados.

10. Se invita a la gente a ir a Mundo de Autos la semana que viene.

Activity 15 brings together all concepts presented.

14 Turista

Hablar/Escribir Estás en Los Ángeles en un hotel donde se habla español. Le preguntas al (a la) recepcionista varias cosas. Dramatiza la conversación con un(a) compañero(a) y luego cambien de papel.

modelo

¿cómo? / llegar al banco desde el hotel

Tú: ¿Cómo se llega al banco desde el hotel?

Compañero(a): Salga del hotel y doble a la derecha…

1. ¿a qué hora? / servir el desayuno
2. ¿a qué hora? / abrir el gimnasio
3. ¿a qué hora? / abrir las tiendas
4. ¿a qué hora? / cerrar las tiendas
5. ¿dónde? / comprar discos compactos mexicanos
6. ¿dónde? / alquilar videos en español
7. ¿dónde? / escuchar música mexicana tradicional
8. ¿dónde? / cambiar cheques de viajero

15 DiscoLandia

Hablar En grupos de tres o cuatro, hablen sobre las tiendas en su ciudad. Expresen sus opiniones.

modelo

Tú: ¿Se venden discos compactos buenos en Música Moderna?

Amigo(a) 1: No. Se venden discos muy tradicionales allí.

Tú: ¿En qué tienda te gusta comprar discos compactos?

Amigo(a) 2: En DiscoLandia.

Tú: ¿Te prestas los discos con tus amigos?

Amigo(a) 3: Sí, nos los prestamos.

More Practice:
Más comunicación p. R5

 **Online Workbook** CLASSZONE.COM

Nota cultural

Los Ángeles Una de las comunidades latinas más importantes de Estados Unidos se encuentra en el Este de Los Ángeles. La población de **Greater Eastside** es mayormente de origen mexicano. En las décadas de los setenta y ochenta, el Eastside creció mucho cuando vinieron muchos inmigrantes no sólo de México, sino también de países de Centroamérica como El Salvador, Guatemala y Nicaragua. En esta comunidad también viven grupos del Medio Oriente y de Asia. Es actualmente una de las zonas manufactureras e industriales más importantes de todo el país.

Refrán

La ocupación constante previene las distracciones.

¿Qué te sugiere el refrán sobre las veces cuando sientes aburrimiento? ¿Qué debes hacer para no aburrirte?

AUDIO

En voces
LECTURA

PARA LEER
STRATEGY: READING
Chart contrasts between dreams and reality in a personal narrative Maintaining a balance between dreams and reality is an important part of our growth. Do you have in your imagination a dream house or a dream room? How would you describe it? Set up two charts to compare the author's **casa de sus sueños** *(dreams)* and **la casa de Mango Street**. Consider each one from these points of view:

Interior	Exterior
Habitaciones	Patio
Tamaño	Jardín
Otro	Otro

What do both houses have in common?

LAS CASAS

la cerca	marca el límite de una propiedad
el escalón	parte de una escalera
el agua corriente	agua en casa
el ladrillo	material de construcción
mudarse	cambiar de casa
el pasto	el césped
el tubo	conductor para agua en forma de cilindro

Sobre la autora

Sandra Cisneros, la autora de *La casa en Mango Street*, nació en Chicago en 1954. Escribe ficción y poesía y vive en San Antonio, Texas.

Introducción

*L*a casa en Mango Street es una novela que narra las experiencias de Esperanza Cordero, una chica que vive en un barrio latino de Chicago. Ella quiere tener una casa y escribir cuentos. En prosa sencilla y colorida, Sandra Cisneros describe los pensamientos de esta joven. Elena Poniatowska, una escritora mexicana famosa, tradujo esta selección del inglés al español.

La casa en Mango Street

Siempre decían que algún día nos mudaríamos a una casa, una casa de verdad, que fuera nuestra para siempre, de la que no tuviéramos que salir cada año, y nuestra casa tendría agua corriente y tubos que sirvieran[1]. Y escaleras interiores propias como las de la tele. Y tendríamos un sótano, y por lo menos tres baños para no tener que avisarle[2] a todo el mundo cada vez que nos bañáramos. Nuestra casa sería blanca, rodeada[3] de árboles, un jardín enorme y el pasto creciendo sin cerca. Ésa es la casa de la que hablaba Papá cuando tenía un billete de lotería y ésa es la casa que Mamá soñaba[4] en los cuentos que nos contaba antes de dormir.

Pero la casa de Mango Street no es de ningún modo como ellos la contaron. Es pequeña y roja, con escalones apretados al frente y unas ventanitas tan chicas que parecen guardar su respiración. Los ladrillos se hacen pedazos[5] en algunas partes y la puerta del frente se ha hinchado[6] tanto que uno tiene que empujar[7] fuerte para entrar. No hay jardín al frente sino cuatro olmos[8] chiquititos que la ciudad plantó en la banqueta.

Afuera, atrás hay un garaje chiquito para el carro que no tenemos todavía, y un patiecito que luce[9] todavía más chiquito entre los edificios de los lados. Nuestra casa tiene escaleras pero son ordinarias, de pasillo, y tiene solamente un baño. Todos compartimos recámaras, Mamá y Papá, Carlos y Kiki, yo y Nenny.

[1] that work	[6] has swollen
[2] to announce	[7] to push
[3] surrounded	[8] elms
[4] dreamed about	[9] appears
[5] to fall apart	

¿Comprendiste?

1. ¿Cómo es la casa que imaginaba Esperanza?
2. ¿Cómo es la casa de Mango Street?
3. ¿Dónde vivió Esperanza antes de mudarse a Mango Street?
4. ¿Quiénes son los miembros de la familia Cordero?

¿Qué piensas?

En tu opinión, ¿qué sentimientos tiene la narradora hacia la casa de Mango Street? Explica tus razones.

Hazlo tú

Eres novelista. Escribe los tres primeros párrafos de tu novela. Narra cómo es tu vida en la casa donde vives ahora y cómo era tu vida en la casa donde vivías antes. Luego describe a tu familia y habla un poco de tus planes e ilusiones. ¡Puedes escribir una combinación de autobiografía y ficción!

 Understand and convey information about music with an emphasis on significant people and events

En colores
CULTURA Y COMPARACIONES

PARA CONOCERNOS

STRATEGY: CONNECTING CULTURES

Interview, report, and value musical influences

Do you know someone who has his or her own musical group? Interview that person. Take notes about instruments, influences, and the type of music they play. After reading «En colores» fill in the same information about Tito Puente.

Músico	T. Puente
Instrumentos	
Influencias	
Tipo de música	
Comentario	

El legendario rey del *mambo*

Tito Puente pasó casi medio siglo[1] uniendo[2] diferentes estilos y culturas musicales y produciendo música de constante calidad. Fue un experto compositor y autor de arreglos[3] musicales como los clásicos «Ran Kan Kan» y «Mambo diablo». Fue también un excelente saxofonista y un experto ejecutante de muy distintos instrumentos de percusión. Además fue el mejor timbalero del mundo, el patriarca de los timbales, esos tambores[4] metálicos cubanos que son el alma[5] de muchos ritmos latinos.

---

[1] century
[2] uniting
[3] arrangements
[4] drums
[5] soul

Ernest Anthony Puente Jr. nació en 1923 en la ciudad de Nueva York, hijo de puertorriqueños recién emigrados a los Estados Unidos. En East Harlem, «El Barrio» de la ciudad de Nueva York, «Ernestito» creció[6] en un mundo de boleros y rumbas que se mezclaban[7] con los grandes conjuntos de swing de la época y la creciente tendencia a la improvisación en el jazz.

Con el tiempo fue reclutado por la Marina[8] de Estados Unidos y esto le dio algunos beneficios muy positivos. Una de las decisiones más acertadas[9] de su vida fue aprovechar[10] la ley de ayuda a los veteranos de guerra[11], que le permitió asistir a la Escuela de Música Juilliard, de Nueva York. Allí estudió teoría, orquestación y dirección.

Tito Puente Jr., el hijo de Tito Puente, también es músico. Sus versiones de «Oye como va» y «Azúcar» se oyen en discos alrededor del mundo.

La carrera de Tito Puente fue una larga lista de éxitos[12]. Además, Tito Puente les dio oportunidades a los jóvenes, otorgándoles becas para continuar sus estudios de música. En esos jóvenes músicos Tito Puente vio la continuación de la forma de arte que él comenzó. Este músico genial falleció en la ciudad de Nueva York el 1 de junio del 2000.

[6] grew up	[9] right	[12] successes
[7] mixed	[10] to take advantage of	[13] scholarships
[8] Navy	[11] war	

CLASSZONE.COM
More About Latinos

¿Comprendiste?

1. ¿Qué instrumentos musicales tocaba Tito Puente?
2. ¿De dónde era Tito Puente? ¿Y su familia?
3. ¿Dónde estudió Puente música? ¿Qué clases tomó?

¿Qué piensas?

1. ¿Cómo contribuyó la vida de El Barrio a la formación musical de Tito Puente?
2. Para ti, ¿cuáles son los aspectos más admirables de la vida de Tito Puente?

Hazlo tú

Busca información sobre un(a) músico que te guste. Escribe un breve artículo sobre la vida y la carrera del músico usando como modelo el artículo sobre Tito Puente.

The best of Tito Puente
El Rey Del Timbal!

ETAPA 3

En uso
REPASO Y MÁS COMUNICACIÓN

OBJECTIVES
- Talk about household chores
- Say what friends do
- Express feelings

Now you can...

- talk about household chores.

To review

- vocabulary for chores see p. 76.

---

1 ¡Ayúdame!

Tu amigo(a) está en ciertas situaciones. Ayúdalo a resolver los problemas.

modelo

La lámpara no funciona.

Tienes que cambiar la bombilla.

1. Tuvimos una fiesta ayer y hoy hay mucha basura.
2. Me voy a volver loco. El teléfono ha estado sonando *(ringing)* todo el día.
3. El televisor no enciende. No sé por qué.
4. Mira este gabinete. ¡Qué desastre!
5. El jardín está lleno de malas hierbas.
6. Tengo que cortar el césped, pero el cortacésped no funciona.
7. ¡Las plantas están muy secas! No ha llovido en dos semanas.

Now you can...

- express feelings.

To review

- reflexive verbs see p. 84.

---

2 Un día en la casa de Alma

Alma describe cómo se sintieron varias personas de su familia el otro día. Usa los verbos de la lista.

aburrirse   cansarse   desanimarse   entusiasmarse   sentirse frustrado(a)

animarse   divertirse   enojarse   ponerse nervioso   preocuparse

modelo

Mi hermanito no tenía nada que hacer. (él)   *Se aburrió.*

1. Papá hizo quehaceres todo el día. (él)
2. La profesora me dijo que había hecho un buen trabajo. (yo)
3. Por la tarde, fuimos a ver una película muy divertida. (nosotros)
4. Ricardo no hizo lo que le pidió mamá. (mamá)
5. Mi mamá tenía mucho trabajo y la computadora no funcionaba. (ella)
6. Mi hermano estudió mucho y de todas maneras sacó una mala nota. (él)
7. Cuando papá dijo que íbamos a salir, nos dio mucho gusto. (nosotros)

Now you can...

• say what friends do.

To review

• reflexives used as reciprocals see p. 86.

3 La telenovela

Una telenovela le encanta a tu abuela y ella te cuenta sobre los personajes. Completa su descripción con los verbos de la lista. Se puede repetir el verbo.

pelearse	conocerse	hablarse	escribirse
telefonearse	odiarse	perdonarse	saludarse
contarse chismes	entenderse	llevarse muy mal	quererse
	verse		ayudarse

Érica y Olivia siempre __1__ cuando tienen problemas. __2__ con todo, con el trabajo de la casa y de la oficina. __3__ muy bien porque han sido amigas desde pequeñas. Casi nunca __4__. Cuando __5__, __6__ de todos los vecinos.

Hay un vecino que no les gusta para nada. Se llama José. José y Érica __7__. __8__ frecuentemente sobre cosas muy tontas. Un día __9__ porque el perro de José corrió por el jardín de Érica y destruyó todas las plantas.

Olivia y José también __10__. Unos años antes, Olivia y José eran novios en la universidad y __11__ mucho. Entonces __12__ todos los días por Internet. __13__ muy bien porque los dos eran atletas y pasaban los días entrenándose. Ahora no __14__, imagínate, ¡ni una palabra! No __15__ desde que rompieron como novios. __16__ muy poco, y cuando se ven, ni __17__. ¡Ay, la juventud!

Now you can...

• say what friends do.

To review

• impersonal constructions with **se** see p. 89.

4 Se hace así

Tu amigo(a) quiere saber cosas sobre tus vacaciones con tu primo(a). ¿Qué te pregunta y cómo le contestas?

modelo

¿a qué hora? / servir el almuerzo (a las doce)

Tú: *¿A qué hora se sirve el almuerzo en la casa de tu primo?*

Compañero(a): *En su casa se sirve el almuerzo a las doce.*

1. ¿a qué hora? / apagar el televisor (a las ocho)
2. ¿cuándo? / hacer los quehaceres (los domingos)
3. ¿cuáles? / hablar idiomas (inglés y español)

4. ¿qué clase? / escuchar música (música clásica)
5. ¿cuándo? / hacer la tarea (temprano)

6. ¿a qué hora? / servir la cena (a las siete)

5 ¿Qué te anima?

STRATEGY: SPEAKING

Identify feelings important in a friendship
Before beginning your interview, identify for yourself those feelings you consider important in a friend. Then review the vocabulary for talking about friendship. Finally, write down the questions you really want to ask to know your classmates better, then conduct your interviews.

Quieres saber qué inspira ciertos sentimientos en tus amigos. Escribe una lista de preguntas para conocerlos mejor. Luego, haz las preguntas a cuatro o cinco compañeros. Anota sus respuestas y haz un resumen de los resultados para la clase.

modelo

Preguntas
¿Qué te pone nervioso(a)? ¿Qué te aburre? ¿Qué te anima?

Resultados
Cuatro personas se ponen nerviosas cuando tienen un examen. Tres personas se aburren en la clase de biología. Dos personas se animan cuando sacan buenas notas.

6 ¡Se odian!

Trabajando en grupos, inventen cuatro personajes que viven en un pueblito y que tienen muchos sentimientos entre sí. Expliquen por qué se sienten así el uno hacia el otro. Escriban por lo menos seis frases que describan estas relaciones complicadas.

modelo

Marcos y Marcelo se odian porque los dos quieren a Gloria. Marcos y Marcelo se pelean frecuentemente.

7 En tu propia voz

ESCRITURA ¡Vas a empezar tu propia compañía! Primero decide qué clase de compañía es y qué servicios va a ofrecer. Dale un nombre a tu compañía y luego escribe un anuncio para poner en el periódico.

modelo

La compañía DiseñoNet

Se ofrecen clases de Internet.

Se diseñan páginas iniciales.

TÚ EN LA COMUNIDAD

La'Donna es alumna en New Jersey. Trabaja de voluntaria en un hospital. Ella habla español con los pacientes para saber si tienen hambre, cómo llegaron al hospital, cuántos años tienen, si viven solos, etc. También habla español con su tía y a veces con sus amigos.

En resumen
REPASO DE VOCABULARIO

Flashcards
CLASSZONE.COM

TALK ABOUT HOUSEHOLD CHORES

Tasks

conectar	to connect
desarmar	to take apart
desconectar	to disconnect, to turn off
desenchufar	to unplug
desyerbar	to weed
encender (e→ie)	to turn on
enchufar	to plug in
esconderse	to hide
regar (e→ie)	to water
reparar	to repair
vaciar	to empty

Objects

el basurero	trash can, wastebasket
la bombilla	lightbulb
el cortacésped	lawnmower
desorganizado(a)	disorganized
el desván	attic
el gabinete	cabinet
las malas hierbas	weeds
el sótano	basement

EXPRESS FEELINGS

animarse	to get encouraged, interested
dedicarse a	to apply oneself to something
desanimarse	to get discouraged
entusiasmarse	to get excited
oponerse a	to oppose
ponerse nervioso(a)	to get nervous
sentirse (e→ie) frustrado(a)	to feel frustrated

OTHER WORDS

 Ya sabes

armar	to assemble
organizado	organized
el televisor	television set

SAY WHAT FRIENDS DO

apoyarse	to support each other
ayudarse	to help each other
conocerse bien/mal	to know each other well/ not very well
contarse (o→ue) chismes	to tell each other gossip
contarse secretos	to tell each other secrets
llevarse bien/mal con	to get along well/badly (with)
odiarse	to hate each other
pelearse/no pelearse frecuentemente	to fight/not to fight often
perdonarse	to forgive each other
quejarse	to complain
saludarse	to greet, to say hello to each other
telefonearse	to phone each other

Impersonal *se*

Se habla español.
Se venden libros.

Juego

¡Tanto que hacer!

Escribe los quehaceres indicados. Luego pon en orden las letras en los círculos para saber cuál es la última cosa que todavía queda por hacer.

1. __ ◯ __ __ __ ◯ __ EL CORTACÉSPED.
2. ◯ __ __ __ __ __ __ __ ◯ __ EL JARDÍN.
3. __ __ __ __ ◯ __ ◯ EL BASURERO.

Create paragraphs
when writing

En tu propia voz

ESCRITURA

Presentaciones personales

Vas a representar a tu club de español en una reunión nacional de clubes en Estados Unidos. Cada participante debe mandar una foto y preparar una descripción personal para presentarse ante los otros delegados. Tienes que escribir una descripción personal para el programa.

Función: Describirse a sí mismo

Contexto: Presentarse ante los delegados de la reunión

Contenido: Información sobre tu personalidad e intereses

Tipo de texto: Cuadro personal

HABLA
Conferencia anual
de estudiantes de español

Los Ángeles, California
14 al 17 de marzo

¡Bienvenido a HABLA!

HABLA es una conferencia para que los estudiantes de español practiquen, aprendan y disfruten juntos al compartir sus experiencias de aprendizaje y las excursiones y eventos de HABLA, nuestra organización.

Agradecemos al alcalde y la ciudad de Los Ángeles por brindarnos su hospitalidad.

PARA ESCRIBIR • STRATEGY: WRITING

Use details to enrich a description Include in your profile a complete description with specific details and facts that communicate personality, interests, and activities. Each paragraph should begin with a clear topic sentence supported by facts and details. Others want information that shows your uniqueness!

Modelo del estudiante

The writer included a **specific anecdote** to support his more generalized statement.

Each paragraph is introduced by a **concise topic sentence** that defines its content.

The author indicates a **personal preference** to add depth to his description and to support the topic sentence.

Mi descripción personal

● ¡Hola! Mi nombre es Javier Gutiérrez, pero mis parientes y amigos me llaman «Hooper» porque me encanta el baloncesto. Me puse un poco nervioso durante los partidos importantes, pero ¡todo resultó bien el año pasado cuando mi equipo ganó el campeonato del estado!

● Además de participar activamente en el club de español, soy miembro del club de ajedrez y toco el violín en la orquesta.
● Me encanta la música clásica, especialmente la de Beethoven, mi compositor favorito.

Aunque tengo muchos intereses académicos y extracurriculares, lo que más me importa es mi relación con mi familia y mis amistades. El verano pasado mis amigos y yo fuimos a las montañas para acampar. Nos divertimos pescando, haciendo montañismo y volando en planeador. Nos sentimos más unidos por la experiencia y yo los aprecio aun más que antes.

**Language Arts
Writing Standard 2.1b**
Write fictional, autobiographical,
or biographical narratives. Locate
scenes in specific places.

Estrategias para escribir

Antes de escribir …

Para crear una descripción completa, es necesario pensar en los
detalles que debes incluir para hacerla más específica. Haz una
lista de categorías, usando la lista de abajo como ejemplo. Luego
usa tu lista para crear una «red de palabras» como la de la
derecha. Usa el gráfico para organizar tu información en
párrafos lógicos con datos interesantes y reveladores.

Música:	Me gusta la música clásica
Deportes:	Mis favoritos son…
Comida:	Prefiero comer…
Características:	Soy… pero no soy…

Revisiones

Después de escribir el primer borrador *(draft)*, compártelo
con un(a) compañero(a) de clase. Pregúntale:

- *¿Qué otras ideas o datos específicos debo incluir?*
- *¿Cómo puedo mejorar la organización de las ideas?*
- *¿Cómo refleja la descripción mi personalidad,
  intereses y actividades?*

La versión final

Antes de crear tu versión final, léela de nuevo y
repasa los siguientes puntos:

- *¿Usé **ser** y **estar** correctamente?*

Haz lo siguiente: Subraya *(Underline)* estos verbos
y verifica que escogiste el verbo correcto y que la
forma concuerda *(agrees)* con el sujeto.

- *¿Están correctas las formas de los verbos **gustar,
  encantar, interesar, importar** y otros parecidos?*

Haz lo siguiente: Subraya el sustantivo *(noun)* y el verbo
para ver si concuerdan. Haz un círculo alrededor del
pronombre para ver si se refiere a la persona correcta.

> **PROOFREADING SYMBOLS**
>
> ∧ Add letters, words,
> or punctuation
> marks.
>
> ∼ Switch the
> position of letters
> or words.
>
> ≡ Capitalize a letter.
>
> ⌐ Take out letters
> or words.
>
> / Make a capital
> letter lowercase.

> Me llamo Martina Ibañez.
> Soy ~~Estoy~~ de Denver, Colorado y soy
> estudiante. ⓜe gustan la nieve y gusta
> ⓜe encanta esquiar. ~~Soy~~ Estoy muy
> contenta cuando estoy en
> las montañas...

UNIDAD 2

STANDARDS

Communication
- Saying what you want to do
- Making requests and suggestions
- Saying what should be done
- Reacting to ecology and nature
- Reacting to others' actions
- Expressing doubt
- Relating events in time

Cultures
- Influential people from Mexico and Central America
- Literacy in the Spanish-speaking world
- Volunteer opportunities
- Natural reserves in Costa Rica

Connections
- Science: Charting recycling efforts
- Science: Promoting the preservation of rainforests

Comparisons
- Ethnic groups in Central America and in the U.S.
- Literacy rates in the Spanish-speaking world and in the U.S.
- Ecotourism in the Spanish-speaking world and in the U.S.

Communities
- Using Spanish in volunteer activities
- Using Spanish with other students

INTERNET Preview

CLASSZONE.COM
- More About Mexico and Central America
- Webquest
- Self-Check Quizzes
- Flashcards
- Writing Center
- Online Workbook
- eEdition Plus Online

¡EL MUNDO ES NUESTRO!

MÉXICO
MÉXICO, D.F.
★

GUATEMALA
GUATEMALA
★

HONDURAS
TEGUCIGALPA
★

NICARAGUA
★ MANAGUA

SAN SALVADOR ★
EL SALVADOR

SAN JOSÉ
COSTA RICA
★

PANAMÁ
PANAMÁ
★

MÉXICO
MARÍA IZQUIERDO (1902–1955)
Es una de las artistas más importantes de México. Sus pinturas tratan costumbres y escenas rurales. ¿Qué temas de la comunidad se ven en esta pintura?

ALMANAQUE CULTURAL

POBLACIÓN: México: 101.879.171, Centroamérica: 37.240.008

ALTURA: 4211m sobre el nivel del mar, Volcán Tajumulco, Guatemala (punto más alto)

CLIMA: (más alta) 82°F (28°C) Panamá, Panamá, (más baja) 63°F (18°C) Ciudad de México, México

COMIDA TÍPICA: Tamales, pupusas, cebiche

GENTE FAMOSA DE MÉXICO Y CENTROAMÉRICA:
María Izquierdo (artista), Óscar Arias (político), Juan José Arreola (escritor), Rigoberta Menchú (activista)

 VIDEO DVD Mira el video para más información.

 CLASSZONE.COM
More About Mexico and Central America

RUINAS DE COPÁN

HONDURAS
RUINAS DE COPÁN En el siglo XX, la O.N.U *(U.N.)* declaró que las Ruinas de Copán son «patrimonio universal de la humanidad». ¿Por qué crees que esta declaración es importante?

CENTROAMÉRICA
¡PROTEGE LA SELVA TROPICAL! Animales como éste y árboles bellos viven en la selva tropical. ¿Cómo puedes protegerlos?

¡Protege la selva tropical!

COSTA RICA
ÓSCAR ARIAS ganó un premio por trabajar para proteger el medio ambiente. ¿Qué otras personas famosas protegen el medio ambiente?

GUATEMALA
TEJIDOS GUATEMALTECOS
Los tejidos de Guatemala se hacen desde hace cientos de años. ¿Qué aspecto de la cultura indígena crees que represente este tejido?

MÉXICO Y CENTROAMÉRICA
CEBICHE MIXTO El cebiche es un plato de camarones y pescado en salsa de limón. Mira el mapa. ¿Por qué crees que el cebiche es popular en México y Centroamérica?

103

¡EL MUNDO ES NUESTRO!

- Comunicación
- Culturas
- Conexiones
- Comparaciones
- Comunidades

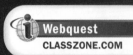
Webquest
CLASSZONE.COM

Explore communication in Mexico and Central America through guided Web activities.

Comunicación en acción Estas personas están discutiendo un tema importante. ¿Crees que están de acuerdo? ¿Por qué?

Comunicación

¿Te gusta expresar tus opiniones? ¿Cómo reacciones al oír opiniones opuestas a las tuyas? ¿Te interesa hacer sugerencias y buscar soluciones a los problemas del mundo?

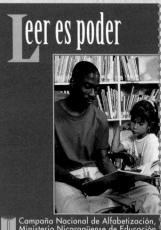

Leer es poder

Campaña Nacional de Alfabetización, Ministerio Nicaragüense de Educación

Comunidades

En esta unidad hablaremos de nuestras oportunidades para servir a la comunidad. Conoceremos a un alumno de Washington que sirve a la comunidad usando el español en su trabajo de voluntario. ¿Cómo ayudan los alumnos de las fotos?

EL RIO ES FUENTE DE VIDA
NO CONTAMINE

joven, ¡participa!
concierto de rock al aire libre
El grupo internacional de artistas
voces del mundo
presenta un concierto para hacer de la Ciudad de México un mejor lugar para vivir y unir la comunidad.

¿Cuándo?
El sábado 20 de junio a las 8:00

¿Dónde?
Ciudad de México, D.F.

nos vemos allí

Comparaciones

En tu opinión, ¿cómo se comparan las preocupaciones de los jóvenes hispanos con las de los jóvenes estadounidenses? ¿Son semejantes o diferentes? ¿Expresan sus opiniones de la misma manera?

Culturas

México y Centroamérica representan su cultura con muchos colores. ¿Piensas que eso está relacionado con su alegría o con la naturaleza? Explica tu opinión.

Conexiones

El español te ayuda a investigar y presentar información relacionada con las ciencias naturales. ¿Qué sabes ya de las selvas tropicales?

Fíjate

Observa las fotos de estas dos páginas. ¿Puedes identificar algunos de los problemas del mundo que vamos a tratar en esta unidad? Haz una lista.

ETAPA

1

Pensemos en los demás

OBJECTIVES

- Say what you want to do

- Make requests

- Make suggestions

¿Qué ves?

Mira la foto. Contesta las preguntas.

1. ¿Desde qué punto de vista piensas que se tomó esta foto?

2. ¿Qué clase de evento crees que es éste?

3. ¿Cómo crees que se conecta con la comunidad?

4. Lee el póster. ¿Conoces a otros artistas que hagan lo mismo?

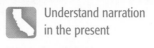
EL CIUDADANO
el periódico estudiantil

Participa en la campaña para embellecer la ciudad. Puedes…

recoger basura

sembrar árboles

juntar fondos

Expresa tu opinión. ¡Vota!

¿Estás a favor de…?   ¿Estás en contra de…?

Sí
- comprar uniformes nuevos para el equipo de fútbol
- tener una fiesta a fines del año

No
- construir un parque pequeño al lado de la escuela
- ayudar a eliminar la pobreza

¿Cómo vamos a usar el dinero que juntamos este año?

¡Mantengamos nuestra ciudad limpia!

¡No seas parte del problema!

¡Sé parte de la solución!

En esta edición:

- Participa en la campaña para embellecer la ciudad.
- ¡Vota!
- Oportunidades para trabajar de voluntario(a)
- Opinión
- Discusión sobre las diferencias culturales

Éstas son las opiniones de dos estudiantes. ¿Qué piensas tú?

«Yo creo que los jóvenes no participamos mucho en la comunidad. Como parte de nuestros estudios, debemos hacer veinte horas de servicio social al mes para ser buenos ciudadanos».

«Estoy muy ocupado, con el fútbol, la tarea, los quehaceres en casa... No quiero usar mi tiempo libre para las causas de la comunidad. ¡Ésas son cosas de adultos!»

Oportunidades para trabajar de voluntario(a): ¿Cuál es tu talento? Puedes trabajar de voluntario(a) en...

El Centro de la Comunidad

Los ancianos valoran tu tiempo y tus atenciones. Puedes conversar, pasear, leer o jugar al ajedrez con ellos.

El Comedor de Beneficencia

La gente sin hogar también valora tu tiempo. Puedes servir la cena, donar ropa o dar clases de inglés.

Vamos a educar al público **sobre:** la importancia de luchar contra el prejuicio y cómo convivir con otras culturas.

¡TODOS SOMOS IGUALES!

Online Workbook
CLASSZONE.COM

¿Comprendiste?

1. ¿Cómo usarías el dinero que juntaron estos estudiantes?
2. ¿Es importante hacer algo para embellecer tu ciudad?
3. ¿Hay un centro de la comunidad donde vives?
4. ¿Qué puedes hacer en el centro de beneficencia de tu comunidad?
5. ¿Qué harías para cambiar los prejuicios contra otra cultura?
6. ¿Crees que los jóvenes deben ser voluntarios unas horas al mes?

AUDIO

SITUACIONES

¡Los candidatos!

Lee sobre dos candidatos para alcalde (*mayor*). Luego escucha un debate entre los dos y decide por quién vas a votar.

 Leer

Lee sobre los candidatos para alcalde de tu ciudad.

Eduardo Herrera Garza

- *Es hombre de negocios.*

- *Sabe juntar fondos y aumentarlos.*

- *Ha servido como presidente de la organización para embellecer la ciudad.*

Vote por
Eduardo Herrera Garza
y tendrá

- una ciudad limpia y bella

- mejor uso de los fondos municipales

- un trabajo para cada individuo

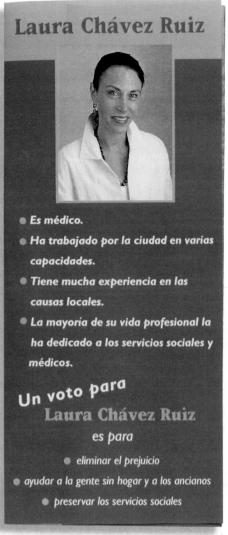

Laura Chávez Ruiz

- *Es médico.*

- *Ha trabajado por la ciudad en varias capacidades.*

- *Tiene mucha experiencia en las causas locales.*

- *La mayoría de su vida profesional la ha dedicado a los servicios sociales y médicos.*

Un voto para Laura Chávez Ruiz *es para*

- eliminar el prejuicio
- ayudar a la gente sin hogar y a los ancianos
- preservar los servicios sociales

 ## 2 Escuchar

Ahora escucha un debate entre la señora Chávez Ruiz y el señor
Herrera Garza. En una hoja aparte, anota las posiciones de los candidatos:
¿De qué cosas están a favor? y ¿de qué cosas están en contra?

Laura Chávez Ruiz	
Está a favor de…	Está en contra de…
_____	_____
_____	_____

Eduardo Herrera Garza	
Está a favor de…	Está en contra de…
_____	_____
_____	_____

3 Escribir

¿Por quién quieres votar? ¿Por qué? De las causas que se mencionan en el debate,
¿por cuáles estás a favor? ¿y en contra? Escribe dos cosas por las cuales estás a
favor y dos cosas por las cuales estás en contra.

En acción

PARTE A

Práctica del vocabulario

Objectives for Activities 2-4
• Say what you want to do • Make requests

1 Las campañas

Escribir Las campañas siempre tienen un lema (*slogan*) publicitario. Escoge frases de la lista para completar los siguientes lemas.

gente sin hogar

trabajar de voluntario(a)

sembrar árboles

votar

convivir

comedores de beneficencia

1. Para embellecer la ciudad, hay que _____ .

2. Para ser buen(a) ciudadano(a), hay que _____ en las elecciones.

3. Para combatir el prejuicio, hay que _____ con vecinos de todas las culturas.

4. Para luchar contra la pobreza, hay que ayudarle a la _____ .

5. Para luchar contra el hambre, hay que construir más _____ .

6. Para apoyar al centro de la comunidad, hay que _____ .

2 ¿Qué vas a hacer?

Hablar/*Escribir* Tu compañero(a) quiere saber qué vas a hacer hoy para mejorar la comunidad. Basándote en los dibujos, dile lo que vas a hacer. Luego, cambien de papel.

modelo

Compañero(a): *¿Qué vas a hacer hoy?*

Tú: *Voy a votar.*

1.

2.

3.

4.

5.

6.

También se dice

Aunque la palabra **anciano** se usa en todo el mundo hispanohablante, también se utilizan las siguientes expresiones:

• **gente grande** (Argentina)

• **abuelitos** (México), aun para personas que no sean familiares

3 ¿Estás a favor de...?

Hablar/Escribir Tú quieres saber si tu compañero(a) está a favor o en contra de estas cosas. ¿Qué le preguntas?

modelo

¿a favor de? / la campaña para embellecer la ciudad

Tú: *¿Estás a favor de la campaña para embellecer la ciudad?*

Compañero(a): *Sí, (No, no) estoy a favor de la campaña para embellecer la ciudad.*

1. ¿a favor de? / construir un parque
2. ¿a favor de? / tener menos horas de clase
3. ¿en contra de? / tener una sola comida en el menú de la cafetería
4. ¿en contra de? / usar uniformes para la escuela
5. ¿en contra de? / recoger basura todos los sábados por la mañana
6. ¿a favor de? / reciclar vidrio, plástico y papel
7. ¿en contra de? / luchar contra el hambre
8. ¿...? / ¿...?

4 Voluntarios

Hablar/Escribir Escribe dos cosas que quieres hacer en la comunidad. Luego, pide ayuda a tu compañero(a). Después, cambien de papel.

modelo

Tú: *¿Puedes hacerme un favor?*

Compañero(a): *Sí, con mucho gusto.*

Tú: *¿Podrías darme una mano con esta ropa? Voy a llevarla al centro de la comunidad.*

Compañero(a): *Lo siento mucho, pero no puedo. Hoy voy a participar en la campaña para embellecer la ciudad.*

Tú: ...

Vocabulario

Para pedir ayuda y responder

¿Cómo puedo ayudarte(lo, la)? *How can I help you?*

¿Podría(s) darme una mano? *Could you give me a hand?*

¿Puede(s) ayudarme? *Can you help me?*

¿Puede(s) hacerme un favor? *Can you do me a favor?*

Estoy agotado(a). *I'm exhausted.*

Lo siento mucho, pero... *I'm sorry, but...*

Me es imposible. *It's just not possible for me.*

No, de veras, no puedo. *No, really, I can't.*

¿Por qué no? *Sure, why not?*

Sí, con mucho gusto. *Yes, gladly.*

Si pudiera, lo haría. *If I could, I would.*

▶ ¿Cómo pides ayuda y cómo respondes?

Objectives for Activities 5–13
• Say what you want to do • Make suggestions • Make requests

REPASO   **Command Forms**

One of the ways to tell someone to do or not to do something is to use command forms. The **Ud.** and **Uds. command forms** are all formed by taking the **yo form** of a verb, dropping the **-o** and adding the appropriate endings.

- For **Ud.** commands add:
    -e for **-ar** verbs
    -a for **-er** or **-ir** verbs

- For **Uds.** commands add:
    -en for **-ar** verbs
    -an for **-er** or **-ir** verbs

El Sr. Arroyuelo **cambia** la bombilla en el desván.
*Mr. Arroyuelo **is changing** the lightbulb in the attic.*

Sr. Arroyuelo, por favor **cambi**e la bombilla.
*Mr. Arroyuelo, please **change** the lightbulb!*

> Regular
> **tú commands** look just
> like the third person
> indicative.

The **tú** command has a different form for **negative** commands.

- For **negative tú** commands add:
    -es for **-ar** verbs
    -as for **-er** or **-ir** verbs

No **com**as ese dulce.
*Don't **eat** that candy.*

If the stem of a verb is **irregular** in the **yo form**, it will be irregular in the command form. The endings will be the same as regular commands.

infinitive	yo form indicative	command form
seguir →	si**g**o →	¡Si**g**a!

Other verbs like this are: **caer, hacer, oír, poner, salir, venir, tener, traer, ofrecer.**

Apoyo para estudiar

Here is an auditory way to help you remember all these commands. Tape record yourself as you carefully say each of the commands with pauses between each one. Then, using the form in Actividad 5, replay the tape. As you hear each command, check the correct form and say it. This reinforces your visual, kinesthetic, and auditory memory.

Remember that verbs ending in **-c**ar, **-g**ar, and **-z**ar require spelling changes to keep the pronunciation consistent.

c	→	qu
g	→	gu
z	→	c

changes before an -e

yo bus**c**o   ¡Bus**qu**e Ud.!   ¡Bus**c**a tú!

no change before an -a

Practice: **Actividades**
5 6 7

Más práctica *cuaderno p. 45*
Para hispanohablantes *cuaderno pp. 43–44*

Online Workbook
CLASSZONE.COM

5 La clase de ejercicio

Escuchar/Escribir El instructor de la clase de ejercicio aconseja a personas de varias edades. Escucha sus consejos. Primero, decide si el mandato es afirmativo o negativo. Luego, decide si él usa **tú**, **usted** o **ustedes**.

modelo

afirmativo _X_ negativo _____ tú _____ usted _____ ustedes _X_

1. afirmativo _____ negativo _____	tú _____ usted _____ ustedes _____
2. afirmativo _____ negativo _____	tú _____ usted _____ ustedes _____
3. afirmativo _____ negativo _____	tú _____ usted _____ ustedes _____
4. afirmativo _____ negativo _____	tú _____ usted _____ ustedes _____
5. afirmativo _____ negativo _____	tú _____ usted _____ ustedes _____
6. afirmativo _____ negativo _____	tú _____ usted _____ ustedes _____
7. afirmativo _____ negativo _____	tú _____ usted _____ ustedes _____

6 Tu amigo(a)

Hablar/Escribir Tienes un(a) amigo(a) que siempre te dice lo que tienes que hacer. Usa los siguientes verbos para expresar lo que te dice. Sigue el modelo.

modelo

ir   *Ve al centro de la comunidad.*

1. votar
2. donar
3. ayudar
4. recoger
5. tirar

6. reciclar
7. convivir
8. gastar
9. sembrar
10. educar

7 Tráemelo

Hablar/Escribir Usa los verbos a continuación para hacer preguntas a tu compañero(a). Sigue el modelo.

modelo

traer

Tú: *¿Traigo los libros?*

Compañero(a): *Sí, tráelos. o*
No, no los traigas.

Nota: Gramática

Remember that pronouns are attached to the end of affirmative commands. In negative commands they come before the verb.

—**Dime**, papá. ¿Traigo el periódico?

—No, gracias. **No lo traigas. Déjalo** en la sala.

llevar	vender
comprar	dar
traer	regalar
preparar	prestar

Activity 8: Converse and listen during face-to-face social interactions

REPASO   Nosotros Commands

> When you want to say *let's do something* or *let's not do something* you use **nosotros commands**. Remember to start with the **yo form** of the verb, drop the **-o** and add the appropriate ending.

- **-ar** verbs end in **-emos**
- **-er** and **-ir** verbs end in **-amos**

Participemos en la campaña para mejorar nuestra ciudad.
Let's participate in the campaign to improve our city.

> If a verb has an irregular **yo form** it also appears in the **nosotros command**.

Irregular yo form:

Yo siempre **dig**o la verdad sobre los problemas en nuestra ciudad.
*I always **tell** the truth about the problems in our city.*

Nosotros command:

Digamos la verdad sobre los problemas en nuestra ciudad.
*Let's **tell** the truth about the problems in our city.*

> Some verbs are **irregular** in the **nosotros command** form and are not created using the **yo form**.
>
> Use the present subjunctive of the **nosotros** form for stem-changing verbs.

dar	demos
estar	estemos
saber	sepamos
ser	seamos

When you want to say *let's go,* use: **vamos**   To say *let's not go,* use: **no vay**amos

> Besides the command form, you already know another way to say *let's do something:*

To say *let's not do something* you must use:

vamos a + **infinitive**

no + **nosotros command**

Vamos a luchar contra el prejuicio en nuestra sociedad.
*Let's **fight** against prejudice in our society.*

No olvidemos a los ancianos ni a los enfermos.
*Let's **not forget** the elderly and the sick.*

> With **reflexive verbs,** you drop the final **s** of the **command form** before attaching the reflexive pronoun **nos**:

levantemo~~s~~ + **nos** ➝ Levantémo**nos**
Let's get up.

Notice the **accent.** It is added to keep the pronunciation consistent.

Practice: **Actividades** 8 9 10 11   **Más práctica** *cuaderno p. 46*
Para hispanohablantes *cuaderno p. 45*

Online Workbook CLASSZONE.COM

8 Hagamos algo

Hablar/Escribir Pasas el fin de semana con un(a) amigo(a). El (Ella) quiere hacer varias cosas, pero a ti no te gustan sus ideas. Sugiere que hagan otra cosa.

modelo

ver la película romántica (la película de acción)

Compañero(a): *¿Por qué no vemos la película romántica?*

Tú: *No, mejor veamos la película de acción.*

1. ir al museo (centro comercial)
2. comprar el helado de chocolate (de fresa)
3. jugar al tenis (al fútbol)
4. visitar a mis primos (a mis amigos)
5. preparar hamburguesas (salchichas)
6. andar por el parque (por la playa)
7. salir por la tarde (por la noche)
8. votar a Eduardo Herrera (a Laura Chávez)
9. pasear por la ciudad (por el campo)
10. donar videos (ropae)

9 El club de voluntarios

Hablar/Escribir Estás en una reunión del club de voluntarios. Todos tienen ideas diferentes de lo que deben hacer. ¿Qué dicen?

modelo

luchar contra el hambre

Luchemos contra el hambre.

1. convivir con nuestros vecinos
2. preservar los derechos humanos
3. resolver el problema de la gente sin hogar
4. consumir menos
5. conservar más
6. cuidar de los ancianos
7. hacer un esfuerzo para unir al pueblo
8. trabajar de voluntarios en el centro de rehabilitación
9. acabar con el racismo
10. crear soluciones realísticas para nuestros problemas

Vocabulario

En la comunidad

colaborar con *collaborate with*
conservar más *to conserve more*
consumir menos *to consume less*
crear *to create*
cuidar de *to take care of*
hacer un esfuerzo *to make an effort*
permitir *to permit*
pertenecer *to belong*
preservar *to preserve*
resolver *to resolve*

el centro de rehabilitación *rehabilitation center*
los derechos humanos *human rights*
el desarrollo *development*
la discriminación *discrimination*
los enfermos *the sick*
los minusválidos *the physically challenged*
el ser humano *human being*

▶ ¿Cómo puedes usar estas palabras para describir tus actividades?

Activity 10: Clarify, ask for and comprehend clarification

10 ¿Qué haremos?

Hablar/*Escribir* Tú y tu compañero(a) son buenos ciudadanos. Conversen sobre sus planes para servir a la comunidad.

modelo

trabajar de voluntarios

Tú: *¿Dónde debemos trabajar de voluntarios?*

Compañero(a): *Trabajemos en el centro de la comunidad.*

Tú: *No, no trabajemos en el centro de la comunidad. Mejor trabajemos en el comedor de beneficencia.*

1. luchar contra
2. hacer un esfuerzo
3. donar
4. colaborar con
5. trabajar de voluntarios
6. cuidar de
7. acabar con
8. educar al público
9. votar
10. embellecer
11. participar
12. convivir con

11 La publicidad

Escribir Estás a cargo de la publicidad para tu club de voluntarios. Escribe tres lemas publicitarios para un folleto (*brochure*) que describa los propósitos de tu club.

modelo

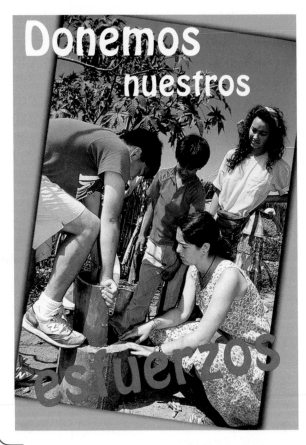

More Practice: Más comunicación *p. R6*

Nota cultural

En México se han formado grupos de jóvenes para ayudar a la comunidad. Se conocen como asociaciones de segundo piso y su propósito es establecer relaciones con hombres y mujeres de negocios para recolectar desechos (*scrap material*) y basura industriales. Luego estos grupos venden los materiales y juntan fondos para programas sociales dirigidos a los ancianos, niños y minusválidos, entre otros.

Speculating with the Conditional

To talk about what you *should, could,* or *would do,* use the **conditional tense.**

The conditional:

- helps you to talk about what would happen under certain conditions.
- is used to make polite requests.

   Yo no **me quejaría** tanto. Yo **me llevaría** bien con todos.
   *I **wouldn't complain** so much. I **would get along** well with everyone.*

Verbs ending with **-ar, -er** and **-ir** all have the same endings in the **conditional.** You add the endings directly to the **infinitive.**

conditional ending

infinitive +	-ía	-íamos
	-ías	-íais
	-ía	-ían

   Yo **estaría** a favor de comprar árboles y flores para embellecer el centro de la comunidad.
   *I **would be** in favor of buying trees and flowers to beautify the community center.*

If a verb has an **irregular stem** in the **future,** you use that same stem to form the conditional. The endings are the **same endings** as in the above chart.

infinitive	irregular future stem	conditional
decir →	diré →	diría

Note: The conditional of **hay** is always **habría.**

infinitive	irregular stem
decir	dir-
hacer	har-
poder	podr-
poner	pondr-
querer	querr-
saber	sabr-
salir	saldr-
tener	tendr-
valer	valdr-
venir	vendr-

Practice: **Actividades**   **Más práctica** *cuaderno pp. 47–48*
Para hispanohablantes *cuaderno p. 46*

 **Online Workbook**
CLASSZONE.COM

Activity 15: Express and understand opinions

12 Costa Rica ♻ 🎧

Escuchar/*Escribir* Cada año, la familia de Fausto va al mismo campo en Costa Rica. Escucha su descripción de las vacaciones anteriores y las de este año. Decide si cada persona en su familia ya participó en una actividad, o si participaría en esa actividad durante el próximo viaje.

modelo

Joaquín: volar en planeador

Δ *Ya voló en planeador.*   ✗ *Volaría en planeador.*

1. Fausto	5. los padres de Fausto
Δ Acampó.	Δ Escalaron montañas.
Δ Acamparía.	Δ Escalarían montañas.
2. Fausto y su hermana	6. las hermanas de Fausto
Δ Navegaron en tabla de vela.	Δ Esquiaron en el agua.
Δ Navegarían en tabla de vela.	Δ Esquiarían en el agua.
3. Juan	7. Ana
Δ Hizo surfing.	Δ Levantó pesas.
Δ Haría surfing.	Δ Levantaría pesas.
4. la mamá de Fausto	8. el papá de Fausto
Δ Hizo alpinismo.	Δ Pescó en alta mar.
Δ Haría alpinismo.	Δ Pescaría en alta mar.

13 La Ciudad de México

Hablar/*Escribir* Imagínate que vas de viaje a la Ciudad de México. ¿Qué harías allí?

modelo

visitar el Palacio de Bellas Artes

Visitaría el Palacio de Bellas Artes.

1. ir a las pirámides de Teotihuacán

2. comprar regalos en la Zona Rosa

3. escuchar música de mariachi

4. asistir a un concierto de Luis Miguel

5. comer en el restaurante del San Ángel Inn

6. mandar tarjetas postales a todos mis amigos

7. buscar el museo de Frida Kahlo

8. pasearme por el Parque de Chapultepec

Nota cultural

En América Latina y España normalmente se habla a la gente mayor con mucho respeto. Sería mal educado (*impolite*) decir a alguien mayor «¡Tráigamelo!» Es más cortés decir «Por favor don Ramón, ¿puede traérmelo?» o «Don Ramón, ¿me hace usted el favor de traerlo?»

Activities 14–15 bring together all concepts presented.

14 Mi comunidad

STRATEGY: SPEAKING

Name social problems, then propose solutions
Here are a few words to help you get started. Issues include **pobreza, discriminación, crimen.** People who need help include **ancianos, gente sin hogar, jóvenes, refugiados.** Organizations that can help include **centros de la comunidad, escuelas, servicios sociales.** One can always give **ayuda, dinero, consejos.** Add your own ideas for being a good citizen.

Hablar/Escribir Eres el (la) presidente(a) de una organización que cuida el medio ambiente. En tu discurso, le dices a la gente lo que debe hacer.

modelo

trabajar de voluntarios en...

Tú: *Trabajen de voluntarios en un programa de reciclaje.*

1. votar por la campaña para…
2. estar a favor de…
3. estar en contra de…
4. participar en…
5. hacer el esfuerzo para…
6. colaborar con…
7. conservar…
8. valorar…
9. cuidar de…
10. resolver…
11. ¿….?

15 En mis sueños

Hablar Imagínate que puedes hacer lo que quieras: que no tienes límites ni de dinero ni de tiempo. ¿Sabes lo que harías? En grupos de tres o cuatro, expresen tres cosas que harían.

modelo

Tú: *Yo pasaría más tiempo con mis amigos y mi familia.*

Amigo(a) 1: *Yo compraría una casa cerca de la playa.*

Amigo(a) 2: *Yo viajaría por toda Latinoamérica.*

Amigo(a) 3: *Yo donaría mucho dinero a las causas de mi comunidad.*

More Practice: **Más comunicación** *p. R6*

Online Workbook
CLASSZONE.COM

Refrán

Haz bien y no mires a quién.

¿Qué quiere decir el refrán? ¿Qué harías tú para «hacer bien» sin «mirar a quién»?

AUDIO

En voces
LECTURA

PARA LEER
STRATEGY: READING

Comprehend complex sentences
Rigoberta Menchú is an admirable international figure. When she writes or speaks, she usually combines several ideas in one paragraph. For example, here are the meaningful units inside one of her sentences:

Tenemos el reto / de construir y consolidar la democracia y la paz, / resolviendo los problemas internos y privilegiando el diálogo.

Focus first on the small units of meaning rather than on the entire sentence. Read the sections aloud. What words connect the units of meaning? Try this technique with other sentences so that you can understand them more easily.

Nota cultural

Rigoberta Menchú habla de **construir la democracia.** Guatemala, al igual que muchos países de Latinoamérica, sufrieron fuertes dictaduras. Al establecerse de nuevo la democracia, todas las personas vuelven a tener derechos.

Rigoberta Menchú

Rigoberta Menchú cuenta una historia tan extraordinaria como su vida. Ella se escapó de la represión del gobierno de Guatemala durante las décadas de los 70 y los 80, cuando muchas personas murieron, incluyendo miembros de la familia de Rigoberta.

Una de las razones de la represión fue la discriminación de la clase alta y la clase media contra los indios, quienes viven en las montañas, lejos de la ciudad, y con muy pocos recursos.

---

[1] Nobel Peace Prize

Rigoberta Menchú decidió organizar a los indios para que se defendieran contra la violencia del gobierno, pero luego tuvo que irse de su país. Fue a París, donde Elizabeth Burgos, una activista de nacionalidad francesa y venezolana, la ayudó a escribir su historia. Por los esfuerzos para mejorar las condiciones de su pueblo, Rigoberta Menchú ganó el Premio Nóbel de la Paz [1] en 1992. Ella habla frecuentemente sobre la paz.

66 El tesoro más grande que tengo en la vida es la capacidad de soñar. En las situaciones más duras y complejas, he sido capaz de soñar con un futuro más hermoso. 99

En su testimonio personal, *Me llamo Rigoberta Menchú y así me nació la conciencia*, Rigoberta Menchú cuenta cómo fue una de las víctimas del prejuicio de pertenecer a un pueblo indígena, donde casi todos hablan su propia lengua, el quiché. Además, en sus entrevistas siempre hace referencia a la unión de los pueblos.

66 La utopía de la interculturalidad debe convertirse en el motor que guíe las relaciones entre pueblos y culturas. 99

¿Comprendiste?

1. ¿Cuál es el sueño más grande de Rigoberta Menchú?
2. ¿De qué habla en su libro?
3. ¿De qué habla en sus entrevistas?

¿Qué piensas?

Para Rigoberta, ¿existen esperanzas de que su pueblo sea aceptado por otras culturas? ¿Con qué palabras expresa ella esta relación?

Hazlo tú

En un mapa de Guatemala busca el Departamento de El Quiché. Estudia los nombres de los pueblos. ¿Cuáles parecen ser de origen indígena y cuáles son nombres castellanos?

ETAPA 1

 Activity 3: Generally choose appropriate vocabulary for familiar topics, but as the complexity of the message increases, there is evidence of hesitation and groping for words, as well as patterns of mispronunciation and intonation

En uso
REPASO Y MÁS **COMUNICACIÓN**

OBJECTIVES

- Say what you want to do
- Make requests
- Make suggestions

Now you can...

- make requests.

To review

- command forms see p. 114.

1 En el centro de la comunidad

Trabajas de voluntario(a) en el centro de la comunidad. Hay letreros que tienen reglas para los estudiantes voluntarios y para las personas que viven allí. ¿Qué dicen?

modelo

a los residentes: no perder sus cosas.

al estudiante: ayudar con el almuerzo

a los dos: escuchar estos anuncios

1. al residente: acostarse antes de la medianoche
2. a la estudiante: llegar antes de las diez
3. a los dos: ver el video sobre el centro
4. al residente: no olvidarse de tomar su medicina
5. a los dos: salir al parque para los ejercicios
6. al estudiante: ir a la cafetería para servir el almuerzo
7. al residente: tener cuidado con el equipo deportivo
8. al estudiante: no hacer tu tarea aquí

Now you can...

- make suggestions.

To review

- **nosotros** commands see p. 116.

2 La reunión municipal

Varias personas en la reunión municipal expresan sus opiniones sobre temas que afectan la ciudad. ¿Qué dice cada persona?

modelo

limpiar nuestra ciudad

Limpiemos nuestra ciudad. **o** *¡No limpiemos nuestra ciudad!*

1. abrir los brazos a la gente sin hogar
2. eliminar la discriminación
3. colaborar con los demás
4. apoyar a nuestros líderes
5. eliminar el hambre
6. participar en las actividades de la comunidad
7. ayudar a la gente sin hogar

124 ciento veinticuatro
Unidad 2

Now you can...

- say what you want to do.

To review

- the conditional see p. 119.

3 Servicio a la comunidad

Usa la frase indicada para empezar una conversación con tu compañero(a) sobre los servicios a la comunidad.

modelo

participar en...

Compañero(a): *¿Participarías en una campaña para embellecer el centro?*

Tú: *Sí, ¿por qué no? Participaría en una campaña para embellecer el centro. o No, lo siento mucho pero no podría hacerlo.*

1. trabajar de voluntario(a)
2. embellecer la ciudad
3. donar ropa
4. juntar fondos para...
5. estar a favor de...
6. estar en contra de...

Now you can...

- say what you want to do.

To review

- the conditional see p. 119.

4 Reacciones

¿Qué harían las siguientes personas con una gran cantidad de dinero?

modelo

Tú: *vender tu carro viejo y comprar un carro nuevo*

Venderías tu carro viejo y comprarías un carro nuevo.

1. yo: poner el dinero en el banco y vivir como siempre
2. tú: gastar todo el dinero y quedarte con nada
3. él: salir todos los días y no preocuparse de nada
4. ella: empacar sus maletas y viajar a Europa
5. usted: construir una casa nueva e invitar a sus amigos
6. ustedes: seguir con sus estudios e ir a la universidad
7. nosotros: divertirnos y también tener tiempo para trabajar de voluntarios

5 El club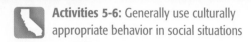

STRATEGY: SPEAKING
Identify the general ideas, then delegate responsibilities
Before delegating the responsibilities, it is good to decide what they are. Agree on a list of things to be done (infinitives). Then decide who is to do them (commands). You can also decide what not to do.

Tú y dos amigos han decidido que van a empezar un club de voluntarios. Quieren tener la primera reunión en tu casa. Todos tienen ideas de qué se debe hacer.

modelo

Tú: *Enrique, haz una lista de las personas que vamos a invitar.*

Amigo(a) 1: *Sandra, tú y Hernán empiecen a llamar a todos en la lista.*

Amigo(a) 2: …

6 En tu lugar

Imagina que tu amigo(a) es famoso(a)—puede ser actor o actriz, artista, escritor(a), atleta, presidente(a) de un país, ¡lo que sea! Trabajen en grupos de tres o cuatro. Primero decidan quién del grupo es famoso(a) y por qué. Luego, los demás dan consejos a la persona famosa empezando con la frase «En tu lugar».

modelo

escritora famosa

Amigo(a) 1: *En tu lugar, yo escribiría una novela de ciencia-ficción.*

Amigo(a) 2: *En tu lugar, yo pediría más dinero por la próxima novela.*

Amigo(a) 3: *En tu lugar, yo descansaría un rato antes de escribir otra novela.*

7 En tu propia voz

ESCRITURA Inventa un producto y escribe un anuncio para venderlo. Los anuncios frecuentemente usan las formas de **usted** o **ustedes** al dar mandatos. Tu anuncio va a salir en Internet, así que asegura que sea ¡interesante y atractivo!

modelo

¡Piense en el futuro de sus hijos! ¡Compre la computadora SúperRápida!

TÚ EN LA COMUNIDAD

James es alumno en Washington. Él trabaja de voluntario con una optómetra en una misión médica en México. Da instrucciones a los pacientes, habla sobre los problemas que tienen y comunica al médico información importante. Cuando está en Washington, habla español con alumnos de otras escuelas.

En resumen

REPASO DE VOCABULARIO

SAY WHAT YOU WANT TO DO

Actions

colaborar con	to collaborate with
conservar más	to conserve more
consumir menos	to consume less
convivir	to live together, to get along
crear	to create
cuidar de	to take care of
donar	to donate
educar al público	to educate the public
embellecer	to beautify
estar a favor de	to be for
estar en contra de	to be against
hacer un esfuerzo	to make an effort
juntar fondos	to fundraise
luchar contra	to fight against
participar	to participate
permitir	to permit
pertenecer	to belong
preservar	to preserve
recoger	to pick up
resolver (o→ue)	to resolve
sembrar (e→ie)	to plant
trabajar de voluntario(a)	to volunteer
valorar	to value
votar	to vote

People, places, and things

los ancianos	the elderly
el árbol	tree
la basura	garbage
la campaña	campaign
el centro de la comunidad	community center
el centro de rehabilitación	rehabilitation center
el (la) ciudadano(a)	citizen
el comedor de beneficencia	soup kitchen
los derechos humanos	human rights
el desarrollo	development
la discriminación	discrimination
los enfermos	the sick
la gente sin hogar	the homeless
los jóvenes	young people
los minusválidos	the physically challenged
la pobreza	poverty
el prejuicio	prejudice
el ser humano	human being
el servicio social	social service
la solución	solution

MAKE REQUESTS

Questions

¿Cómo puedo ayudarte(lo, la)?	How can I help you?
¿Podría(s) darme una mano?	Could you give me a hand?
¿Puede(s) ayudarme?	Can you help me?
¿Puede(s) hacerme un favor?	Can you do me a favor?

Responses

Estoy agotado(a).	I'm exhausted.
Lo siento mucho, pero…	I'm sorry, but…
Me es imposible.	It's just not possible for me.
No, de veras, no puedo.	No, really, I can't.
¿Por qué no?	Sure, why not?
Sí, con mucho gusto.	Yes, gladly.
Si pudiera, lo haría.	If I could, I would.

MAKE SUGGESTIONS

Nosotros commands

Trabajemos de voluntarios.
Ayudemos a los demás.

Juego

¿Qué podemos hacer?

Mira las fotos de varios lugares de México. ¿Qué servicio para la comunidad asocias con cada una?

1.

2.

3.

a. preservar el medio ambiente
b. sembrar árboles
c. trabajar de voluntario(a)

UNIDAD 2

ETAPA **2**

Un planeta en peligro

OBJECTIVES

- Say what should be done

- React to ecology

- React to others' actions

¿Qué ves?

Mira la foto. Contesta las preguntas.

1. ¿Qué cosas llevan los jóvenes? ¿En qué lugar las están usando?

2. ¿Qué hacen? ¿Crees que llevan el equipo correcto?

3. ¿Llevarías otras cosas? ¿Por qué?

4. ¿Crees que sería útil un libro como *El manejo de la iguana verde*? ¿Por qué?

El Manejo de la
Iguana Verde

TOMO VIII

La Iguana Verde
en zonas
de amortiguamiento

En contexto

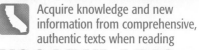

Acquire knowledge and new
information from comprehensive,
authentic texts when reading

VOCABULARIO

Descubre

¿Puedes adivinar el significado
de las siguientes frases? Escoge de la
lista de abajo. ¡Fácil!

1. la capa de ozono
2. el desperdicio
3. el derrame de petróleo
4. los recursos naturales
5. las zonas de reserva ecológica
6. la contaminación del aire

   a. *oil spill*
   b. *conservation land*
   c. *ozone layer*
   d. *air pollution*
   e. *natural resources*
   f. *waste*

Nuestro
Planeta

¡Es el único
que tenemos!

A nivel personal

Problema ecológico:
la contaminación del aire

¿Qué puedes hacer tú?

➤ Compartir el carro con amigos y colegas.
➤ ¡Caminar!
➤ Usar el transporte público.

Problema ecológico:
la destrucción de la capa de ozono

¿Qué puedes hacer tú?

➤ Limitar el uso de **aerosoles**.

Problema ecológico:
el desperdicio

¿Qué puedes hacer tú?

➤ ¡No **eches** los siguientes productos a la basura!
➤ ¡Recíclalos! **Instituye** programas de **reciclaje**.

botellas de vidrio   **plástico**   **cartón**

latas

A nivel oficial

Problema ecológico:
los derrames de petróleo
Soluciones posibles:

➤ **Desarrollar** otras formas de energía.

Problema ecológico:
la destrucción de los **recursos naturales**
Soluciones posibles:

➤ Declarar **zonas de reserva ecológica**.

➤ **Prohibir** el uso de **contaminantes**

➤ No comprar productos que usan **químicos dañinos**.

¡A todos nos toca proteger el **medio ambiente!** ¡No contamines tu **planeta!**

Online Workbook
CLASSZONE.COM

¿Comprendiste?

1. ¿Hay un programa de reciclaje en tu ciudad?
2. En tu casa, ¿se reciclan todos los productos de aluminio, vidrio y papel? ¿Quién lo hace? ¿Qué tienen que hacer: llevar los productos a un sitio específico o poner los productos con la basura?
3. ¿Cómo llegas al colegio? ¿Tratas de caminar o ir en bicicleta cuando puedes?
4. ¿Crees que nos toca a todos cuidar del medio ambiente?
5. ¿Crees que el gobierno está haciendo su parte para la conservación del medio ambiente? ¿Qué más crees que debe hacer?

Understand and convey information about the environment

En vivo

AUDIO
SITUACIONES

PARA ESCUCHAR • STRATEGIES: LISTENING

Pre-listening Are there public service campaigns about protecting the environment in your community? How do you find out about them (television? radio? posters? billboards?)

Inventory efforts to save the environment
First, make a list of efforts in your community to improve the environment. Then, check **Sí** beside those that are mentioned in the campaign in Costa Rica. Finally, comment on similarities and differences between your local campaign and the one heard in **Escuchar.** This chart will help you get started:

Problemas locales	Sí
agua	
aire	
animales	
árboles/plantas	
basura	
energía	
minerales	
tierra	

¡Viva el medio ambiente!

Trabajas en una agencia de publicidad que está creando anuncios con el fin de promover (*to promote*) la protección del ambiente. Mira las imágenes de estos anuncios. Luego, escucha los anuncios de radio y decide a qué imagen corresponden.

① Mirar

Éstas son seis imágenes que van a salir en los anuncios. Estúdialas y piensa sobre qué tema tratan.

A.

B.

C.

D.

E.

❷ Escuchar 🎧

Ahora escucha anuncios de radio que corresponden a las seis imágenes que viste. Escribe el número del anuncio que mejor va con cada imagen en la siguiente tabla.

Dibujo A _____
Dibujo B _____
Dibujo C _____
Dibujo D _____
Dibujo E _____
Dibujo F _____

❸ Hablar/Escribir 👥

Haz una lista de cinco cosas que tú puedes hacer hoy para proteger tu ciudad y el planeta. Compara tu lista con la de dos compañeros. Conversen sobre sus ideas y traten de pensar sobre modos nuevos de proteger el medio ambiente. Si les interesa, monten una campaña ecológica para su colegio.

F.

En acción

Práctica del vocabulario

For Activities 1–4
• Say what should be done • React to ecology • React to others' actions

1 La composición

Escribir Andrés tuvo que escribir una composición sobre el medio ambiente. Completa su composición usando las palabras de la lista.

plástico	cartón
destrucción	contaminación del aire
reciclaje	botellas
latas	desarrollen
desperdicio	medio ambiente
químicos dañinos	reciclar
echar	vidrio
petróleo	capa de ozono

Hay mucho __1__ en el mundo. No deberíamos __2__ todo a la basura. Hay muchas cosas que podemos __3__: las __4__, las botellas de __5__, los productos de __6__, el __7__, los periódicos y las revistas. Si no hay un programa de __8__ en tu ciudad, debes ayudar a empezar uno.

Es importante conservar el __9__. La __10__ se puede controlar. El __11__ que usamos cuando viajamos en nuestros carros echa __12__ al aire. Esos químicos contribuyen a la __13__ de la __14__. Tenemos que pedirles a nuestros políticos y científicos que __15__ otras formas de energía.

2 El horario de Ángela

Hablar/Escribir Ángela vio un documental sobre el medio ambiente y decidió que tenía que hacer algo cada día para conservarlo. Con un(a) compañero(a), conversen sobre su horario.

modelo

usar el metro

Tú: *¿Cuándo va a usar Ángela el metro?*

Compañero(a): *Ángela va a usar el metro para ir al colegio el lunes y el miércoles.*

lunes	
8:00	usar el transporte público para ir al colegio
martes	
4:00	trabajar de voluntaria en el programa de reciclaje
miércoles	
8:00	usar el transporte público para ir al colegio
jueves	
7:00	caminar al colegio
5:00	juntar y recoger las latas para llevarlas a reciclar
viernes	
4:00	llevar cartón, periódicos y revistas a reciclar
sábado	
10:00	pedir a los vecinos que no usen sus carros hoy
2:00	participar en la limpieza del parque municipal
domingo	
10:00	sembrar árboles en el centro
2:00	investigar la política de los candidatos sobre el medio ambiente

1. caminar
2. reciclar
3. investigar
4. hablar con los vecinos
5. participar en la limpieza
6. trabajar de voluntaria

3 La perezosa

Hablar/*Escribir* Antes, Luci no hacía mucho para cuidar del medio ambiente. Describe lo que ocurre en cada dibujo. ¿Cambió Luci su modo de pensar o no?

modelo

Tú: *Había muchas cosas en su casa que podía reciclar, pero antes Luci no lo quería hacer.*

Compañero(a): *¡Sí! Había botellas de plástico y de vidrio…*

Antes

Ahora

4 ¿Qué podemos hacer?

Hablar Tú y tu compañero(a) hablan sobre problemas ecológicos. ¿Qué soluciones hay?

modelo

Tú: *¿Cómo podemos eliminar la contaminación del aire?*

Compañero(a): *Pues, es muy complicado. Si cada familia usa su carro sólo tres veces por semana…*

destrucción de la capa de ozono

contaminación del aire

extinción de las especies

los derrames de petróleo

los combustibles

Vocabulario

El medio ambiente

el combustible *fuel*

complicado(a) *complicated*

descubrir *to discover*

los efectos *effects*

increíble *incredible*

inútil *useless*

¡Qué lío! *What a mess!*

el permiso *permission*

la población *population*

por todas partes *all around*

proteger las especies *to protect the species*

reducir *to reduce*

respetar *to respect*

separar *to separate*

el smog *smog*

la tierra *land*

▶ ¿Cómo puedes usar estas palabras para describir un problema ecológico en tu región?

Práctica: gramática y vocabulario

Objectives for Activities 5–15
• Say what should be done • React to the ecology • React to other's actions

 REPASO | **The Present Subjunctive of Regular Verbs**

¿RECUERDAS? *pp. 114, 116*
Remember when you learned to tell someone not to do something using **negative command forms?**

¡No **habl**es mucho!
¡No **habl**e mucho!
¡No **habl**emos mucho!

The same endings are used when you want to express your opinion or point of view using the subjunctive.

Es importante que uses el transporte público.
It's important that you use public transportation.

The present subjunctive

	-ar hablar	-er comer	-ir escribir
yo	**habl**e	**com**a	**escrib**a
tú	**habl**es	**com**as	**escrib**as
él, ella, usted	**habl**e	**com**a	**escrib**a
nosotros(as)	**habl**emos	**com**amos	**escrib**amos
vosotros(as)	**habl**éis	**com**áis	**escrib**áis
ellos, ellas, ustedes	**habl**en	**com**an	**escrib**an

Remember that you have to change the spelling for some verbs to keep the pronunciation the same.

bus**c**ar → bus**qu**e   pa**g**ar → pa**gu**e

cru**z**ar → cru**c**e   reco**g**er → reco**j**a

Practice: Actividades
5 6 7

Más práctica
cuaderno p. 53
Para hispanohablantes
cuaderno p. 51

 **Online Workbook**
CLASSZONE.COM

5 La ecóloga

Escuchar/Escribir Escucha lo que dice la ecóloga en la radio sobre el medio ambiente y qué podemos hacer para protegerlo. Empieza tus oraciones con las siguientes frases.

modelo
Es triste que...
Es triste que nosotros no respetemos el medio ambiente.

1. Es una lástima que…
2. Es importante que…
3. Es mejor que…
4. Es malo que…
5. Es lógico que…
6. Es necesario que…

También se dice

¡Hay muchos ambientes! Cuando hables del **ambiente,** especifica a cual te refieres.

• Si le preguntas a un(a) amigo(a) **¿cómo está el ambiente?** te refieres a la onda de un lugar.
• Ambiente también significa **atmósfera.**

6 Protejamos el medio ambiente

Hablar/Escribir Los expertos quieren que hagamos ciertas cosas para proteger el medio ambiente. ¿Qué quieren que hagamos?

modelo

no contaminar el medio ambiente

Quieren que no contaminemos el medio ambiente.

> **Nota: Gramática**
>
> Don't forget that verbs that end in **-uir**, like **contribuir**, include a **y** in their subjunctive form: **contribuya, contribuyas**, etc.

1. reciclar los productos de plástico
2. usar el transporte público
3. compartir el carro con amigos y colegas
4. no contribuir al desperdicio
5. andar en bicicleta
6. respetar el medio ambiente
7. no echar las botellas de vidrio a la basura
8. no contaminar el aire con químicos dañinos

7 Tu amigo(a) perezoso(a)

> **STRATEGY: SPEAKING**
>
> **Consider the effect of words and tone of voice**
> What is your purpose: to persuade, irritate, prompt action, work on sense of responsibility? What is a probable response to these two sentences: **¡Es ridículo que no recicles los periódicos!** and **Es una lástima que no recicles los periódicos.** Choose your words and tone of voice to fit your purpose.

Hablar/Escribir Quieres convencer a tu amigo(a) de que debería proteger el medio ambiente. Usa las siguientes expresiones impersonales para empezar un diálogo. Luego, cambien de papel.

modelo

Tú: *¡Es ridículo que no recicles los periódicos!*

Compañero(a): *Sí, lo sé.*

Tú *¡Es importante que los recicles hoy!*

Compañero(a) *Está bien. Más tarde voy al programa de reciclaje.*

Vocabulario

Ya sabes

es bueno que…	es posible que …
es importante que…	es probable que …
es lógico que…	es raro que …
es malo que …	es ridículo que …
es mejor que …	es triste que …
es necesario que …	es una lástima que …
es peligroso que …	

▶ ¿Qué otras opiniones puedes expresar?

Activity 10: Converse and listen during face-to-face social interactions

REPASO **The Present Subjunctive of Irregular Verbs**

♻ **¿RECUERDAS?** *p.136* You have already learned to form the **subjunctive** of regular verbs to express your opinion or point of view.

> **Es importante que** uses el transporte público.
> *It's important that you use public transportation.*

▶ Some verbs have **irregular forms** in the **subjunctive**.

dar	estar	ir	saber	ser
dé	esté	vaya	sepa	sea
des	estés	vayas	sepas	seas
dé	esté	vaya	sepa	sea
demos	estemos	vayamos	sepamos	seamos
deis	estéis	vayáis	sepáis	seáis
den	estén	vayan	sepan	sean

Es bueno que vayas a la escuela en autobús.
It's good that you go to school by bus.

Es malo que no estén de acuerdo.
It's too bad that they don't agree.

The subjunctive of **haber** is **haya.**

Hay mucha basura.	**Es malo que** haya mucha basura.
There is a lot of trash.	*It's bad that there is a lot of trash.*

▶ Verbs with **yo forms** that end in **-go** or **-zco** in the present indicative use the same **irregular stem** in the subjunctive.

decir → digo

diga	digamos
digas	digáis
diga	digan

conocer → conozco

conozca	conozcamos
conozcas	conozcáis
conozca	conozcan

Other verbs like these are:

caer, hacer, oír, poner, salir, venir, tener, traer, ofrecer

Practice: Actividades ❽ ❾ ❿

Más práctica *cuaderno p. 54*
Para hispanohablantes *cuaderno p. 52*

🌐 **Online Workbook** CLASSZONE.COM

---

❽ Un viaje a Honduras

Hablar/*Escribir* Vas a ir a visitar a tus abuelos en Tegucigalpa, Honduras. Tu abuelita te da consejos. ¿Qué te dice?

> **modelo**
> *hacer las reservaciones*
> *Es importante que hagas las reservaciones.*

1. tener tu pasaporte
2. salir temprano para el aeropuerto
3. estar en el mostrador dos horas antes de la salida
4. saber el número de tu vuelo
5. oír los anuncios para los vuelos
6. ser cortés con los otros pasajeros
7. traer suficiente ropa para dos semanas
8. conocer la ciudad de Tegucigalpa
9. leer sobre la historia de Honduras
10. ¡venir a vernos!

9 El guía turístico

Hablar/Escribir Un grupo de tu colegio ha viajado a Guatemala con su profesor(a) de español. El guía turístico les recomienda diferentes cosas a todos. ¿Qué les recomienda?

modelo

tú / ir al museo

Es mejor que tú vayas al museo.

1. yo/ver las pirámides
2. nosotros/ser turistas responsables
3. usted/dar una donación al pueblo
4. ustedes/estar en el hotel a las siete
5. tú/traer tu cámara para sacar fotos

10 Los consejos

Hablar/Escribir Dale consejos a tu amigo(a) usando los verbos en la lista y una expresión impersonal de la página 137. ¡Trata de dar consejos útiles! Luego cambien de papel.

modelo

conocer

Tú: *Es importante que conozcas los problemas ecológicos.*

Compañero(a): *Tienes razón. Buscaré información.*

1. poner   5. tener
2. ir   6. salir
3. dar   7. hacer
4. ver   8. escribir

REPASO **The Present Subjunctive of Stem-Changing Verbs**

When you use the **present subjunctive** of **-ar** and **-er** **stem-changing verbs**, remember to make the same stem-changes as in the present indicative.

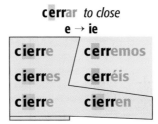

cerrar *to close*
e → ie

cierre	cerremos
cierres	cerréis
cierre	cierren

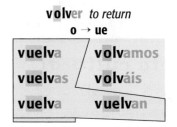

volver *to return*
o → ue

vuelva	volvamos
vuelvas	volváis
vuelva	vuelvan

Notice that **-ir** verbs change their stems differently. The stem of **mentir** alternates between **ie** and **i**, and **dormir** alternates between **ue** and **u**.

The verb **pedir** also has a stem change. The stem changes from **e** to **i** in all forms in the subjunctive.

mentir *to lie*
(e → ie and i)

mienta	mintamos
mientas	mintáis
mienta	mientan

dormir *to sleep*
(o → ue and u)

duerma	durmamos
duermas	durmáis
duerma	duerman

pedir *to ask for, to order*
(e → i)

pida	pidamos
pidas	pidáis
pida	pidan

Practice: **Actividades** 11 12 13   **Más práctica** *cuaderno p. 55*   **Para hispanohablantes** *cuaderno p. 53*

Online Workbook
CLASSZONE.COM

Activity 13: Use strings of related sentences when speaking

11 ¿Es bueno o malo?

Hablar/Escribir Tú y tu compañero(a) no siempre están de acuerdo. ¿Qué dicen? Sigan el modelo.

modelo

nosotros: pedir dinero a nuestros padres

Tú: *Es lógico que pidamos dinero a nuestros padres.*

Compañero(a): *Es malo que les pidamos dinero. Es importante que ganemos nuestro propio dinero.*

1. él/ella: pensar en su novia(o) todo el tiempo
2. tú: vestirte de estilo *grunge*
3. ustedes: volver tarde a casa
4. usted: servir la cena antes de las ocho
5. los niños: jugar videojuegos todo el día
6. nosotros: acostarnos temprano

Nota cultural

Muchas de las unidades monetarias de Centroamérica muestran personas o cosas históricas. En Guatemala la moneda oficial es el **quetzal**, también nombre de un pájaro nativo de Centroamérica que tenía gran significado para los mayas. En Costa Rica y El Salvador se usa el **colón**, nombrado así en honor al famoso explorador Cristóbal Colón. El **balboa** es la moneda oficial de Panamá. Su nombre se refiere a Vasco Núñez de Balboa, uno de los primeros exploradores de esa región. Las otras monedas de esta región son el **peso** (México), la **córdoba** (Nicaragua) y el **lempira** (Honduras).

12 Consejos para proteger el planeta

Hablar/Escribir Hablas con tu primo(a) que es menor que tú. Quieres darle buenos consejos para que comience a proteger el planeta. ¿Qué le dices?

modelo

importante / empezar

Es importante que empieces a pensar en el ambiente.

1. importante/ empezar
2. bueno/ pensar
3. necesario/ entender
4. lógico/ resolver
5. mejor / pedir
6. peligroso/ perder
7. posible / querer
8. ¿...?

13 Recomendaciones

Hablar Tú y tu compañero(a) conversan sobre el medio ambiente. Hablen sobre los problemas que vean y algunas soluciones posibles.

modelo

Tú: *Es necesario que resolvamos el problema de la contaminación del aire. ¿No crees?*

Compañero(a): *Sí, claro. Por eso es importante que busquemos otras formas de energía. El petróleo es malo para el medio ambiente.*

More Practice: **Más comunicación** *p. R7*

¿RECUERDAS? *p. 46* You have already learned how to form the **present perfect** in the **indicative.**

> **present** tense of the
> auxiliary verb **haber,** *to have*   +   **past participle** of the verb.

The subjunctive also has a present perfect tense. To form it you use:

> present subjunctive
> of **haber**   +   **past participle** of the verb.

haya **lleg**ado	**hay**amos **lleg**ado
hayas **lleg**ado	**hay**áis **lleg**ado
haya **lleg**ado	**hay**an **lleg**ado

You use the present perfect subjunctive to indicate that the action of the subordinate clause took place in the past. Compare these sentences.

Las ruinas de Mitla

present subjunctive

Es posible que Juan **visit**e Mitla.
*It's possible that Juan is **visiting/will visit** Mitla.*

present perfect subjunctive

Es posible que Juan **hay**a **visit**ado Mitla.
*It's possible that Juan **has visited/visited** Mitla.*

present subjunctive

Es bueno que **hag**as eso.
*It's good that **you're doing/will do** that.*

present perfect subjunctive

Es bueno que **hay**as **hecho** eso.
*It's good that **you've done/you did** that.*

Notice how the meanings of two subjunctives contrast with each other.

Practice: **Actividades**   **Más práctica** *cuaderno p. 56*    **Online Workbook** CLASSZONE.COM
14 15   **Para hispanohablantes** *cuaderno p. 54*

Activity 15: Understand most spoken language when the message is deliberately and carefully conveyed by a speaker accustomed to dealing with learners when listening

14 El colegio

Hablar/*Escribir* Todos tienen mucho que hacer para sus clases en el colegio. Tú y tu compañero(a) conversan sobre qué han hecho y no han hecho los estudiantes de su clase. ¿Qué dicen?

modelo

yo: estudiar para el examen toda la noche (es importante)

Tú: *Yo estudié para el examen toda la noche.*

Compañero(a): *Es importante que (tú) hayas estudiado para el examen toda la noche.*

1. Marta y Martín: no hacer la tarea (es increíble)
2. Pedro: no ir a la biblioteca todavía (es posible)
3. tú: trabajar toda la noche (es bueno)
4. nosotros: olvidar el examen (es malo)
5. Arturo: comprender la tarea (es importante)
6. Fumiko y Kai: no entender el capítulo (es posible)

Nota cultural

El movimiento ecológico mexicano empezó en serio a principios de los años ochenta, con la constitución del **Grupo de los Cien**. Este grupo incluyó a los artistas, intelectuales, académicos y políticos más famosos de México. Actualmente existen diversos grupos ambientalistas como **Biodiversidad** y **Red Ambiental Joven de México**.

15 El club de ecología

Escuchar/Hablar/*Escribir* Escucha lo que dice el presidente del club de ecología sobre lo que hizo cada miembro esta semana. Luego, di si lo que han hecho es bueno o malo. Sigue el modelo.

modelo

Armando y Laura (es bueno que)

Es bueno que Armando y Laura hayan participado en la campaña para embellecer la ciudad.

1. Juan (es bueno que)
2. Esperanza y Arnoldo (es bueno que)
3. usted (es malo que)
4. ustedes (es malo que)
5. nosotros (es bueno que)
6. tú (es malo que)

Vocabulario

Nuestro planeta

la altura *height, altitude*

el bosque *forest*

el cielo *sky*

el clima *climate*

la colina *hill*

diverso(a) *diverse*

el ecosistema *ecosystem*

la fauna silvestre *wild animal life*

la flora silvestre *wild plant life*

la naturaleza *nature*

la piedra *rock*

la selva *jungle, forest*

la sequía *drought*

el valle *valley*

▶ ¿Qué palabras usarías para describir la región donde vives?

Activities 16–17 bring together all concepts presented.

16 Sierra Madre

Hablar/Leer Tú y tu compañero(a) ven este anuncio para la empresa Sierra Madre en una revista. Léanlo y juntos contesten las siguientes preguntas.

1. ¿Para qué trabaja el grupo Sierra Madre?

2. ¿Dónde tiene programas?

3. ¿Qué hace el grupo Sierra Madre en la sociedad?

4. ¿De qué depende en gran parte el futuro de los mexicanos?

5. Según lo que aprendiste en este capítulo, ¿qué clase de grupo es Sierra Madre?

6. ¿Qué te gustaría hacer si trabajaras en Sierra Madre?

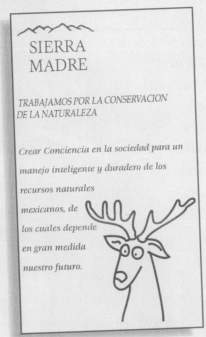

SIERRA MADRE

TRABAJAMOS POR LA CONSERVACIÓN DE LA NATURALEZA

Crear Conciencia en la sociedad para un manejo inteligente y duradero de los recursos naturales mexicanos, de los cuales depende en gran medida nuestro futuro.

17 La naturaleza

Hablar En grupos de tres o cuatro, conversen sobre la naturaleza de su pueblo. Mencionen los efectos del clima y todo lo que se relaciona con la conservación de la naturaleza.

modelo

Tú: *Por fin llovió este mes.*

Amigo(a) 1: *Sí. Es bueno que la sequía se haya acabado.*

Amigo(a) 2: *Tienes razón. El mes pasado los ríos habían bajado.*

Amigo(a) 3: *Y las plantas silvestres se estaban muriendo.*

More Practice: **Más comunicación** *p. R7*

Online Workbook
CLASSZONE.COM

Refrán

El que planta árboles ama a otros además de a sí mismo.

¿Qué quiere decir el refrán? En tu opinión, ¿por qué es sembrar árboles un acto de beneficencia? ¿Puedes pensar en otras actividades que benefician a la comunidad y a sus ciudadanos?

En colores
CULTURA Y COMPARACIONES

UNIDOS podemos hacerlo

PARA CONOCERNOS
STRATEGY: CONNECTING CULTURES Gather and analyze information about **literacy** Look up the word literacy. Then use the word **alfabetización** to add to that definition. Below is a chart showing literacy rates in three countries. In the library, look up two or three other countries and add them to the chart. What do you think accounts for differences among countries?

País	Por ciento
Argentina	96%
E.E.U.U.	97%
Honduras	73%

Nicaragua es uno de los países más hermosos de América Central, pero ha tenido que hacer muchos cambios para avanzar su desarrollo económico. A principios de los años 80, casi 50% de los nicaragüenses no sabían leer ni escribir. Desde entonces, Nicaragua ha organizado campañas de alfabetización para enseñar a leer y escribir a sus ciudadanos. Ahora, cerca del 65% de estos saben leer.

Vas a leer una página del diario de Rafaela Dávila, una estudiante que ayudó a la comunidad sirviendo como profesora en la campaña de alfabetización[1] en Nicaragua. Ella era estudiante de preparatoria[2] en Masaya cuando pasó el verano en el pueblo de Santo Tomás del Norte, situado en las montañas cerca de Honduras.

---

[1] literacy
[2] preparatory school

Santo Tomás del Norte, martes 30 de junio

Ya llevo una semana en Santo Tomás del Norte y me encuentro muy a gusto[3] aquí. Vivo en casa de los Rodríguez, una familia que tiene cinco hijos. La señora Rodríguez, Doña Rosa, sabe leer un poco, pero el señor Rodríguez, Don Mario, no sabe ni una letra del abecedario[4]. Es carpintero y él y los otros hombres del pueblo se han juntado para reparar una vieja casa abandonada que nos servirá de[5] escuela.....

Noto que hay mucho entusiasmo en el pueblo. Esta gente nunca ha tenido la oportunidad de estudiar nada y tiene muchas ganas de aprender. Un caso en especial me pareció muy conmovedor[6]. Ayer en la calle me habló una anciana. Dicen en el pueblo que tiene ochenta y cinco años. Ella me mostró su lápiz y su cuaderno y me preguntó que si las clases comenzaban ese mismo día. Le dije que hoy no, que pasado mañana. —Bueno, me contestó con una sonrisa, —hace más de ochenta años que espero. Puedo esperar dos días más.

---

[3] comfortable, happy
[4] alphabet
[5] will be used as
[6] moving

Campaña Nacional de Alfabetización,
Ministerio Nicaragüense de Educación

CLASSZONE.COM
More About Central America

¿Comprendiste?

1. ¿Quién es Rafaela Dávila?
2. ¿Qué hace en Santo Tomás?
3. ¿Qué hacen los hombres de Santo Tomás del Norte para ayudar a Rafaela?
4. ¿Qué quería saber la anciana que habló con Rafaela?

¿Qué piensas?

¿Por qué están tan entusiasmados por aprender los residentes de Santo Tomás del Norte?

Hazlo tú

Piensa en lo que significa ser voluntario(a). ¿Qué conocimientos y cualidades personales debe tener esa persona? Elige un campo en el cual te interese ser voluntario(a) (la alfabetización, el servicio a la comunidad, la política, u otro). Haz un póster o escribe un anuncio para atraer a otros voluntarios.

Activity 1: Generally use culturally appropriate behavior in social situations

En uso
REPASO Y MÁS COMUNICACIÓN

OBJECTIVES
- Say what should be done
- React to the ecology
- React to others' actions

Now you can...

- say what should be done.

To review

- the present subjunctive of regular verbs see p. 136.
- the present subjunctive of irregular verbs see p. 138.

1 ¿Crees que...?

Tu compañero(a) quiere hacerte algunas preguntas sobre el medio ambiente. ¿Qué te pregunta y cómo le respondes?

modelo

nosotros / proteger el planeta

Compañero(a): *¿Debemos proteger el planeta?*

Tú: *Sí, es importante que protejamos el planeta.*

1. nosotros / ir en autobús
2. yo / reciclar las botellas de vidrio
3. las comunidades / limitar el uso de aerosoles
4. nosotros / no echar todo a la basura
5. las ciudades / instituir un programa de reciclaje
6. nosotros / no destruir la capa de ozono
7. yo / saber más sobre el medio ambiente
8. los políticos / declarar zonas de reserva ecológica

Now you can...

- say what should be done.

To review

- the present subjunctive of regular verbs see p. 136.
- the present subjunctive of stem-changing verbs see p. 139.

2 ¡Un robo!

Dos personas vieron un robo. ¿Qué es necesario que hagan?

modelo

ayudar a las víctimas

Es necesario que ayuden a las víctimas.

1. llamar a la policía
2. pedir información de las víctimas
3. volver a la escena del crimen
4. contar lo que pasó a la policía
5. describir a los criminales
6. recordar qué pasó

Now you can...

• react to the ecology.

To review

• the present subjunctive of regular verbs see p. 136.

• the present subjunctive of irregular verbs see p. 138.

3 Una entrevista

Tú eres el (la) nuevo(a) presidente(a) del club para proteger el medio ambiente. Te hacen una entrevista para el periódico de la escuela sobre tus opiniones.

modelo

¿Es necesario hacer todo lo posible para preservar nuestro planeta?

¡Claro! Es necesario que hagamos todo lo posible para preservar nuestro planeta.

1. ¿Es posible reducir la contaminación del aire?
2. ¿Es bueno reciclar los productos de plástico?
3. ¿Es importante buscar otras formas de energía?
4. ¿Es malo contaminar el aire con combustibles?
5. ¿Es lógico limitar el uso de aerosoles?
6. ¿Es malo no respetar el medio ambiente?
7. ¿Es triste no comprender la importancia de proteger la fauna silvestre?
8. ¿Es una lástima no saber conservar los recursos naturales?

Now you can...

• react to others' actions.

To review

• the present perfect subjunctive see p. 141.

4 Mi reacción

¿Qué piensas de las actividades de los demás? Escribe una oración que describa tu reacción.

modelo

bueno: Los científicos desarrollaron otras formas de energía.

Es bueno que los científicos hayan desarrollado otras formas de energía.

1. bueno: Tú reciclaste las latas.
2. lógico: Mis vecinos participaron en la limpieza del parque.
3. malo: Mi amigo no fue a la reunión del club de ecología.
4. lástima: Nosotros destruimos la capa de ozono.
5. triste: Ustedes contribuyeron a la contaminación del aire.
6. bueno: Las ciudades prohibieron el uso de contaminantes.
7. lógico: Los políticos instituyeron un programa de reciclaje.
8. malo: Tú echaste las revistas a la basura.

5 Y yo, ¿qué puedo hacer?

STRATEGY: SPEAKING
Express support or lack of support Listen to your group members' plans and decide if you want to support their efforts. Use impersonal expressions (p. 133) to state your position or to ask for clarification from your classmates.

En grupos de dos o tres, conversen sobre lo que pueden hacer hoy para proteger el medio ambiente.

modelo

Tú: *Yo voy a caminar al colegio.*

Amigo(a) 1: *Yo voy a reciclar mis revistas.*

Amigo(a) 2: …

6 ¿Cuál es el problema?

¿Cuál es el problema ambiental más grave en tu ciudad? ¿Qué sugieres para resolverlo? En grupos de dos o tres, conversen sobre los problemas en su comunidad y las soluciones posibles.

modelo

Amigo 1: *Yo creo que el desperdicio es el problema más grave que tenemos.*

Amigo 2: *Es importante que reciclemos todos los productos de plástico, papel y vidrio.*

Amigo 3: *Yo creo que la contaminación del aire es el problema más grave.*

Amigo 1: …

7 En tu propia voz

ESCRITURA Te han elegido para escribir y diseñar un folleto (*brochure*) que se va a distribuir en tu colegio. El propósito del folleto es educar a los estudiantes sobre el medio ambiente y cómo protegerlo. Escribe dos o tres oraciones que expliquen la importancia de proteger el medio ambiente. Ilustra tus ideas si quieres.

modelo

¿Tienes bicicleta? ¡Úsala! Las bicicletas no echan combustibles al aire. Cada vez que te subes a un carro, estás destruyendo la capa de ozono. ¿Lo sabías?

CONEXIONES

Las ciencias Haz una investigación sobre el reciclaje en tu comunidad. ¿Quién recicla y cuánto? Haz una gráfica de lo que reciclas tú y lo que reciclan tu familia, tu escuela y tu comunidad en una semana. Incluye la cantidad. Haz un reportaje de lo que aprendas. Compara el porcentaje de lo que echan a la basura y lo que reciclan. ¿Es necesario que tu escuela y comunidad hagan más? ¿Es posible que tú puedas hacer más?

El reciclaje de...	el papel/el cartón	el vidrio	las latas	el plástico
mi casa			10 latas	2 botellas de jugo
mi escuela	papel de computadora			
mi comunidad	periódicos/revistas		latas de sopa	

En resumen
REPASO DE VOCABULARIO

REACT TO THE ECOLOGY

Environment

el combustible	fuel
los efectos	effects
el medio ambiente	environment
el planeta	planet
la población	population
por todas partes	all around
los recursos naturales	natural resources
la tierra	land

Problems

el aerosol	aerosol
la capa de ozono	ozone layer
complicado(a)	complicated
la contaminación del aire	air pollution
el contaminante	pollutant
dañino(a)	damaging
el derrame de petróleo	oil spill
el desperdicio	waste
la destrucción	destruction
echar	to throw away
inútil	useless
¡Qué lío!	What a mess!
el químico	chemical
el smog	smog

Solutions

¡A todos nos toca!	It's up to all of us!
la botella	bottle
el cartón	cardboard
desarrollar	to develop
descubrir	to discover
increíble	incredible
instituir	to institute
la lata	can
el permiso	permission
el plástico	plastic
el programa de reciclaje	recycling program
prohibir	to prohibit

Solutions (continued)

proteger	to protect
reducir	to reduce
respetar	to respect
separar	to separate
el vidrio	glass
las zonas de reserva ecológica	conservation land

Nature

la altura	altitude
el bosque	forest
el cielo	sky
el clima	climate
la colina	hill
diverso(a)	diverse
el ecosistema	ecosystem
las especies	species
la fauna silvestre	wild animal life
la flora silvestre	wild plant life
la naturaleza	nature
la piedra	rock
la selva	jungle, forest
la sequía	drought
el valle	valley

SAY WHAT SHOULD BE DONE

 Ya sabes

Es...	
bueno que	It's good that
importante que	It's important that
lógico que	It's logical that
malo que	It's bad that
mejor que	It's better that
necesario que	It's necessary that
peligroso que	It's dangerous that
posible que	It's possible that
probable que	It's probable that
raro que	It's rare that
ridículo que	It's ridiculous that
triste que	It's sad that
una lástima que	It's a pity that

SAY WHAT SHOULD BE DONE

Subjunctive with regular verbs

Es malo que **contaminemos** el aire y el agua.

Subjunctive with irregular verbs

Es bueno que **sepas** más sobre el medio ambiente.

Subjunctive with stem-changing verbs

Es importante que **recuerden** que el planeta es para todos.

REACT TO OTHERS' ACTIONS

Present perfect subjunctive

Es lógico que **hayas empezado** un programa de reciclaje.

Juego

Sopa de letras

Pon en orden las letras siguientes para saber qué es lo que no se puede ver, pero que nos protege todos los días. (¡Ojo! Son cuatro palabras enteras.)

ETAPA
3

La riqueza natural

OBJECTIVES

- React to nature

- Express doubt

- Relate events in time

¿Qué ves?

Mira la foto. Contesta las preguntas.

1. ¿Qué cosas ven los turistas?

2. ¿Dónde se encuentran?

3. ¿Cómo crees que se sentirán el chico de la mochila azul y el chico de la gorra? ¿Cómo te sentirías tú?

4. ¿Sabes de qué país viene la foto pequeña? ¿Cómo podrías averiguarlo?

BIENVENIDO
PARQUE NACIONAL VOLCAN POAS

HORARIO DE INGRESO
DE 8:00 A.M A 3:30 P.M

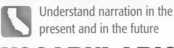
EcoTurista

Lo mejor de dos mundos: la ecología y el turismo

Seis recomendaciones para el EcoTurista que quiere conocer CENTROAMÉRICA.

En COSTA RICA

Los ríos Sarapiquí, Corobicí y Pacuare Costa Rica cuenta con buenos ríos para la **navegación por rápidos**. Este deporte, a veces **peligroso**, es una aventura. El que esté atento verá iridiscentes **mariposas** azules y los colores llamativos de los **tucanes**.

En EL SALVADOR

Parque Nacional Montecristo
Si tienes suerte, aquí verás monos, **zorrillos** y **osos hormigueros**. Hay 87 especies de pájaros como los **picaflores** y los **búhos**.

En GUATEMALA

Biotopo Cerro Cahuí
Aquí verás los colores brillantes de las muchas variedades de tucanes, **loros** y **halcones**. En la selva hay más de 20 especies de animales, tales como los **monos araña**, los **venados** y los **ocelotes**. En el agua hay **tortugas** y **serpientes**.

En PANAMÁ

El Refugio de Vida Silvestre Isla Iguana

Este refugio es famoso no sólo por las **iguanas** que le dan su nombre, sino también por las **ballenas jorobadas** que visitan sus mares de junio a noviembre. En Isla Iguana puedes **bucear** y **nadar con tubo de respiración** si quieres ver los **peces** tropicales, pescar o solamente descansar en la playa.

En HONDURAS

Refugio de Vida Silvestre Cuero y Salado

En este refugio encontrarás una selva llena de animales **salvajes:** monos, **jaguares** y **boas constrictoras**. Entre las 196 especies de pájaros hay tucanes, loros y **pelícanos**.

En NICARAGUA

Isla de Ometepe

Esta isla es la más grande del mundo en un lago **de agua dulce**. Los únicos **tiburones** de agua dulce del mundo viven en el Lago de Nicaragua. ¡Atrévete!

Online Workbook CLASSZONE.COM

¿Comprendiste?

1. ¿Has ido alguna vez a una reserva ecológica? ¿Cuál?
2. ¿Has visto algunos de los animales mencionados en la revista EcoTurista? ¿Cuáles?
3. ¿Te gustaría conocer uno de los lugares recomendados por la revista EcoTurista? ¿Cuál? ¿Por qué?
4. ¿Crees que serías un(a) buen(a) EcoTurista? ¿Por qué o por qué no?
5. ¿Piensas que es importante que los países protejan su flora y fauna? ¿Crees que el ecoturismo es una buena idea para realizar eso?

AUDIO

En vivo

SITUACIONES

PARA ESCUCHAR • STRATEGY: LISTENING

Pre-listening What memories do you have of visits to the zoo? What are the pros and cons of taking animals out of their natural habitat?

Determine your purpose for listening You will hear a guide give a tour of a zoological park. Before listening, decide whether you are on a field trip or a visit with friends. Then, make a list of listening strategies you consider most appropriate and use them while listening.

How effective were your purpose and strategies? Would a different purpose and strategies change your understanding and memory?

El Parque SalvaNatura

Estás en el Parque SalvaNatura y escuchas al guía turístico, quien explica las secciones y la vida silvestre del parque.

➊ Leer

Cuando llegas al Parque SalvaNatura, la primera cosa que te da el guía turístico es un mapa del parque. El parque tiene seis secciones que representan la flora y fauna de cada país en Centroamérica. Estudia el mapa para ver qué hay en cada uno.

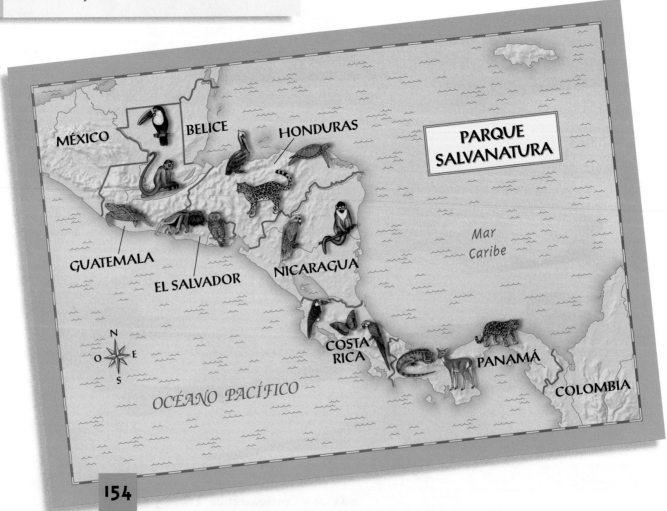

② Escuchar

El guía turístico te explica qué animales hay en cada sección del parque. Escucha y decide si esos animales están (sí) o no están (no) en cada sección.

En la sección guatemalteca		
• tucanes	Sí	No
• monos araña	Sí	No
• serpientes	Sí	No

En la sección hondureña		
• jaguares	Sí	No
• venados	Sí	No
• tortugas	Sí	No

En la sección salvadoreña		
• boas constrictoras	Sí	No
• mariposas	Sí	No
• osos hormigueros	Sí	No

En la sección nicaragüense		
• tiburones	Sí	No
• monos	Sí	No
• serpientes	Sí	No

En la sección costarricense		
• mariposas	Sí	No
• loros	Sí	No
• jaguares	Sí	No

En la sección panameña		
• iguanas	Sí	No
• venados	Sí	No
• jaguares	Sí	No

③ Hablar/Escribir

Ahora tú eres el (la) guía turístico(a) de un parque zoológico. ¿Qué animales escogerías para tu parque? Con dos o tres compañeros, diseñen un parque, decidan qué animales traerían a su parque y dibujen un mapa para los turistas con el país de origen de cada animal. Den un nombre imaginativo a su parque.

Parque Nacional Corcovado Costa Rica

En acción

PARTE A — Práctica del vocabulario

Objectives for Activities 1-4
• React to nature

1 Los animales

Hablar/Escribir ¿Qué oración describe a cada dibujo?

I.

2.

3.

4.

5.

6. ¡Buenos días!

a. El tucán es un pájaro con un pico de muchos colores.

b. Los jaguares son de la familia de los gatos.

c. Hay muchas mariposas en esta selva.

d. Dicen que las tortugas son muy lentas.

e. Los loros pueden imitar el habla de la gente.

f. ¡Mira la ballena jorobada! ¡Qué enorme!

2 El hábitat natural

Hablar/Escribir ¿Cuál es el hábitat natural de estos animales? Pregúntale a tu compañero(a).

modelo

ballena

  a. río   b. mar   c. tierra

Tú: *¿Cuál es el hábitat natural de la ballena?*

Compañero(a): *Las ballenas viven en el mar.*

I. venado
  a. río   b. mar   c. bosque

2. jaguar
  a. selva   b. mar   c. río

3. tortuga
  a. palmera   b. montañas   c. río

4. mono araña
  a. río   b. mar   c. selva

5. pelícano
  a. mar   b. bosque   c. montañas

6. tiburón
  a. montañas   b. mar   c. río

También se dice

Hay muchas maneras de nombrar los animales.

- **chango** = mono
- **perico, cotorra** = loro
- **víbora** = serpiente
- **colibrí** = picaflor
- **mofeta** = zorrillo

3 **¿Has visto...?**

Hablar/Escribir Quieres saber qué animales tu compañero(a) ha visto y qué piensa de ellos. Hazle preguntas sobre los animales en las fotos. Luego cambien de papel.

modelo

Tú: *¿Has visto un mono araña alguna vez?*

Compañero(a): *Sí, lo vi en el zoológico.*

Tú: *¿Qué piensas de los monos?*

Compañero(a): *Pues, son muy inteligentes y cómicos.*

I.

2.

3.

4.

5.

6.

4 **¡Vamos a Costa Rica!**

Hablar/Escribir Vas a viajar a Costa Rica. Pregúntale a tu amigo(a) costarricense dónde puedes ver las cosas que te interesan. Usa el mapa como guía. Luego, cambien de papel.

modelo

Tú: *Me interesan mucho las tortugas. ¿Adónde debo ir?*

Compañero(a): *Debes ir al Parque Nacional Tortuguero. Allí hay muchas tortugas.*

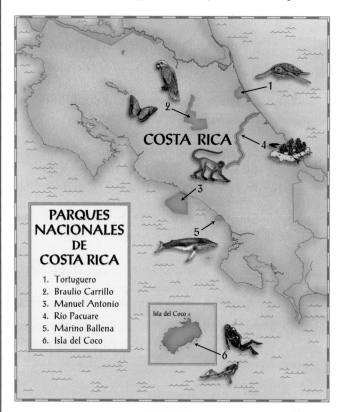

PARQUES NACIONALES DE COSTA RICA

1. Tortuguero
2. Braulio Carrillo
3. Manuel Antonio
4. Río Pacuare
5. Marino Ballena
6. Isla del Coco

Objectives for Activities 5–14

• React to nature • Express doubt • Relate events in time

REPASO | **The Subjunctive with Expressions of Emotion**

▶ Remember that you use the **subjunctive** after expressions of emotion such as
I'm happy and *I'm sad.*

The vocabulary box at the right lists some **expressions of emotions.**

> **Me alegro de que** te **guste** la naturaleza.
> *I'm happy that you like nature.*

> **Espero que podamos** hacer algo pronto.
> *I hope that we can do something soon.*

▶ You can use either the **present subjunctive** or the
present perfect subjunctive after expressions of emotion.
Just remember that:

• the **present subjunctive** refers to **present** or **future** time.

• the **present perfect subjunctive** refers to the **past**.

> **Es triste que haya** tanta contaminación.
> *It's sad that there is so much pollution.*

> Sí. **Ojalá que puedan** reducirla. **Tengo miedo
> de que sea** muy peligrosa
> *Yes. I hope that they can reduce it. I'm afraid that it is very dangerous.*

> **Siento que** no **haga** buen tiempo.
> *I'm sorry the weather is not good.*

> **Siento que** no **haya hecho** buen tiempo.
> *I'm sorry the weather wasn't good.*

Vocabulario

♻ **Ya sabes**

Es ridículo que...

Es triste que...

Es una lástima que...

Espero que...

Me alegro de que...

Ojalá que...

Siento que...

Tengo miedo de que...

Practice: | **Actividades** ⑤ ⑥ ⑦ | **Más práctica** *cuaderno p. 61*
Para hispanohablantes *cuaderno p. 59* | **Online Workbook**
CLASSZONE.COM

Nota cultural

La isla de Ometepe se considera una de las islas más bellas del
mundo, localizada en el Lago de Nicaragua. El lago es casi del mismo
tamaño(*size*) del lago Titicaca en Perú. La isla de Ometepe tiene dos
grandes montañas gemelas. Los aventureros salen de las montañas en
planeadores o avionetas pequeñas para ver la belleza natural de esta
isla, uno de los centros turísticos más bellos de Centroamérica.

Activity 7: Identify, state, and understand feelings and emotions

5 Reunión familiar

Hablar/Escribir Tu tía llama a toda la familia para hablar de sus esperanzas para la reunión familiar. ¿Qué le dice a cada uno?

> **modelo**
>
> a ustedes (llegar a tiempo)
>
> *Espero que lleguen a tiempo.*

1. a ti (gustar la comida)
2. a usted (divertirse)
3. a mi prima (invitar a tu novio)
4. a mis papás (bailar mucho)
5. a mí (comer suficiente)
6. a ustedes (vestirse bien)
7. a tu tío (llegar temprano)
8. sus amigas (traer el postre)

6 Juan

Escuchar/Escribir Escucha lo que les dice Juan a varios amigos y parientes. Luego escribe cómo se siente Juan en cada situación. Primero escucha el modelo.

> **modelo**
>
> Adriana
>
> *Juan espera que Adriana vaya con él a la fiesta.*

7 El mundo de hoy ♻ 👥

STRATEGY: SPEAKING

Gain thinking time before speaking Sometimes ideas do not come to us as quickly as we would like. One way to gain time is to restate what was just said which may in turn trigger a fresh idea. Example: **Sí, es una lástima. Espero que se proteja la selva también.**

Hablar/Escribir Tú y tu compañero(a) hablan sobre el estado del mundo de hoy. Usa las frases de la primera columna para expresar cómo te sientes sobre alguna situación en la segunda columna.

> **modelo**
>
> **Tú:** *Es una lástima que los jaguares estén en peligro de extinción.*
>
> **Compañero(a):** *Tienes razón, es muy triste. Espero que…*

Me alegro de que…	países: (no) declarar zonas de reserva ecológicas
Siento que…	jaguares: (no) estar en peligro de extinción
Tengo miedo de que…	ciudades: (no) instituir programas de reciclaje
Es una lástima que…	gente: (no) reciclar el plástico
Ojalá que…	nosotros: (no) proteger las especies
Espero que…	nosotros: (no) destruir la capa de ozono
Es triste que…	gente: (no) respetar el planeta
Es ridículo que…	carros: (no) contaminar el aire (no) haber derrames de petróleo ¿…?

More Practice: Más comunicación *p. R8*

Activity 8: Express and understand opinions

REPASO **The Subjunctive to Express Doubt and Uncertainty**

▶ Remember that you use the **subjunctive** in the dependent clause after **expressions of doubt** and **uncertainty** such as those in the vocabulary box.

> **Dudo que** tus primos **quieran** acampar con nosotros.
> *I doubt that your cousins **want** to camp with us.*

> ¿Quién sabe? **Quizás** les **interese** la idea.
> *Who knows? **Maybe** the idea **will interest** them.*

> **No creo que** Pedro **haya visto** un animal salvaje en toda su vida.
> *I don't think that Pedro **has seen** a wild animal in his whole life.*

> **Tal vez quiera** ir con nosotros al refugio de vida silvestre.
> *Maybe he **would want** to come with us to the wildlife preserve.*

Vocabulario

 Ya sabes

Dudo que…

No creo que…

No es cierto que…

No es seguro que…

No es verdad que…

Quizás…/ Quizá…

Tal vez…

▶ You can use the **present subjunctive** or the **present perfect subjunctive** after **expressions of doubt** and **uncertainty.**

* the **present subjunctive** refers to **present** or **future** time

> **No es cierto que** Julio y Vera **naden** con tubo de respiración.
> *It's not true that Julio and Vera **are going** to snorkel.*

* the **present perfect subjunctive** refers to the **past**

> **No es cierto que** Julio y Vera **hayan nadado** con tubo de respiración.
> *It's not true that Julio and Vera **have** snorkeled.*

▶ Normally, you don't use the **subjunctive** after the following expressions because they express **certainty,** not doubt.

> *expresses certainty*
>
> Yo **no dudo que** él ya sabe navegar por rápidos.
> *I **don't doubt that** he already knows how to white water raft.*

no dudo que…

creo que…

es cierto que…

es verdad que…

es seguro que…

Practice: **Actividades** **8 9 10**

Más práctica *cuaderno p. 62*
Para hispanohablantes *cuaderno p. 60*

 **Online Workbook**
CLASSZONE.COM

8 Dudo que...

Hablar/Escribir Tienes algunas dudas sobre los animales de la selva y el mar. Exprésalas.

> **modelo**
>
> *(no) dudo que: haber / tiburones en la playa*
>
> *Dudo que haya tiburones en la playa.*

1. (no) creo que: las mariposas / vivir solamente en el bosque
2. (no) dudo que: el ecoturismo / resolver todos los problemas ecológicos
3. (no) es verdad que: los venados / ser peligrosos
4. (no) es seguro que: los monos / entender a los humanos
5. (no) dudo que: los tucanes / poder hablar
6. (no) creo que: las serpientes / comer peces
7. (no) es cierto que: las ballenas / existir sólo en los ríos
8. (no) es cierto que: las tortugas / estar en peligro de extinción
9. (no) es verdad que: la isla de Ometepe / ser muy bella
10. (no) creo que: los osos hormigueros / comer fruta

La tortuga

Es uno de los animales más apreciados por el hombre, de nosotros depende su conservación. No adquiera ni consuma productos de tortuga, ayudemos a preservar con ella la biodiversidad de México.

9 No te creo

Hablar/Escribir Tu amigo(a) siempre exagera y dice que ha hecho cosas increíbles. Tú nunca le crees. ¿Cómo le respondes cuando dice que ha hecho las siguientes cosas?

> **modelo**
>
> *Fui a...*
>
> **Amigo(a):** *Fui a Centroamérica.*
>
> **Tú:** *No creo que hayas ido a Centroamérica.*

1. Hice alpinismo en…
2. Navegué por rápidos en…
3. Vi un jaguar en…
4. Buceé en…
5. Nadé con tubo de respiración en…
6. Pesqué en alta mar…
7. Piloté una avioneta en…
8. Conocí a Rigoberta Menchú…

Nota cultural

Recientemente muchos de los países de Centroamérica, particularmente Guatemala, Honduras, Nicaragua, Costa Rica y Panamá, han designado varias áreas como **reservas naturales** destinadas a la preservación de la fauna y flora de la región.

Activity 10: Converse and listen during
face-to-face social interactions

10 Conversaciones diarias

Hablar A veces tenemos dudas sobre las situaciones en que nos encontramos. Con un(a) compañero(a), expresa algunas dudas que tengas usando las expresiones de la página 160. Tu compañero(a) te contesta usando *tal vez* o *quizás*. Luego, cambien de papel.

modelo

Tú: *No creo que papá esté en casa.*

Compañero(a): *Tal vez esté en la oficina.*

Compañero(a): *Dudo que mi hermana me preste el dinero.*

Tú: *¿Quién sabe? Quizás se sienta generosa.*

comprar	tener
hablar	venir
invitar	¿?

GRAMÁTICA   The Subjunctive with **cuando** and Other Conjunctions of Time

▶ You use the subjunctive after certain **conjunctions of time** to show that you are **not sure when** or **if** something will happen.

You use the indicative with the same **conjunctions of time**, if the **main clause** refers to the present or the past. Using the indicative shows that you are already certain about the outcome of the action described in the subordinate clause.

Vocabulario

El tiempo

cuando *when*

en cuanto *as soon as*

hasta que *until*

tan pronto como *as soon as*

Subjunctive	Indicative
not sure of outcome	certain of outcome

Bucearán **hasta que** anochezca.
They will scuba dive until it gets dark.

Bucearon **hasta que** anocheció.
They were scuba diving until it got dark.

Avísame **cuando** sepas.
Let me know when you find something out.

Siempre me avisas **cuando** sabes.
You always let me know when you find something out.

Practice: **Actividades**   11 12 13 14

Más práctica *cuaderno pp. 63–64*
Para hispanohablantes *cuaderno pp.61–62*

 Online Workbook
CLASSZONE.COM

11 El Refugio

Hablar/Escribir Un grupo de tu colegio va al refugio. ¿Qué dicen que van a hacer y hasta cuándo?

modelo

seguir el sendero / acabarse

Vamos a seguir el sendero hasta que se acabe.

> ### Nota: Gramática
>
> Remember that when you form the subjunctive of verbs that end in **-cer** (such as **conocer, oscurecer, anochecer, amanecer,** etc.), the subjunctive forms include a **z: conozca, conozcas, conozca, conozcamos, conozcan.**

1. caminar / no haber luz
2. nadar / oscurecer
3. dormir / amanecer
4. hacer alpinismo / anochecer
5. disfrutar de la naturaleza / (nosotros) cansarse
6. observar la flora y fauna / (nosotros) tener hambre
7. buscar monos / (nosotros) ver uno
8. quedarnos / (ellos) cerrar el refugio

12 Tan pronto como...

Hablar/Escribir Tú y tu hermano(a) menor van a acampar en un parque nacional y él (ella) te hace muchas preguntas. ¿Qué le dices?

modelo

recoger la leña

Tu hermano(a): *¿Por qué no recoges la leña?*

Tú: *Recogeré la leña tan pronto como lleguemos al campamento.*

1. hacer el fuego
2. abrir la tienda de campaña
3. acostarte en el saco de dormir
4. darme la manta
5. poner la linterna
6. abrir las latas
7. bajar la almohada del carro
8. traer agua
9. buscar un animal raro
10. cocinar sopa

Vocabulario

Vamos a acampar

el abrelatas *can opener*

la almohada *pillow*

(el) amanecer *dawn; to start the day*

(el) anochecer *nightfall; to get dark*

(el) atardecer *late afternoon; to get dark*

el campamento *camp*

descubrir *to discover*

el fósforo *match*

el fuego *fire*

la leña *firewood*

la linterna *flashlight*

la luz *light*

la manta *blanket*

la navaja *jackknife*

oscurecer *to get dark*

la oscuridad *darkness*

el saco de dormir *sleeping bag*

el sendero *path*

la tienda de campaña *tent*

▶ ¿Qué objetos necesitas para acampar?

Activity 13: Understand most spoken language when the message is deliberately and carefully conveyed by a speaker accustomed to dealing with learners when listening

13 Los quehaceres

Escuchar/Escribir ¡Pobre Imelda! Tiene muchos quehaceres hoy. Escucha lo que dice. Di si ya hizo el quehacer que menciona, o si todavía le falta hacerlo. Luego escribe una oración usando **en cuanto, hasta que** o **tan pronto como**.

modelo

No ha cortado el césped. Lo cortará en cuanto pueda.

1. planchar
2. cortar
3. limpiar
4. barrer
5. pasar
6. regar
7. hacer
8. sacar

14 El tiempo

Hablar/Escribir Pregúntale a tu compañero(a) si quiere hacer algo para disfrutar de la naturaleza. Él (Ella) te contesta de acuerdo a cómo está el tiempo. Luego, cambien de papel.

modelo

quitarse la neblina

Tú: *¿Quieres ir a pescar? o ¿Quieres ir a bucear?*

Compañero(a): *Sí, vamos a pescar (bucear) en cuanto se quite la neblina*

1. parar los truenos
2. irse el huracán
3. estar soleado
4. terminar el aguacero
5. irse las nubes
6. bajar la temperatura
7. pasar el relámpago
8. parar la llovizna

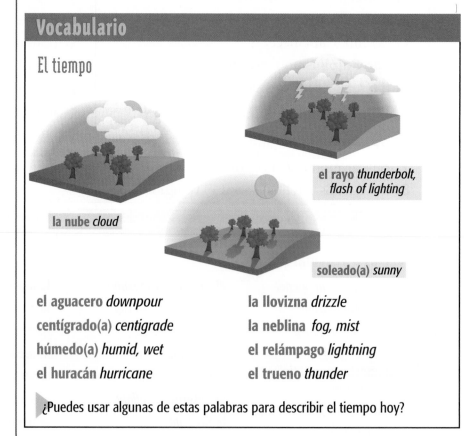

Vocabulario

El tiempo

el rayo *thunderbolt, flash of lighting*

la nube *cloud*

soleado(a) *sunny*

el aguacero *downpour*
centígrado(a) *centigrade*
húmedo(a) *humid, wet*
el huracán *hurricane*

la llovizna *drizzle*
la neblina *fog, mist*
el relámpago *lightning*
el trueno *thunder*

▶ ¿Puedes usar algunas de estas palabras para describir el tiempo hoy?

Activity 15 brings together all concepts presented.

15 La finca de mariposas

Hablar/Leer En grupos de tres o cuatro, hablen sobre este folleto de La Finca de Mariposas en La Guácima de Alajuela, Costa Rica.

modelo

Tú: ¿Te gustan las mariposas?

Amigo(a) 1: Sí. Pero no sé cómo atraerlas a mi jardín.

Amigo(a) 2: Pues, aquí dice que …

Amigo(a) 3: ¿Sabes cuántas especies de mariposas hay…?

Refrán

A cada pájaro le gusta oír su propio canto.

¿Qué quiere decir el refrán? ¿Estás de acuerdo? ¿Crees que preferimos las cosas que conocemos bien a las que no conocemos?

La Finca de Mariposas
GUÍA

P: ¿Cómo puedo atraer mariposas a mi jardín?
R: Sembrando tanto plantas ricas en néctar, como las plantas hospederas necesarias para las especies locales de mariposas.

P: ¿Cuántas especies de mariposas hay?
R: Hay aproximadamente 20,000 especies en el mundo. De éstas, cerca del 5%, o sea, 1000 especies existen en Costa Rica.

P: ¿Qué hace a una mariposa un insecto?
R: Como todos los insectos, las mariposas tienen 6 patas y 3 partes en su cuerpo: cabeza, tórax y abdomen. Las mariposas pertenecen al orden de insectos llamado lepidóptera (alas con escamas).

P: ¿Cuánto tiempo vive una mariposa?
R: El ciclo completo dura aproximadamente 2½ meses. El huevo dura unos 3 a 5 días, la larva 3 a 4 semanas y la pupa 1 ó 2 semanas. Luego el adulto vivirá otras 2 ó 3 semanas.

P: ¿Cuál es la función de las mariposas en la naturaleza?
R: Su principal función es servir como alimento a muchos depredadores. Ejemplos serían, hormigas, arañas, avispas, culebras, pájaros, monos, moscas parásitas y otros. Estos atacarán en uno o varios estados de su ciclo de vida (huevo, larva, pupa y adulto).

Fundada en 1983, mayor exportador de mariposas en occidente. Todos los visitantes a la finca reciben un tour de 1½ hora acerca de estos fascinantes y bellos insectos y sus hábitats naturales. Para más información llame o escriba a
La Finca de Mariposas, Frente Club Campestre Los Reyes, La Guácima de Alajuela, C.R.-Tel/fax 48-01-15.

More Practice: Más comunicación *p. R8*

 **Online Workbook** CLASSZONE.COM

AUDIO

En voces

LECTURA

PARA LEER

STRATEGY: READING

Recognize uses of satire, parody, and irony What advertisements, TV shows, or movies do you see that use humor about someone or something? Their humor is often based on these three devices:

Satire: use of sarcasm to make fun of someone or something

Parody: a satirical imitation of a serious piece of writing

Irony: use of language whose meaning is the opposite of what is intended

Give examples of satire, parody, or irony in *Baby H.P.*

EN LA CASA

ama de casa	mujer que trabaja en la casa
el agobiante ajetreo hogareño	actividad energética de la casa
la rabieta	cuando un niño llora y grita
los vástagos	niños

Sobre el autor

Juan José Arreola, cuentista mexicano, es uno de los escritores más originales y más importantes de su generación. Nació en Ciudad Guzmán en el estado de Jalisco en 1918. Publicó sus primeros cuentos en unas revistas de Guadalajara durante los años 40. Las piezas cortas escritas por Arreola son cuentos, fábulas, viñetas o simplemente piezas cortas. Arreola se sirve del humor para satirizar ciertas características de la sociedad y del ser humano.

Introducción

«Baby H.P.», escrita en 1952, es una pieza satírica que trata del mundo de la publicidad y los anuncios. El autor parodia los anuncios dirigidos a las amas de casa, describiendo un aparato que se pone al niño para conservar su energía y convertirla después en electricidad.

Baby H.P.

Señora ama de casa: convierta usted en fuerza motriz[1] la vitalidad de sus niños. Ya tenemos a la venta[2] el maravilloso Baby H.P., un aparato que está llamado a revolucionar la economía hogareña.

. . . .

El Baby H.P. es una estructura de metal muy resistente y ligera que se adapta con perfección al delicado cuerpo infantil, mediante cómodos cinturones, pulseras, anillos y broches[3]. Las ramificaciones de este esqueleto suplementario recogen[4] cada uno de los movimientos del niño, haciéndolos converger en una botellita de Leyden que puede colocarse en la espalda o en el pecho, según necesidad.

. . . .

De hoy en adelante usted verá con otros ojos el agobiante ajetreo de sus hijos. Y ni siquiera perderá la paciencia ante una rabieta convulsiva, pensando en que es una fuente[5] generosa de energía.

---

[1] power, moving force
[2] for sale
[3] fasteners, clips
[4] to collect
[5] source
[6] leftover, surplus
[7] storage battery
[8] overflowing

Las familias numerosas pueden satisfacer todas sus demandas de electricidad, instalando un Baby H.P. en cada uno de sus vástagos, y hasta realizar un pequeño y lucrativo negocio, transmitiendo a los vecinos un poco de la energía sobrante[6].

. . . .

Los niños deben tener puesto día y noche su lucrativo H.P. Es importante que lo lleven siempre a la escuela, para que no se pierdan las horas preciosas del recreo, de las que ellos vuelven con el acumulador[7] rebosante[8] de energía.

Online Workbook
CLASSZONE.COM

¿Comprendiste?

1. ¿Cómo son las obras de Arreola?
2. ¿Para qué sirve el aparato Baby H.P.?
3. ¿Cómo es el aparato?
4. ¿Qué ventaja tienen las familias numerosas?

¿Qué piensas?

1. ¿Qué te parece la idea de Arreola? ¿Por qué es buena? ¿Por qué no?
2. ¿Crees que él habla en serio? ¿Por qué? Cita frases del texto en tu respuesta.

Hazlo tú

Inventa un aparato y descríbelo, imitando el estilo de Arreola. Menciona cómo es, para qué sirve y los beneficios que tiene.

En colores

CULTURA Y COMPARACIONES

Un país de encanto

PARA CONOCERNOS

STRATEGY:
CONNECTING CULTURES
Analyze the advantages and
disadvantages of ecotourism

This brochure makes
ecotourism very appealing by
providing new experiences for
tourists and economic benefits
for local citizens. Analyze the
positive and possible negative
consequences of ecotourism.
Present your findings in a
chart that might be used to
inform both the Department
of Tourism and the Chamber
of Commerce. What is your
personal position as a result
of this analysis?

	ventajas	desventajas
ecoturismo	1.	1.
	2.	2.
	3.	3.

Are able to understand and retain most key ideas
and some supporting detail when reading

Costa Rica es muy conocido por sus
parques nacionales, reservas biólogicas
y refugios naturales. Miles de turistas
visitan este país para conocer sus
ecosistemas y disfrutar de su belleza.

Los aficionados de la
naturaleza encontrarán en los
bosques de Costa Rica más de
850 especies de aves[1] como
halcones, tucanes y pelícanos.

Al norte de San
José se encuentra el Parque
Nacional Braulio Carrillo,
creado para proteger el bosque
tropical de la construcción de la
carretera que va desde San José a
Puerto Limón.

---

[1] birds

El parque tiene tres tipos de vegetación según la altura del terreno [2] y varias especies de animales. Puedes explorar muchos senderos, como el Sendero Botella, donde hay una hermosa catarata [3]. Allí también viven monos, osos hormigueros, tortugas, serpientes, jabalíes [4], una variedad de mariposas y hasta coyotes y tapires.

En la costa del Pacífico, el Parque Nacional Manuel Antonio llama la atención por sus playas. ¡Ojo! En este parque los monos y las iguanas no tienen miedo de la gente. Suelen acercarse [5] a los turistas en las hermosas playas de Espadilla Sur, Manuel Antonio y Puerto Escondido. Pero no des de comer a los monos porque se ponen muy pesados. ¡Hasta se meten [6] en las bolsas!

---

[2] altitude of terrain
[3] waterfall
[4] wild boars
[5] they often come up to
[6] get into

CLASSZONE.COM
More About Central America

¿Comprendiste?

1. ¿Qué animales se pueden ver en los parques nacionales de Costa Rica?
2. ¿Cómo es el Parque Nacional Braulio Carrillo?
3. ¿Qué parque nacional tiene playas hermosas?
4. ¿Cuál de los parques tiene una catarata?
5. ¿Qué animales llaman la atención en el Parque Nacional Manuel Antonio? ¿Qué hacen?

¿Qué piensas?

¿Qué piensas de los parques zoológicos en contraste con los refugios de vida silvestre? ¿Cuál prefieres? ¿Por qué?

Hazlo tú

Estudia un mapa de Costa Rica y planea una visita a unos de sus parques nacionales. Indica lo que quieres ver en cada parque.

Activity 2: Generally choose appropriate vocabulary for familiar topics, but as the complexity of the message increases, there is evidence of hesitation and groping for words, as well as patterns of mispronunciation and intonation

En uso
REPASO Y MÁS COMUNICACIÓN

OBJECTIVES
- React to nature
- Express doubt
- Relate events in time

Now you can...
- react to nature.

To review
- subjunctive with expressions of emotion see p. 158.

1 El fin de semana

Vas al Parque Nacional con tu familia este fin de semana. Combina las siguientes frases.

modelo

Vamos al Parque Nacional este fin de semana. Me alegro.

Me alegro de que vayamos al Parque Nacional este fin de semana.

1. Mi amigo no viene con nosotros. Es una lástima.
2. Mi hermano no tiene tiempo para acompañarnos. Siento mucho.
3. Siempre llueve durante nuestras vacaciones. Es ridículo.
4. Papá viene con nosotros. Me alegro.
5. Siempre nos perdemos en el parque. Mi hermanita tiene miedo.
6. Hay tantas especies de mariposas en el parque. Me alegro mucho.
7. Nos divertimos mucho. Ojalá.

Now you can...
- express doubt.

To review
- the subjunctive to express doubt and uncertainty see p. 160.

2 El campamento

Tú y tu compañero(a) están en un campamento. Él (Ella) dice algunas cosas que tú dudas. ¿Qué te dice y cómo le respondes?

modelo

¡Hay serpientes en el campamento! (no es cierto)

Compañero(a): *¡Hay serpientes en el campamento!*

Tú: *No es cierto que haya serpientes en el campamento.*

Compañero(a): *¿Entonces qué hay?*

Tú: *Hay pájaros y hormigas.*

1. Veré un jaguar. (dudo que)
2. Haré un fuego con leña. (no creo que)
3. Tengo fósforos. (no es verdad que)
4. Escucharé un disco compacto. (no es seguro que)
5. Veremos un mono araña. (no creo)
6. Sacaré fotos de los osos hormigueros. (dudo que)
7. Usaré un saco de dormir. (no creo que)

Now you can...

• relate events in time.

To review

• the subjunctive with **cuando** and other expressions of time see p. 162.

3 Primera vez

Es la primera vez que Miguel va a acampar. ¿Qué dice que va a hacer? Usa **cuando** en sus oraciones.

modelo

abrir la tienda de campaña (llegar al campamento)

Abriré la tienda de campaña cuando llegue al campamento.

1. juntar leña (necesitar más)
2. empezar el fuego (tener frío)
3. abrir las latas de comida (encontrar el abrelatas)
4. guardar la comida (terminar de comer)
5. acostarme en el saco de dormir (cansarme)
6. despertarme (salir el sol)

Now you can...

• relate events in time.

To review

• the subjunctive with **cuando** and other expressions of time see p. 162.

4 Después de clases

Le describes a un(a) amigo(a) qué van a hacer tú y tu hermano(a) hoy después de clases. ¿Qué le dices?

modelo

ir a la biblioteca / en cuanto / salir de la última clase

Iremos a la biblioteca en cuanto salgamos de la última clase.

1. quedarse en el colegio / hasta que / venir mamá por nosotros
2. hacer la tarea / en cuanto / llegar a casa
3. preparar la cena / tan pronto como / terminar la tarea
4. encender el televisor / tan pronto como / lavar los platos
5. ver la tele / hasta que / oscurecer
6. acostarse / en cuanto / apagar el televisor
7. dormirse / tan pronto como / acostarse
8. no despertarse / hasta que / amanecer

5 ¡Somos ecoturistas!

STRATEGY: SPEAKING

Reassure others When people are planning an extensive trip together, it is important that they express hopes, feelings, concerns, and doubts. How can you reassure them? Here are some ways: **¡No te pongas triste! ¡No te preocupes! ¡Sé más optimista!** Can you think of other ways?

En grupos de dos o tres, conversen sobre un viaje que van a hacer como ecoturistas en Centroamérica. Primero decidan adónde van, y entonces digan cómo se sienten.

modelo

Tú: *Me alegro de que vayamos a acampar este fin de semana.*

Compañero(a) 1: *Es triste que Carlos no pueda venir con nosotros.*

Compañero(a) 2: *¡Sí, pobrecito! Ojalá que mañana se sienta mejor.*

6 Cuando haga esto...

En grupos de dos o tres, conversen sobre los planes que realizarán cuando terminen otras cosas.

modelo

Amigo(a) 1: *Yo voy a buscar trabajo en cuanto termine este año escolar.*

Amigo(a) 2: *Yo no. Yo voy a viajar por Centroamérica cuando pueda.*

Amigo(a) 3: *Yo no sé qué voy a hacer. Quizás me vaya a Nueva York.*

Amigo(a) 1: …

7 En tu propia voz

ESCRITURA Escoge uno de los animales de Centroamérica y haz una investigación sobre ese animal. Escribe una composición que describa el animal: cuál es su hábitat natural, qué come, cómo es, etc. Si prefieres, escribe un cuento de ficción sobre un animal centroamericano.

modelo

El jaguar vive en la selva. Sólo sale al amanecer…

CONEXIONES

Las ciencias Las selvas tropicales producen una gran parte del oxígeno que necesita el mundo para vivir. Investiga algunos productos que vienen de la selva. ¿Crees que son necesarios? ¿Cuáles son algunos cambios que podrían pasar al medio ambiente si desaparecen las selvas tropicales? Puedes buscar en la biblioteca o en Internet. Comparte tu información en un ensayo o en un póster.

En resumen
REPASO DE VOCABULARIO

In the wild

el agua dulce	freshwater
(el) amanecer	dawn; to start the day
(el) anochecer	nightfall; to get dark
(el) atardecer	late afternoon; to get dark
bucear	to scuba-dive
el campamento	camp
descubrir	to discover
el (la) ecoturista	ecotourist
la luz	light
nadar con tubo de respiración	to snorkel
navegar por rápidos	to go white-water rafting
oscurecer	to get dark
la oscuridad	darkness
peligroso(a)	dangerous
el refugio de vida silvestre	wildlife refuge
salvaje	wild
el sendero	path

Animals, birds, and insects

la ballena jorobada	humpback whale
la boa constrictora	boa constrictor
el búho	owl
el halcón	falcon
la iguana	iguana
el jaguar	jaguar
el loro	parrot
la mariposa	butterfly
el mono araña	spider monkey
el ocelote	ocelot
el oso hormiguero	anteater
el pelícano	pelican
el pez	fish
el picaflor	hummingbird
la serpiente	snake
el tiburón	shark
la tortuga	turtle
el tucán	toucan
el venado	deer
el zorrillo	skunk

Camping

el abrelatas	can opener
la almohada	pillow
el fósforo	match
el fuego	fire
la leña	firewood
la linterna	flashlight
la manta	blanket
la navaja	jackknife
el saco de dormir	sleeping bag
la tienda de campaña	tent

Weather

el aguacero	downpour
centígrado	centigrade
húmedo(a)	humid, wet
el huracán	hurricane
la llovizna	drizzle
la neblina	fog, mist
la nube	cloud
el rayo	thunderbolt, flash of lightning
el relámpago	lightning
soleado(a)	sunny
el trueno	thunder

Ya sabes

Es ridículo que...	It's ridiculous that…
Es triste que...	It's sad that…
Es una lástima que...	It's a shame that…
Espero que...	I hope that…
Me alegro de que...	I'm happy that…
Ojalá que...	I hope that…
Siento que...	I'm sorry that…
Tengo miedo de que...	I'm afraid that…

Ya sabes

Dudo que...	I doubt that…
No creo que...	I don't think that…
No es cierto que...	It's not certain that…
No es seguro que...	It's not sure that…
No es verdad que...	It's not true that…
quizás / quizá	maybe
tal vez	perhaps, maybe

cuando	when
en cuanto	as soon as
hasta que	until
tan pronto como	as soon as

Juego

La selva misteriosa

¿Qué hay en la selva? ¿Puedes encontrar dos animales cuyos nombres empiecen con la letra **m**?

Create paragraphs
when writing

En tu propia voz

ESCRITURA

¡A todos nos toca!

Tu amiga costarricense es candidata para presidenta de su clase. Te pide que le escribas un discurso sobre la preservación del medio ambiente. Tienes que escribir un discurso persuasivo dando tres razones por las cuales los alumnos deben votar por ella.

Función: Persuadir a votar   **Contenido:** La protección del medio ambiente
Contexto: Informar a los alumnos   **Tipo de texto:** Discurso persuasivo

PARA ESCRIBIR • STRATEGY: WRITING

Persuade by presenting solutions to problems A persuasive political speech convinces voters that a candidate can identify important issues, define shared goals, and offer specific solutions to problems. It also highlights the benefits of solving the problems.

Modelo del estudiante

The writer clearly **identifies the problems** the candidate intends to solve.

The writer involves listeners in the speech directly by using **nosotros** as opposed to **yo** forms, as well as **nosotros commands**.

As the speech is developed, the writer **establishes a clear relationship** between the problems, the solutions, and the future benefits.

> Estudiantes votantes: ¿Valoran el futuro? ¿Quieren contribuir con nuestra comunidad? Todos deseamos mejorar el mundo en que vivimos.
>
> Sin embargo, día tras día tenemos que confrontar la creciente contaminación de nuestro aire y agua. ¿Podemos evitar los malos efectos de la destrucción de nuestras áreas verdes? ¿Es posible reducir la contaminación, consumir menos y conservar más?
>
> ¡Yo creo que sí! Si nuestra clase me elige presidenta, verán lo que puede hacer una líder dedicada a la preservación de la ecología. ¡Trabajemos juntos para luchar contra la degradación de nuestro medio ambiente! ¡Colaboremos desde nuestra escuela para nuestra comunidad!
>
> Como presidenta de la clase, crearé y pondré en práctica programas efectivos. Les prometo organizar una campaña de sembrar árboles y flores alrededor de la escuela para embellecerla y devolver el oxígeno al aire. Lucharé por grupos de estudiantes para recoger basura, para que no se ensucie más nuestra vecindad. Organizaré un programa de reciclaje más completo para conservar papel, latas y plásticos de la escuela.
>
> Su voto por mí es un voto por el futuro. Estar a favor de mi presidencia es estar a favor del medio ambiente. Ojalá que juntos podamos cumplir con mis esperanzas y planes para nuestra clase. ¿Me pueden ayudar? ¡Protejamos los beneficios del aire limpio, de un paisaje bello y de un futuro más seguro! No se olviden... ¡A todos nos toca!

 **Language Arts
Writing Standard 1.1**
Demonstrate an understanding of the elements of discourse
(e.g., purpose, speaker, audience, form) when completing narrative,
expository, persuasive, or descriptive writing assignments

 **Writing Center**
CLASSZONE.COM

Estrategias para escribir

Antes de escribir...

Piensa en varios problemas ecológicos que se deben
mencionar. ¿Qué problemas puedes solucionar con
programas en la escuela? Crea una tabla como la
de la derecha. Determina qué problemas son más
importantes para los alumnos e identifica
soluciones generales. Después, piensa en programas
que puedan implementar las soluciones de una
manera concreta.

Problema general	Solución general	Programas específicos de la candidata
contaminación del aire	devolver el oxígeno	sembrar árboles y plantas
consumo de recursos naturales	conservar	reciclar papel, latas, plástico y más
una vecindad sucia	limpiar	recoger la basura

Revisiones

Después de escribir el primer borrador, pídele a un(a) compañero(a)
que lo lea en voz alta. Pregúntale:

- *¿Qué más debo hacer para identificar los problemas de
  una manera clara?*
- *¿Cómo puedo explicar la relación entre los problemas,
  las soluciones y los beneficios?*
- *¿Qué más necesito hacer para que los estudiantes tomen
  un interés personal?*

La versión final

Antes de crear tu versión final, léela de nuevo y repasa
los siguientes puntos:

- *¿Usé el subjuntivo con expresiones de duda, emoción
  o deseo?*

Haz lo siguiente: Subraya todas las expresiones de duda,
emoción o deseo. Subraya dos veces los verbos que se
usan con estas expresiones. Míralos para ver si debes usar
el subjuntivo o el indicativo en cada caso. ¡Ojo! No te
olvides de usar el indicativo con expresiones de certeza.

- *¿Incluí mandatos con la forma nosotros en el discurso?*

Haz lo siguiente: Haz un círculo alrededor de todos los mandatos con
la forma nosotros. Míralos y corrígelos si es necesario. Recuerda que
estos mandatos tienen la misma forma nosotros en el subjuntivo, con
la excepción de los irregulares.

La naturaleza es la responsabilidad de
todos. ¡Hacemos (Hagamos) nuestra parte! Es
peligroso que no la hemos (hayamos) protegido más
y que no hayamos descubierto soluciones
para preservarla. ¡Tengo miedo de que la
destrucción del medio ambiente va (vaya) a ser
desastroso (a) para todos! Es evidente que
tengamos (tenemos) que sacrificar y conservar para
mantener el equilibrio ecológico. ¡No
contaminámos! (contaminemos) ¡Sobrevivamos!

ciento setenta y cinco
México y Centroamérica Unidad 2   **175**

UNIDAD
3

STANDARDS

Communication
- Describing celebrations, holidays, and historic events
- Saying what people want
- Linking events and ideas
- Hypothesizing
- Expressing doubt, disagreement, and emotion
- Making suggestions and wishes
- Stating cause and effect

Cultures
- Regional vocabulary
- Celebrations, holidays, and historic events in the Spanish-speaking Caribbean world
- The history and culture of the Spanish-speaking Caribbean world

Connections
- Art: Caribbean art style
- Social Studies: Investigating independence days in the Spanish-speaking world

Comparisons
- Celebrations, holidays, and historic events in the Spanish-speaking Caribbean world and in the U.S.
- Music in the Spanish-speaking Caribbean world and in the U.S.

Communities
- Using Spanish in the workplace
- Using Spanish with friends at school

INTERNET Preview
CLASSZONE.COM
- More About the Caribbean
- Webquest
- Self-Check Quizzes
- Flashcards
- Writing Center
- Online Workbook
- eEdition Plus Online

CELEBRACIÓN DE MI MUNDO

CUBA
LOS MUÑEQUITOS DE MATANZAS
Este grupo de música afrocubana utiliza tradiciones e instrumentos de África. También cantan en yoruba, un lenguaje africano. ¿En cuáles otros países del Caribe se puede ver la influencia africana?

POBLACIÓN: Puerto Rico: 3.839.810
República Dominicana: 8.581.477
Cuba: 11.184.023

ALTURA: 3175 m sobre el nivel del mar (punto más alto: Pico Duarte, República Dominicana)

CLIMA: 80°F (27°C) San Juan (Temp. más alta); 77°F (25°C) Habana (Temp. más baja)

COMIDA TÍPICA: chicharrones, sancocho, tostones

GENTE FAMOSA DEL CARIBE: José Martí (escritor), Pedro Martínez (atleta), Juan Luis Guerra (músico), Rosario Ferré (escritora)

VIDEO   DVD   Mira el video para más información.

CLASSZONE.COM
More About the Caribbean

PUERTO RICO
ROSARIO FERRÉ es una de las escritoras más prominentes de América Latina. Ha escrito y publicado libros en inglés y español. ¿Cómo crees que la historia de Puerto Rico ha influido en esto?

EL CARIBE
FRUTAS TROPICALES Las guayabas, quenepas y piñas son algunas de las frutas que se cultivan en el clima tropical del Caribe. Se hacen muchos dulces y refrescos con éstas. ¿Conoces otras frutas que vengan de un clima tropical?

LA REPÚBLICA DOMINICANA
JUAN LUIS GUERRA Este músico es uno de los grandes del merengue. El merengue es la música más popular de la República Dominicana. Es conocida a través del mundo latinoamericano. ¿Sabes qué otra cosa significa **merengue**?

PUERTO RICO
PARQUE CEREMONIAL TAÍNO, UTUADO
Los taínos, las personas que vivían en Puerto Rico cuando llegaron los españoles, celebraban el batey o juego de pelota en este parque ceremonial. Según la foto, ¿a qué deporte crees que se parece este juego?

EL CARIBE
MARACAS Este instrumento de percusión es muy popular en el Caribe. Es de origen africano. ¿Qué otros instrumentos conoces de origen africano?

Utuado, P.R.

3

CELEBRACIÓN DE MI MUNDO

- Comunicación

- **Culturas**

- Conexiones

- Comparaciones

- Comunidades

Explore cultures in the Caribbean through guided Web activities.

Culturas

La cultura del Caribe es única. ¿Cómo y por qué se distinguen las culturas de diferentes regiones y países?

Culturas en acción **Observa estas tres fotos. ¿Te dicen algo especial sobre el Caribe? ¿Te recuerdan otros países? Explica.**

Monumento a Colón, Santo Domingo, República Dominicana

Comunicación

Nos gusta conversar mucho cuando celebramos alguna ocasión especial. Imagina la conversación de esta familia. En tu opinión, ¿qué están diciendo?

Comparaciones

En todos los países del mundo se celebran algunos eventos históricos. ¿Cómo se comparan las celebraciones de la foto? ¿En cuál te gustaría participar?

Conexiones

En esta unidad tendrás la oportunidad de investigar el arte caribeño y los días de independencia de varios países. ¿Cuál de estas dos áreas te interesa más? ¿Por qué?

Comunidades

En esta unidad conocerás a un alumno que habla español en el restaurante donde trabaja. ¿Dónde hablas español en tu comunidad?

Fíjate

Hay diferentes tipos de celebraciones. En la siguiente tabla escribe dos celebraciones que compartas con tu familia o amigos, y dos celebraciones que compartas en la escuela.

Celebraciones con familia o amigos	Celebraciones en la escuela
_____	_____
_____	_____

ETAPA

1

¡Al fin la graduación!

OBJECTIVES

- Describe personal celebrations

- Say what people want

- Link events and ideas

¿Qué ves?

Mira la foto de la celebración. Contesta las preguntas.

1. ¿Qué se está celebrando?

2. ¿Quiénes son las personas en la foto?

3. ¿Dónde es la celebración?

4. Según el diploma, ¿en qué país sucede este evento?

181

El día de mi graduación

 Descubre

Adivina el significado de las palabras
en azul según el contexto.

1. Estás en una ceremonia de
graduación. Ahora entra el **desfile
de graduandos.**

2. En su **discurso** el profesor les dice
algunas palabras a los graduandos.

3. Les **desea mucho éxito** a todos
los estudiantes.

4. Los padres del graduando están
muy **orgullosos** de él.

5. El graduando ha **llevado a cabo**
sus estudios.

6. Antes de comer hay un **brindis** por
el graduando.

7. El graduando les da las gracias a
sus padres y a sus **padrinos** por
su ayuda.

8. Les dice «**se la agradezco** mucho».
   a. wishes them success
   b. toast
   c. speech
   d. accomplished
   e. procession of graduates
   f. is grateful to them
   g. proud
   h. godparents

Estamos entrando al auditorio
en el desfile de graduandos.
¡Todos estamos muy emocionados!
Es un día muy importante para
nosotros. ¡Por fin vamos a
graduarnos!

Entrada de los graduandos

El profesor Julio
Capetillo León dio la
bienvenida. Su discurso
fue muy inspirador. Sus
últimas palabras fueron:
«Les deseo mucho
éxito a todos los
graduandos. Espero que
disfruten de la vida y
que su educación los
lleve a lo mejor que la
vida puede ofrecer.»

La ceremonia de graduación

¡Aquí estoy yo con
mi diploma! Me veo
muy sofisticada en
mi toga y birrete.
¡Estoy lista para
conquistar el mundo!

El diploma

¡Felicitaciones!

Aquí estoy con mis padrinos. Nos damos la enhorabuena. ¡Qué felices estamos, y qué orgullosos! ¡Por fin! En este momento podemos ver que vale la pena trabajar duro y seguir nuestros sueños.

La celebración entre familia

Luego todos fuimos a casa para celebrar. En esta foto brindamos por mi graduación y futuro. Yo pude apreciar el apoyo que me ha dado mi familia: «Mil gracias por su generosidad. Se la agradezco mucho. Verdaderamente han sido muy generosos. Por esto voy a poder llevar a cabo mis sueños.»

Online Workbook
CLASSZONE.COM

¿Comprendiste?

1. ¿Alguien en tu familia se ha graduado de la secundaria? ¿Quién?
2. ¿Has ido a una ceremonia de graduación? ¿Dónde fue?
3. ¿Hubo discursos? ¿Qué pensaste de ellos?
4. ¿Hubo una celebración después? ¿Qué hicieron?
5. ¿Cómo se sintieron los padres del graduando?
6. ¿Esperas tu día de graduación con emoción? ¿Cómo crees que te vas a sentir ese día? Explica.

Centro de Estudios José Reyes

Ceremonia de graduación
4 de junio
Auditorio Duarte

En vivo

AUDIO

SITUACIONES

PARA ESCUCHAR

STRATEGY: LISTENING

Pre-listening Think ahead to your own graduation. What are the main parts of the ceremony? What events do you think family members might videotape?

Listen and recognize major transitions As Jorge narrates his sister's graduation, how does he indicate changes of activity and location? Write a title for each scene he describes.

La graduación de Rosanna

Jorge, el hermano de Rosanna, grabó *(taped)* la graduación de su hermana con una videocámara. Vas a ver unas imágenes del video. Luego escucha la narración de Jorge durante ese día especial.

❶ Mirar

Estudia las siguientes imágenes del video que hizo Jorge de la graduación de Rosanna.

② Escuchar

Primero, estudia el programa de la ceremonia.
Luego, escucha la narración de Jorge mientras
describe lo que está grabando. Decide si las
siguientes oraciones son ciertas o falsas. Si son
falsas, corrígelas.

1. La ceremonia de graduación es en el gimnasio.

2. Hay mucha gente en el auditorio.

3. Los graduandos entran con sus padres.

4. Jorge graba a Rosanna mientras entra con sus
   padrinos.

5. Tocan el Himno Nacional después del primer
   discurso.

6. Jorge decide que no va a grabar el discurso del
   profesor Julio C. León.

7. Jorge no graba el momento cuando su hermana
   recibe el diploma.

8. Jorge está orgulloso de su hermana.

9. Rosanna no quiere hablar ante la cámara.

PROGRAMA

1.- Entrada de graduandos junto con sus padrinos.

2.- Himno Nacional.

3.- Bienvenida y presentación, a cargo del
    Prof. Julio C. León.

4.- Palabras a cargo del presidente Luis Rivas

5.- Acto de investidura:
    a) Entrega de Diplomas a los graduandos.
    b) Juramento de graduandos.
    c) Recibimiento de graduandos.

6.- Reconocimiento a la directiva y profesores.

7.- Palabras de gracias a cargo del relacionador
    Jhonson Morillo.

8.- Despedida de la Promoción
    (Comité Pro-Graduación)

9.- Acto sorpresa a cargo del Grupo los Magníficos.

10.- Himno Nacional

③ Hablar/Escribir

¿Piensas mucho en el día de tu graduación? Con
dos o tres compañeros, conversen sobre ese
día futuro. Hagan planes detallados para todo
el día, desde por la mañana hasta por la noche.

En acción

Objectives for Activities 1-4
Describe personal celebrations

1 ¡Qué emoción!

Escribir Rosanna le escribió esta carta a su prima Cristina en Nueva York para describir su ceremonia de graduación. Completa la carta con las siguientes palabras.

aprecio	toga	graduación
desfile	discurso	generosidad
~~me gradué~~	birrete	orgulloso
felicitar	~~diploma~~	llevado a cabo
graduandos	~~generosa~~	se emocionaron
~~mil gracias~~	éxito	

Querida Cristina,

¡Qué emoción! Ayer fue mi ceremonia de ___1___. ¿Puedes creerlo? Por fin ___2___ de la secundaria. Me hubieras visto en mi ___3___. ¡Me veía muy intelectual! Mis padres ___4___ mucho al verme recibir mi ___5___.

Déjame describirte todo. Primero hicimos el ___6___ de ___7___. Entonces el profesor León dio un ___8___ muy inspirador. Él nos deseó mucho ___9___ en nuestras carreras y dijo que estaba muy ___10___ de todos los graduandos.

¡ ___11___ por el regalo que me mandaste! Siempre has sido muy ___12___: no sabes cuánto ___13___ tu ___14___. Por favor escríbeme y dime cómo fue tu graduación. Te quiero ___15___ por haber ___16___ tu educación secundaria.

Cuídate y muchos abrazos,

tu prima Rosanna

2 ¡Enhorabuena!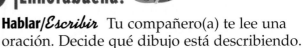

Hablar/*Escribir* Tu compañero(a) te lee una oración. Decide qué dibujo está describiendo.

f.

modelo
Tú: *«¡Enhorabuena, hija! Te felicitamos».*

Compañero(a): *Es el dibujo f.*

a.    b.    c.

d.   e.

1. «Mis padrinos brindaron porque me gradué segunda en mi clase.»

2. «Los graduandos llevan toga y birrete.»

3. «¡Los discursos me aburren! ¡Qué fatal!»

4. «Fuimos a la ceremonia de graduación de Rosanna. ¡Qué bonito estuvo todo!»

5. «¡Mira! Puse mi diploma en la pared para que lo vieran todos.»

3 Mi graduación

Hablar/Escribir Imagina que te vas a graduar de la secundaria. Usa las siguientes palabras para dramatizar esta situación con tu compañero(a). Luego, cambien de papel.

modelo

Tú: *Hoy es la ceremonia de mi graduación.*

Compañero(a): *¡Te felicito!*

Tú: *Mil gracias, te lo agradezco.*

graduando
toga y birrete
diploma
generosidad
desfile
felicitar
brindar
graduarse
desfilar
apreciar
enhorabuena
emocionarse
llevar a cabo
mil gracias
ceremonia de graduación
Te lo agradezco.
Te deseo mucho éxito.
¿…?

4 Un día importante

Hablar/Escribir Conversa con un(a) compañero(a) sobre lo que ocurre en cada dibujo. Describan cada escena con todo el detalle que puedan.

modelo

Tú: *Es una ceremonia de graduación.*

Compañero(a): *Sí. Los graduandos llevan togas y birretes.*

Tú: *Están entrando en el desfile de graduandos…*

1.

2.

3.

4.

Práctica: gramática y vocabulario

Objectives for Activities 5–14
• Say what people want • Link events and ideas

REPASO   The Subjunctive for Expressing Wishes

▶ Remember that you learned to use the **subjunctive** with **impersonal expressions** such as **es importante que** and **es necesario que.**

You know how to use the subjunctive after verbs like **querer que** and **preferir que** to indicate that one person wants someone else to do something.

> Rosanna **quiere que** sus padrinos la **acompañen.**
> *Rosanna **wants** her godparents to **accompany** her.*

▶ You also use the **subjunctive** after other verbs, like the ones below, that express wishes.

> Rosanna **espera que** no **llueva.**
> *Rosanna **hopes** it **doesn't rain.***

▶ You only use the **subjunctive** with these **verbs** when there is a change of **subject.**

Compare the following sentences.

> Yo **quiero que** tú **asistas** a la ceremonia.
> *I want **you** to attend the ceremony.*

> Yo **quiero** **asistir** a la ceremonia.
> *I want to attend the ceremony.*

> When there is no change of **subject,** you use the **infinitive** instead of the subjunctive.

Vocabulario

Otros verbos

dejar *to allow*
exigir *to demand*
insistir en *to insist*
oponerse a *to oppose*
prohibir *to prohibit*
rogar (o→ue) *to beg*
suplicar *to ask, plead*

♻ **Ya sabes**

aconsejar
desear
esperar
mandar
pedir (e→i)
permitir
querer (e→ie)
recomendar (e→ie)
sugerir (e→ie)

Practice: **Actividades** ⑤ ⑥ ⑦   **Más práctica** *cuaderno p. 69*
Para hispanohablantes *cuaderno p. 67*

 Online Workbook
CLASSZONE.COM

También se dice

El (la) **graduando(a)** es el (la) estudiante que está en la ceremonia de graduación y se está graduando. Se dice **graduado(a)** a la persona que ya se graduó y tiene su diploma de bachiller.

5 Pedro

Hablar/Escribir Hoy es el día de graduación de Pedro. Su madre le dice que quiere que haga ciertas cosas. ¿Qué le dice?

modelo

querer que / estar listo a las cinco

Quiero que estés listo a las cinco.

> **Nota: Gramática**
>
> Verbs that end in **-ger** and **-gir** change **g** to **j** in the subjunctive: **exija, exijas,** etc.

1. insistir en que / cortarse el pelo
2. recomendar que / comprar zapatos nuevos
3. sugerir que / recoger tu cuarto
4. esperar que / sentirse orgulloso
5. rogar que / no ponerse esa camisa
6. pedir que / llegar a tiempo

6 Consejos a los graduandos

Escuchar/Escribir ¿Qué consejos tienen los amigos y familiares de los graduandos? Escucha y escribe cada consejo.

modelo

Armando y Nydia   Les sugiero que vayan a la universidad.

1. Ana Luisa
2. Juan y María
3. Arturo
4. Martín y Laura
5. Eduardo
6. Joaquín
7. Lisa y Dina
8. Rogelio

7 Por favor

> **STRATEGY: SPEAKING**
>
> **Accept or Reject Advice** Here are ways to respond to well-meant advice or requests:
>
> **a.** Accept and seek more information or show a positive reaction: **¿Cuándo? Me alegro de que…**
>
> **b.** Indicate uncertainty or a condition: **Dudo que… Puedo salir en cuanto…**
>
> **c.** Reject and give reasons: **No puedo… Es imposible que… No es lógico que…**

Hablar/Escribir Es la última semana antes de la graduación y quieres que tu compañero(a) haga ciertas cosas contigo. ¿Qué le dices y cómo te responde? Usa los verbos de la lista. Luego, cambien de papel.

modelo

querer que

Tú: *Quiero que estudies conmigo para los exámenes finales.*

Compañero(a): *Está bien. ¿Qué vamos a estudiar?*

> desear que
> sugerir que
> suplicar que
> aconsejar
> insistir en que
> mandar que
> rogar que
> recomendar que
> prohibir que
> pedir que

Activity 10: Express and understand opinions

 **REPASO** **The Subjunctive with Conjunctions**

 **¿RECUERDAS?** *p. 162* Remember that you use the **subjunctive** with **cuando** and **conjunctions of time** in order to express that you are not sure when or if something will happen.

▶ The following **conjunctions** also express degrees of doubt or certainty about events.

Iremos a la fiesta **a menos que** no nos **inviten**.
*We will go to the party **unless** they **don't invite us**.*

Vendrán a la fiesta **con tal de que** bailen.
*They will come to the party **as long as** they **can dance**.*

Trae el mapa **en caso de que** nos perdamos.
*Bring the map in case **we get lost**.*

Acérquense todos **para que** brindemos.
*Come close **so that** we **can make a toast**.*

> **Vocabulario**
>
> **Conjunciones**
>
> a menos que *unless*
>
> con tal (de) que *provided that, as long as*
>
> en caso de que *in case*
>
> para que *so that*
>
> **Ya sabes:** antes (de) que

▶ The adverbial expression **antes de que** requires the subjunctive. However, the expression **antes de** requires the infinitive. Compare these sentences.

Yo necesito felicitar al graduando **antes de que** él **salga**.
*I need to congratulate the graduate before **he leaves**.*

Yo necesito felicitar al graduando **antes de salir**.
*I need to congratulate the graduate before **I leave**.*

> One exception is **a menos que** which always requires the subjunctive.

Practice: **Actividades** **8** **9** **10**   **Más práctica** *cuaderno p. 70*
Para hispanohablantes *cuaderno p. 68*

Online Workbook
CLASSZONE.COM

Nota cultural

En la República Dominicana, como en muchos países de Latinoamérica, existe la tradición de que grupos de amigos hagan juntos un viaje de graduación después de la ceremonia formal de la graduación. Por lo regular, si viven en pueblos del interior o de la costa, van a las ciudades principales como Santo Domingo, Santiago o Puerto Plata. Si son de las ciudades, van a la costa o al campo. Para conseguir dinero para el viaje, hacen rifas *(raffles)* y fiestas durante el año escolar.

8 Para que...

Hablar/Escribir Todos van a hacer algo para que otros puedan hacer otra cosa. ¿Qué van a hacer?

modelo

Tú vas a llamar a tus amigos. (ellos venir a tu casa)

Tú vas a llamar a tus amigos para que ellos vengan a tu casa.

1. Yo voy a comprar los boletos. (ustedes ir al concierto)
2. Ricardo va a preparar una paella. (nosotros probarla)
3. Mis hermanos van a salir a jugar. (yo poder estudiar)
4. Nosotros lavaremos el coche. (papá sentirse orgulloso)
5. Pepe va a escribir a sus padres. (ellos saber que está bien)
6. María va a hablar con el profesor. (él explicarle la tarea)

9 ¿Cuándo?

Hablar/Escribir Pregúntale a tu compañero(a) cuándo va a hacer ciertas cosas. Él o ella responde usando **antes de que** o **después de que** con el subjuntivo. Luego, cambien de papel.

modelo

Tú: *¿Cuándo vas a hacer la tarea?*

Compañero(a): *Voy a hacer la tarea antes de que salgamos.*

limpiar	llamar
estudiar	felicitar
cenar	agradecer
comprar	¿...?
salir	

10 Mis opiniones

Escribir Piensa en una celebración personal reciente. ¿Qué opiniones tienes sobre el evento? Escribe cuatro oraciones usando las siguientes frases.

modelo

Me encantan las fiestas de cumpleaños con tal de que pueda comer pastel.

a menos que

con tal de que

en caso de que

para que

More Practice: **Más comunicación** *p. R9*

Activity 12: Understand most spoken language when the message is deliberately and carefully conveyed by a speaker accustomed to dealing with learners when listening

GRAMÁTICA The Imperfect Subjunctive

▶ You already know the **present** and **present perfect** **subjunctive.** There are also **past** forms of the subjunctive. Use the **imperfect subjunctive** instead of the present subjunctive when the context of the sentence is in the **past.**

Compare the following pairs of sentences.

Present context

Los padrinos quieren **que** felicitemos al graduando.

*The godparents **want** us to **congratulate** the graduate.*

La madre de la graduanda sugiere **que** hagamos un brindis.

*The mother of the graduate **suggests** we **make** a toast.*

Past context

Los padrinos querían **que** felicitáramos al graduando.

*The godparents **wanted** us to **congratulate** the graduate.*

La madre de la graduanda sugirió **que** hiciéramos un brindis.

*The mother of the graduate **suggested** we **make** a toast.*

▶ You form the **imperfect subjunctive** by removing the **-ron** ending of the **ellos/ellas/Uds.** form of the **preterite** and adding a special set of endings . The endings are the same for **-ar, -er,** and **-ir** verbs.

hablar ⟶ habla**ron** ⟵ endings

habla**ra**	hablá**ramos**
habla**ras**	habla**rais**
habla**ra**	habla**ran**

Notice the **accent** in the **nosotros** form.

Los padres de la graduanda querían **que** nosotros comié**ramos** con ellos.
*The graduate's parents **wanted** us **to eat** with them.*

Los padrinos querían mucho **que** Rosanna recibie**ra** su diploma.
*The godparents really **wanted** Rosanna **to receive** her diploma.*

▶ If a verb is **irregular** in the **ellos/ellas/Uds.** form of the preterite, like the verb **ir (fueron),** it will also be **irregular** in the **imperfect subjunctive (fuera).**

Practice: **Actividades**

Más práctica *cuaderno pp. 71–72*
Para hispanohablantes *cuaderno pp. 69–70*

Online Workbook
CLASSZONE.COM

11 Los padres

Hablar/*Escribir* Los padres siempre quieren que los hijos sean perfectos. ¿Qué querían los padres de Alma que hicieran ella y sus hermanos?

modelo

estudiar mucho

Querían que estudiaran mucho.

1. sacar buenas notas en todas las clases
2. no salir mucho
3. llegar temprano
4. limpiar sus cuartos
5. poner su educación ante todo
6. entender la importancia de una buena educación
7. darles las gracias a sus abuelos por su apoyo
8. ser estudiantes modelos
9. comer con toda la familia
10. hablar de su futuro

12 ¿En qué insistían?

Escuchar/*Escribir* Estos estudiantes acaban de llegar a la universidad. Están en una reunión y leen apuntes sobre su niñez. ¿En qué insistían sus padres? Completa las oraciones según lo que dicen.

modelo

Mi papá insistía en que yo **fuera a la universidad** .

1. Mi mamá insistía en que mis hermanos y yo _____.
2. Mi papá insistía en que yo _____.
3. Mi mamá insistía en que mis hermanos y yo _____.
4. Mis padres insistían en que mis hermanos y yo _____.
5. Mi papá insistía en que yo _____.
6. Mi mamá insistía en que mis hermanos _____.
7. Mis padres insistían en que mis hermanos y yo _____.
8. Mis padres insistían en que mis hermanos _____.
9. Mi papá insistía en que yo _____.

Nota cultural

Fiesta de graduación En la República Dominicana, no se celebra la graduación con un baile como el «prom». Generalmente los padres y los padrinos dan una fiesta para el (la) graduando(a). A veces dan una fiesta para un grupo de graduandos que son amigos en la casa de su(s) familias o en un restaurante u hotel.

UNA INVITACIÓN ESPECIAL

Querido Pedro

Te invito a ti y a tu familia a que vengan a la
FIESTA DE GRADUACIÓN.

DÓNDE: Salón principal,
Hotel Mar Azul, Santo Domingo

CUÁNDO: domingo 13 de mayo
12:00 p.m. a 7:00 p.m.

Habrá baile y comida.

¡Te espero allí!

Un abrazo y felicidades.

Tu amiga, Estefanía

Activity 15: Express and understand opinions

13 Los chismes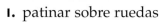

Hablar/Escribir Hablas con un(a) amigo(a) y le cuentas los chismes que otro(a) amigo(a) exagerado(a) te contó la semana pasada. Tu compañero(a) sabe que tu amigo(a) es muy exagerado(a) y te dice lo que ustedes ya se imaginan. Sigue el modelo y usa las siguientes expresiones.

modelo

comprar un coche nuevo

Tú: *Me dijo que se compró un coche nuevo.*

Compañero(a): *Y no era cierto que se comprara un coche nuevo, ¿no?*

No era cierto que	Era mentira que
No era verdad que	Era dudoso que

1. conocer a Michael Jordan
2. ganar la lotería
3. ir de viaje a Nepal
4. leer cien libros en una semana
5. escribir una composición de quinientas páginas
6. actuar en un programa de televisión
7. aprender a hacer alpinismo en una hora
8. llegar a casa a las siete todos los sábados
9. hacer un brindis con Eva
10. asistir a tu graduación

14 Permiso

Hablar/Escribir Necesitamos el permiso de nuestros padres para participar en varias actividades. En grupos de tres o cuatro, conversen sobre sus planes para la semana después de la graduación. Digan qué piensan sus padres.

modelo

Tú: *¿Van a poder salir esta noche?*

Amigo(a) 1: *Mi papá se opuso a que yo saliera esta noche.*

Amigo(a) 2: *Mi mamá me pidió que yo volviera temprano.*

Amigo(a) 3: *Mi papá recomendó que mejor saliéramos el viernes.*

1. patinar sobre ruedas   2. remar

3. esquiar en el agua   4. ir al teatro

5. acampar   6. hacer alpinismo

Activity **15** brings together all concepts presented.

15 Camino al éxito

Leer/Escribir Lee la introducción en el programa de la graduación de Rosanna Lisette Cruz de la Rosa. Luego contesta las preguntas.

1. ¿Quién crees que camina hacia el futuro?
2. ¿Importa su posición social o económica?
3. Al avanzar, ¿de qué están seguros los graduandos?
4. ¿Cuáles necesidades vencen siempre?
5. ¿Qué sienten al avanzar?
6. ¿Cuál es la gran satisfacción que obtienen?
7. ¿Cómo siguen adelante?
8. ¿Hacia qué siguen avanzando?
9. ¿Crees que la introducción inspira a los graduandos? ¿Por qué?

Vocabulario

El futuro

avanzar *to advance*	**el orgullo** *pride*
el camino *path*	**el tropiezo** *setback*
imponer *to impose*	**vencer** *to defeat, triumph*

 **Ya sabes:** la meta

▶ ¿Relacionas estas palabras con tu futuro?

INVITACIÓN Y PROGRAMA
ACTO DE INVESTIDURA

PROMOCIÓN
"EXITUS 00"
"CAMINO AL ÉXITO"

INTRODUCCIÓN

Cuando caminamos hacia el futuro, sin importar nuestra posición social o económica, avanzamos con la seguridad de llegar a nuestras metas, sin importar los tropiezos y obstáculos que encontremos en el camino.

Venciendo siempre las necesidades que nos impone la sociedad y la vida misma, como pago a nuestro esfuerzo y al de nuestros padres; por eso siempre seguiremos adelante sin importar las barreras de la vida, ni los derrumbes de vicios que obstaculizan nuestro camino, avanzando con orgullo y con la frente en alto, para obtener la gran satisfacción que sentimos al llegar a la meta deseada y seguir adelante con respeto, disciplina, trabajo y esfuerzo, para seguir avanzando...camino al éxito.

J. H. M. A.

More Practice: Más comunicación *p. R9*

Online Workbook
CLASSZONE.COM

Refrán

La prosperidad hace amigos, pero la adversidad los prueba.

¿Qué quiere decir el refrán? ¿Crees que si alguien no está a tu lado durante los tiempos difíciles es verdaderamente un(a) amigo(a)? ¿Por qué? ¿Cuáles son algunas características de un(a) buen(a) amigo(a)?

AUDIO

En voces

LECTURA

PARA LEER • STRATEGY: READING

Interpret metaphors A metaphor is an implicit comparison between two things, such as "a sea of troubles." It may also be a symbol in which one thing represents another. **Ébano** *(Ebony)* is defined as **un árbol cuya madera es dura y negra**. **Real** can mean both royal and real. Read **"Ébano real"** and listen to it read aloud. What do you think this tree represents to the poet?

EL ÁRBOL

duro	muy denso
el ébano	tipo de árbol de color muy oscuro
la madera	material que viene del árbol
el tronco	parte vertical del árbol

Sobre el autor

Nicolás Guillén tal vez sea el poeta cubano más conocido. Nació en Camagüey, Cuba, en 1902 y viajó a México y luego a España. En aquel entonces estaba prohibido tocar el son, un tipo de música que combina baile y cantos de estilo africano con romances castellanos de España. Guillén adoptó los ritmos[1] del son en sus poemas, creando un estilo nuevo de poesía que honra la cultura de sus compatriotas afroamericanos. Se llama «poesía negra» a la poesía de Guillén y otros poetas del Caribe. Esta poesía se basa en los ritmos y los temas folklóricos de la gente de descendencia africana.

Introducción

En el poema «Ébano real», Guillén describe un árbol viejo y majestuoso. La repetición de las frases y las palabras con sonidos africanos, como **arará** y **sabalú**, contribuyen a la musicalidad del son cubano en los versos.

---

[1] rhythms

Ébano real

Te vi al pasar, una tarde,
ébano, y te saludé;
duro entre todos los troncos,
duro entre todos los troncos,
tu corazón[2] recordé.
    Arará, cuévano,
    arará, sabalú.
—Ébano real[3], yo quiero un barco,
ébano real, de tu negra madera…
Ahora no puede ser,
espérate, amigo, espérate,
espérate a que me muera.
    Arará, cuévano,
    arará, sabalú.
—Ébano real, yo quiero un cofre[4],
ébano real, de tu negra madera…
Ahora no puede ser,
espérate, amigo, espérate,
espérate a que me muera.
    Arará, cuévano,
    arará, sabalú.
—Ébano real, yo quiero un techo[5],
ébano real, de tu negra madera…
Ahora no puede ser,
espérate, amigo, espérate,
espérate a que me muera.
    Arará, cuévano,
    arará, sabalú.

—Quiero una mesa cuadrada
y el asta[6] de mi bandera[7];
quiero mi pesado lecho[8]
quiero mi lecho pesado,
ébano, de tu madera…
Ahora no puede ser,
espérate, amigo, espérate,
espérate a que me muera.
    Arará, cuévano,
    arará, sabalú.
Te vi al pasar, una tarde,
ébano, y te saludé;
duro entre todos los troncos,
duro entre todos los troncos,
tu corazón recordé.

---

[2] heart	[5] roof	[7] flag
[3] royal	[6] flagpole	[8] heavy bed
[4] chest		

Online Workbook
CLASSZONE.COM

¿Comprendiste?

1. ¿Qué relación tiene la forma musical del son con la poesía de Guillén?
2. ¿Contra qué protesta la poesía de Guillén?
3. ¿Qué cosas le pide el autor al árbol?

¿Qué piensas?

¿Cuáles son las palabras repetidas del poema? ¿Qué efecto tienen?

Hazlo tú

Piensa en una cosa que ves a menudo. ¿Qué palabras puedes usar para «saludarla»? Describe qué representa este objeto para ti en unos versos.

Activities 1–4: Generally uses culturally appropriate behavior in social situations

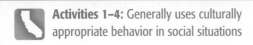

En uso
REPASO Y MÁS COMUNICACIÓN

OBJECTIVES

- Describe personal celebrations
- Say what people want
- Link events and ideas

Now you can...

- describe personal celebrations.

- say what people want

To review

- the subjunctive for expressing wishes see p. 188.

1 Sugiero que...

Tu compañero(a) hace el papel del (de la) graduando(a) en su día de graduación. Tú haces el papel de su mamá o de su papá. ¿Qué le dices?

modelo

(sugerir que) llegar temprano a la ceremonia de graduación

Tú: *Sugiero que llegues temprano a la ceremonia de graduación.*

Compañero(a): *Está bien, mamá (papá). Llegaré temprano.*

1. (querer que) vestirse bien
2. (querer que) darles las gracias a tus padrinos
3. (recomendar que) dar un discurso inspirador
4. (sugerir que) no emocionarse demasiado
5. (aconsejar que) no ponerse la toga y el birrete antes de la ceremonia
6. (sugerir que) felicitar a tus compañeros
7. (esperar que) disfrutar de la ceremonia

Now you can...

- say what people want.

To review

- the subjunctive for expressing wishes see p. 188.

2 Mi familia

Las familias siempre tienen mucho que decir sobre las actividades de los familiares jóvenes. Haz oraciones que expresan estos deseos usando frases de las dos columnas.

modelo

mi madre querer / yo asistir a la ceremonia de graduación

Mi madre quiere que yo asista a la ceremonia de graduación.

1. mis abuelos esperar / mis hermanos saludar a los padrinos
2. mi tío prohibir / mi prima salir con los amigos después de la celebración
3. mi padre insistir en / yo agradecer la generosidad de los invitados
4. mis primos pedir / mis hermanos y yo llegar temprano a la fiesta
5. mi madre exigir / yo vestirme muy elegante

Now you can...

• link events and
  ideas.

To review

• the subjunctive
  with conjunctions
  see p. 190.

③ ¡Ay, abuela!

Hoy es tu día de graduación y tu abuela tiene muchas cosas que decirte. ¿Qué te dice?

modelo

(tú: pensar ir con tus amigos) Ven conmigo a la ceremonia a menos que _____.

Ven conmigo a la ceremonia a menos que pienses ir con tus amigos.

1. (el discurso: ser largo) Voy a hacer un video de toda la ceremonia a menos que _____.
2. (tú: olvidarse) Aquí está tu bolso antes de que _____.
3. (la ceremonia: no terminar muy tarde) Vamos a quedarnos hasta el final con tal de que _____.
4. (tú: ponerse la toga) Péinate antes de que _____.
5. (tú: recibir el diploma) Mira la cámara antes de que _____.
6. (llover) La ceremonia va a ser en el auditorio en caso de que _____.
7. (tú: decidir acompañarme) Voy a esperarte después en caso de que _____.
8. (tú: necesitarme) Voy a estar allí en caso de que _____.
9. (tú: descansar) Siéntate un rato para que _____.

Now you can...

• say what people
  want.

To review

• the imperfect
  subjunctive
  see p. 192.

④ El profesor

Tu profesor favorito dio un discurso a la clase de graduandos antes de la ceremonia. ¿Qué les dijo a ustedes?

modelo

no ponerse nerviosos

Nos dijo que no nos pusiéramos nerviosos.

1. disfrutar del tiempo libre
2. brindar con nuestros padres
3. sentirse orgullosos
4. apreciar a nuestros profesores
5. apoyarse el uno al otro
6. dedicarnos a nuestros estudios
7. dar las gracias a nuestra familia
8. felicitar a nuestros compañeros de clase
9. llevar a cabo nuestros sueños
10. avanzar hacia nuestras metas

5 El (La) graduando(a)

STRATEGY: SPEAKING

Give advice and best wishes Think of the different suggestions and recommendations you might make to the graduate. You can offer the usual phrases, be socially correct, or you might capture the graduate's attention by saying something unusual and unorthodox. Use your imagination!

En grupos de tres o cuatro, escojan a una persona que sea el (la) graduando(a). Los demás deben hacer el papel de parientes y felicitarlo(la) y darle consejos.

modelo

Tú: *¡Te felicito en tu graduación!*

El graduando: *Muchas gracias, se lo agradezco.*

Amigo(a) 1: *¡Enhorabuena, hijo! Estamos muy orgullosos de ti.*

Amigo(a) 2: *Te deseo mucho éxito, y sugiero que…*

6 En tu propia voz

ESCRITURA Imagina el día de tu graduación. ¿Cómo va a ser? ¿Cómo te vas a sentir? ¿A quiénes vas a invitar? ¿Dónde y cómo van a celebrar? Escribe tres leyendas detalladas para tu álbum de fotos.

TÚ EN LA COMUNIDAD

Jordan es alumno en Arkansas. Su familia vivió en la República Dominicana por diez años y todos aprendieron a hablar español. Jordan trabaja en un restaurante de Arkansas. Habla español con los clientes que no hablan inglés y les ayuda a pedir la comida. También habla español con sus compañeros de la escuela de vez en cuando. ¿Hablas español cuando tienes la oportunidad?

En resumen
REPASO DE VOCABULARIO

DESCRIBE PERSONAL CELEBRATIONS

Graduation

agradecer	to thank
el apoyo	support
apreciar	to appreciate
el birrete	cap
brindar	to make a toast
el brindis	toast
la ceremonia de graduación	graduation ceremony
el desfile	parade, procession
el diploma	diploma
el discurso	speech
emocionarse	to be thrilled, touched
la enhorabuena	congratulations
el éxito	success
la generosidad	generosity
generoso(a)	generous
el (la) graduando(a)	graduate
graduarse	to graduate
llevar a cabo	to accomplish
Mil gracias.	Many thanks.
orgulloso(a)	proud
los padrinos	godparents
la toga	gown
valer la pena	to be worthwhile

The future

avanzar	to advance
el camino	path, road
imponer	to impose
el orgullo	pride
el tropiezo	setback
vencer	to defeat, to overcome

SAY WHAT PEOPLE WANT

The subjunctive for expressing wishes

dejar	to allow
exigir	to demand
insistir en	to insist on
oponerse a	to oppose
prohibir	to prohibit
rogar (o→ue)	to beg
suplicar	to ask, to plead

♻ Ya sabes

aconsejar	to advise
desear	to want
esperar	to hope, wait
mandar	to order, send
pedir (e→i)	to request
permitir	to permit
querer (e→ie)	to want
recomendar (e→ie)	to recommend
sugerir (e→ie)	to suggest

LINK EVENTS AND IDEAS

Conjunctions

a menos que	unless
con tal de que	provided that, as long as
en caso de que	in case
para que	so that

♻ Ya sabes

antes (de) que	before

Juego

Consejos para el graduando

¿Cuál de estas palabras no se aplica al dibujo?

a. emocionados

b. prohibir

c. enhorabuena

d. llevar a cabo

ETAPA 2

¡Próspero Año Nuevo!

OBJECTIVES

- **Talk about holidays**

- **Hypothesize**

- **Express doubt and disagree**

- **Describe ideals**

¿Qué ves?

Mira la foto. Contesta las preguntas.

1. ¿Qué observas en la foto?

2. ¿Crees que es una noche regular o una noche especial en este lugar? ¿Por qué?

3. ¿Qué época del año es? ¿Cómo te sientes durante esa época del año?

4. ¿Qué otras cosas puede decir la tarjeta?

¡Próspero Año Nuevo!

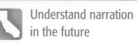
¡Gánate una FIESTA continua!

De **KQ 102** y Milton Canales, **el disc-jockey** que conoce tus gustos...

¿Cómo vas a recibir el **Año Nuevo** en San Juan? KQ **102** quiere ser tu **anfitrión**. Por ese **motivo**, ¡KQ **102** va a **patrocinar** un **concurso** de baile! La **pareja** que gane el concurso va a **festejar** la **despedida del año** en una **fiesta continua** por todo San Juan. ¡Van a **pasarlo muy bien**!

Premios para la pareja ganadora...

Como huéspedes del Hotel Caribe, la pareja primero va a disfrutar de una cena fabulosa con todas las comidas **típicas** de Puerto Rico: **lechón asado**, **arroz con gandules** y **pasteles**. ¡Buen provecho!

La noche sigue con una **gala** en el Hotel Caribe, donde van a oír la música de la gran **orquesta** de Gilberto Santa Rosa. Algunos de los mejores **músicos** de todo San Juan tocan en esta orquesta formidable.

Radioemisora KQ102

¡Donde se oye salsa todo el día!

Después una limosina los llevará a El Morro, donde verán un *show* espectacular de **fuegos artificiales**. ¡Qué **ruido** hacen los **cohetes**! Y la gente también, por supuesto.

A la medianoche la pareja se encontrará en el Viejo San Juan. Allí se comerán las doce uvas tradicionales: tienen que comérselas una por una sincronizadas con las **campanadas** de la Catedral. No se olviden de brindar por un ¡**Próspero Año Nuevo**!

En la **madrugada**, la pareja irá a la playa del Condado para ver el espectáculo que es el amanecer puertorriqueño.

¡Anímate! ¡Participa en el concurso de baile!

Gánate una despedida de año **inolvidable** cortesía de tu estación favorita, KQ102.

Online Workbook
CLASSZONE.COM

¿Comprendiste?

1. ¿Celebras el Año Nuevo? ¿Cómo? ¿Con quiénes?
2. ¿Hay fuegos artificiales en tu ciudad para el Año Nuevo? ¿Vas a verlos o los ves por la televisión?
3. ¿Te quedas despierto(a) hasta que cambia el año? ¿Por qué?
4. ¿Tienes tradiciones personales para el Año Nuevo? ¿Cuáles son?

Listen to
audio texts

AUDIO

En vivo
SITUACIONES

PARA ESCUCHAR

STRATEGY: LISTENING

Pre-listening Do you listen to or watch talk-show interviews? How does the host move the conversation along and keep it interesting? What kinds of questions does he/she ask?

Observe interview techniques As you listen to the winners' responses, notice the kinds of questions Milton Canales asks. Check the frequency with which he uses these question types:

Tipo de pregunta	Muchas veces	A veces	Casi nunca
sí/no			
dos respuestas			
respuesta corta			
varias respuestas			

Which one elicits the most interesting information?

¡Próspero Año Nuevo!

Estás en casa y ves un anuncio en el periódico sobre el concurso de salsa y la pareja ganadora. Luego, escuchas una entrevista entre ellos y el disc jockey de KQ102, Milton Canales.

❶ Leer

Lee el anuncio de KQ102 en el periódico.

LA PAREJA GANADORA DEL CONCURSO DE SALSA DE

KQ 102

EMILIA RUEDAS

Y

ALEX ORTIZ

¿Cómo pasaron la despedida de año Emilia y Alex?

Si quieres saber, pon la radio en tu estación favorita,  *a las doce en punto. Milton Canales va a entrevistar a la pareja ganadora.*

RESTAURANTE EL COQUÍ

¡GRACIAS POR SU VISITA!

EL MORRO

❷ Escuchar

Milton Canales, el disc-jockey de KQ102, habla con Emilia
Ruedas y Alex Ortiz, la pareja ganadora del concurso de
salsa. Escucha y escoge la respuesta correcta.

1. La despedida de año
   a. Lo pasaron muy bien.
   b. Fue una despedida de año horrible para la pareja.

2. La orquesta de Gilberto Santa Rosa
   a. Alex cree que Gilberto Santa Rosa no es buen músico.
   b. Alex cree que Gilberto Santa Rosa es excelente.

3. La cena
   a. La pareja comió muy bien.
   b. La pareja no disfrutó de la cena.

4. Los fuegos artificiales
   a. Emilia dice que El Morro no es un buen sitio para
      ver los fuegos artificiales.
   b. Emilia dice que no hay otro lugar como el Morro
      para ver los fuegos artificiales.

5. Las campanadas
   a. Se pudieron comer las uvas a tiempo con cada campanada.
   b. No pudieron comerse las uvas a tiempo con cada campanada.

6. El amanecer
   a. La pareja cree que el amanecer en la playa Condado es ordinario.
   b. La pareja cree que el amanecer en la playa Condado es inolvidable.

❸ Hablar

En grupos de dos o tres, conversen sobre la despedida de año ideal.
¿Qué harían? ¿Adónde irían? ¿Dónde cenarían? ¿Les gustaría pasar la
despedida de año en Puerto Rico? Comparen las celebraciones en
Estados Unidos con la celebración de la pareja ganadora en
San Juan, Puerto Rico.

En acción

Objectives for Activities 1-4
• Talk about holidays

1 ¿Quién habla?

Hablar/*Escribir* Lee las oraciones. ¿Quién habla: el anfitrión de una celebración, un invitado o cualquiera de los dos?

modelo

Bienvenidos, pasen, pasen. Denme sus abrigos.

el anfitrión

1. Su casa es bella, señora Ruiz.
2. Gracias por venir a festejar con nosotros.
3. Un brindis para nuestros invitados de honor.
4. La orquesta es buenísima, ¿no crees?
5. Le traje un regalito. Espero que le gusten las rosas.
6. ¡Buen provecho! Ojalá que les guste el lechón asado.
7. ¿Vinieron en auto?
8. Es medianoche. ¡Próspero Año Nuevo!
9. ¡Vamos a bailar!
10. Lo pasamos muy bien. Gracias por invitarnos.

2 ¿Cuál es?

Hablar/*Escribir* Léele una oración a tu compañero(a). Él (Ella) te va a decir qué dibujo describes. Luego, cambien de papel.

modelo

Tú: *Dicen que ese cantante es increíble.*

Compañero(a): *Estás describiendo el dibujo e.*

a.    b.

c.    d.

e.    f.

1. La gente lo está pasando muy bien en la gala.
2. ¡Próspero Año Nuevo!
3. ¡Mira los fuegos artificiales! ¡Qué bonitos!
4. Es la madrugada. ¡No hay nada como el amanecer!
5. Es la orquesta más famosa de San Juan.

3 No hay de qué

Hablar/Escribir Habla con tu compañero(a) sobre el año nuevo.

modelo

el Año Nuevo

Tú: *Gracias por venir a celebrar el Año Nuevo con nosotros.*

Compañero(a): *No hay de qué, es un placer para mí.*

Tú: *Espero que lo pases muy bien.*

1. orquesta
2. músico
3. festejar
4. fuegos artificiales
5. pasteles
6. fiesta continua

Vocabulario

Dar las gracias

Muy amable. *That's kind of you.*
No hay de qué. *It's nothing.*

 Ya sabes

De nada.
Es un placer…
Gracias.
Mil gracias.
Se lo agradezco.

▶ ¿Cuándo usas estas frases?

4 ¡Bienvenidos!

STRATEGY: SPEAKING

Socialize as host or guest As the host at your own party, you will want to suggest choices of food, drink, or entertainment. **(Tú podrás…, Recomiendo…, Sugiero…, ¿Te gustaría…?, Sería buena idea…, Quizás…)** As a guest you will want to accept or decline by expressing your own preferences. **(Prefiero…, Es posible…, Me gustaría…, Quisiera…, Se puede…)**

Hablar/Escribir Tú eres el (la) anfitrión(a) de una gala para celebrar el Año Nuevo. Con dos o tres compañeros, dramaticen esta situación.

modelo

Tú: *Pasen, pasen. Bienvenidos a mi casa.*

Compañero(a): *Gracias, muy amable.*

Tú: *Hay mucho que hacer. Si tienen ganas de bailar…*

Compañero(a): …

Objectives for Activities 5–15
• Express doubt and disagree • Hypothesize • Describe ideals

GRAMÁTICA   Subjunctive with Nonexistent and Indefinite

▶ If you want to say that something **may not exist,** you use the **subjunctive.**

may not exist

No hay orquesta que me **guste.**
There is no orchestra that I like.

may not exist

No conozco a nadie que lo **pase** bien.
I don't know anyone who is having a good time.

> The **thing** or **person** probably doesn't exist: *there is no orchestra…*

▶ **Expressions** that trigger this use of the **subjunctive** include:

No hay… que
No hay nadie que…
No hay nada que…
No hay ningún/ninguna… que…

▶ A related way to use the **subjunctive** is in subordinate clauses that are **indefinite** or **uncertain:**

> **Uncertain:** we don't know if these musicians exist or not.

subordinate clause

Buscamos músicos **que sepan** tocar música bailable.
We're looking for musicians who know how to play dance music.

▶ **Words** and **expressions** that trigger this use of the **subjunctive** include:

Buscar/Querer/Necesitar… que
¿Hay algo/alguien que… ?
¿Conoces a alguien que… ?
¿Tienes algo que… ?

Practice:
Actividades
5 **6** **7**

Más práctica
cuaderno p. 77
Para hispanohablantes
cuaderno p. 75

Online Workbook
CLASSZONE.COM

---

5 **En la comunidad** ♻

Escribir Quieres saber quién participa en actividades para la comunidad. Escribe diez preguntas que puedes hacerle a tu clase.

modelo

participar en la campaña para embellecer la ciudad

¿Hay alguien que participe en la campaña para embellecer la ciudad?

Nota: Gramática

Remember that **sembrar** is an **e→ie** stem-changing verb. **Recoger** and **educar** have spelling changes in most of their subjunctive forms; **g→j** and **c→qu** respectively.

1. trabajar de voluntario(a) en un comedor de beneficencia
2. juntar fondos para la comunidad
3. estar en contra de los servicios sociales
4. donar ropa a la gente sin hogar
5. sembrar árboles en la comunidad
6. recoger basura en el vecindario
7. pasar tiempo con ancianos
8. educar al público sobre los problemas sociales
9. querer colaborar con alimentos
10. conocer un centro de reciclaje

6 **¿Existe o no?**

Escuchar/*Escribir* Copia la siguiente tabla y escucha las oraciones de varias personas. Si la persona o cosa indicada existe en la vida de la persona que habla, marca **sí**. Si en este momento esa cosa o persona no existe, marca **no**.

modelo

¿Existe alguien que sepa reparar computadoras?

Sí	No
x	

¿Existe…?	Sí	No
1. un señor que hable francés		
2. alguien que pueda tocar la guitarra		
3. una orquesta que no cueste mucho		
4. un músico que cante muy bien		
5. un apartamento que tenga jardín		
6. unos estudiantes que puedan trabajar los fines de semana		
7. unos amigos que sepan bailar salsa		
8. una radio que funcione		

7 **¿Conoces a alguien…?**

Hablar/*Escribir* Pregunta a tu compañero(a) si él (ella) conoce a varias personas que puedan hacer las cosas indicadas. Luego, cambien de papel.

modelo

Tú: *¿Conoces a alguien que celebre el Año Nuevo con sus padres?*

Compañero(a): *Sí, conozco a alguien que celebra el año nuevo con sus padres. o No, no conozco a nadie que celebre el año nuevo con sus padres.*

> **bailar muy bien**
> **tocar en una orquesta**
> **ser músico**
> **saber preparar comida puertorriqueña**
> **tener una limosina**
> **querer comprar un vestido muy elegante**
> **dar clases de salsa**

Nota cultural

Salsa En Puerto Rico, la música **salsa** tiene muchos entusiastas y grandes exponentes como Tito Puente y Willie Colón. A través de toda la isla, hay salones de baile donde los **salseros** (los entusiastas de la salsa) pueden bailar y divertirse.

Activity 9: Express and understand opinions

REPASO The Subjunctive for Disagreement and Denial

Another way to use the **subjunctive forms** you have already learned is to express **doubt** or **disagreement**. You already know many ways to express doubt or to disagree with someone.

Jorge: ¿Sabes si Mamá invitó a doña Laura?

Do you know if Mom invited doña Laura?

Rosanna: Yo creo que sí, pero **es improbable que venga.** Está enferma.

*I think so, but **it's unlikely that** she **will come.** She's sick.*

Vocabulario

 Ya sabes

Dudar que…
Es imposible que…
Es improbable que…
no creer que…
no pensar (e→ie) que…

No es cierto que…
No es seguro que…
No es verdad que…
no estar seguro (de) que…
no opinar que…

Practice: Actividades
8 9 10 11

Más práctica
cuaderno p. 78
Para hispanohablantes
cuaderno p.76

Online Workbook
CLASSZONE.COM

---

8 Las dudas de Enrique

Hablar/Escribir Enrique, tu mejor amigo, siempre duda de lo que dices. ¿Qué le dices y cómo te responde?

modelo

los García: *dar una gala*

Tú: *Yo creo que los García van a dar una gala.*

Enrique: *Dudo que los García den una gala.*

1. mis papás: festejar hasta la madrugada
2. nosotros: pasarlo muy bien en esa fiesta
3. Andrés: ponerse un traje
4. la orquesta: ser muy buena
5. los fuegos artificiales: ser magníficos
6. la fiesta: acabarse a la medianoche

Nota cultural

En Puerto Rico se celebran muchos días festivos. Por ser un Estado Libre Asociado, muchos de esos días son los mismos que se celebran en Estados Unidos, como el Día de la Independencia de Estados Unidos (4 de julio) y el Día del Trabajo (primer lunes de septiembre). Además se celebran fiestas nacionales como el Descubrimiento de Puerto Rico (19 de noviembre), el Día de la Abolición de la Esclavitud (22 de marzo), y el nacimiento de héroes de la independencia puertorriqueña, como Eugenio María de Hostos (10 de enero), y de poetas, como Luis Muñoz Rivera (18 de julio).

9 No estoy seguro(a)

Hablar/Escribir Tú y tu amigo(a) comentan sobre una fiesta, pero él (ella) no está seguro(a) de lo que tú dices. Dramaticen la situación.

modelo

Tú: *La música es divertida.*

Compañero(a): *No estoy seguro(a) de que la música sea divertida.*

1. Los invitados se divierten.
2. La anfitriona cocina muy bien.
3. La orquesta toca toda la noche.
4. Los fuegos artificiales hacen mucho ruido.
5. Oímos las campanadas de la catedral a la medianoche.
6. ¿...?

10 ¡Es imposible!

Hablar/Escribir Le sugieres ideas a tu padre para celebrar el Año Nuevo. ¿Cómo te responde?

modelo

Tú: *Papá, ¿por qué no damos una gala para despedir el año?*

Papá: *¡Es imposible que demos una gala!*

1. invitar
2. cocinar
3. buscar
4. celebrar
5. aprender
6. comprar
7. festejar
8. ir
9. brindar
10. llamar

11 Los días festivos

Hablar/Escribir Usa palabras de la lista y comenta sobre cómo van a celebrar los días festivos.

modelo

la Navidad

Dudo que celebremos la Navidad en casa de mis tíos.

1. el Día de Acción de Gracias
2. el Día del Trabajo
3. el Día de las Madres o de los Padres
4. el Día de la Amistad
5. el Día de la Raza
6. el Día de la Independencia

no pensar	no es seguro
no es cierto	es improbable
no creer	dudar

Vocabulario

Los días festivos

el Día de Acción de Gracias *Thanksgiving*

el Día de la Amistad *Valentine's Day*

el Día de la Independencia *Independence Day*

el Día de las Madres / los Padres *Mother's / Father's Day*

el Día de la Raza *Columbus Day*

Hanuka *Hanukkah*

la Navidad *Christmas*

las Pascuas *Easter*

la quinceañera *fifteenth birthday*

▶ ¿Cómo celebras estos días festivos en casa?

More Practice: Más comunicación *p. R10*

Activity 13: Understand most spoken language when the message is deliberately and carefully conveyed by a speaker accustomed to dealing with learners when listening

GRAMÁTICA **Conditional Sentences**

▶ In Spanish, many sentences are composed of a si-clause (*if-clause*) and a main clause.

To predict a future result based on an initial action, use:
the **present tense** in the si-clause and the **future** in the main clause.

present tense → **si-clause main clause** ← *future tense*

Si **vienes,** lo **pasarás** bien.
If you **come,** **you will have** *a good time.*

▶ In order to say what things would be like if circumstances were different, you use:
the **imperfect subjunctive** in the si-clause and the **conditional** in the main clause.

imperfect subjunctive → **si-clause main clause** ← *conditional*

Si **vinieras,** lo **pasarías** bien.
If you **came (could come),** **you would have** *a good time.*

▶ Compare these two sentences.

In the first example, your friend might come to the party. So her **future** (*having a good time*) will happen based on her **initial action** (*coming to the party*).

Si **vienes,** lo **pasarás** bien.
If you **come, you will have** *a good time.*

In the second, you know that your friend is probably not coming to the party. If **circumstances were different,** (*if she came*) you want her to know **what it would be like** (*she'd have a good time*).

Si **vinieras,** lo **pasarías** bien.
If you **came, you would have** *a good time.*

▶ In both of these cases, the order of the clauses can be switched.

Lo **pasarás** bien si **vienes.**   Lo **pasarías** bien si **vinieras.**

Practice: **Actividades** **Más práctica** *cuaderno pp. 79–80*   **Online Workbook**
          ⑫ ⑬ ⑭ ⑮ **Para hispanohablantes** *cuaderno pp. 77–78*   CLASSZONE.COM

12 Los sueños

Hablar/*Escribir* Todos tenemos sueños de qué haríamos bajo ciertas condiciones. ¿Qué dicen las siguientes personas?

modelo

hablar francés (viajar a Francia)

Si hablara francés, viajaría a Francia.

1. ser actor o actriz (irse para Los Ángeles)
2. estar en la universidad (estudiar informática)
3. trabajar (guardar mi dinero)
4. tener mucho dinero (no trabajar)
5. poder hacerlo todo (conocer a Europa)
6. manejar (comprar un carro deportivo)
7. vivir en Puerto Rico (vivir en San Juan)
8. saber tocar un instrumento (tocar en una orquesta)

Nota cultural

Chayanne Nacido en Río Piedras, Puerto Rico, este joven es uno de los cantantes más populares de Latinoamérica. Ganó un premio en el Festival de la Canción de Viña del Mar, en Chile, quizá el concurso más importante de música popular de Latinoamérica.

13 ¿Qué va a hacer?

Escuchar/*Escribir* Gustavo dice que va a hacer algunas cosas y también dice que haría otras cosas si pudiera. Di bajo qué condiciones haría esas cosas.

modelo

Comprar un traje.

a. Lo va a hacer.

b. Lo haría si tuviera dinero.

1. Ir a una gala.
   a. Lo va a hacer.
   b. Lo haría si _____.
2. Preparar un lechón asado.
   a. Lo va a hacer.
   b. Lo haría si _____.
3. Empezar un grupo musical.
   a. Lo va a hacer.
   b. Lo haría si _____.
4. Participar en un concurso de baile.
   a. Lo va a hacer.
   b. Lo haría si _____.
5. Ir a ver los fuegos artificiales.
   a. Lo va a hacer.
   b. Lo haría si _____.
6. Ir a ver el amanecer en la playa.
   a. Lo va a hacer.
   b. Lo haría si _____.

Activity 15: Express and understand opinions

14 **Las profesiones**

Hablar/Escribir Tú y tu compañero(a) conversan sobre sus planes para después de la graduación. Di qué harían si estuvieran en ciertas profesiones.

modelo

Compañero(a): *Si fuera músico(a), escribiría mis propias canciones.*

Tú: *Si yo fuera músico(a), escribiría canciones románticas.*

actor (actriz)
cantante
escritor
artista
doctor(a)
mesero
director(a)
dependiente(a)
profesor(a)

vender
servir
actuar
hacer
empezar
irse
cantar
enseñar
curar
jugar

¿?

También se dice

Puerto Rico también se conoce por el nombre **Borinquen o Boriquén**, el nombre taíno de la isla. Los taínos vivían en la región al llegar los españoles. Algunos puertorriqueños también usan la palabra **boricua** para indicar que son de Puerto Rico.

15 **¿Qué harías?**

Hablar/Escribir En grupos de tres o cuatro, conversen sobre qué harían en ciertos días festivos.

modelo

Tú: *¿Qué harías si fuera la quinceañera de tu prima?*

Amigo(a) 1: *Si fuera la quinceañera de mi prima, iríamos a la casa de mis abuelos.*

Amigo(a) 2: *Nosotros iríamos a una gala en un hotel elegante.*

Amigo(a) 3: *Mi familia y yo compraríamos regalos como....*

1.

2.

3.

4.

5.

6.

7.

8.

Activity **16** brings together all concepts presented.

16 Planes para celebrar

Leer/*Escribir* Piensas celebrar el Año Nuevo con tu familia en el restaurante Casa Borinquen. Mira el menú. Luego explica qué deseos tienes, qué comida pedirías y qué dudas tienes sobre la celebración.

modelo

Me gustaría comer con toda mi familia. Pediría pasteles para todos. Si viniera mi hermano, comería mucho budín. No creo que mis tíos puedan venir.

Vocabulario

Comidas típicas

el arroz con dulce *dessert dish of rice, cinnamon, and coconut milk*

el arroz con gandules *rice and pigeon peas*

el arroz con leche *dessert dish of sweet rice and milk*

el budín *pudding*

el coquito *eggnog with coconut and condensed milk*

los guineítos en escabeche *small green bananas in a garlic, vinegar, and oil sauce with red pepper*

el lechón asado *suckling pig*

los pasteles *tamale-like item of plantain, yuca, and meat*

el pavo *turkey*

el tembleque *coconut-milk custard*

▶ ¿Incluyes algunas de estas comidas en tu dieta?

CASA BORINQUEN

Platos principales
Lechón asado
Pavo
Arroz con gandules
Pasteles
Guineítos

Bebidas
Coquito
Refrescos
Batidos de fruta

Postres
Arroz con dulce
Arroz con leche
Budín
Tembleque

More Practice: **Más comunicación** *p. R10*

Online Workbook
CLASSZONE.COM

Refrán

Este mundo es un fandango y, bien o mal, hay que bailarlo.

En tu opinión, ¿qué dice este refrán sobre la vida y las celebraciones? ¿Crees que si tienes que hacer algo por lo menos debes divertirte? ¿Por qué?

En colores

CULTURA Y COMPARACIONES

Una tradición de Puerto Rico

PARA CONOCERNOS

STRATEGY: CONNECTING CULTURES

Recognize and describe uses of disguise As children, did you and your friends like to take on new identities by disguising yourselves? What did you do to change your appearance? How did this change make you feel?

Where in the adult world do you find disguises? Masks are one type of disguise. Think of social events, holidays, characters in literature. Make a list of examples. How long a list can you make?

EVENTOS SOCIALES	FIESTAS	LITERATURA
	carnaval	

En Puerto Rico, la fabricación y el uso de máscaras [1] es una importante tradición que continúa hasta hoy. Según la mayor parte de los antropólogos, el uso de las máscaras en el Puerto Rico de hoy viene de las tradiciones españolas de la Edad Media. Las máscaras se usaban en las fiestas religiosas.

En Puerto Rico la manera de hacer máscaras varía de una ciudad a otra. La ciudad de Ponce, ubicada [2] en la costa sur de la Isla, se conoce por las máscaras de cartón que se fabrican allí. Estas caretas [3] se ponen para Carnaval y se admiran por su brillante colorido.

---

[1] masks
[2] located
[3] masks

para la temporada de Carnaval. Los mascareros aprenden el arte de sus padres y abuelos y transmiten su arte a sus hijos. Hoy día muchos de estos mascareros gozan de[8] prestigio y fama en Puerto Rico y fuera de la Isla. Las máscaras no sólo se usan para Carnaval. También se venden en las galerías de arte, los museos y las tiendas especializadas y turísticas. Se consideran objetos de arte popular que tienen valor histórico y decorativo.

Niñas en el Taller de Máscaras del Residencial Roosevelt, Mayagüez, Puerto Rico

[8] enjoy, have

Los mascareros[4] hacen las caretas del carnaval ponceño con papel, pintura[5] y engrudo. El engrudo es una pasta blanca que se consigue al cocinar harina de trigo[6] con agua.

Pocos artesanos se dedican a la fabricación de estas caretas a tiempo completo[7]. Muchos trabajan en otro oficio y fabrican las máscaras

[4] maskmakers
[5] paint
[6] wheat flour
[7] full time

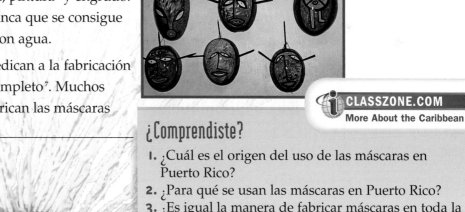

Máscaras en una tienda en Bayamón

CLASSZONE.COM
More About the Caribbean

¿Comprendiste?

1. ¿Cuál es el origen del uso de las máscaras en Puerto Rico?
2. ¿Para qué se usan las máscaras en Puerto Rico?
3. ¿Es igual la manera de fabricar máscaras en toda la isla? ¿Qué ciudad se conoce por sus máscaras?
4. ¿Ha cambiado el modo de trabajar de los mascareros?

¿Qué piensas?

¿Por qué crees que las máscaras tienen valor histórico y decorativo?

Hazlo tú

¿En qué fiestas norteamericanas se usan máscaras? ¿Qué máscara crearías tú para esa fiesta? ¿Con qué materiales la fabricarías? Dibuja la máscara y explica lo que representa.

 Activity 3: Tend to become less accurate as the task or message becomes more complex, and some patterns of error may interfere with meaning

En uso
REPASO Y MÁS **COMUNICACIÓN**

OBJECTIVES

- Talk about holidays
- Hypothesize
- Express doubt and disagree
- Describe ideals

Now you can...

- describe ideals.

To review

- the subjunctive with nonexistent and indefinite see p. 210.

1 La gala

Todos están haciendo planes para una gala que van a dar para el Año Nuevo. ¿Qué buscan o necesitan?

modelo

nosotros: buscar unos músicos / saber tocar salsa

Buscamos unos músicos que sepan tocar salsa.

1. nosotros: buscar un lugar / no costar mucho para alquilar
2. yo: necesitar una limosina / ser bastante grande para todos los invitados
3. Marta y Juan: buscar unos cohetes / no ser caros
4. nosotros: necesitar un cocinero / preparar lechón asado
5. yo: necesitar un camarógrafo / hacer un video de la gala
6. Elena: buscar unos fuegos artificiales/ no hacer mucho ruido
7. nosotros: buscar un lugar donde / poderse ver el amanecer
8. yo: buscar un cantante / tener una voz fenomenal

Now you can...

- express doubt and disagree.

To review

- the subjunctive for disagreement and denial see p. 212.

2 ¡Es muy improbable!

Le dices a tu compañero(a) qué harás con otras personas. Él (Ella) no te cree. ¿Qué le dices y cómo te responde?

modelo

Voy a celebrar el día de Acción de Gracias en San Juan. (dudar que)

Compañero(a): *Dudo que celebres el día de Acción de Gracias en San Juan.*

1. Le voy a comprar un carro a mi novio(a). (es improbable que)
2. Mis hermanos y yo le daremos un viaje a Puerto Rico a nuestro papá. (no es seguro que)
3. Mi hermana y yo limpiaremos la casa. (no creer que)
4. Mi familia y yo invitaremos a cien personas a casa. (es imposible que)
5. Yo voy a festejar las Pascuas con mis primos en España. (dudar que)
6. Yo tendré una fiesta para mi cumpleaños. (es improbable que)
7. Mis primos no van al colegio el Día de la Raza. (es imposible que)

③ Si fuera verdad

Eres muy imaginativo(a) y estás pensando en una fiesta que quisieras dar. Pero lamentablemente las cosas no son como las sueñas. Di cómo serían las cosas bajo ciertas condiciones.

Now you can...

• hypothesize.

To review

• conditional sentences see p. 214.

modelo

Va a haber una fiesta. Yo voy a preparar los pasteles.

Si hubiera una fiesta, yo prepararía los pasteles.

1. Vamos a ser los anfitriones. Vamos a invitar a muchas personas.
2. Tengo que traer un postre. Voy a traer budín.
3. Nydia sabe cocinar. Va a hacer unos guineítos en escabeche.
4. Vamos a invitar a todas las clases de español. Nos vamos a divertir mucho.
5. Mucha gente va a venir a la fiesta. Vamos a comprar mucha comida.
6. Vamos a probar el lechón asado. Probablemente nos va a gustar.
7. Los chicos pueden tocar salsa. Los vamos a invitar a tocar en la fiesta.
8. Las chicas van a festejar hasta la madrugada. Pueden ver el amanecer.

④ Los días festivos

Ya sabes mucho de cómo se celebran los días festivos, pero todavía quieres saber más. Usando frases de las dos columnas, haz oraciones que expresen lo que dudas o lo que te gustaría saber.

Now you can...

• talk about holidays.

To review

• the subjunctive with nonexistent and indefinite see p. 210.

• the subjunctive with disagreement and denial see p. 212.

modelo

Las Pascuas / organizar un concurso de salsa

No creo que nadie organice un concurso de salsa para las Pascuas. o:
¿Hay alguien aquí que organice un concurso de salsa para las Pascuas?

1. las Pascuas / desyerbar el jardín
2. el día de Acción de Gracias / hacer un viaje en limosina
3. el día de la Amistad / tener una orquesta formal en casa
4. el día de la Independencia / preparar una cena para cien invitados
5. el día de las Madres/ los Padres / volar en planeador sobre el océano
6. el día de la Raza / comer dos lechones asados
7. la quinceañera / cantar una ópera entera con los amigos

5 ¿Hay alguien que...?

STRATEGY: SPEAKING

Encourage participation Write the names of two or three classmates and one thing that person can do and one thing he or she probably cannot do. Then pool the papers. Each person selects one paper at random and begins the conversation about what people may or may not be able to do for the class party.

Van a dar una fiesta para celebrar un día festivo en su clase de español. Tienen que decidir cómo van a participar todos. En grupos de tres o cuatro, conversen sobre los varios talentos de todos los estudiantes. Claro, ¡siempre hay personas que dudan de los demás!

modelo

Tú: *¿Hay alguien aquí que sepa cocinar comida puertorriqueña?*

Amigo(a) 1: *Dudo que haya alguien en la clase que sepa hacerlo.*

Amigo(a) 2: *¡Yo sé preparar arroz con gandules!*

Amigo(a) 3: *¡No es cierto que sepas preparar arroz con gandules!*

6 Para celebrar

Pregunta a tu compañero(a) qué hizo para celebrar varios días festivos. Luego pregúntale qué haría si pudiera hacer lo que quisiera. Después, cambien de papel.

modelo

Tú: *¿Qué hiciste para celebrar el Año Nuevo?*

Compañero(a): *Me quedé en casa y vi unos videos.*

Tú: *¿Qué harías si pudieras hacer lo que quisieras?*

Compañero(a): *Si alguien me invitara, iría a una gala.*

7 En tu propia voz

ESCRITURA Imagina que pasaste la semana de Año Nuevo con tus primos. En tu diario, escribe lo que hicieron cada día.

modelo

El lunes fuimos a...

El martes visitamos...

El miércoles celebramos...

CONEXIONES

El arte Crea una tarjeta para desear a un(a) amigo(a) un Próspero Año Nuevo. Investiga el arte del Caribe y haz una tarjeta con arte al estilo caribeño. Explica a la clase cómo tu tarjeta representa el arte caribeño.

En resumen
REPASO DE VOCABULARIO

TALK ABOUT HOLIDAYS

Actions

festejar	to celebrate
pasarlo bien	to have a good time
patrocinar	to sponsor

Expressions

¡Buen provecho!	Enjoy! (your meal)
¡Próspero Año Nuevo!	Happy New Year!

Give thanks

Muy amable.	That's kind of you.
No hay de qué.	It's nothing.

Foods

el arroz …	
con dulce	rice-coconut milk dessert
con gandules	rice and pigeon peas
con leche	sweet rice-milk dessert
el budín	pudding
el coquito	eggnog
los guineítos en escabeche	small green bananas in garlic vinegar, red pepper, and oil
el lechón asado	roast suckling pig
el pastel	tamale-like mixture of plantain, yuca, and meat
el pavo	turkey
el tembleque	coconut-milk custard

Holidays

El día de …	
Acción de Gracias	Thanksgiving
la Amistad	Valentine's Day
la Independencia	Independence Day
las Madres / de los Padres	Mother's / Father's Day
la Raza	Columbus Day
la Hanuka	Hanukkah
la Navidad	Christmas
las Pascuas	Easter
la quinceañera	fifteenth birthday

People and things

el anfitrión	host
la anfitriona	hostess
el Año Nuevo	New Year
la campana	bell
la campanada	tolling of the bell
el cohete	firecracker
el concurso	contest
la despedida del año	New Year's Eve
la fiesta continua	party in stages
los fuegos artificiales	fireworks
la gala	big, formal party
inolvidable	unforgettable
la madrugada	early morning, dawn
el motivo	purpose
el (la) músico(a)	musician
la orquesta	orchestra
la pareja	couple
la radioemisora	radio station
el ruido	noise
típico(a)	typical, regional

Ya sabes

De nada.	You're welcome.
Es un placer…	It's a pleasure…
Gracias.	Thank you.
Mil gracias.	Many thanks.
Se lo agradezco.	It's really appreciated.

EXPRESS DOUBT AND DISAGREE

Ya sabes

dudar que	to doubt that
Es imposible que	It's impossible that
Es improbable que	It's improbable that
no creer que	to not think that
no es cierto que	it's not true that
no es verdad que	it's not true that
no estar seguro(a) (de) que	to not be sure that
no opinar que	to not be of the opinion that
no pensar (e→ie) que	to not think that

DESCRIBE IDEALS

Ya sabes

buscar	to look for
no hay nada	there is nothing
no hay nadie	there is nobody
¿Hay algo…?	Is there anything…?
¿Hay alguien…?	Is there anyone…?

HYPOTHESIZE

Use conditional sentences

Si vas a la fiesta, te divertirás.
Si fueras a la fiesta, te divertirías.

¿Qué hay en la mesa? ¿Puedes encontrar dos tipos de comida cuyos nombres empiecen con la letra **p**?

UNIDAD 3

ETAPA 3

Celebraciones de patria

OBJECTIVES

- Describe historic events

- Make suggestions and wishes

- Express emotion and doubt

- State cause and effect

¿Qué ves?

Mira la foto. Contesta las preguntas.

1. ¿Quiénes son estas personas? ¿Cómo crees que se sienten?

2. ¿Qué celebración crees que sea? ¿Has estado en celebraciones parecidas?

3. Mira el sello. ¿Cómo puedes averiguar en qué país están?

En contexto  VOCABULARIO

EL DIARIO   23-30 marzo

Dos ensayos patrióticos

Estos chicos caribeños participaron en una competencia de ensayos patrióticos. Tuvieron que escribir un breve ensayo que describe el día festivo que tiene más importancia en su familia.

Día de la Abolición de la Esclavitud en Puerto Rico

PRIMER LUGAR

Descubre

Usa tu intuición y lo que ya sabes para decidir el significado de cada palabra. Escoge del segundo grupo de palabras.

1. costumbre	a. fight, struggle
2. acudir	b. ancestors
3. antepasados	c. custom
4. lucha	d. to honor
5. enfrentar	e. essays
6. honrar	f. slaves
7. esclavos	g. to commemorate
8. conmemorar	h. to confront
9. competencia	i. to attend
10. ensayos	j. contest

22 de marzo

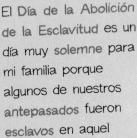

El Día de la Abolición de la Esclavitud es un día muy solemne para mi familia porque algunos de nuestros antepasados fueron esclavos en aquel entonces. Tenemos varias costumbres para honrar su memoria. Empezamos el día con unos momentos de silencio alrededor de la mesa. Papá dice unas palabras sobre la injusticia que sufrieron nuestros antepasados y la lucha que enfrentaban todos los días contra el opresor. Este día fue una victoria justa contra los proponentes de la esclavitud.

Luego vamos al Parque de Bombas para celebrar. Me gusta ir a la Plaza para oír la bomba y plena que tocan los conjuntos afrocaribeños. Allí todos compartimos y recordamos juntos. Es una ocasión alegre.

Emilio Hernández de la Cruz

Monumento a la Abolición de la Esclavitud, Ponce, Puerto Rico

Pleneros celebran el Día de la Abolición

Día de la Raza en la República Dominicana

El día en que Cristóbal Colón descubrió las Américas.

Los dominicanos tienen una relación especial con la familia Colón. El **Almirante** Cristóbal fue el primer europeo que llegó a la isla en diciembre de 1492. El hermano de Cristóbal, Bartolomé, fundó Santo Domingo poco después de la fundación de una colonia española en la isla. Y el hijo de Cristóbal, Diego, se estableció en Santo Domingo por unos años.

En la zona colonial de Santo Domingo hay varios monumentos históricos para **conmemorar** el **descubrimiento** de las Américas. Cada 12 de octubre, miles de estudiantes dominicanos **acuden** a estos monumentos para celebrar el día de la Raza. Hay **procesiones** de estudiantes y **bandas** y **suenan** las campanas de las iglesias. En general, es un día de mucha alegría.

Las escuelas organizan desfiles por la zona colonial de Santo Domingo y en el **Faro** de Colón, con **banderas** dominicanas y con flores de diversos tipos, sobre todo rosas. Es un día que no olvido, no importa en donde esté. En Nueva York o en Santo Domingo, el 12 de octubre será un día en el cual siempre estaré consciente de mi herencia. ¡Dominicana hoy y siempre!

Ángela Beatriz Corona

Monumento a Colón, Santo Domingo, República Dominicana

Colón llega al Caribe

Online Workbook
CLASSZONE.COM

¿Comprendiste?

1. ¿Hay un día festivo que tenga importancia especial en tu familia?
2. Algunos días festivos celebran relaciones personales, como el Día de las Madres y de los Padres, o el Día de la Amistad. Pero otros conmemoran días nacionales, como el Día de la Independencia. ¿Cuál prefieres? ¿Por qué?
3. ¿Piensas mucho en tu nacionalidad? ¿Hay días festivos en los cuales piensas más en tu nacionalidad que en otros?
4. ¿Crees que es importante tener días festivos nacionales? ¿Por qué?
5. ¿Qué sabes de Cristóbal Colón? ¿Tienes la misma relación con él y su familia como la que tiene Ángela? ¿Por qué?

AUDIO

En vivo

SITUACIONES

PARA ESCUCHAR

STRATEGY: LISTENING

Pre-listening Here you will listen to a formal presentation about Columbus. What differences do you anticipate between an oral report and an informal conversation?

Listen and take notes Quick comprehension of numbers is often one of the last listening skills we master. Write down each date when you hear it. Afterward, go back and jot down what happened on that date.

Los viajes del Almirante

Estudias los viajes de Cristóbal Colón en tu clase de historia. Primero vas a mirar un mapa de sus viajes y luego escucharás a un compañero dominicano que da un informe oral sobre Colón.

➊ Mirar

Estudia el mapa que sigue los viajes de Cristóbal Colón en el año 1492.

El Viaje de Colón

Océano Atlántico

Guanahaní

Cuba

La Española

❷ Escuchar 🎧

En tu clase de historia, Miguel, un estudiante dominicano, escribió su ensayo sobre Cristóbal Colón. Primero, lee las oraciones a continuación. Luego, escucha mientras Miguel lee su ensayo. Ordena las oraciones según la información que da Miguel.

_____ Colón establece la primera ciudad española y la nombra «La Isabela».

_____ Los monarcas Fernando e Isabel le dieron el dinero a Colón para hacer su viaje.

_____ Colón murió en España, convencido de que su expedición llegó a las Indias.

_____ El tres de agosto, 1492, la expedición de Colón salió de España en la Niña, la Pinta y la Santa María.

_____ Colón llegó a una bella isla que nombró «La Española».

_____ Colón llegó a una isla llamada Guanahaní.

_____ El hermano menor de Colón, Bartolomé, fundó Santo Domingo.

Puerto de Palos

❸ Hablar/Escribir 👥

En grupos de dos o tres, conversen sobre la vida de Cristóbal Colón. ¿Creen que su descubrimiento fue importante? ¿Saben algo más sobre sus viajes? ¿Pueden añadir información al ensayo de Miguel? Busquen más información sobre Colón en Internet o en una enciclopedia. Cada persona del grupo debe traer un dato importante sobre el hombre.

Activity 4: Narrate in the present

En acción

PARTE A

Práctica del vocabulario

Objectives for Activities 1-4
• Describe historic events

1 Descripciones

Leer/Escribir Lee las descripciones. Decide qué oración mejor describe cada foto.

a.

b.

c.

d.

1. La procesión de los estudiantes a la estatua de Colón es una costumbre del día de la Raza.

2. El día de los Veteranos es un día solemne en el cual honramos la memoria de los soldados de nuestro país.

3. La bandera es un símbolo que representa la historia de un país.

4. Muchas personas acudieron para apoyar su discurso.

2 El ensayo patriótico

Hablar/Escribir Completa las oraciones con las palabras de la lista.

acuden

banda

costumbres

antepasados

ensayo patriótico

injusticia

honrar

1. Voy a escribir un _____ sobre la lucha para nuestra independencia.

2. En los desfiles, siempre hay una _____ que toca el himno nacional.

3. En mi familia tenemos varias _____ para celebrar el día de la Raza.

4. Durante el día de la Raza, miles de personas _____ a la plaza central de la ciudad para oír discursos y celebrar el descubrimiento de las Américas.

5. Es importante _____ a los héroes que dieron sus vidas para luchar contra la esclavitud.

6. No debemos tolerar la _____ contra las personas sin hogar.

7. Nuestros _____ lucharon para nuestra independencia, algo que olvidamos fácilmente.

3 Conversación

Hablar/*Escribir* Quieres saber qué cosas tu compañero(a) ha hecho o visto relacionadas con el patriotismo. Conversen sobre los temas a continuación.

modelo

ensayo patriótico

Tú: *¿Has escrito un ensayo patriótico alguna vez?*

Compañero(a): *Sí, lo escribí sobre el día de la Independencia de Estados Unidos.*

procesión patriótica

enfrentar una injusticia

celebrar una victoria

conmemorar antepasados

bailar bomba y plena

una banda en desfile

costumbres de la familia para el día de...

Vocabulario

El orgullo nacional

la patria *mother country*

el (la) patriota *patriot*

patriótico(a) *patriotic*

el patriotismo *patriotism*

▶ ¿Crees que tenemos mucho patriotismo en este país? ¿Por qué?

4 Las costumbres

STRATEGY: SPEAKING

Describe celebrations There are many aspects in a description of a celebration: **el lugar, la gente, la ropa, la comida, las acciones.** Think also about the time frame you want to use: **Por lo general, vamos a... vemos a..., llevamos..., comemos..., hacemos....** What if you changed the time frame? What tenses would you use if you said: **Pero el año pasado...** or **Cuando era niño(a)**...

Hablar Quieres saber si tu compañero(a) tiene algunas costumbres para ciertos días festivos. Conversen sobre dos o tres días festivos de su ciudad, estado o país. Mira la página 213 para repasar los días festivos.

modelo

Tú: *¿Cómo celebran el día de Acción de Gracias en tu familia?*

Compañero(a): *Pues, tenemos varias costumbres. Primero... Y tu familia, ¿cómo celebra el día de...?*

Práctica: gramática y vocabulario

Objectives for Activities 5–15
• Make suggestions and wishes • Express emotion and doubt • State cause and effect

REPASO Summary of the Subjunctive (Part 1)

As you have learned, you use the **subjunctive** in Spanish in subordinate clauses when the **main clause** expresses…

• **wishes:** querer, recomendar, insistir en, aconsejar, etc.

> **Queremos que vengas** a la procesión.
> *We want you to come to the procession.*

> ¿Qué **recomienda Ud. que hagamos**?
> *What do you recommend that we do?*

• **emotion:** alegrarse, sentir, esperar, ojalá, es bueno/malo/mejor que, etc.

> **Me alegro de que Uds. conozcan** la bomba y plena.
> *I am happy that you know the bomba and plena.*

• **doubt, disagreement,** and **denial:** no creer/pensar, dudar, no es cierto/verdad que, etc.

> **Dudábamos de que** ellos **enfrentaran** el problema.
> *We doubted that they would face the problem.*

··

When you have a sentence with a subordinate clause and the subjunctive is required, the main clause will be in the indicative and the subordinate clause in the subjunctive. The tense you use in the main clause will help you determine which tense to use in the subordinate clause.

Main Clause Indicative	Subordinate Clause Subjunctive
if **present, future, present perfect**	use present
Siento que **termine** la celebración. *I'm sorry that the celebration is ending.*	
if **present**	use present perfect (if action has taken place)
Siento que **se haya terminado** la celebración. *I'm sorry that the celebration has ended.*	
if **preterite, imperfect, conditional, past perfect**	use imperfect
Sentía que **terminara** la celebración. *I was sorry that the celebration was ending.*	

Practice:
5 6 7 8

Más práctica
cuaderno p. 85
Para hispanohablantes
cuaderno pp. 83–84

Online Workbook
CLASSZONE.COM

5 El desfile

Escuchar/*Escribir* Tu familia te habla sobre el desfile que van a ver hoy. Escucha y di qué quieren que tú hagas.

modelo
mi abuela / ir al desfile
Mi abuela quiere que yo vaya al desfile con ella.

1. mi abuela / ponerse ropa de verano
2. mi abuela / llegar al desfile temprano
3. mi abuela / traer una silla
4. mi papá / invitar a dos amigos
5. mis papás / venir a la casa dos horas antes del desfile
6. mis hermanos / quedarse en el parque todo el día

Apoyo para estudiar

Tenses of the subjunctive
This chart will help you remember what tenses to use:

Main Clause Indicative	Subordinate Subjunctive
present, future, present perfect	present
present	present perfect (if action has taken place)
preterite, imperfect, conditional, past perfect	imperfect

6 ¡Me alegro!

Escribir Vas a ir a la República Dominicana para celebrar el día de la Raza con tu amiga dominicana Susana. Ella te escribe esta carta. Completa la carta con la forma correcta de los verbos indicados.

> **Nota: Gramática**
>
> Don't forget that verbs that end in **-cer** add a **z** to their subjunctive forms. Verbs that end in **-gar** add a **u** to their subjunctive forms.

Querido(a) amigo(a),

¡No puedo creer que la próxima semana vas a estar aquí en Santo Domingo! Ya sabes que vamos a ir a la celebración del día de la Raza el sábado. Me alegro de que tú ___1___ (querer) venir con nosotros. Quiero que tú ___2___ (ver) como celebramos nosotros los dominicanos. No es cierto que ___3___ (ser) un día solemne. Hay muchas cosas divertidas que hacer. Siento que tu hermana no ___4___ (poder) venir contigo. Dudo que (nosotros) ___5___ (quedarse) todo el día, porque queremos llegar a casa antes de que ___6___ (oscurecer). ¡Espero que ___7___ (divertirse)! Recomiendo que tú ___8___ (llegar) a nuestra casa como a las diez de la mañana. ¡Qué lindo será verte!

Un abrazo,

Susana

7 Santo Domingo

Hablar Tu compañero(a) va a celebrar el día de la Raza en Santo Domingo. ¿Qué le recomiendas?

modelo

Compañero(a): *Voy a celebrar el día de la Raza en Santo Domingo.*

Tú: *¡Recomiendo que hagas tus reservaciones hoy!*

Compañero(a): *Pienso viajar en avión.*

Tú: *Recomiendo que…*

> **Nota: Gramática**
>
> Remember that verbs that end in **-ger** change the **g** to a **j** in their subjunctive forms.

hacer tus reservaciones

comprar tus boletos

llamar a la agencia de viajes

escoger un hotel cerca del centro

llevar ropa de verano

comprar cheques de viajero

no ir solo(a)

visitar los monumentos históricos

ser parte del desfile estudiantil

Activity 8: Clarify, ask for and comprehend clarification

8 ¡Vamos a celebrar!

Hablar/Escribir Ayer fue el día de la Raza en Santo Domingo. ¿Qué querían todos que hicieras?

modelo

Compañero(a): ¿Qué querían tus padres que hicieras?

Tú: Mis padres querían que fuera al desfile con mi hermana.

tu mejor amigo(a)

tu tío(a)

tu primo(a)

tu profesor(a)

tu hermano(a)

tus amigos(as)

Nota cultural

El naufragio de la Santa María
En diciembre de 1492, la Santa María, uno de los tres barcos de Colón, naufragó (shipwrecked) cerca de la isla La Española. Con la madera y otros materiales rescatados del naufragio, Colón y sus acompañantes construyeron un fuerte al que dieron el nombre de La Navidad. En su segundo viaje, en 1493, Colón llegó al fuerte que encontró vacío. Abandonóel fuerte y estableció la colonia de Isabela cerca del Cabo Isabela en lo que hoy se conoce como la República Dominicana.

REPASO **Summary of the Subjunctive (Part 2)**

You have also learned to use the **subjunctive** after the **nonexistent** or **indefinite** antecedents:

No hay discurso **que me interese.**
There's no speech that interests me.

Busco una banda **que sepa tocar** el himno nacional.
I'm looking for a band that knows how to play the national anthem.

No hay nada/nadie que…
Busco/Necesito/Quiero… algo/alguien que…
¿Hay algo/alguien que… ?
¿Conoces a alguien que… ?

Use the **subjunctive** with these **conjunctions of time,** but only if the main clause has a **command** or refers to the **future**:

Nos iremos cuando termine la fiesta.
We'll leave when the party ends.

Quédense aquí **hasta que empiece** el desfile.
Stay here until the parade begins.

cuando
en cuanto
después (de) que
tan pronto como
hasta que

You do not use the subjunctive if the **conjunction** is in a past-tense context.

Estaba lloviendo **cuando** empezó el discurso.
It was raining when the speech began..

Use the **subjunctive** with these **conjunctions:**

Te lo digo/Te lo diré **para que** te des cuenta del problema.
I'm telling you/I will tell you so that you'll realize the problem.

Te lo dije **para que** te dieras cuenta del problema.
I told you so that you'd realize the problem.

antes (de) que
con tal (de) que
a menos que
para que
en caso (de) que

Practice: Actividades
9 10 11 12

Más práctica
cuaderno p. 86
Para hispanohablantes
cuaderno pp. 83–84

Online Workbook
CLASSZONE.COM

9 La comunidad ♻

Leer/Escribir Eres el (la) presidente(a) del comité para mejorar la ciudad. Creas varios pósters que pones por toda la ciudad. Le explicas a un(a) compañero(a) el propósito *(purpose)* de cada cartel. ¿Qué le dices?

modelo

buscar voluntarios / mantener limpia la ciudad

Buscamos voluntarios que mantengan limpia la ciudad.

1. *buscar / músicos*
hacer audición para…

2. *necesitar / candidato*
ser conservador

3. *querer / voluntarios*
mantener…

4. *buscar / personas*
luchar contra…

5. *querer / ciudadanos*
respetar leyes de la ciudad

6. *necesitar / personas*
resolver problemas…

Activity 12: Use strings of related sentences when speaking

10 El gobierno

Hablar/Escribir La política es muy complicada. ¿Qué necesitan los ciudadanos y las personas en el poder?

modelo

gente / querer / gobierno / escribir leyes justas

La gente quiere un gobierno que escriba leyes justas.

1. pueblo / querer / alcalde / entender los problemas de la ciudad
2. monarquía / insistir en que / ciudadanos / obedecer las leyes
3. gente / querer / gobierno / ser democrático
4. país / querer / presidente(a) / tener una ideología honorable
5. gente / querer / líder / saber tomar decisiones
6. alcalde / necesitar / ciudadanos / apoyarlo
7. estado / querer / gobernador(a) / no olvidar sus responsabilidades
8. rey / buscar / ejército / estar siempre listo
9. ciudadanos / querer / elecciones / ser limpias
10. políticos / necesitar / gente / votar por ellos

Vocabulario

La política

el (la) alcalde(sa) *mayor*

conservador(a) *conservative*

la constitución *constitution*

la democracia *democracy*

democrático(a) *democratic*

el derecho *the (legal) right*

el ejército *army*

el (la) gobernador(a) *governor*

el gobierno *government*

la ideología *ideology*

la ley *law*

el (la) líder *leader*

liberal *liberal*

la monarquía *monarchy*

el poder *power*

el (la) presidente(a) *president*

la reina *queen*

el rey *king*

¿Qué tipo de gobierno tenemos en este país? En tu opinión, ¿cómo es? ¿Quiénes son los líderes más importantes? ¿Cómo son?

11 Las celebraciones

Hablar/Escribir ¿Qué hacen todos antes y después de la celebración? Combina palabras y frases de las dos columnas para explicarlo. Usa el adverbio **cuando** y los detalles que necesites en tus oraciones.

modelo

Ellos se irán a casa cuando termine la celebración.

> **Nota: Gramática**
>
> Remember that verbs that end in **-zar** change the **z** to a **c** in their subjunctive forms.

yo	terminar la celebración
tú	empezar el desfile
él/ella	subir la bandera
ellos	visitar el monumento
nosotros(as)	tocar el himno nacional
todos	sonar las campanas
	acabarse el discurso

12 ¿Qué van a hacer?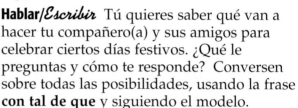

Hablar/Escribir Tú quieres saber qué van a hacer tu compañero(a) y sus amigos para celebrar ciertos días festivos. ¿Qué le preguntas y cómo te responde? Conversen sobre todas las posibilidades, usando la frase **con tal de que** y siguiendo el modelo.

modelo

Tú: *¿Van a participar en el desfile?*

Compañero(a): *Sí. Vamos a participar en el desfile con tal de que no tome mucho tiempo.*

More Practice: Más comunicación *p. R11*

También se dice

A los dominicanos también se les llama **quisqueyanos.**

Nota cultural

El Himno Nacional de la República Dominicana fue escrito en 1883. El poeta y educador Emilio Prud'homme compuso la letra y el maestro José Reyes la música. Reyes vio el himno de Argentina en un periódico parisino y decidió escribir uno para la República Dominicana. Invitó a su amigo Prud'homme a escribir la letra. El himno se cantó en febrero de 1884 al llevar a la República los restos de Juan Pablo Duarte, el libertador del país, quien murió en 1876 en Venezuela.

> **El Himno Nacional**
>
> Quisqueyanos valientes, alcemos Nuestro canto con viva emoción, Y del mundo a la faz ostentemos Nuestro invicto, glorioso perdón.
>
> ¡Salve! el pueblo que, intrépido y fuerte, A la guerra a morir se lanzó, Cuando en bélico reto de muerte sus cadenas de esclavo rompió.

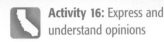

Activity 16: Express and understand opinions

GRAMÁTICA Subjunctive vs. Indicative

▶ Use the **subjunctive:**

- after expressions of **doubt**

  No creo que sepan la respuesta.

  I don't think that they know the answer.

- to make **suggestions** or **recommendations**

  Marta dijo que viéramos fuegos artificiales.

  *Marta said **we should see** the fireworks.*

Use the *indicative:*

- to express **certainty**

  Creo que *saben* la respuesta.

  I think they know the answer.

- to **report** actions

  Marta dijo que *vimos* fuegos artificiales.

  *Marta said **we saw** the fireworks.*

Vocabulario

 Ya sabes

dudar	pensar
no creer	creer
es dudoso	es cierto
es improbable	es verdad
no estoy seguro(a)	estoy seguro(a)

Practice:
Actividades

Más práctica
cuaderno pp. 87–88
Para hispanohablantes
cuaderno pp. 85–86

Online Workbook
CLASSZONE.COM

13 Opiniones opuestas

Hablar/*Escribir* Di si estás de acuerdo con estas ideas.

modelo
ese candidato: ser liberal
Creo que ese candidato es liberal.
No creo que ese candidato sea liberal.

1. el gobierno: ser democrático
2. el alcalde: tener dinero
3. el desfile: ser solemne
4. el (la) presidente(a): dar un discurso
5. el (la) gobernador(a): venir a la celebración
6. la ley: ser justa
7. el (la) presidente(a): llegar temprano

14 El (La) dudoso(a)

Escuchar/*Escribir* Escucha las oraciones y di lo que dudas.

modelo
Dudo que el presidente sea eficiente.

15 ¿Qué dijo?

Hablar/*Escribir* Tu hermano fue a una celebración nacional y te contó todo lo que pasó. Ahora le cuentas los eventos a tu compañero(a). Primero habla sobre los discursos de los políticos durante la celebración y luego cuenta qué pasó, según tu hermano. Sigue el modelo.

modelo

el alcalde / ver la bandera

Compañero(a): *¿Qué dijo el alcalde que hicieran?*

Tú: *Dijo que viéramos la bandera.*

Compañero(a): *¿Qué dijo tu hermano que hicieron?*

Tú: *Dijo que vieron la bandera.*

el alcalde	ver la bandera
la gobernadora	celebrar el día de la Raza
el presidente	subir la bandera
el político liberal	tocar el himno nacional
la política conservadora	luchar por la independencia
	ver el desfile del ejército
	sonar las campanas

16 La entrevista

Hablar Tú eres reportero(a) y dos o tres de tus compañeros son candidatos para alcalde de la ciudad. Hazles preguntas sobre sus campañas.

modelo

Tú: *¿Por qué quiere ser alcalde?*

Compañero(a) 1: *Es importante que participemos en la política de la ciudad.*

Compañero(a) 2: *Yo también voy a hacer campaña. Yo creo que es necesario que…*

Compañero(a) 3: *Cuando sea alcalde…*

More Practice: **Más comunicación** *p. R11*

 **Online Workbook** CLASSZONE.COM

Refrán

El que quiere ser cabeza, que sea puente.

¿Qué quiere decir el refrán?

¿Crees que los políticos pueden juntar varios grupos de gente? En tu opinión, ¿es mejor ser el (la) líder o parte del grupo?¿Quién debe establecer las metas para todos?

Read
poems

AUDIO

En voces
LECTURA

PARA LEER
STRATEGY: READING

Observe what makes poetry Poems are meant to be spoken and are often sung. Here is a poem in which the language is simple but the thought and form are complex. Four basic characteristics of a poem are: rhythm, rhyme, metaphor, and inverted word order.

Rhythm Read the poem aloud. Can you tap a steady beat?

Rhyme Scan the sounds of the last word of each line. Is there a pattern?

Metaphor Find a comparison between two things or a person and a thing. For example in line 8, Martí says «**En los montes, monte soy.**» What do you think he means?

Inverted word order To make everything work together, sometimes the poet changes natural word order. Can you find an example?

LA NATURALEZA

las alas	lo que usa el pájaro para volar
el alma	el espíritu
crecer	vivir y florecer
la lumbre	luz
las yerbas	el césped

Sobre el autor

José Martí, poeta, escritor y patriota cubano, nació en La Habana en 1853 cuando Cuba era todavía una colonia española. Escribió y habló a favor de la independencia de Cuba y fue exiliado a España por sus actividades revolucionarias. Luego fundó el Partido Revolucionario Cubano en 1892 y murió en una batalla por la independencia de Cuba en 1895. Martí murió como vivió, al servicio de la libertad de su patria.

Introducción

La poesía de Martí es directa y sincera. Entre sus poesías más famosas se destacan los *Versos libres* e *Ismaelillo*, escritos alrededor de 1882. Aquí tienes unos versos de su libro más conocido, *Versos sencillos*, escrito en 1891.

Nota cultural

El poema «Versos sencillos» fue la inspiración para la famosa canción "Guantanamera". ¿La conoces?

240 doscientos cuarenta
Unidad 3

de *Versos sencillos*: I.

Yo soy un hombre sincero
De donde crece la palma,
Y antes de morirme quiero
Echar¹ mis versos del alma.

Yo vengo de todas partes,
Y hacia todas partes voy:
Arte soy entre las artes,
En los montes, monte soy.

---
¹ to send out

Yo sé los nombres extraños²
De las yerbas y las flores,
Y de mortales engaños³,
Y de sublimes dolores⁴.

Yo he visto en la noche oscura
Llover sobre mi cabeza
Los rayos de lumbre⁵ pura
De la divina belleza.

---
² strange
³ tricks, deceits
⁴ pains
⁵ light

Alas nacer vi en los hombros
De las mujeres hermosas:
Y salir de los escombros⁶,
Volando las mariposas.

Todo es hermoso y constante,
Todo es música y razón,
Y todo, como el diamante,
Antes que luz es carbón.

---
⁶ rubble, debris

Online Workbook
CLASSZONE.COM

¿Comprendiste?

1. ¿Cómo participó Martí en la lucha por la independencia de Cuba?
2. ¿Cuáles son las imágenes que usa Martí? ¿Qué piensas de ellas?
3. ¿En qué líneas del poema habla Martí de su origen?

¿Qué piensas?

1. ¿Cómo trata Martí los temas de la naturaleza y el patriotismo?
2. ¿Qué significan estos versos en el contexto de la poesía?
   a. Yo vengo de todas partes / Y hacia todas partes voy.
   b. Y todo, como el diamante, / Antes que luz es carbón.

Hazlo tú

Escribe un poema parecido a éste. Empieza con **yo soy…, de donde…, y antes de morirme…**.

En colores

CULTURA Y COMPARACIONES

Una historia única

PARA CONOCERNOS

STRATEGY: CONNECTING CULTURES

Analyze national celebrations You have both read about and experienced national celebrations. Pick a particular celebration, perhaps July 4 in the United States. What did the celebration originally represent? What has it become in popular culture? Make a chart to compare your observations:

DÍA DE LA INDEPENDENCIA EE.UU.

Importancia histórica	Cómo lo conmemoramos
1.	1.
2.	2.
Etc.	Etc.

What conclusions would you make about your observations?

esfiles, disfraces[1], festejos… El día de la Independencia en la República Dominicana es una ocasión especial por su historia única.

Historia

En 1492, Cristóbal Colón llegó a una isla que llamó La Española. Los españoles luego se instalaron en la parte este de la isla, pero en la parte oeste, los franceses crearon la colonia de Haití. En 1804, los haitianos ganaron su independencia de Francia y en 1822 ocuparon la ciudad española de Santo Domingo.

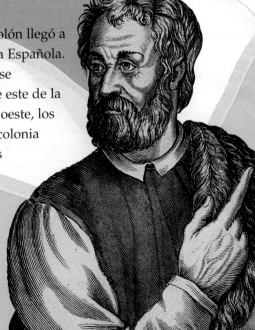

[1] costumes

Independencia

Después de la ocupación, el patriota dominicano Juan Pablo Duarte fundó un movimiento de resistencia. La noche del 27 de febrero de 1844, los dominicanos declararon la independencia de Haití. Poco después, los haitianos tuvieron que irse de la nueva nación, la República Dominicana.

Celebraciones

En la República Dominicana hay eventos oficiales durante este día patriótico y también hay eventos populares. El 27 de febrero, la gente disfruta del comienzo del Carnaval disfrazándose de payaso[2], animal u otras cosas. En Santo Domingo, orquestas tocan música bailable a lo largo del hermoso malecón[3] de la capital. Los dominicanos celebran hasta altas horas de la noche, conmemorando el día de la Independencia con la alegría del Carnaval.

[2] clown
[3] boardwalk

CLASSZONE.COM
More About the Caribbean

¿Comprendiste?

1. ¿Qué dos países colonizaron La Española?
2. ¿Qué día se celebra la independencia dominicana? ¿Qué evento histórico conmemora?
3. ¿Qué hacen los dominicanos en el malecón en la noche del 27 de febrero?

¿Qué piensas?

1. ¿Cuáles son los dos elementos que dan a la celebración del Día de la Independencia de la República Dominicana un aspecto especial?
2. Compara esta celebración con la de Estados Unidos. ¿Qué elementos comparten?¿Cómo son diferentes?

Hazlo tú

Escribe una breve descripción del día de la Independencia de Estados Unidos. ¿Qué eventos históricos conmemora? ¿Cómo celebra el pueblo estadounidense su independencia? ¿Cómo celebran el día tú y tu familia?

 Activity 3: Generally use culturally appropriate behavior in social situations

En uso
REPASO Y MÁS COMUNICACIÓN

OBJECTIVES

- Describe historic events
- Make suggestions and wishes
- Express emotion and doubt
- State cause and effect

Now you can...

- describe historic events.
- make suggestions and wishes.

To review

- the subjunctive, see pp. 232, 234.

1 La celebración

Vas a tener una celebración para el día de la Raza, pero tienes muchos problemas. Conversa con un(a) compañero(a) sobre qué necesitan para la celebración.

> **modelo**
>
> *El conjunto no sabe tocar bomba y plena.*
>
> **Tú:** *Tenemos un problema.*
>
> **Compañero(a):** *¿Qué?*
>
> **Tú:** *Necesitamos un conjunto que sepa tocar bomba y plena.*

1. La banda no toca el himno nacional.
2. El cocinero no prepara comida dominicana.
3. Los músicos no cantan canciones tradicionales.
4. La candidata no quiere dar un discurso patriótico.
5. Los estudiantes no quieren participar en el desfile.
6. El lugar no es grande.

Now you can...

- make suggestions and wishes.

To review

- the subjunctive, see pp. 232, 234.

2 Santo Domingo

Piensas visitar a tu amigo(a) dominicano(a). Él te escribe una carta con sus planes. Complétala con el subjuntivo de los verbos entre paréntesis para saber qué te dice.

Querido Esteban:

Siento que se _____ (haber terminado) el verano pero me alegro de que _____ (venir) a visitarme a Santo Domingo. Yo dudaba de que _____ (poder) venir. Quiero que _____ (conocer) nuestras playas. Voy a llamar a alguna amiga que _____ (saber) bailar merengue para que te _____ (enseñar). No dejaré que _____ (regresar) a Boston hasta que _____ (aprender). Cuando _____ (bailar) como yo, serás casi dominicano. Para que te _____ (dar cuenta), aquí baila todo el mundo.

Hasta pronto,

Rubén

Now you can...

• express emotion and doubt.

To review

• subjunctive vs. indicative, see p. 238.

③ El debate

Estás en una reunión de tu comunidad. Tú y tu vecino siempre tienen opiniones opuestas sobre la política. ¿Qué le dices y cómo te responde?

modelo

alcalde: *ser conservador*

Tú: *Pienso que el alcalde es conservador.*

Compañero(a): *Dudo que el alcalde sea conservador.*

1. la gobernadora: tener mucho poder
2. el presidente: resolver los problemas de los ciudadanos
3. todos los países: necesitar un ejército
4. el rey: colaborar con la gente

5. el gobierno: preservar los derechos humanos
6. la constitución: necesitar cambios
7. la presidenta: apoyar el desarrollo de los países pobres
8. la reina: tener una ideología liberal

Now you can...

• state cause and effect.

To review

• the subjunctive, see pp. 232, 234.

④ Con tal de que...

Tú dices que vas a hacer algo bajo ciertas condiciones. ¿Qué dices?

modelo

Voy a votar por el candidato liberal con tal de que _____ (resolver los problemas de la ciudad)

Voy a votar por el candidato liberal con tal de que resuelva los problemas de la ciudad.

1. Voy a donar mucho dinero a esa organización con tal de que _____ (honrar la memoria de los veteranos)
2. Vamos a visitar a ciertos países con tal de que _____ (tener un gobierno democrático)
3. Vamos a apoyar a ese gobierno con tal de que _____ (colaborar con los gobiernos de otros países)

4. Voy a votar por el candidato liberal con tal de que _____ (organizar a la gente del pueblo)
5. Voy a participar en la campaña para juntar fondos con tal de que _____ (usarse para mantener limpia la ciudad)
6. Voy a trabajar de voluntario(a) para esa organización con tal de que _____ (luchar contra la pobreza)

5 Mi comunidad

STRATEGY: SPEAKING

Express yourself In this discussion of politics, express your wishes, hopes, emotions, doubts, uncertainties, and concerns about the actions of your political leaders. How can you use what you have learned in this unit?

En grupos de dos o tres, conversen sobre los políticos de su ciudad, su estado y su país. Si es necesario, lean sobre los políticos en el periódico para saber más de su ideología.

modelo

Tú: *El alcalde de mi ciudad es bastante liberal. Apoya los servicios sociales y …*

Amigo 1: *La gobernadora de mi estado es conservadora…*

Amigo 2: *El presidente de nuestro país es…*

6 Vamos a celebrar

En grupos de dos o tres, planeen una celebración para un día patriótico o histórico. Primero, escojan un día que quieran celebrar. Luego, preparen los planes para la fiesta.

modelo

Tú: *¿Por qué no celebramos el día que fundaron nuestra ciudad?*

Amigo 1: *Buena idea. Podemos buscar datos…*

Amigo 2: *Y luego podemos…*

7 En tu propia voz

ESCRITURA Imagínate que estás en Puerto Rico para el día de la Abolición de la Esclavitud. Escribe una tarjeta postal a tu familia describiendo qué hiciste y qué viste.

CONEXIONES

Los estudios sociales En Estados Unidos celebramos nuestra independencia de Inglaterra el 4 de julio. Escoge tres países hispanohablantes y haz una investigación para descubrir lo siguiente: ¿Celebra el país un día de la independencia? Si lo celebra, ¿de quién ganó la independencia y cuándo se celebra? Si no celebra un día de la independencia, ¿por qué no? Escribe un reportaje y compara los resultados con dos o tres compañeros de la clase.

En resumen
REPASO DE VOCABULARIO

DESCRIBE HISTORIC EVENTS

Columbus Day

acudir a	to attend
el (la) almirante	admiral
la banda	band
la bandera	flag
el descubrimiento	discovery
el faro	lighthouse
la procesión	procession
sonar (o→ue)	to sound, to ring

Abolition of slavery

los antepasados	ancestors
la bomba	Afro-Caribbean dance
el conjunto	musical group
conmemorar	to commemorate
la costumbre	custom
el día de la Abolición de la Esclavitud	Abolition Day
enfrentar	to confront
el (la) esclavo(a)	slave
honrar	to honor
la injusticia	injustice
justo(a)	just, fair
la lucha	fight
el (la) opresor(a)	oppressor
la plena	Afro-Caribbean dance
el (la) proponente	supporter
solemne	solemn
la victoria	victory

Government

conservador(a)	conservative
la constitución	constitution
la democracia	democracy
democrático(a)	democratic
el derecho	the (legal) right
el ejército	army
el gobierno	government
la ideología	ideology
la ley	law
liberal	liberal
la monarquía	monarchy
el poder	power

Leaders

el (la) alcalde(sa)	mayor
el (la) gobernador(a)	governor
el (la) líder	leader
el (la) presidente(a)	president
la reina	queen
el rey	king

Patriotism

la competencia	competition
el ensayo	essay
la patria	mother country
el (la) patriota	patriot
patriótico(a)	patriotic
el patriotismo	patriotism

EXPRESS EMOTION AND DOUBT

♻ Ya sabes

dudar	to doubt
es cierto	it's certain
es dudoso	it's doubtful
es improbable	it's improbable
es verdad	it's true
estoy seguro(a)	I'm sure
no estoy seguro(a)	I'm not sure
no creer	to not believe
creer	to believe
pensar (e→ie)	to think

STATE CAUSE AND EFFECT

The subjunctive

Vamos a la celebración a menos que no **haya** tiempo.
Podemos quedarnos hasta que la banda **empiece** a tocar.

MAKE SUGGESTIONS AND WISHES

The subjunctive

Busco un conjunto que **sepa** tocar la plena y la bomba.
El gobernador quiere que **participemos** en la procesión.

Juego

Saludo a la gente pero no tengo manos. Me comunico con la gente pero no tengo boca. Soy alto pero no necesito ropa especial.
Doy apoyo y protección a la gente. ¿Qué soy?

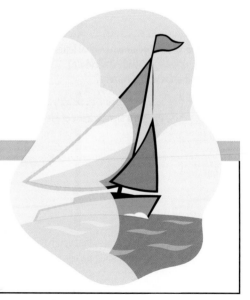

Create paragraphs
when writing

En tu propia voz
ESCRITURA

¡Les deseamos mucho éxito!

¡Felicitaciones! Te vas a graduar este año y tu clase quiere
expresar sus deseos para los estudiantes del futuro. Están
preparando una «cápsula de tiempo» donde van a incluir una
carta dirigida a estos estudiantes.

Función: Expresar deseos

Contexto: Escribirles a los
estudiantes del futuro

Contenido: Experiencias en la escuela

Tipo de texto: Carta personal

PARA ESCRIBIR • STRATEGY: WRITING

Use transitions to make text flow smoothly Using transition words
helps you organize the major points of your letter and move
logically from one point to another. Some Spanish transition
words are **primero, segundo, tercero, al principio, entonces,
luego, además, por último,** and **también.**

Modelo del estudiante

The salutation for a personal letter
usually begins with **Querido(a)**
and ends with a comma.

Primero signals a list of ideas.

Para concluir indicates that you have
reached the concluding paragraph.

Queridos estudiantes del futuro,

Estamos a punto de graduarnos y pensamos en el futuro. ¡Ojalá que
pudiéramos ver esta escuela en 100 años y hablar con los estudiantes del
siglo veintidós! Tenemos muchas esperanzas para el futuro.

Primero, ¡sería mejor que en el futuro los estudiantes no tuvieran que
pasar tanto tiempo haciendo la tarea! ¿No sería mucho mejor si tuvieran
más tiempo para las actividades y para estudiar para exámenes?

Segundo, nos gustaría que hubiera unos cambios en nuestra
comunidad. ¿No sería buena idea permitir a los jóvenes conducir un coche
a la edad de catorce años en vez de a los dieciséis? ¡Entonces los
otros miembros de la familia no tendrían que gastar su tiempo llevándonos
a nuestras actividades!

También, esperamos que en el futuro las posibilidades de empleo sean
mejores para todos los estudiantes. ¡Necesitamos más oportunidades!

Para concluir, ¡les deseamos todo lo bueno de nuestra época y
esperamos que todo lo malo haya mejorado al llegar a la suya!

Atentamente,

Los estudiantes del siglo veintiuno

Typical closings for a personal letter are
Atentamente, Cordialmente, Un abrazo,
etc., followed by a comma and a signature.

Estrategias para escribir

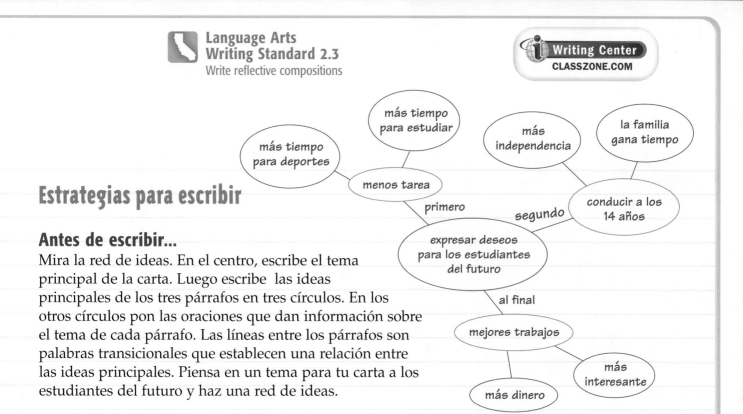

Antes de escribir...

Mira la red de ideas. En el centro, escribe el tema principal de la carta. Luego escribe las ideas principales de los tres párrafos en tres círculos. En los otros círculos pon las oraciones que dan información sobre el tema de cada párrafo. Las líneas entre los párrafos son palabras transicionales que establecen una relación entre las ideas principales. Piensa en un tema para tu carta a los estudiantes del futuro y haz una red de ideas.

Revisiones

Después de escribir el primer borrador, pídele a un(a) compañero(a) que la compare con su red de ideas. Pregúntale:

- *¿Cuáles son las palabras y expresiones transicionales que usé para establecer una relación entre los tres párrafos?*
- *¿Cuál es la relación entre cada párrafo y el tema principal?*
- *¿Cómo se relacionan las oraciones de cada párrafo con la idea principal del párrafo?*

La versión final

Para completar tu carta, léela de nuevo y repasa los siguientes puntos:

- *¿Usé alguna expresión de duda, incertidumbre, juicio, no-existencia o emoción que requiere el subjuntivo?*

Haz lo siguiente: Subraya los verbos en el presente del indicativo o en el subjuntivo. Determina por qué el contexto requiere estos tiempos. Corrige los verbos.

- *¿Usé el tiempo verbal condicional correctamente?*

Haz lo siguiente: Confirma la aplicación apropiada del condicional. Haz un círculo alrededor de los verbos y las formas. ¿Está conjugado correctamente el verbo condicional? ¿Deberías usar subjuntivo para expresar una situación hipotética?

Espero que los estudiantes del
siglo veintiuno tienen mucho éxito.
(tengan)
Primero, quiero que disfruten de
los años de estudio en la escuela
superior. ¡Se van muy rápido!
Después
Antes, no pueden olvidarse de sus
amigos. Los amigos son una parte
importante de la vida.

UN FUTURO BRILLANTE

STANDARDS

Communication
- Describing your studies
- Asking questions
- Saying what you are and were doing
- Talking about careers
- Confirming, denying, and hypothesizing
- Expressing emotions
- Clarifying meaning
- Expressing possession
- Expressing past probability

Cultures
- The culture of the Southern Cone countries
- Fields of study in schools in Latin America
- Careers in Latin America
- Latin America economics

Connections
- Social Studies: Job requirements
- Social Studies: International organizations

Comparisons
- High school students' future goals and plans
- Economic situations
- How language reflects culture

Communities
- Using Spanish in the workplace
- Using Spanish in volunteer activities

URUGUAY
RAFAEL GUARGA inventó un método eficaz para proteger las frutas de las temperaturas frías. ¿Qué datos crees que tomó en cuenta para su invención?

CHILE
LA UNIVERSIDAD DE CHILE es una de las más prestigiosas de América del Sur. ¿Qué cosas crees que se estudian allí?

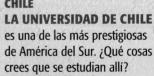

INTERNET Preview
CLASSZONE.COM
- More About the Southern Cone
- Webquest
- Self-Check Quizzes
- Flashcards
- Writing Center
- Online Workbook
- eEdition Plus Online

ALMANAQUE CULTURAL

POBLACIÓN: Argentina: 37.384.816; Chile: 15.328.467; Paraguay: 5.734.139; Uruguay: 3.360.105

ALTURA: 6.959 m sobre el nivel del mar, Cerro Aconcagua (punto más alto)

TEMPERATURA: (más alta) 74ºF (24ºC) Asunción, Paraguay. (más baja) 48ºF (11ºC) Bariloche, Argentina

COMIDAS: mate, parrillada, dulce de leche, puchero

GENTE FAMOSA: Isabel Allende (escritora), Gabriel Batistuta (futbolista), Mario Benedetti (poeta), Adolfo Pérez Esquivel (pacifista)

 Mira el video para más información.

CLASSZONE.COM
More About the Southern Cone

ANTONIO BERNI

ARGENTINA
ANTONIO BERNI (1905–1981) Muchas veces decidimos nuestro futuro durante nuestra niñez. Este pintor argentino celebra estos momentos en su pintura *El club atlético de Chicago,* 1937. Según lo que ves, ¿qué serán estos niños en el futuro?

EL CONO SUR
MATE El mate es una bebida parecida al té. Se toma en un envase (*container*), también llamado mate, con un objeto especial llamado bombilla. ¿Conoces otras comidas o bebidas de América del Sur?

PARAGUAY
EL ARPA es uno de los instrumentos típicos de Paraguay. ¿Qué cultura crees que desarrolló este instrumento?

ARGENTINA
LA BOLSA Éste es uno de los centros de comercio principales de Latinoamérica. ¿Qué países crees que participan en ella?

UN FUTURO BRILLANTE

c@fe INTERNET

capacitación
mantenimiento

e-mail
web
irc

09-554-137

- Comunicación

- Culturas

- **Conexiones**

- Comparaciones

- Comunidades

252

Conexiones

Al estudiar español, aprendemos más de otras materias, como el arte, las ciencias, las matemáticas, la música y los estudios sociales. ¿Cuál es tu materia favorita? ¿Qué has aprendido de esa materia por medio del español?

Conexiones en acción Identifica la materia representada por cada foto y explica cómo podrías usar el español para aprender más.

Comparaciones

Es fascinante comparar los idiomas que se hablan en el mundo. ¿Conoces estas palabras? ¿Sabes cuál es su origen?

chocolate

papa

maíz

llama

huracán

Comunicación

Usamos varias formas de comunicación todos los días. Por ejemplo, nos comunicamos por teléfono y por medio de los anuncios. ¿Cuántas formas de comunicación puedes identificar en la foto?

Culturas

La cultura distinta de los países del Cono Sur se refleja en sus productos, costumbres y perspectivas. ¿Cuánto sabes ya de la cultura de Chile, Argentina, Paraguay y Uruguay? ¿Sabes algo de su economía?

Fíjate

¿Esperas tener un futuro brillante? Explica la relación que puede tener una de las fotos de estas páginas con tu futuro. Si no encuentras una foto adecuada, haz un dibujo que represente tus planes para el futuro.

Comunidades

En esta unidad conocerás a Toño, un alumno que usa el español en su comunidad. ¿Cómo puede Toño ayudar a una estudiante hispanohablante que no comprende su tarea?

ETAPA

1

El próximo paso

OBJECTIVES

- Describe your studies
- Ask questions
- Say what you are doing
- Say what you were doing

¿Qué ves?

Mira la foto. Contesta las preguntas.

1. ¿En qué clase estarán estos estudiantes? ¿Cómo lo sabes?

2. ¿Tienen esta clase en tu escuela? ¿La has tomado?

3. ¿Cómo crees que se sienten los estudiantes? ¿Por qué?

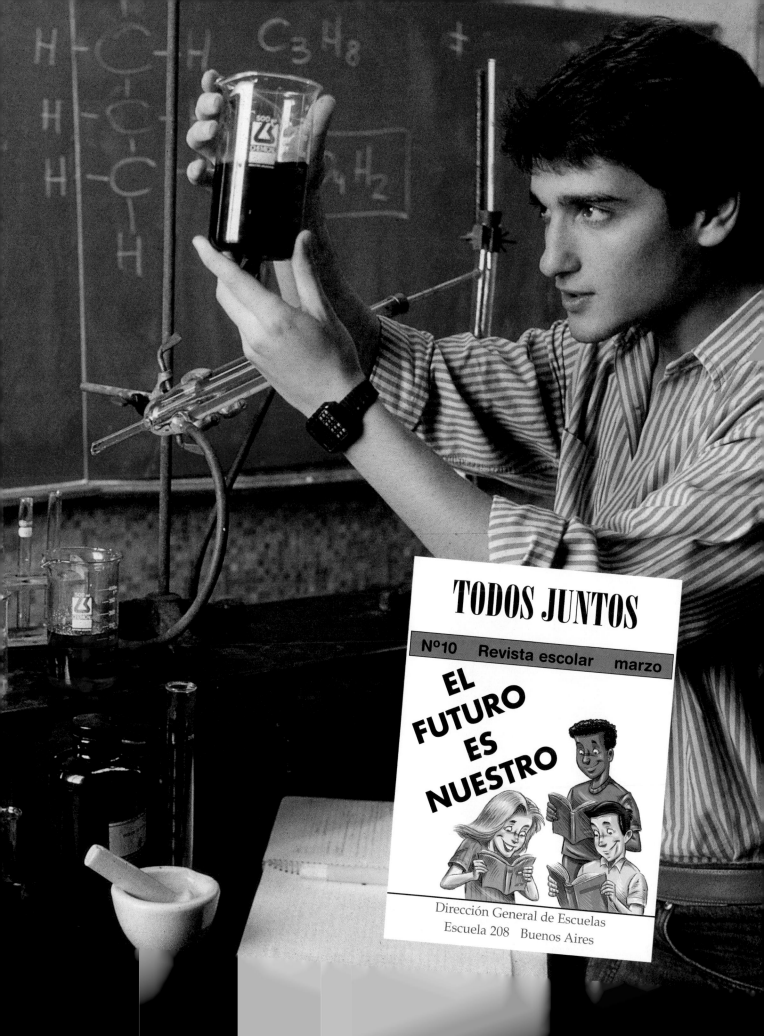

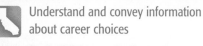
¿Qué quieres hacer?

Es importante pensar en tu futuro. ¿Qué campo de estudio te interesa después del colegio?

Administración de empresas

Administración de empresas
Te llevas bien con la gente. Crees que podrías manejar el personal de un negocio.

Comercio
Quieres ser un hombre o mujer de negocios para ver cómo funcionan los negocios grandes.

Contabilidad
Tienes gran habilidad para las matemáticas. ¡Lo cuentas todo!

Finanzas
Lo que más te interesa es cómo aumentar fondos.

Informática
¡Quieres hacer que la computadora haga lo que tú quieras!

Mercadeo
Te gustaría saber qué estrategias convierten un producto en un éxito.

Ventas
¡Tienes poder de persuasión! Convences a los consumidores sobre los productos que deben comprar.

Agronomía

Agronomía
Te interesa cultivar vegetales, frutas, o cualquier producto de la tierra.

Educación

Siempre has querido ser maestro o maestra. Te encantaría enseñar a otros todo lo que sabes.

Humanidades

Te interesa todo lo que tiene que ver con la gente y la cultura: la filosofía, la literatura, las bellas artes.

Educación

Ingeniería civil

Diseño

Eres muy creativo(a) y artístico(a). Ya sean libros o revistas o anuncios, tú los quieres diseñar.

Publicidad

Tú quieres crear los anuncios que atraen la atención del público.

Relaciones públicas

Para tener una buena relación con el público, las compañías emplean a gente que se dedica solamente a eso.

Dibujo técnico

Te gusta dibujar y eres muy exacto(a). Te encanta pasar horas dibujando con mucha precisión.

Ingeniería civil

¿Cómo se construyen las ciudades? Quieres diseñar y construir **carreteras** y puentes.

Ingeniería mecánica

Este campo se dedica a la producción, el diseño y el uso de las máquinas.

Diseño

Online Workbook
CLASSZONE.COM

¿Comprendiste?

1. ¿Qué campo te interesa? ¿Por qué?
2. ¿Qué tienen en común los campos de publicidad, mercadeo y ventas?
3. ¿Te interesa seguir la profesión de alguien que conoces? ¿Por qué?
4. ¿En qué campo de estudio crees que podrías desarrollar tus habilidades y talentos?
5. ¿Explorarás varios campos antes de seguir uno? ¿Sabes cuál vas a seguir? Explica.

Listen to audio texts

En vivo

AUDIO

SITUACIONES

PARA ESCUCHAR

STRATEGIES: LISTENING

Pre-listening Scan the fields of study on p. 256. In addition to knowledge, each one requires certain personal qualities. Which ones require the ability to get along with others? Which ones are best suited for those who think for themselves and set personal goals?

Evaluate recommendations: Read **Datos personales** and analyze Emilio's interests and skills. What courses would you recommend for him? Then listen to his interview with the counselor. How does your suggestion compare with hers? How do you explain the differences?

¡Eso sí me interesa!

Esperas hablar con la consejera del colegio y ves el formulario de otro estudiante. Luego, escuchas una entrevista entre este estudiante y la consejera sobre los planes que él tiene para el futuro.

1 Leer

Encuentras este formulario en la oficina mientras estás esperando a la consejera. Estúdialo para saber más sobre el estudiante que lo dejó.

DATOS PERSONALES

Nombre: _Emilio García Ávila_

Dirección: _____

No. de teléfono: _____ **Fecha de nacimiento** _____

Marca los campos de estudio que te interesan:

☒ **Agricultura** *Me parece un poco aburrido...*

☐ **Administración de empresas**

☐ **Comercio**

☐ **Contabilidad**

☑ **Dibujo técnico**

☐ **Diseño**

☐ **Educación**

☐ **Finanzas**

☒ **Humanidades**

☐ **Informática** *¡Eso no es para mí!*

☐ **Ingeniería civil**

☐ **Ingeniería mecánica**

☐ **Mercadeo**

☐ **Publicidad** *Soy demasiado tímido para trabajar con la gente. ¡Olvida Ventas y Relaciones públicas!*

☐ **Relaciones públicas**

☒ **Ventas**

---

2 Escuchar

Emilio está en la oficina de la consejera para hablar con
ella de los campos de estudio que le interesan. Escucha
la entrevista. Marca **sí** si el campo de estudio le interesa a
Emilio. Marca **no** si no le interesa. Marca **no se menciona**
si ese campo de estudio no se menciona en la entrevista.

	Sí	No	No se menciona
Agricultura	⬭	⬭	⬭
Comercio	⬭	⬭	⬭
Contabilidad	⬭	⬭	⬭
Dibujo técnico	⬭	⬭	⬭
Diseño	⬭	⬭	⬭
Educación	⬭	⬭	⬭
Finanzas	⬭	⬭	⬭
Humanidades	⬭	⬭	⬭
Ingeniería civil	⬭	⬭	⬭
Ingeniería mecánica	⬭	⬭	⬭
Mercadeo	⬭	⬭	⬭
Publicidad	⬭	⬭	⬭
Relaciones públicas	⬭	⬭	⬭
Tecnología	⬭	⬭	⬭

3 Hablar

En grupos de dos o tres, conversen sobre los campos de estudio que más les
interesan. Hablen de sus talentos y habilidades. Escojan dos o tres campos y
digan por qué les interesan. Luego escojan otros dos o tres campos y digan por
qué no les interesan.

En acción

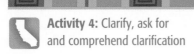

PARTE A

Práctica del vocabulario

Objectives for Activities 2–4
• Describe your studies • Ask questions

1 El (La) consejero(a)

Escribir Eres el (la) consejero(a) de un colegio. Varios estudiantes quieren tu opinión. Según sus intereses, diles qué deberían estudiar.

Diseño Informática Finanzas

Comercio Educación Publicidad

Ingeniería mecánica Relaciones públicas

1. Quiero ser maestro en una escuela secundaria.
2. Soy muy social. Quiero un puesto en algo que tenga que ver con la gente.
3. Me fascinan las máquinas. Me gusta desarmarlas para ver cómo funcionan.
4. Soy muy artística. Me gustaría diseñar libros.
5. El mundo de los negocios me interesa mucho.
6. ¡Quiero hacerme rico! Saber invertir el dinero es muy importante para vivir bien.
7. Soy muy creativa. Creo que puedo inventar anuncios para la tele.
8. Tengo tres computadoras. ¡No puedo vivir sin mis computadoras!

2 La especialización

Hablar/Escribir Quieres saber en qué se especializaron las siguientes personas. Sigue el modelo.

hermano mayor

Tú: ¿En qué se especializó tu hermano(a) mayor?

Compañero(a): *Se especializó en dibujo técnico.*

1.
Luis Pablo

2.

tío Ernesto

3.

prima Aurora

4.
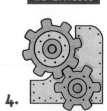
Mercedes

5.

doña Carmen

6.

Rodolfo

3 La solicitud

Hablar/Escribir Tú y tu compañero van a llenar una solicitud de empleo para los siguientes departamentos de una empresa. Hablen sobre la información que tienen que dar usando expresiones del vocabulario.

modelo

mercadeo

Tú: *En la solicitud preguntan para qué estás capacitado(a).*

Compañero(a): *Me especialicé en mercadeo. Me gusta tomar decisiones. ¿Y tú?…*

1. relaciones públicas
2. finanzas
3. educación
4. diseño

Vocabulario

Tus habilidades

adaptarse *to adapt oneself*

capacitado(a) *qualified*

correr riesgos *to take risks*

desempeñar un cargo *to carry out a responsibility*

emprendedor(a) *enterprising*

encargarse de *to be in charge of*

especializarse *to specialize*

estar dispuesto(a) a *to be willing to*

la formación *training, education*

superarse *to get ahead, excel*

tomar decisiones *to make decisions*

▶ ¿Qué palabras usarías para describirte?

4 ¿Cuál te interesa?

Hablar/Escribir Quieres saber qué campo de estudio le interesa a tu compañero(a) y cuáles son las habilidades que tiene para ese campo. Luego, él (ella) quiere saber lo mismo sobre ti.

modelo

Tú: *¿Qué campo de estudio te interesa?*

Compañero(a): *Me interesa mucho la ingeniería mecánica.*

Tú: *¿De veras? ¿Por qué?*

Compañero(a): *Desde niño(a) me han fascinado las máquinas. También quiero diseñar equipo mecánico.*

Práctica: gramática y vocabulario

Objectives for Activities 5–16
• Ask questions • Say what you are doing • Say what you were doing

REPASO Interrogative Words

▶ There are many words you can use to ask questions. Many of the ones you already know are reviewed in the vocabulary box below.

▶ When you ask someone to repeat what they just said, use **¿Cómo?** rather than **¿Qué?** It is more polite.

—**Mañana no hay clases.**
Tomorrow there are no classes.

—**¿Cómo?**
What (did you say)?

▶ As you know, the meaning of **¿Cómo?** changes depending on whether you use it with **ser** or **estar** :

asks about appearance, character, and personality

asks about her health

¿Cómo es Laura?
What does Laura look like?
What is Laura like?

¿Cómo está Laura?
How is Laura?

Don't forget to write Spanish question words with an **accent mark.**

Vocabulario

♻ **Ya sabes**

¿Adónde?	**¿De quién(es)?**
¿A quién(es)?	**¿Dónde?**
¿Cómo?	**¿Para qué?**
¿Cuál(es)?	**¿Por qué?**
¿Cuándo?	**¿Qué?**
¿Cuánto(s)/Cuánta(s)?	**¿Quién(es)?**
¿De dónde?	

Practice: Actividades
5 6 7

Más práctica
cuaderno p. 94
Para hispanohablantes
cuaderno p. 91

🛈 **Online Workbook**
CLASSZONE.COM

5 Preguntón

Hablar/*Escribir* Estás sentado(a) en el jardín de la Universidad de Chile. Un estudiante te ve y empieza a hacerte muchas preguntas. ¿Qué te pregunta? Usa las palabras interrogativas del vocabulario para completar sus preguntas.

modelo

Yo soy de Santiago.
Y tú, ¿de dónde eres?

1. ¿_____ te graduaste de la secundaria?

2. ¿_____ es tu clase favorita?

3. ¿_____ te gusta la ingeniería mecánica?

4. ¿_____ libros necesitas comprar para esa clase?

5. ¿_____ es tu profesor de dibujo mecánico?

6. ¿_____ aprendiste a dibujar tan bien?

7. ¿Ya te vas? ¿_____ vas?

8. ¿_____ está la biblioteca?

9. ¿_____ es esa bicicleta?

10. ¿Tomaste apuntes en clase? ¿_____ le prestaste

6 Los datos

Escuchar/*Escribir* Escucha una entrevista entre la señora Madrigal y un empleado de la universidad. Luego completa el formulario.

Solicitud de empleo

Nombre: _____ Fecha de nacimiento: _____

Estado civil: _____ Campo de estudio: _____

Educación:

Licenciatura: _____ Maestría: _____ Doctorado: _____

Fecha de solicitud: _____

Vocabulario

Dar información por escrito

el currículum vitae *resumé*

los datos *facts; information*

el doctorado *doctorate*

el estado civil *civil status (married, divorced, single)*

la estatura *height*

la fecha de nacimiento *date of birth*

la firma *signature*

la licenciatura *university degree*

la maestría *master's degree*

el paquete *package*

el sobre *envelope*

solicitar *to request, to apply for*

la solicitud *application*

▶ ¿Qué información has dado por escrito y para qué?

7 Un(a) nuevo(a) amigo(a)

STRATEGY: SPEAKING

Establish closer relationships In developing a new friendship, you not only want to find out biographical information, but also know about that person's plans, hopes, beliefs, and feelings. These verbs will help you: **pensar, esperar, creer, sentirse.**

Hablar/*Escribir* Imagina que tú y tu compañero(a) se acaban de conocer. Hazle preguntas primero. Luego, cambien de papel.

¿Adónde? ¿Para qué?

¿Cuánto?/¿Cuántos?

¿Dónde? ¿Cuánta?/¿Cuántas?

¿De dónde? ¿Por qué? ¿Qué?

¿Cuál?/¿Cuáles? ¿Quién?/¿Quiénes?

¿Cómo? ¿De quién?/¿De quiénes?

¿A quién?/¿A quiénes? ¿Cuándo?

modelo

Tú: *¡Hola! Yo me llamo… Y tú, ¿cómo te llamas?*

Compañero(a): *Me llamo… ¿De dónde eres tú?*

Tú: *Yo soy de….*

More Practice: Más comunicación *p. R12*

Activity 8: Converse in simple transactions on the phone

REPASO **The Present Progressive**

The **present progressive** tense is only used for an action that is going on at the time of the sentence. The present progressive is like the **-ing** form (gerund) of a verb in English.

*What are you do**ing**?*
*I am study**ing**.*

To form the **present progressive tense** use:

present tense of estar + **present participle**

You already know the forms of **estar**. To form the **present participle**, drop the ending of the infinitive and add the appropriate ending.

-ar verbs   estudi **ar** + **ando** ➡ estudi**ando**

-er, ir verbs   com **er** + **iendo** ➡ com**iendo**

leer, oír, creer   le **er** + **yendo** ➡ le**yendo**

—Buenas tardes, señora. ¿José Antonio **está comiendo**?
*Good afternoon, Ma'am. **Is** José Antonio **eating** (now)?*

—No, ya comió. **Está estudiando**.
*No, he already ate. **He's studying** (at this moment).*

....

When you have **object pronouns** or **reflexive pronouns** with the present progressive, you can either:

before

- place the **pronouns** before **estar**   Los estudiantes **se están adaptando** a la vida universitaria.
*The students **are adapting** to university life.*

attaches

- or attach them to the **present participle.**   Los estudiantes **están adaptándose** a la vida universitaria.
*The students **are adapting** to university life.*

> When you attach the **pronouns**, you add an **accent mark** to the **a** or **e** before **-ndo**.

Practice: Actividades 8 9 10

Más práctica *cuaderno p. 93*
Para hispanohablantes *cuaderno p. 92*

 **Online Workbook** CLASSZONE.COM

---

8 Las llamadas

Hablar/*Escribir* Haces unas llamadas para hablar con varios amigos. Pero no pueden hablar contigo porque están haciendo otra cosa. ¿Qué están haciendo?

modelo
Ariel / ducharse
Tú: *¿Puedo hablar con Ariel?*
Compañero(a): *No, lo siento. Se está duchando. (Está duchándose.)*

Nota: Gramática
-ir stem-changing verbs change
e → i and **o → u** in the present participle.
v**e**stirse → v**i**stiéndose
d**o**rmir → d**u**rmiendo

1. Esteban: afeitarse
2. Mónica: arreglarse el pelo
3. Arturo: bañarse
4. Marta: correr
5. Cristina: maquillarse
6. Armando: peinarse
7. Raquel: vestirse
8. Beto: secarse el pelo
9. Joaquín: escribir una carta
10. Jimena: acostarse

9 La limpieza ♻

Escribir La familia Márquez está muy ocupada hoy. Es el primer fin de semana de la primavera y ya es hora de hacer la limpieza anual. ¿Qué están haciendo?

modelo

El Sr. Márquez está barriendo el suelo.

10 ¡No puede!

Hablar/Escribir Buscas a varias personas para salir. ¡Pero nadie puede porque todos están ocupados! Pregunta a tu compañero(a) dónde están cinco amigos. Luego, él (ella) te pregunta por otros cinco. Sigue el modelo.

modelo

Tú: *¿Dónde está Miguel? ¿Puede ir al cine con nosotros?*

Compañero(a): *No, no puede. Está estudiando para el examen de español.*

More Practice:

Más comunicación *p. R12*

Nota cultural

Estados Unidos: «¡Excelente!»
Uruguay: «O.K.»

Estados Unidos: «O.K.»

Uruguay: «Lo dudo.»

Activity 13: Use strings of related sentences when speaking

 ¿RECUERDAS? *p. 264* You already know that you use the **present progressive** tense to say what you are doing right now. You can also use the verbs *ir, andar,* and *seguir* instead of **estar** with the **present participle.**

▶ Each of these verbs has a special meaning in progressive constructions.

ir + **present participle**

Anita **se va adapt ando** a su puesto.
*Anita **is (slowly but surely) adjusting** to her job.*

andar + **present participle**

Isabel **anda busc ando** trabajo.
*Isabel **is going around looking** for work.*

seguir + **present participle**

Isabel **sigue busc ando** trabajo.
*Isabel **is still looking for** work.*

Practice:
Actividades
11 **12** **13** **14**

Más práctica
cuaderno p. 95
Para hispanohablantes
cuaderno p. 93

Online Workbook
CLASSZONE.COM

Nota cultural

En muchos países hispanohablantes se usa el título profesional. Por ejemplo: Ingeniero Fernández, Arquitecta Herrera Valentín, etc. Recuerda que los hispanohablantes usan dos apellidos, el del padre y el de la madre. Las mujeres casadas usan el apellido de su padre y el de su esposo.

SPMNET

Ingeniero Suárez Santini

Calle Mar de Plata
#167890 Unidad 28
1040 Buenos Aires

tel. 1/788-5468
fax 1/788-5788

---

11 Un amigo de Paraguay

Hablar/*Escribir* Ramiro, tu amigo de Paraguay, está hablando por teléfono con sus padres. Les está contando cosas de su vida en Estados Unidos. ¿Qué les dice?

> *modelo*
> *estudiar tres horas al día*
> *Sigo estudiando tres horas al día.*

1. trabajar los fines de semana

2. aprender más idiomas

3. tocar la guitarra

4. correr por las mañanas

5. hacer ejercicio todos los días

6. buscar un carro barato

7. correr riesgos en mi trabajo

8. tomar decisiones para mi futuro

9. superarse en mi campo de estudio

10. adaptarse a las costumbres norteamericanas

También se dice

El **mercadeo** también se llama:
- **mercadotecnia**
- **marketing**

---

12 ¿Es cierto?

Escuchar/Escribir Ruiz, el gerente, habla sobre la vida de sus empleados. Di si las oraciones son ciertas o falsas según Ruiz.

modelo

Gómez trabaja en la compañía.
Cierto.

1. Ribeira sigue superándose en su puesto.
2. Escobar nunca comparte sus ideas con otros compañeros de trabajo.
3. Prado va perdiendo mucho dinero.
4. Durán sigue estudiando mercadeo y relaciones públicas.
5. Vega no se está adaptando bien a la vida profesional.
6. Silva anda corriendo riesgos innecesarios.

Nota cultural

En Estados Unidos es común llamar a alguien por su nombre de pila *(first name)*. En la mayoría de los países hispanohablantes se usa el título y el apellido. Por ejemplo, «Sr. Blázquez, ¿está listo el folleto?» Solamente familia, buenos amigos o niños usan el nombre de pila.

13 ¿Qué sabes de...?

Hablar/Escribir Con un(a) compañero(a), hablen sobre amigos que acaban de ver. Haz una lista de cinco amigos. Usa la construcción progresiva con **ir, andar** o **seguir.** ¡Sé creativo(a)!

modelo

Tú: *Oye, ¿has hablado con Enrique? ¿Qué sabes de él?*

Compañero(a): *Sí, hablé con él el otro día. Anda disfrutando de las vacaciones. ¿Viste a Marta? Vino a visitar a sus padres ayer.*

Tú: *Sí, la vi. Me dice que está estudiando mucho.*

> acostumbrarse a las tradiciones de…
> adaptarse a su puesto
> buscar trabajo
> correr riesgos sin tener por qué
> darse cuenta de que [campo de estudio] no es nada fácil
> decidir su carrera
> disfrutar de las vacaciones
> inventar proyectos para el verano
> preocuparse más y más sobre…
> prepararse para la universidad
> tomar decisiones sin pensar
> tratar de convencer a sus padres de que…
> vender su carro

14 Verano

Hablar/Escribir Conversa con un(a) compañero(a) sobre lo que estás haciendo en el verano. Luego, escribe tus ideas.

modelo

Tú: *¿Sigues buscando trabajo para el verano?*

Compañero(a): *Sí, estoy entrevistándome en varias compañías.*

Activity 16: Converse and listen
during face-to-face social interactions

GRAMÁTICA Past Progressive Tenses

▶ You use the **past progressive** to emphasize that an action was in progress at a
particular time in the past. It is usually formed by using:

> *The action was
> in progress at a specific
> time in the past.*

imperfect form of estar + present participle

—¿Qué estabas **haciendo**
a las nueve de la mañana?
*What **were you doing** at nine A.M.?*

—Estaba **desayun ando** y **estudi ando** para
el examen de mercadeo.
*I **was eating breakfast** and **studying** for the marketing exam.*

▶ You can also use the **past progressive** to emphasize that an action continued in the past for
a specific period of time until it came to an end. To form this use of the past progressive use:

preterite of estar + present participle

> *It is clear
> that the action
> has ended.*

Estuvimos **escrib iendo** toda la mañana.
***We were writing** all morning. (**We spent** the whole morning **writing**.)*

Practice: **Actividades** 15 16 **Más práctica** *cuaderno p. 96*
Para hispanohablantes *cuaderno p. 94*

 Online Workbook CLASSZONE.COM

---

15 **Ayer ellos estuvieron...**

Hablar/*Escribir* ¿Qué estuvieron haciendo ayer?

modelo
Gabriel: escribir un ensayo para la clase
Gabriel estuvo escribiendo un ensayo para la clase.

1. yo: diseñar un folleto
2. Paula y Carlos: hacer un dibujo mecánico
3. Herlinda: leer para la clase de educación
4. Gustavo: hacer ventas por teléfono
5. Clara: leer la sección de negocios
6. María y Consuelo: estudiar un programa
   de computación
7. Leonardo: buscar trabajo
8. Nosotros: leer artículos periodísticos

16 **A las dos, a las tres, etc.**

Hablar/*Escribir* Conversen sobre lo que tú y
tu compañero(a) estaban haciendo a ciertas
horas ayer.

modelo
Tú: *Ayer a las dos, yo estaba tomando el examen
de informática. ¿Qué estabas haciendo tú?*

Compañero(a): *Yo estaba estudiando en la biblioteca.*

tomar el examen de...
enviar mi solicitud
llenar el formulario para el puesto de...
escribir un trabajo para la clase de...
trabajar en la oficina de...
hacer la tarea para la clase de...

Activity **17** brings together all concepts presented.

17 Paramedia

Leer/*Escribir* Lee el folleto de la compañía Paramedia y contesta las preguntas.

1. ¿Qué clase de compañía es Paramedia?

2. ¿Cuáles son las cuatro áreas en que se especializa Paramedia?

3. ¿Hacen anuncios publicitarios? ¿Para qué medios de comunicación?

4. ¿En qué países tienen contactos?

5. Si fueras artístico(a), ¿en qué área trabajarías?

6. Si te gustara escribir, ¿en qué área trabajarías?

7. Si estudiaras relaciones públicas, ¿dónde crees que te pondrían a trabajar?

8. ¿Qué les dirías de tus habilidades para que te dieran un puesto? ¿En qué lugares estuviste trabajando antes?

AUDIOVISUAL

Especializados en Spots publicitarios de 35 mm. Realizadores nacionales e internacionales. Equipos de producción. Proyectos multimedia. Vídeo interactivo, corporativo e industrial. Posproducciones digitales y grafismo electrónico. Fotografía. Contactos en Francia, Italia, Finlandia, Holanda, Inglaterra y Estados Unidos.

COMUNICACIÓN

Transmita lo que desee, cuando y como precise, a quien usted quiera. Estrategias planificadas. Convocatorias informativas. Presencia en los medios: Prensa, radio, TV... Convenciones y presentaciones. Relaciones públicas. Desde el posicionamiento hasta el target, cuidamos su imagen y su proyección corporativa.

EDITORIAL

De la concepción y gerencia del producto a la entrega del mismo totalmente acabado. Diseño, diagramación, maquetación. Selección de soportes y artes gráficas. Libros, revistas, posters, folletos, catálogos, papelería corporativa. A la vanguardia en la utilización de papeles especiales para todos los sistemas de impresión.

CREATIVIDAD

Manuales de Identidad Corporativa. Logotipos y aplicaciones. Arte publicitario. Bocetos, ilustraciones, grafismo, dibujo. Entorno Macintosh. Artes finales. Maquetas en volumen. Aerografía. Desde el concepto hasta su desarrollo y total ejecución. Contamos con elementos humanos y técnicos para hacer realidad su mejor idea.

PARAMEDIA

- CREATIVIDAD
- EDITORIAL
- AUDIOVISUAL
- COMUNICACIÓN

❸

More Practice: **Más comunicación** *p. R12*

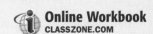
Online Workbook
CLASSZONE.COM

Refrán

Con paciencia se gana lo imposible.

¿Qué quiere decir el refrán? ¿Puedes pensar en una situación en que tuviste que tener mucha paciencia? En tu opinión, ¿es importante tener paciencia?

En voces

AUDIO
LECTURA

Analyze the role of identity and fantasy Movies and television often tell stories about people who are uncertain about their identity. Think about a T.V. story or movie where you have seen this theme. What elements does the character see as fact? Which ones does she or he see as fiction? After the reading, list the elements from Borges' life that he uses in his works. Why do you think he chose those? Would you choose the same? Explain.

EL AUTOR

el hogar	lugar donde uno vive
el tigre	gato salvaje
mudarse	cambiarse de casa
por su cuenta	por sí mismo
paterno(a)	del padre
materno(a)	de la madre
reconocido(a)	famoso(a)

Nota cultural

Borges pasó los últimos años de su vida casi ciego, pero su ceguera *(blindness)* no le impidió seguir escribiendo. Contó con el apoyo de su esposa, María Kodama, quien lo ayudó mucho. Ella hacía los trabajos que él no podía.

Jorge Luis Borges

Los laberintos y sueños[1], la fantasía, las identidades misteriosas y la suspensión del tiempo… todos son temas importantes en las obras de Jorge Luis Borges, uno de los autores latinoamericanos más reconocidos del siglo XX.

Borges nació en Buenos Aires en 1899 y vivió allí hasta 1914. Comenzó a escribir a la edad de nueve años, cuando publicó una traducción al español del cuento *The Happy Prince* de Oscar Wilde. Muchas de sus primeras lecturas fueron en inglés porque su hogar era bilingüe, ya que su abuela era inglesa. A los trece años, publicó su primer cuento original sobre tigres. Desde entonces, los tigres fueron un símbolo importante en la obra de Borges.

En 1914, su familia se mudó a Suiza y en 1919, se trasladó[2] a España, donde Borges publicó «Himno al mar», su primer poema en español. Regresó a Buenos Aires en 1921, fundó revistas y publicó su primera colección de poemas, *Fervor de Buenos Aires* (1923). Publicó poesía a lo largo de[3] su vida.

---

[1] dreams   [2] moved   [3] throughout

Elogio de la sombra (1969), *El oro de los tigres* (1972) y *La rosa profunda* (1975) son otros libros de poemas conocidos. En estos libros, Borges trata los temas de la historia de su familia, una que participó en varias etapas de la historia de Argentina. Su abuelo paterno participó en la guerra civil de Argentina; su abuelo materno también fue soldado. Borges se veía muy distinto a ellos, como dice en «Soy», un poema de *La rosa profunda*:

" Soy… el que no fue una espada[4] en la guerra. "

Borges no luchó con una espada de verdad, pero libró batallas de la imaginación[5] que resultaron en una obra voluminosa. Además de poemas, publicó varias colecciones de cuentos. Entre las más importantes se encuentran *Ficciones* (1944) y *El Aleph* (1949). En sus cuentos, Borges explora el límite entre la realidad y la fantasía y cómo a veces estas cosas se confunden en nuestras vidas.

El sentido del ser—quiénes somos y cómo formamos nuestra identidad—es otro de los temas importantes en la obra de Borges. Él veía su identidad como escritor aparte de su identidad como hombre. Pero Borges el escritor es el que captura finalmente la esencia de Borges el hombre. Hablando de sí mismo como escritor dijo:

" …todas las cosas quieren perseverar en su ser[6]; la piedra eternamente quiere ser piedra y el tigre un tigre. Yo he de quedar en Borges, no en mí (si es que alguien soy)… "

---

[4] sword
[5] fought battles of the imagination
[6] persevere in being themselves

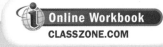

¿Comprendiste?

1. ¿Cómo comenzó la carrera literaria de Borges? ¿Qué lo hizo famoso?
2. ¿Cuáles son unos temas importantes de sus obras?
3. ¿Qué tipos de obras literarias escribió Borges? ¿Cómo es el estilo de Borges?

¿Qué piensas?

1. ¿Cómo crees que la historia de la familia de Borges influyó sus escritos?
2. ¿Por qué crees que la naturaleza forma una parte importante de la obra de Borges?

Hazlo tú

Piensa en las personas y cosas que hacen que tú seas la persona que eres: tu familia, el lugar donde vives, tus intereses, las cosas que has estudiado y tus sueños para el futuro. Luego, escribe un poema o cuento que incluya aspectos importantes de tu relación con estas personas o cosas. También puedes buscar otro poema o cuento de Borges y escribir una opinión corta.

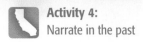

Activity 4:
Narrate in the past

En uso
REPASO Y MÁS COMUNICACIÓN

OBJECTIVES

• Describe your studies
• Ask questions
• Say what you are doing
• Say what you were doing

Now you can...

• describe studies.

• ask questions.

To review

• interrogative words
see p. 262.

1 Inés

Completa las preguntas para saber más sobre la amiga de
tu compañero(a).

modelo

Tú: *¿Quién es tu amiga?*

Compañero(a): *¿Ella? Es Inés de la Cruz.*

1. —¿_____ estudia Inés? —Estudia en la Universidad de Buenos Aires.
2. —¿_____ estudia Inés? —Estudia publicidad.
3. —¿_____ estudia publicidad? —Trabajará en una agencia de publicidad.
4. —¿_____ toma clases? —Toma clases de lunes a viernes.
5. —¿_____ es su clase favorita? —Es «Publicidad para la televisión».
6. —¿_____ es el profesor de esa clase? —El señor Chávez.
7. —¿_____ estudiantes hay en esa clase? —Hay sesenta estudiantes.
8. —¿_____ clases toma por semestre? —Toma seis clases por semestre.

Now you can...

• say what you
are doing.

To review

• the present
progressive
see p. 264.

2 Los estudiantes

En este momento todos los estudiantes están haciendo algo
para sus clases. ¿Qué están haciendo?

modelo

Ángela: estudiar para su examen de comercio

Está estudiando para su examen de comercio.

1. Tito: diseñar un folleto publicitario
2. Berta: entrevistar a un candidato
3. Enrique: enviar un paquete
4. Alicia: escribir un ensayo para la clase de humanidades
5. Clara: hablar con su profesor de dibujo técnico
6. Tomás: ver un video para su clase de publicidad
7. Bárbara: tratar de hacer una venta
8. Eliseo: completar la solicitud para el puesto

Now you can...

• say what you are doing.

To review

• the progressive with **ir**, **andar**, and **seguir** see p. 266.

3 La familia

Algunos miembros de tu familia están en varias situaciones en sus carreras. Explica sus situaciones. Sigue el modelo.

modelo

mi hermana Eugenia: ir / prepararse para la universidad

Mi hermana Eugenia se va preparando para la universidad.

1. mi primo Hernán: seguir / pensar en ser ingeniero
2. mi prima Amelia: ir / adaptarse a su puesto en la agencia de publicidad
3. mi tío: ir / acostumbrarse a la vida de un ingeniero civil
4. mi tía: seguir / trabajar en relaciones públicas
5. mi hermano menor: andar / tratar de decidir en su carrera
6. mi primo Rolando: seguir / enviar solicitudes a varias compañías

Now you can...

• say what you were doing.

To review

• the past progressive see p. 268.

4 ¿Qué estaba haciendo?

Ésta es la agenda de Lorena. Ayer pasó el día investigando varios campos de estudio en la universidad. Di qué estaba haciendo a ciertas horas.

modelo

8:00 / revisar el folleto de la universidad

Tú: *¿Qué estabas haciendo a las ocho de la mañana?*

Compañero(a): *Estaba revisando el folleto de la universidad.*

miércoles

9:00	llegar a la universidad
10:00	entrevistarse con el profesor de contabilidad
10:30	ver los anuncios publicitarios en la clase de diseño
12:00	almorzar con un estudiante de informática
4:00	leer varios ensayos para la clase de educación
5:00	completar la solicitud para el programa de verano
7:00	visitar la clase de ventas
7:30	regresar a la casa

5 A mí me interesa...

STRATEGY: SPEAKING

Extend a conversation By now, you have had lots of experience in brief question-and-answer conversations. Here you can put together all that you know in discussing different areas of study. You can state opinions and preferences, give reasons, ask clarifying questions, give personal or emotional reactions, express doubts or concerns. These are what you have been learning and practicing!

En grupos de dos o tres, conversen sobre los campos de estudio que les interesan y los que no les interesan.

modelo

Tú: *A mí me interesa la finanza. Creo que es importante saber manejar el dinero.*

Amigo(a) 1: *Yo quiero estudiar la publicidad porque me encantaría hacer anuncios para la tele.*

Amigo(a) 2: …

6 Su propia compañía

En grupos de tres, imaginen que van a tener su propia compañía. Primero decidan qué tipo de compañía será. Luego, decidan quién se encarga de cada departamento y por qué.

modelo

Amigo 1: *¿Por qué no hacemos una compañía de publicidad?*

Amigo 2: *Buena idea. Tenemos que darle un nombre.*

Amigo 3: *Y tenemos que decidir quién se encarga de las finanzas…*

7 *En tu propia voz*

ESCRITURA Imagínate que eres el (la) gerente de una empresa. Vas a tener que entrevistar a varias personas para un puesto en tu compañía. Escribe seis preguntas para los candidatos. Luego, hazles la entrevista a dos compañeros.

modelo

- ¿En qué se especializó?
- ¿Cuáles son sus habilidades?
- ¿Por qué cree que usted sería el mejor candidato para este puesto?
- ¿…?

TÚ EN LA COMUNIDAD

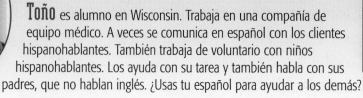

Toño es alumno en Wisconsin. Trabaja en una compañía de equipo médico. A veces se comunica en español con los clientes hispanohablantes. También trabaja de voluntario con niños hispanohablantes. Los ayuda con su tarea y también habla con sus padres, que no hablan inglés. ¿Usas tu español para ayudar a los demás?

En resumen
REPASO DE VOCABULARIO

DESCRIBE YOUR STUDIES

Coursework

la administración de empresas	business administration
la agronomía	agronomy
el campo de estudio	field of study
la carretera	road, highway
el comercio	business
la contabilidad	accounting
el dibujo técnico	technical drawing
el diseño	design
la educación	education
las finanzas	finance
las humanidades	humanities
la informática	computer science
la ingeniería civil	civil engineering
la ingeniería mecánica	mechanical engineering
el mercadeo	marketing
la publicidad	publicity
las relaciones públicas	public relations
las ventas	sales

Abilities and experience

adaptarse	to adapt oneself
capacitado(a)	qualified
correr riesgos	to take risks
desempeñar un cargo	to carry out a responsibility
emprendedor(a)	enterprising
encargarse de	to take charge of
especializarse	to specialize
estar dispuesto(a) a	to be willing to
la formación	training, education
superarse	to get ahead, to excel
tomar decisiones	to make decisions

Written information

el currículum vitae	résumé
los datos	facts, information
el doctorado	doctorate
el estado civil	civil status
la estatura	height
la fecha de nacimiento	date of birth
la firma	signature
la licenciatura	university degree
la maestría	master's degree
el paquete	package
el sobre	envelope
solicitar	to request, to apply for
la solicitud	application

ASK QUESTIONS

♻ **Ya sabes**

¿Adónde?	Where to?
¿Dónde?	Where?
¿De dónde?	From where?
¿Cómo?	How?
¿Cuándo?	When?
¿Cuánto(s)/Cuánta(s)?	How much/many?
¿Cuál(es)?	Which? (choosing items)
¿Para qué?	For what purpose?
¿Por qué?	Why?
¿Qué?	What?
¿Quién(es)?	Who?
¿A quién(es)?	Whom?
¿De quién(es)?	Whose?

SAY WHAT YOU ARE DOING

The Present Progressive Tense

En este momento **estoy estudiando** para el examen.
Mi hermano **sigue buscando** trabajo.

SAY WHAT YOU WERE DOING

The Past Progressive Tense

El viernes a las ocho **estábamos asistiendo** a la clase.

Juego

¡A trabajar!

¿Qué oración va mejor con el dibujo? ¿Por qué?

1. A Magda le gusta trabajar en una oficina.

2. A Magda le gusta correr riesgos.

3. Magda no presta atención a su trabajo.

ETAPA 2

¿Cuál será tu profesión?

OBJECTIVES

- Talk about careers

- Confirm and deny

- Express emotions

- Hypothesize

¿Qué ves?

Mira la foto. Contesta las preguntas.

1. ¿Qué detalles te dan una clave del lugar donde están estas personas?

2. ¿Crees que son amigos o trabajan juntos? ¿Por qué?

3. ¿Qué cosas crees que hace el chico?

4. ¿Qué conocimientos crees que necesitas para ser un(a) diseñador(a) gráfico(a) web?

SPMNET

DISEÑADOR GRÁFICO WEB

☑ Empresa líder en Internet SPMNET requiere personas con conocimientos sobre Internet.

☑ Experiencia en diseño y páginas-Web. Plataforma Mac o PC.

Llamar a Raúl Cerviño, tel. 1/788-9150, fax 1/788-9158

La décima reunión

¡Qué gusto ver a todos! Dicen que yo—Maité Martínez—lleno el requisito perfecto para escribir un resumen de las noticias—¡Soy la más chismosa de todos!

Descubre

Usa las palabras dentro de las palabras para adivinar las profesiones. Mira los dibujos y escoge el equivalente.

1. **deportista** (deporte)
2. **bailarina** (bailar)
3. **diseñador(a) gráfico(a)** (diseño)
4. **cartero** (carta)

a.

b.

c.

d.

Josefina Álvarez es abogada en un bufete en Buenos Aires. Nos presentó a su esposo que también es abogado. Son empleados en la misma oficina. ¡Felicidades, Josefina!

¡No van a creer quién se hizo bailarín! Jorge Valdez me contó que ahora vive en la gran ciudad de Montevideo donde es miembro del grupo de ballet del Teatro Solís.

¿Qué noticias tenemos de la famosa deportista de nuestra clase? Andrea, la gran tenista de Santiago de Chile, viaja por todo el mundo con su raqueta.

La arquitecta Ramona Díaz me contó que ha diseñado varios edificios comerciales en la ciudad de Bariloche.

Lorenzo Godoy ahora es **agricultor** en su ciudad natal en la pampa argentina. En las Granjas Godoy, se producen todo tipo de vegetales y granos. ¡Qué delicia!

Le pregunté al **cartero** Víctor Benedetti si había notado alguna disminución en el número de cartas a causa de Internet, ¡pero parece que él todavía tiene mucho que hacer!

¿Se acuerdan como Tito Villarreal le cortaba el pelo a todo el mundo? Ahora es **peluquero** con un salón exclusivo en Buenos Aires. ¡Pasen a verlo y saldrán con un elegante corte de pelo!

A Carmen Rossi siempre le han encantado los **coches,** así que no me sorprendió nada saber que ahora es **mecánica** en varios talleres de reparación de autos en la ciudad de Valparaíso. ¡Tiene **conocimiento** de todo tipo de coche!

Elsa Jiménez tiene el **puesto** de **diseñadora gráfica** web. Me parece muy natural… siempre estaba haciendo dibujos en la computadora. ¡Y lo sigue haciendo!

¿Y yo? ¿Qué hago yo? Soy **secretaria** ejecutiva bilingüe para una **empresa** multinacional muy importante. ¡Aprender inglés fue la mejor decisión de mi carrera!

Abrazos y saludos a todos.
¡Hasta la próxima!

Maité

Online Workbook
CLASSZONE.COM

¿Comprendiste?

1. De todas las profesiones, ¿cuál te interesa más? ¿Por qué?
2. ¿Conoces a alguien que esté en una de las profesiones que menciona Maité? ¿Quién? ¿Le gusta su trabajo?
3. ¿Qué habilidades crees que debería tener una persona que quiere ser abogado(a)? ¿deportista? ¿diseñador(a) gráfico(a)? ¿agricultor(a)?
4. ¿Has pensado en tu futuro? ¿Crees que es importante saber cuál profesión te interesa antes de la graduación? ¿Por qué?

En vivo

AUDIO
SITUACIONES

PARA ESCUCHAR

STRATEGY: LISTENING

Pre-listening Have you ever applied for a job? What information did the employer want? Write down some key words.

Identify key information for careers Employment ads have two purposes: (1) to encourage qualified applicants, and (2) to discourage unqualified applicants. List the information that should be included to attain those goals.

Información a incluir
1.
2.
3.
etc.

Listen to the radio job bank and decide if the announcements meet your guidelines. Are there major differences?

Y yo, ¿qué quiero ser?

Estás buscando trabajo en Buenos Aires. Vas a ver unos anuncios clasificados y luego escucharás un programa de radio que describe trabajos posibles.

❶ Leer

Ves estos anuncios clasificados en el periódico *La Nación*. Léelos.

SE BUSCA ABOGADO

- Edad de 25 a 35 años

- Dispuesto a viajar por todo el país

Interesados llamar al 1/312-3612 para pedir solicitud y dirección. Deben presentarse con currículum vitae y foto.

MULTINACIONAL

SE SOLICITAN

AUXILIARES DE CONTABILIDAD

Departamento de Ventas

Requisitos:
- Licencia en contabilidad
- Mínimo de 2 años de experiencia
- Edad de 23 a 28 años

INBURSA

Por favor comunicarse con el 1/371-2939 o presentarse en Lavalle 1444 con su currículum vitae entre las 8.00-14.00 horas.

SPMNET

DISEÑADOR GRÁFICO WEB

☑ *Empresa líder en Internet SPMNET requiere personas con conocimientos sobre Internet.*

☑ *Experiencia en diseño y páginas Web. Plataforma Mac o PC.*

Llamar a Raúl Cerviño, tel. 1/788-9140, fax 1/788-9158

❷ Escuchar

Después de leer los anuncios en el periódico, pones la radio. Escuchas un programa que se dedica a dar anuncios clasificados para las empresas que buscan empleados. Copia la tabla siguiente. Luego, escucha los anuncios. Para cada uno, completa la tabla con la información que falta. Si el anuncio no da esa información, deja el espacio en blanco.

Anuncio:	
Puesto:	
Compañía:	
Años de experiencia:	
Edad:	
Requisitos:	
Número de teléfono:	

❸ Hablar/Escribir

En grupos de dos o tres, conversen sobre las profesiones que les interesan. ¿Tienes alguna idea de qué quieres ser? Pregúntales a tus compañeros si ellos saben a qué profesión quieren entrar. Hagan una lista de posibilidades y comparen sus listas con los otros grupos en la clase. ¿Cuántas personas en su clase quieren ser abogados? ¿secretarios? ¿diseñadores? Den los resultados al (a la) profesor(a).

SECRETARIOS EJECUTIVOS BILINGÜES

Requisitos:
➤ inglés 80%
➤ presencia excelente
➤ manejar computadora ambiente Windows
➤ experiencia mínima de 5 años
➤ carrera comercial

Empresa Internacional

Interesados llamar al 1/806-2433

En acción

Práctica del vocabulario

Objectives for Activities 1–4
• Talk about careers

1 Deberías ser...

Hablar/Escribir Tu compañero(a) no sabe qué quiere ser. ¿Qué le dirías si te dijera estas cosas?

modelo

Compañero(a): *Me gusta hablar con la gente y andar por los barrios de la ciudad. No me importa el mal tiempo.*

Tú: *Deberías ser cartero(a).*

1. La música me encanta. Siempre quiero bailar cuando oigo música.

2. Me gusta el arte. También me encanta organizar elementos visuales para crear un concepto total.

3. Me fascinan los edificios: las casas, los edificios de apartamentos, los edificios comerciales.

4. Practico varios deportes: el fútbol, el béisbol, el tenis y el golf.

5. Me encanta desarmar los motores de los carros, encontrar el problema y armarlos de nuevo.

6. Mi familia ha tenido granjas desde hace muchos años. Cultivamos vegetales, granos y frutas.

2 ¿Cuál es su profesión?

Hablar/Escribir Tus padres tienen amigos en varias profesiones. ¿Cuáles son sus profesiones?

modelo

La señora Ibáñez trabaja en un bufete donde protegen los derechos legales de la gente. La señora Ibáñez es abogada.

1. El señor Gómez trabaja en una compañía que diseña páginas para la red.

2. La señorita Campoy hace planos para edificios nuevos.

3. La señora Botero trabaja en un taller donde se reparan automóviles.

4. El señor Varo crea peinados muy modernos. También se especializa en teñir el pelo.

3 ¿Qué quieres ser?

Hablar/Escribir Habla con un(a) compañero(a) sobre qué les gustaría ser y por qué. Luego, cambien de papel.

modelo

Tú: ¿Qué quieres ser? ¿Por qué?

Compañero(a): ¿Yo? Yo quiero ser abogado(a). Me gustaría ayudar a las personas con problemas legales.

Profesiones	¿Por qué?
abogado(a)	ayudar a las personas con problemas legales
arquitecto(a)	diseñar edificios y supervisar la construcción
niñero(a)	cuidar a los niños pre-escolares
peluquero(a)	saber cortar el pelo al estilo preferido
técnico(a) de sonido	trabajar en una radioemisora
veterinario(a)	trabajar con animales
otra	¿…?

Vocabulario

Más profesiones

el (la) **artesano(a)** *artisan*

el (la) **asistente** *assistant*

el **bombero** *firefighter*

el (la) **contador(a)** *accountant*

el (la) **dueño(a)** *owner*

el (la) **entrevistador(a)** *interviewer*

el (la) **gerente** *manager*

el (la) **ingeniero(a)** *engineer*

el (la) **jardinero(a)** *gardener*

el (la) **juez(a)** *judge*

el (la) **niñero(a)** *baby sitter*

el (la) **obrero(a)** *worker*

el (la) **operador(a)** *operator*

el (la) **taxista** *taxi driver*

el (la) **técnico(a)** *technician*

el (la) **veterinario(a)** *veterinarian*

▶ ¿Conoces a personas que tengan estas profesiones?

4 La agencia de empleos

STRATEGY: SPEAKING

Anticipate what others want to know Rehearse in order to be well-prepared and confident for a job interview. Here are some areas you should be ready to talk about: **intereses, estudios, trabajos, cualidades personales y ambiciones para el futuro.**

Hablar/Escribir Una agencia de empleos te entrevista. Tu compañero(a) hace el papel del agente. Luego, cambien de papel.

modelo

Compañero(a): ¿En qué se especializó?

Tú: En diseño gráfico.

Compañero(a): ¿Qué experiencia tiene?

Tú: He trabajado en una agencia de publicidad por tres años. o No tengo experiencia todavía, pero aprendo rápidamente.

Objectives for Activities 5–14

• Talk about careers • Confirm and deny • Express emotions • Hypothesize

REPASO   Affirmative and Negative Expressions

You have learned many words that you can use in negative and affirmative sentences. Here are some you already know in addition to a few more.

• Remember that Spanish uses a double negative: when a negative word follows the **verb**, use **no** before the **verb.**

follows the verb

No estoy haciendo **nada** ahora.
*I'm **not** doing **anything** now.*

• But when you use a negative word before the **verb,** omit **no**:

before the verb

Nunca trabajo los domingos.
*I **never** work on Sundays.*

Affirmative and negative **adjectives** agree with the nouns that they modify.

agrees

algun**as** empres**as**

agrees

ningun**a** muchach**a**

Alguno and **ninguno** change to **algún** and **ningún** when they come before a **masculine singular noun.**

Estoy buscando **algún** trabaj**o**, pero no encuentro **ninguno.**
*I'm looking for **some kind of** a job, but I'm **not** finding **one**.*

Vocabulario

Palabras afirmativas y negativas

a menudo, muchas veces *often*

a veces *sometimes*

ni…ni *neither…nor*

o…o *either…or*

 Ya sabes

algo	nada
alguien	nadie
alguno(a)	ninguno(a)
siempre	nunca, jamás
también	tampoco

Practice: Actividades

5 6 7

Más práctica
cuaderno p. 101
Para hispanohablantes
cuaderno p. 99

Online Workbook
CLASSZONE.COM

5 Preguntas 🎧

Escuchar/*Escribir* Elena, la nueva amiga argentina de Jorge, quiere saber más sobre su pasado. Escucha sus preguntas y escoge la palabra o frase que mejor describe la respuesta de Jorge.

modelo

Elena: *¿Has conocido a un ingeniero alguna vez?*

Jorge: *Sí, hace años conocí a un ingeniero.*

☒ *alguna vez*   ❏ *nunca*

1. ❏ alguna vez
   ❏ nunca
2. ❏ algo
   ❏ nada
3. ❏ alguien
   ❏ nadie
4. ❏ siempre
   ❏ nunca
5. ❏ también
   ❏ tampoco
6. ❏ alguna idea
   ❏ ninguna idea

También se dice

Cuando una persona se gradúa se dice que recibe su **diploma** o su **título.** En Argentina también se dice que la persona "se recibe".

Activity 5: Understand most spoken language when the message is deliberately and carefully conveyed by a speaker accustomed to dealing with learners when listening

6 Necesitas saber ♻ 👥👤

Hablar/Escribir Es tu primer día en la Universidad de Buenos Aires. Le haces muchas preguntas a tu nuevo(a) amigo(a) argentino(a). Conversa con tu compañero(a) sobre los temas en la lista. Luego cambien de papel.

modelo

Tú: *¿Conoces a alguien que estudie diseño?*

Compañero(a): *No, no conozco a nadie que estudie diseño. o Sí, sí conozco a alguien que estudia diseño.*

> conocer a alguien que estudia diseño
> estudiar finanza alguna vez
> saber mucho del mercadeo
> tener algún interés en estudiar dibujo técnico
> conocer a alguien que estudia ingeniería
> saber algo de informática
> ver algún anuncio publicitario para la educación
> ¿...?

7 El día de profesiones 👥👤

Hablar/Escribir Tú y tus compañeros(as) forman parte de un comité que organizará el día de las profesiones en su escuela. Ustedes van a decidir a quiénes van a invitar. Primero, investiguen las profesiones que les interesan a sus compañeros de clase. Entonces, decidan a qué profesionales van a invitar.

modelo

¿Hay alguien en la clase que quiera ser abogado?

sí_____   no ⠀‖‖‖ ‖‖‖‖

No hay nadie en la clase que quiera ser abogado.

¿Hay alguien en la clase que sepa algo de informática?

sí ‖‖‖‖ ‖   no _____ ‖‖‖

Hay seis personas en la clase que saben algo de informática.

More Practice: **Más comunicación** *p. R13*

Nota cultural

La escuela secundaria es el equivalente del *high school* estadounidense. Al finalizar el último año de estudios secundarios, el (la) estudiante recibe un diploma, con el cual puede conseguir trabajo y seguir sus estudios universitarios. En la mayoría de los países latinoamericanos los estudiantes toman un examen de aptitud para poder entrar en la universidad. Algunas escuelas secundarias requieren un año de clases adicionales. De esta manera, los estudiantes aseguran su plaza universitaria sin tener que tomar el examen de aptitud.

Gabriela Columbaro

Calle 9 N° 380, (1900) La Plata, Pcia. de Buenos Aires
Tel./Fax: (021) 89-8783
e-mail: gcolumbaro@edu.com

EXPERIENCIA
- **Médica general,** *Hospital de Emergencias de La Plata* (1996 / Presente)
- **Asistente del Departamento de Psiquiatría,** *Clínica Dr. Eugenio, Necochea* (Abril 1993 / 1996).
- **Asistente médica,** *Seguro de Salud Cruz Azul, Buenos Aires* (1990 / 1993)
- **Empleada administrativa,** *Clínica Yoli, Salto* (1988 / 1990)

ESTUDIOS
- **Universidad de La Plata:** Doctorado en medicina general (1993)
- **Universidad de Belgrano:** Curso: Investigación en psiquiatría (1992)
- **Cruz Roja Argentina:** Curso: Primeros auxilios (1990)

CONOCIMIENTOS
- Computación: procesador de textos y base de datos.
- Idiomas: Inglés, italiano y francés.

PASATIEMPOS • Lectura, música, turismo.

Referencias disponibles.

Activity 8: Converse and listen
during face-to-face social interactions

GRAMÁTICA   Past Perfect Subjunctive

▶ You can use the past perfect subjunctive to say that you wish that things had happened differently than they did. For example, use it after **ojalá que** to express a wish about something that didn't happen:

Ojalá que hubiera llamado.
I wish I had called. (But I didn't.)

▶ To form the past perfect subjunctive use:

past subjunctive
of **haber**   +   **past participle** of the verb.

hubiera llamado	**hubiéramos llamado**
hubieras llamado	**hubierais llamado**
hubiera llamado	**hubieran llamado**

▶ Here are some **irregular past participles:**

abrir → **abierto,** cubrir → **cubierto,** decir → **dicho,** escribir → **escrito,**
hacer → **hecho,** morir → **muerto,** poner → **puesto,** resolver → **resuelto,**
romper → **roto,** ver → **visto,** volver → **vuelto.**

▶ You can also use the past perfect subjunctive, like the present perfect subjunctive, to say that one action took place before another action. You use the past perfect subjunctive when the verb of the main clause is in the **imperfect** or the **preterite**.

I don't
know if they did.

Compare these sentences:

Espero que te hayan dado el puesto.
I hope they gave you the job.

Esperaba que te hubieran dado el puesto.
*I hoped they **had given** you the job.*

I had hoped they
would have given you the
job, but they didn't.

Practice: **Actividades**   **Más práctica** *cuaderno pp. 102–103*   **Online Workbook**
8 9 10 11   **Para hispanohablantes** *cuaderno pp. 100–101*   **CLASSZONE.COM**

8 La celebración ♻ 👥

Hablar/Escribir Hubo una celebración para el Año Nuevo. Todos tuvieron experiencias distintas esa noche. Describe qué hicieron según las indicaciones. Luego, tu compañero(a) te contesta. Cambien de papel.

modelo

yo: no ir a la gala

Tú: *No fui a la gala.*

Compañero(a): *Ojalá que hubieras ido.*

1. nosotros: no festejar el Año Nuevo
2. tú y Paulina: no ver los fuegos artificiales
3. nosotros: no dar las gracias a la anfitriona
4. tú: no volver temprano
5. nosotros: no celebrar hasta la madrugada
6. tú y Aída: no pasarlo bien en la fiesta
7. yo: no gustar el lechón asado
8. ustedes: no llamar a su profesor(a)
9. Hilda: no comer pan dulce
10. ellos: no tocar la guitarra

9 Después de la entrevista

Hablar/Escribir Tuviste una entrevista para un puesto que querías conseguir. Estás nervioso(a) después de la entrevista y le cuentas a tu madre cómo te fue. ¿Cómo te responde ella?

modelo

Mamá, no me dijeron los requisitos para el puesto antes de la entrevista.

Esperaba que te los hubieran dicho.

Nota: Gramática

Remember that when you have both **direct** and **indirect object pronouns**, the indirect object pronoun goes first.

Esperaba que te los hubieran dicho.

1. Mamá, no me hicieron muchas preguntas.
2. Mamá, no me pidieron el currículum vitae.
3. Mamá, no me pidieron recomendaciones.
4. Mamá, no me preguntaron sobre mi licenciatura.
5. Mamá, no me ofrecieron entrenamiento.
6. Mamá, no me explicaron los beneficios.
7. Mamá, no me dijeron el resultado de la entrevista.
8. Mamá, no sé si me dieron el puesto.

Vocabulario

Un puesto nuevo

aumentar *to increase*

los beneficios *benefits*

el contrato *contract*

el entrenamiento *training*

las habilidades *capabilities*

jubilarse *to retire*

la puntualidad *punctuality*

requerir (e→ie) *to require*

el seguro (médico) *(health) insurance*

el sueldo *salary*

Activity 10:
Narrate in the past

10 ¿Supiste?

Hablar/Escribir Pasaron muchas cosas en el trabajo y se las cuentas a tu compañero(a). Di lo que pasó para que él (ella) te responda. Luego, cambien de papel.

modelo

dar el contrato

Tú: *¿Supiste que me dieron el contrato?*

Compañero(a): *Sí, me alegré muchísimo de que te hubieran dado el contrato.*

> aumentar el sueldo
>
> dar el contrato
>
> dar recomendaciones positivas
>
> jubilarse (mi jefe)
>
> ofrecer beneficios muy buenos
>
> ofrecer entrenamiento
>
> pagar ayer
>
> ¿...?

11 Las reacciones de la familia

Escribir Andas buscando trabajo y pasas por muchas experiencias buenas y también difíciles. ¿Cómo reacciona tu familia? Primero escribe cinco cosas que te pasaron (te fue bien/mal en la entrevista, te dieron o no te dieron el puesto, etc.). Luego, para cada cosa que te pasó, escribe una oración describiendo las reacciones de tu familia. Usa los verbos de la lista.

modelo

No me dieron el puesto.

Papá dudaba que me hubieran dado el puesto. o Abuela sentía que no me hubieran dado el puesto. o Mi hermana esperaba que me hubieran dado el puesto.

> alegrarse de que
>
> esperar que
>
> sentir que
>
> ojalá que
>
> dudar que

Nota cultural

Las carreras tradicionales en los países hispanos siempre fueron medicina, abogacía y administración de empresas. Por lo general, las carreras humanísticas, como arte o lenguaje, han estado en segundo lugar. En los últimos años, la computación es muy popular y muchas personas se dedican a estudiarla. Es necesaria para conseguir trabajo.

▶ You use the **conditional perfect** to say that you *would have done* something:

Yo **habría trabajado**.
I would have worked.

Le **habría ofrecido** el puesto.
I would have offered him (her) the job.

To form the **conditional perfect** tense use:

conditional of **haber**   +   **past participle** of the verb.

habría trabajado	habríamos trabajado
habrías trabajado	habríais trabajado
habría trabajado	habrían trabajado

....

▶ The **conditional perfect** is most commonly used with a **si clause** to say what might have been if things had been different. In these sentences you use the **past perfect subjunctive** and the **conditional perfect** together.

Si **hubieras sabido** hablar español, **te habrían dado** el puesto.
If you had known how to speak Spanish, they would have given you the job.

Habríamos podido trabajar en finanzas si **hubiéramos estudiado** economía.
We would have been able to work in finance if we had studied economics.

....

▶ Contrast the meaning of the three types of sentences with **si clauses** that you have learned:

present tense   →   *future tense*
Si **solicitas** el empleo, lo **conseguirás**.
If you apply for the job you will get it.

imperfect subjunctive   →   *conditional*
Si **solicitaras** el empleo, lo **conseguirías**.
If you were to apply for the job (which you aren't), you would get it.

past perfect subjunctive   →   *conditional perfect*
Si **hubieras solicitado** el empleo, lo **habrías conseguido**.
If you had applied for the job you would have gotten it.

Practice: | **Actividades** | **Más práctica** *cuaderno p. 104*
Para hispanohablantes *cuaderno p. 102* |  **Online Workbook** CLASSZONE.COM

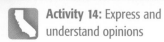

Activity 14: Express and understand opinions

12 Tampoco o también

Hablar/Escribir Muchos de tus amigos están tomando decisiones en relación a sus carreras. Di qué habrías o no habrías hecho tú en las mismas circunstancias. Sigue el modelo.

modelo

Juan no estudió para ser abogado.

Yo tampoco habría estudiado para ser abogado.

1. María no aceptó el trabajo a tiempo parcial.
2. Joaquín cambió de carrera.
3. Arturo no quería un trabajo a tiempo completo.
4. Ana buscó trabajo en publicidad.
5. El señor Miranda no se jubiló hasta que cumplió setenta años.
6. Hernán llegó a todas sus entrevistas a tiempo.

Vocabulario

El trabajo

la carrera *career*

la desventaja *disadvantage*

emplear *to employ*

el empleo *job*

ganarse la vida *to earn a living*

el trabajo a tiempo completo *full-time job*

el trabajo a tiempo parcial *part-time job*

la ventaja *advantage*

▶ ¿Puedes usar algunas de estas palabras y frases para describir tu empleo, si tienes uno?

13 ¡Por eso!

Escuchar/Escribir El consejero de la agencia de empleos les habla a varias personas para explicar por qué no tuvieron éxito en sus entrevistas. Escucha lo que dice para luego decir lo mismo de una manera más cortés.

modelo

¡No sabes hablar español! Por eso no te dieron el puesto.

Si hubieras sabido hablar español, te habrían dado el puesto.

1. Si hubieras tenido las destrezas necesarias, te _____.
2. Si hubieras llegado a la entrevista a tiempo, te _____.
3. Si hubieras llevado tu currículum vitae, te _____.
4. Si hubieras pedido recomendaciones, _____ ese puesto.
5. Si hubieras estudiado ingeniería, _____ el puesto.
6. Si te hubieras puesto una corbata, _____ una buena impresión.

También se dice

Aunque **trabajo a tiempo parcial** es la frase más común en el mundo hispanohablante, también se utilizan las siguientes expresiones:

- **trabajo de medio tiempo** (Argentina)
- **trabajo de media jornada** (España)

Activity 15 brings together all concepts presented.

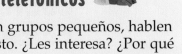

14 Cómo prepararse

Hablar/Escribir Estudia los dibujos. Con un(a) compañero(a), conversen sobre los chicos de los dibujos. ¿Cómo crees que les va a ir en la entrevista? Den todos los detalles que puedan.

modelo

Tú: *Si el chico se hubiera acostado temprano, habría llegado a la entrevista a tiempo.*

Compañero(a): *Es verdad. Probablemente le van a dar el puesto a la chica porque ella llegó a tiempo.*

Cómo prepararse para la entrevista

Cómo vestirse para la entrevista

Cómo comportarse en la entrevista

15 Operadores telefónicos

Hablar/Escribir En grupos pequeños, hablen del siguiente puesto. ¿Les interesa? ¿Por qué sí o por qué no? ¿Qué requisitos se necesitan? ¿Qué les gustaría que pasara en la entrevista?

modelo

Tú: *Ofrecen trabajos a tiempo completo o parcial.*

Compañero(a) 1: *Me interesan los trabajos a tiempo parcial. Así puedo seguir estudiando. Piden conocimiento de PC. Ojalá hubiera encontrado este aviso antes.*

MERKNTEL

Empresa solicita
Operadores telefónicos

Requisitos
- Edad mínima de 20 años
- Preparatoria o equivalente; aceptamos estudiantes de licenciatura
- Buena presentación, excelente ortografía y conocimiento de PC

Ofrecemos
- Sueldo base más bonos
- Trabajos a tiempo completo o parcial
- Ambiente agradable de trabajo
- Capacitación desde tu ingreso

Comunicarse al 2-97765 de 9 a 14 y de 16 a 19 horas o presentarse en: Rambla O'Higgins 5306, Montevideo

More Practice: Más comunicación *p. R13*

Online Workbook
CLASSZONE.COM

Refrán

El mal obrero culpa las herramientas.

¿Qué quiere decir el refrán? Cuando hay problemas con un proyecto, ¿dónde crees que está la causa del problema?

En colores
CULTURA Y COMPARACIONES

Los jóvenes y el futuro

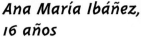 **Chile**

PARA CONOCERNOS

STRATEGY: CONNECTING CULTURES

Formulate plans for the future Think about your future after high school, then write down what you need to do to meet your goals: **escribir, estudiar, ganar, preparar, solicitar, tomar decisiones,** etc. Also acknowledge your feelings: **alegre, dudoso(a), frustrado(a), nervioso(a), preocupado(a), seguro(a)** about each task.

Mis metas:

Para hacer	Lo que siento
1.	
2.	
3.	

With which person in *Los jóvenes y el futuro* do you most identify?

Ana María Ibáñez, 16 años

Yo estudio en un colegio de monjas[1]. Es un internado — eso significa que las chicas viven allí. Ahora estoy cursando[2] mi último año y preparándome para la Prueba de Aptitud Académica, que también se llama la P.A.A. Quiero estudiar en la Universidad Católica, pero para eso necesito sacar más de 740 en la P.A.A. Me interesa estudiar ingeniería comercial. Pero me da un poco de miedo dejar el colegio. ¡Creo que voy a echarlo de menos[3]!

---

[1] nuns
[2] I'm enrolled in, I'm taking
[3] to miss it

Paraguay

**Alfredo Zubizarreta,
17 años**

Estoy en el último año de colegio y pienso mucho en el futuro. Quiero ir a la universidad, pero tengo que pasar el examen de ingreso[4]. Tengo buenas notas, sobre todo en castellano y en literatura, pero dicen que ese examen es muy difícil. Hay pocos puestos en la universidad y muchos estudiantes que quieren estudiar. Por eso algunos salen del país. Si me aceptan en la universidad aquí, voy a estudiar derecho[5], ¡porque los abogados ganan un buen sueldo!

[4] entrance, admission
[5] law

Uruguay

**Miguel Corteggiani,
15 años**

Estudio en un colegio público. El año que viene será el último año de secundaria. Mis padres quieren que vaya a la universidad pero yo dudo que vaya. Preferiría estudiar en una escuela técnica. Me fascinan los carros y me interesa mucho ser mecánico. Algún día quisiera tener mi propio taller. Yo creo que uno tiene que seguir sus intereses. ¿No estás de acuerdo?

CLASSZONE.COM
More About the Southern Cone

¿Comprendiste?

1. ¿En qué tipo de colegio estudia Ana María Ibáñez?
2. ¿Por qué no sabe Alberto Zubizarreta si podrá estudiar en la universidad?
3. ¿Qué campo le interesa a Miguel Corteggiani? ¿Qué piensan sus padres?

¿Qué piensas?

Estos estudiantes no están completamente seguros de sus decisiones. ¿Por qué?

Hazlo tú

Compara las dudas y los miedos de estos jóvenes sudamericanos con los de los jóvenes norteamericanos. ¿Comprendes estos sentimientos? ¿Los tienes también? Escribe un ensayo sobre tus planes para el futuro.

ETAPA **2**

Activity 3:
Narrate in the past

En uso

REPASO Y MÁS COMUNICACIÓN

OBJECTIVES

- Talk about careers
- Confirm and deny
- Express emotions
- Hypothesize

Now you can...

- talk about careers.
- confirm and deny.

To review

- affirmative and negative expressions see p. 284.

1 Algún día

Íñigo va a graduarse pronto y no sabe qué quiere estudiar. Completa esta entrada en su diario con las palabras afirmativas o negativas apropiadas.

algo   jamás   nada   ni...ni   nunca

alguien   ninguno(a)

algún   tampoco

algunos(as)   nadie   siempre

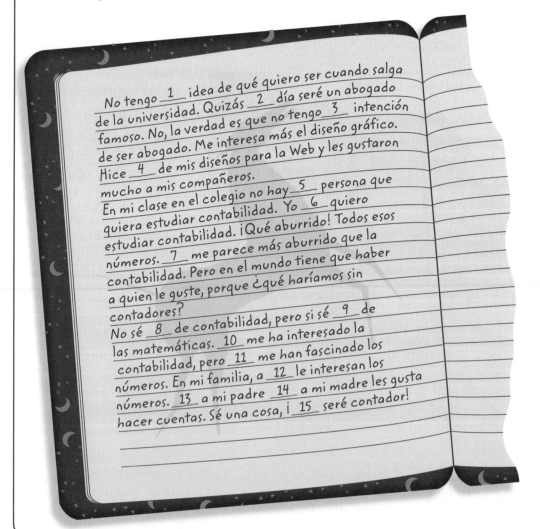

No tengo __1__ idea de qué quiero ser cuando salga de la universidad. Quizás __2__ día seré un abogado famoso. No, la verdad es que no tengo __3__ intención de ser abogado. Me interesa más el diseño gráfico. Hice __4__ de mis diseños para la Web y les gustaron mucho a mis compañeros.

En mi clase en el colegio no hay __5__ persona que quiera estudiar contabilidad. Yo __6__ quiero estudiar contabilidad. ¡Qué aburrido! Todos esos números. __7__ me parece más aburrido que la contabilidad. Pero en el mundo tiene que haber a quien le guste, porque ¿qué haríamos sin contadores?

No sé __8__ de contabilidad, pero sí sé __9__ de las matemáticas. __10__ me ha interesado la contabilidad, pero __11__ me han fascinado los números. En mi familia, a __12__ le interesan los números. __13__ a mi padre __14__ a mi madre les gusta hacer cuentas. Sé una cosa, ¡ __15__ seré contador!

Now you can...

• express emotions.

To review

• past perfect subjunctive see p. 286.

2 Después de la reunión

Vas a una reunión de tu clase y tienes algunas opiniones de la gente que ves allí. Después de la reunión, le escribes a tu mejor amigo sobre cada una de las personas que viste. Cambia los verbos para expresar tus opiniones. Sigue el modelo.

modelo

Yo espero que Felipe haya viajado por todo el mundo.

Yo esperaba que Felipe hubiera viajado por todo el mundo.

1. No estoy seguro(a) de que él haya estudiado arquitectura.
2. Espero que Juan Felipe haya ido a la universidad.
3. No creo que Julia haya sacado su licencia en ingeniería.
4. Ojalá que Teresa y su novio se hayan casado.
5. No creo que Pedro se haya graduado de la universidad.
6. Dudo que Vanessa haya querido hacerse bombero.
7. Espero que Sara se haya preparado para ser veterinaria.
8. Dudo que José Armando haya tenido su propio negocio.

Now you can...

• hypothesize.

To review

• conditional perfect see p. 289.

3 Las posibilidades

Siempre pensamos en lo que pudo ser pero no fue. Un(a) amigo(a) te dice qué habría pasado si hubieras hecho las cosas de otra manera. ¿Qué te dice?

modelo

No tenía experiencia. No me dieron el puesto.

Si hubieras tenido experiencia, te habrían dado el puesto.

1. No fui a la universidad. No recibí la licenciatura.
2. No estudié finanzas. No he ganado mucho dinero.
3. No trabajé a tiempo parcial. Tuve tiempo para mis estudios.
4. No trabajé a tiempo completo. No recibí un buen sueldo.
5. No leí el contrato. No sé qué beneficios me ofrecían.
6. No recibí el entrenamiento. No pude hacer el trabajo.
7. No tenía el conocimiento necesario. No me gustó el trabajo.

4 El cuestionario

STRATEGY: SPEAKING

Conduct an interview How can you handle the social and emotional aspects of an interview? One way is to ask a few open-ended questions that encourage personal expression by the person you are interviewing:

Dígame algo de…

¿Por qué se interesa en…?

¿Qué haría si…?

Si tuviera la oportunidad de…

Remember to use the **usted** form.

En grupos de tres o cuatro, escriban un cuestionario para entrevistar a unos candidatos para un puesto. Primero, decidan cuál es el puesto, los requisitos, etc. Luego, háganse la entrevista. Usen estas ideas.

- puesto
- requisitos
- habilidades
- trabajo a tiempo completo o parcial
- recomendaciones
- sueldo
- beneficios

5 En tu propia voz

ESCRITURA Escribe un anuncio clasificado para un puesto. Primero decide qué ofrece tu compañía. Decide cuál es el puesto y escribe una lista de requisitos. También decide lo que ofrece tu compañía para los empleados (sueldo, horas, beneficios, etc.) ¡Sé creativo(a)!

Empresa solicita

TÉCNICO DE COMPUTADORAS

Requisitos:

✓ mínimo de experiencia: 5 años

✓ conocimiento de varios sistemas

✓ buena presencia y habilidad para trabajar con distintos departamentos

Ofrecemos:

✓ beneficios y salario competitivo

✓ horarios flexibles

✓ bonos

Interesados llamar al 1/495-98710 o enviar su currículo a

Calle Tapes 98 Montevideo, Uruguay

CONEXIONES

Los estudios sociales ¿Qué profesiones te interesan? En tu escuela, entrevista a tres maestros(as) que enseñan tres materias diferentes. Pregunta sobre las profesiones en las cuales hay que saber mucho de las mátematicas (o el arte, las ciencias, etc.). Pregunta también cómo saber español te ayudaría en esa profesión. Comparte lo que descubras con la clase.

En resumen
REPASO DE VOCABULARIO

TALK ABOUT CAREERS

Professions

el (la) abogado(a)	lawyer
el (la) agricultor(a)	farmer
el (la) arquitecto(a)	architect
el (la) artesano(a)	artisan
el (la) asistente	assistant
el bailarín/ la bailarina	dancer
el bombero	firefighter
el (la) cartero(a)	mail carrier
el (la) contador(a)	accountant
el (la) deportista	athlete
el (la) diseñador(a) gráfico(a)	graphic designer
el (la) dueño(a)	owner
el (la) empleado(a)	employee
el (la) entrevistador(a)	interviewer
el (la) gerente	manager
el (la) ingeniero(a)	engineer
el (la) jardinero(a)	gardener
el (la) juez(a)	judge
el (la) mecánico(a)	mechanic
el (la) niñero(a)	baby sitter
el (la) obrero(a)	worker
el (la) operador(a)	operator
el (la) peluquero(a)	hairstylist
el (la) secretario(a)	secretary
el (la) taxista	taxi driver
el (la) técnico(a)	technician
el (la) veterinario(a)	veterinarian

Personal background

el conocimiento	knowledge
el entrenamiento	training
las habilidades	capabilities
la puntualidad	punctuality

In the workplace

aumentar	to increase
los beneficios	benefits
el bufete	law office
la carrera	career
el coche	car
chismoso(a)	gossipy
el contrato	contract
la desventaja	disadvantage
el empleo	job
la empresa	business
ganarse la vida	to earn a living
jubilarse	to retire
el puesto	position
requerir (e→ie)	to require
el requisito	requirement
el sueldo	salary
el seguro médico	health insurance
trabajo a tiempo...	
completo	full-time job
parcial	part-time job
la ventaja	advantage

CONFIRM AND DENY

Affirmative/Negative expressions

a menudo, muchas veces	often
a veces	sometimes
ni…ni	neither…nor
o…o	either…or

♻ Ya sabes

algo	something
alguien	someone, somebody
alguno(a)	some
jamás	never
nada	nothing
nadie	no one, nobody
ninguno(a)	no, not any
nunca	never
siempre	always
también	also
tampoco	neither

EXPRESS EMOTIONS

Past perfect subjunctive

Sentía que no me **hubieran ofrecido** el puesto.

HYPOTHESIZE

Conditional perfect

Si hubiera tenido el entrenamiento, me **habrían ofrecido** el puesto.

Juego

Completa las frases con las palabras apropiadas. Luego pon en orden las letras de los círculos para saber qué es lo bueno de envejecer.

Después de trabajar toda la vida, Carlos quiere

◯ __ __ ◯◯ __ __ .

Él va a recibir muchos

◯ __ __ __ ◯ __ __ ◯ .

La cantidad de dinero que recibe va a

◯ __ __ __ ◯◯◯ __ con los años.

UNIDAD 4

ETAPA 3

Un mundo de posibilidades

OBJECTIVES

- Learn about Latin American economics

- Clarify meaning

- Express possession

- Express past probability

¿Qué ves?

Mira la foto. Contesta las preguntas.

1. ¿Qué cosas ves en la foto?

2. ¿Crees que es un lugar divertido o serio? ¿Cómo lo sabes?

3. ¿Por qué iría alguien a un lugar como éste?

4. ¿Cuáles son algunos(as) profesionales que podrían trabajar aquí?

Países en conferencia

Descubre

A. Los cognados Adivina el significado de los cognados.

1. perfil económico
2. petróleo
3. turismo
4. exportación
5. importación
6. textiles
7. productos forestales
8. principal

B. Palabras similares Ya sabes muchas palabras que terminan en -ía: **frutería, heladería, juguetería, carnicería, panadería.** ¿Qué crees que quieren decir las siguientes palabras?

1. industria pesquera
2. refinería
3. minería
4. ganadería

El representante

¡Hola! Soy Ramón Fuentes del Castillo y voy a representar a mi escuela durante una conferencia del Comité de Economía y Desarrollo Social de la NUMAS (Naciones Unidas Modelo de América del Sur). Me enfocaré en Argentina, Chile, Paraguay y Uruguay.

AMÉRICA economía
Edición 19
Las mayores empresas de América Latina
...Y las 50 globales más competitivas

AMÉRICAS
TRANSFORMACIONES VISUALES DE FRANCISCO TOLEDO

Unidades monetarias: Las unidades monetarias de los países que estudiaré son el peso argentino, el peso chileno, el guaraní y el peso uruguayo.

La investigación

Creo que NUMAS va a ser interesante. Puedes conocer a mucha gente... pero antes ¡hay que estudiar mucho! La economía internacional no es un tema de todos los días, al menos que seas un presidente o un experto en el tema. Antes de la conferencia investigué el **perfil económico** de cada país para estar preparado para los debates del comité: las **industrias,** los **productos principales** de **importación** y **exportación,** las **unidades monetarias**...

Una vez en la conferencia...

¡Por suerte traje una computadora! Así puedo estudiar la información que aprendimos en la clase de geografía y compararla con los datos que obtenga en esta conferencia. Además, en Internet pude conseguir información importante como ésta:

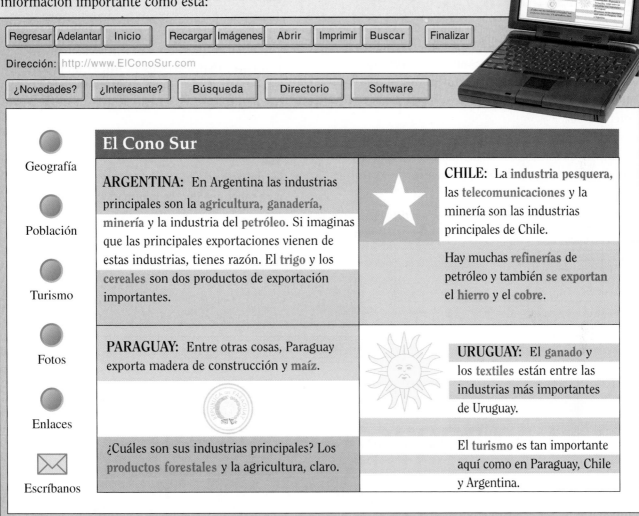

| Regresar | Adelantar | Inicio | | Recargar | Imágenes | Abrir | Imprimir | Buscar | | Finalizar |

Dirección: http://www.ElConoSur.com

| ¿Novedades? | ¿Interesante? | Búsqueda | Directorio | Software |

El Cono Sur

- Geografía
- Población
- Turismo
- Fotos
- Enlaces
- Escríbanos

ARGENTINA: En Argentina las industrias principales son la **agricultura, ganadería, minería** y la industria del **petróleo**. Si imaginas que las principales exportaciones vienen de estas industrias, tienes razón. El **trigo** y los **cereales** son dos productos de exportación importantes.

CHILE: La **industria pesquera**, las **telecomunicaciones** y la minería son las industrias principales de Chile.

Hay muchas **refinerías** de petróleo y también **se exportan** el **hierro** y el **cobre**.

PARAGUAY: Entre otras cosas, Paraguay exporta madera de construcción y **maíz**.

¿Cuáles son sus industrias principales? Los **productos forestales** y la agricultura, claro.

URUGUAY: El **ganado** y los **textiles** están entre las industrias más importantes de Uruguay.

El **turismo** es tan importante aquí como en Paraguay, Chile y Argentina.

Online Workbook
CLASSZONE.COM

¿Comprendiste?

1. ¿Qué categorías puedes incluir en una descripción del perfil económico de Estados Unidos?
2. Indica dos industrias de cada país mencionado.
3. ¿Han trabajado tú o tu familia en una de estas industrias? Da todos los detalles que puedas.
4. ¿Qué problemas debatirías en un club modelo de las Naciones Unidas? ¿Por qué?

Listen to audio texts

AUDIO

En vivo

🎧 SITUACIONES

PARA ESCUCHAR

STRATEGY: LISTENING

Pre-listening Predict what countries are the largest producers of the world's resources. Do you think these are also favorites with tourists? Think of countries in each category and write your predictions.

Use statistics to evaluate predictions Write down the countries as directed in **Escuchar,** then evaluate your predictions. How well did you identify the countries where major world producers and industries are located? Discuss your insights with your classmates.

Alimentos	Minerales	Turismo
1.		
2.		
3.		

¡Encuéntralo por Internet!

Tienes que escribir un informe sobre la producción mundial (*worldwide*) de varios productos. Encuentras información en Internet. Primero ves la información en la página-web. Luego escuchas más información por audio.

❶ Leer

Encontraste esta página en Internet. Lee la página para saber qué tipo de información tiene.

LA PRODUCCIÓN MUNDIAL

Agricultura   Minería   Telecomunicaciones

Ganadería   Petróleo   Textiles

Maderas   Industria pesquera   Turismo

❷ Escuchar

Tienes que informarle a la clase cuáles países del mundo son los que producen la mayor cantidad de ciertos productos. Escucha la información de la página-web «La producción mundial» y escribe los países en el orden correcto.

AGRICULTURA: maíz

País #1: _____
País #2: _____
País #3: _____
País #4: _____
País #5: _____

GANADERÍA: vacas

País #1: _____
País #2: _____
País #3: _____
País #4: _____
País #5: _____

MINERÍA: cobre

País #1: _____
País #2: _____
País #3: _____
País #4: _____
País #5: _____

PETRÓLEO CRUDO

País #1: _____
País #2: _____
País #3: _____
País #4: _____
País #5: _____

INDUSTRIA PESQUERA

País #1: _____
País #2: _____
País #3: _____
País #4: _____
País #5: _____

TURISMO

País #1: _____
País #2: _____
País #3: _____
País #4: _____
País #5: _____

❸ Hablar/Escribir

En grupos de dos o tres, conversen sobre el perfil económico de su ciudad, estado o país. ¿Cuál es la industria más importante de su estado o región? ¿Conocen a alguien que trabaje en esa industria? Entrevisten a esa persona o busquen datos por Internet o en la biblioteca. Escriban un informe en español que explique la importancia de esa industria en su región. También incluyan ideas sobre el desarrollo futuro de sus regiones.

Activity 4: Express and understand opinions

En acción

PARTE A

Práctica del vocabulario

Objectives for Activities 1–4
• Learn about Latin American economics

1 Las compañías

Hablar/*Escribir* Tu compañero(a) quiere saber a qué se dedican varias compañías de Buenos Aires. Como sabes un poco de Argentina, tú le contestas sus preguntas.

modelo

Compañero(a): ¿A qué se dedica esa compañía?

Tú: *Esa compañía se dedica a la industria pesquera.*

1.   2.

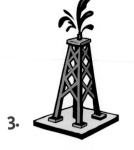

3.   4.

5.   6.

2 Los productos

Escribir En tu clase de geografía, tienes que hacer una tabla que indica los productos que van bajo cada categoría. Copia la tabla y complétala.

Agricultura	cereales			
Ganadería				
Minería				
Industria pesquera				
Textiles				

aceitunas	lana
algodón	maíz
arroz	oro
atún	ovejas *(sheep)*
caballos	pieles de cuero
cabras *(goats)*	plata
café	ropa
calamares	seda
cereales	suéteres
cobre	toallas
frutas	trigo
gallinas	trucha *(trout)*
hierro	vacas

3 Internet

Leer/Hablar Conversa con tu compañero(a) sobre las estadísticas sobre los visitantes a las páginas web de cada país.

modelo

Tú: *Este año, la página de Paraguay recibió un millón doscientos mil visitantes.*

Compañero(a): *De esos visitantes, casi la mitad fue hispanohablante.*

Estadísticas sobre los visitantes

Paraguay: 1.200.000 visitantes
Edad promedio: 32
Hispanohablantes: 50%
Inglés: 25%
Otros idiomas: 25%

Chile: 1.500.800 visitantes
Edad promedio: 25
Hispanohablantes: 33%
Inglés: 33%
Otros idiomas: 33%

Argentina: 2.350.700 visitantes
Edad promedio: 22
Hispanohablantes: 20%
Inglés: 40%
Otros idiomas: 40%

Uruguay: 850.000 visitantes
Edad promedio: 45
Hispanohablantes: 70%
Inglés: 10%
Otros idiomas: 20%

Vocabulario

Comparaciones numéricas

comparar *to compare*
las estadísticas *statistics*
mil millones *a billion*
un millón de millones *a trillion*
la mitad de *one half of*
el por ciento *percent*
el porcentaje *percentage*
el promedio *average*
el quinto *one fifth*

sumar *to add*
el tercio *one third*

 Ya sabes

un cuarto
un décimo
la mayoría
medio(a)

▶ ¿Puedes usar estas palabras para hablar sobre tu ciudad o estado?

4 Tu estado

STRATEGY: SPEAKING

Guess cognates Spanish and English share many words derived from Latin. Try adding a Spanish ending to an English word and it might be a correct word in Spanish. Look at these cognates for discussing your state's industries: **construcción, cinematografía, energía nuclear, radiodifusión, fuerzas armadas.**

Hablar Tú y tu compañero(a) tienen que preparar un reporte sobre tu estado para la clase de estudios sociales. Antes de ir a la biblioteca tienen que decidir qué tipo de información económica necesitan buscar.

modelo

Tú: *La agricultura es muy importante para la economía de Texas.*

Compañero(a): *Tienes razón. También necesitamos información sobre la ganadería.*

Práctica: gramática y vocabulario

Objectives for Activities 5–15
• Clarify meaning • Express possession • Express past probability

---

REPASO Subject and Stressed Object pronouns

Most of the time you do not use **subject pronouns** in Spanish, because the verb ending shows who the subject is. When you do include them it is because you wish to add emphasis, clarify, or make a contrast.

• to show emphasis

Yo le di las estadísticas, no Roberto.
I gave him the statistics, not Roberto.

• to make a comparison or clarify

Él salió. Ella se quedó en casa.
He went out. She stayed home.

You use the prepositional **a** + **subject pronouns** to clarify who the object of a preposition is, except in the case of **yo** and **tú.** Here **special object pronouns** are used (**mí, ti**).

El profesor les dio el reportaje **a ellos.**
*The teacher gave the report **to them.***

No me lo dio **a mí.**
*He didn't give it **to me.***

Vocabulario

 Ya sabes

Subject	a + Pronoun
yo	a mí
tú	a ti
usted	a usted
él	a él
ella	a ella
nosotros	a nosotros
vosotros	a vosotros
ustedes	a ustedes
ellos	a ellos
ellas	a ellas

Practice:
Actividades
5 6 7

Más práctica
cuaderno pp. 109–110
Para hispanohablantes
cuaderno p. 107

Online Workbook
CLASSZONE.COM

---

5 ¿Quién?

Escuchar/*Escribir* Estás en una reunión familar. Contesta las preguntas de tu tío sobre los intereses de todos.

modelo

¿Quién es abogado? él

¿Quién es médica? ella

1. ¿Quién estudió ingeniería?
_____

2. ¿Quién quiere ser veterinario? _____
¿Y bombero? _____

3. ¿Quién fue a la Universidad de Buenos Aires? _____ ¿Quién fue a la Universidad de Chile?
_____

4. ¿Quién estudió para ser arquitecto? _____
¿E ingeniera? _____

5. ¿A quién le interesa el mercadeo? _____ ¿A quién le interesa la publicidad?
_____

6. ¿Quién es bailarina? _____
¿Quién es deportista?

También se dice

Trabajar y **trabajo** son términos universales en todo el mundo de habla española. Pero en México y Colombia se dice también **chambear** y **chamba** para referirse al trabajo. En Argentina se usa la palabra **changa** y en Puerto Rico **chiripa** para referirse a un trabajo pequeño.

Activity 5: Understand most spoken language when the message is deliberately and carefully conveyed by a speaker accustomed to dealing with learners when listening

6 Ganándose la vida

Hablar/Escribir Conversa con tu compañero(a) sobre las profesiones de las personas de la lista y de otras personas que conocen. ¿Cómo se ganan la vida?

Buenos días, Buon giorno, Guten Tag

modelo

la Sra. Martínez

Tú: *¿Cómo se gana la vida la Sra. Martínez?*

Compañero(a): *Ella es intérprete.*

I. el Sr. Martínez

2. Ángel

3. el Sr. Beltrán

4. Susana

5. los Sres. Gutiérrez

6. el Sr. Henares

7. un(a) amigo(a)

8. un(a) pariente

9. un(a) vecino(a)

10. tú

Vocabulario

Carreras con el español

el (la) **académico(a)** *academic*

el (la) **agente de ventas** *sales agent*

el (la) **banquero(a)** *banker*

el (la) **bibliotecario(a)** *librarian*

el (la) **corresponsal** *correspondent*

el (la) **diplomático(a)** *diplomat*

el (la) **financiero(a)** *financial expert*

el (la) **intérprete** *interpreter*

el (la) **trabajador(a) social**
social worker

el (la) **traductor(a)** *translator*

▶ ¿Conoces a alguien que trabaje en una de estas profesiones?

Activity 7: Clarify, ask for and comprehend clarification

7 **¿Los conoces?**

Hablar/Escribir Un alumno nuevo acaba de llegar a tu escuela. Te toca informarle sobre la escuela y los otros alumnos. Están en la clase de español y él te pregunta sobre los alumnos y el (la) maestro(a).

modelo

Compañero(a): *¿Quién es él?*

Tú: *Él es el maestro de español.*

Compañero(a): *¿Y aquellos muchachos allí?*

Tú: *Él es Toño y ella es Ryoko.*

More Practice:

Más comunicación *p. R14*

Nota cultural

En Latinoamérica, buscar y conseguir trabajo no es tan fácil como en Estados Unidos. Si se encuentra un anuncio interesante en el periódico, se debe ir a una entrevista para presentar el currículum personalmente. Frecuentemente hay muchas personas esperando turno y es necesario esperar mucho. Luego se espera la confirmación telefónica y puede haber otra entrevista antes de obtener el trabajo.

REPASO **Possessive Pronouns**

You use **possessive** adjectives and **pronouns** to express possession.

Possessive adjective:

Aquí están **mis** datos.
*Here are **my** facts.*

Aquí está **mi** reportaje.
*Here is **my** report.*

Allí está **tu** reportaje.
*There is **your** report.*

Possessive pronoun:

Los míos están en el libro.
***Mine** are in the book.*

El mío está en la mesa.
***Mine** is on the table.*

Ese reportaje es **el tuyo.**
*That report is **yours**.*

Note that **possessive** adjectives are used with **nouns,** while **possessive** pronouns replace them:

replaced with

Tu carrera es interesante.
***Your career** is interesting.*

Sí, pero **la tuya** es más interesante que **la mía.**
*Yes, but **yours** is more interesting than mine.*

Vocabulario

 Ya sabes

mi	mío(a)
tu	tuyo(a)
su	suyo(a)
nuestro(a)	nuestro(a)
vuestro(a)	vuestro(a)
su	suyo(a)

Practice:
Actividades
8 9 10 11

Más práctica
cuaderno p. 111
Para hispanohablantes
cuaderno p. 108

Online Workbook
CLASSZONE.COM

8 Los productos de América Latina

Hablar/*Escribir* Tú y tu compañero(a) compraron varios productos y comidas de América Latina. Compara tus productos con los de tu compañero(a).

modelo

Tú: *Mi anillo es de oro. ¿Y el tuyo?*

Compañero(a): *El mío es de plata.*

oro/plata

I. Colombia/Oaxaca   **2.** cuero/lana

3. plata/cobre   **4.** Perú/México

5. cuero   **6.** madera

9 ¿De Argentina o de Chile?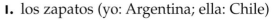

Hablar/*Escribir* Entre tus amigos, todos compraron estas cosas en Argentina o en Chile. ¿De qué país son las cosas que compraron?

modelo

la chaqueta (yo: Argentina; tú: Chile)

La mía es de Argentina.

La tuya es de Chile.

I. los zapatos (yo: Argentina; ella: Chile)
2. el collar (ella: Argentina; tú: Chile)
3. los muebles (nosotros: Argentina; tú: Chile)
4. la camisa (él: Argentina; ella: Chile)
5. las sillas (nosotros: Argentina; ellos: Chile)
6. ¿...?

10 ¿Y el tuyo?

Hablar/*Escribir* Tú y tus compañeros tienen que hacer un informe sobre la economía latinoamericana, pero nadie puede escoger el mismo tema. En grupos de tres o cuatro, hablen del tema que va a tratar el reporte de cada uno.

modelo

Tú: *El informe de Ricardo es sobre la ganadería en Argentina. ¿Y los suyos?*

Compañero(a): *El mío es sobre el turismo en Chile…*

Activity 12: Understand most spoken language when the message is deliberately and carefully conveyed by a speaker accustomed to dealing with learners when listening

11 Mi favorito(a)

Hablar/*Escribir* Comenta tus preferencias con tu compañero(a) mientras él (ella) te pregunta sobre las tuyas. Escoge de la lista o usa tus propias ideas.

modelo

Tú: *Mi clase favorita es el dibujo técnico. ¿Y la tuya?*

Compañero(a): *La mía es la informática porque…*

pasatiempo

poema

estrella de cine

cantante

película

clase

campo de estudio

programa de televisión

libro

¿…?

GRAMÁTICA **The Future Perfect Tense**

▶ You use the **future perfect tense** to express what will have happened by a certain time. To form this tense use:

future of **haber** + **past participle** of the verb

habré terminado	**habremos terminado**
habrás terminado	**habréis terminado**
habrá terminado	**habrán terminado**

—Llegaremos a las dos.
We will arrive at two.

—Pero a esa hora, nosotros ya **habremos salido**.
*But at that time, we **will have** already **left**.*

—Pasaré para recoger el informe a las tres.
I'll come by to pick up the report at three.

—No sé si **habré terminado**.
*I don't know if **I will have finished** (by then).*

▶ The **future perfect tense** is often used with **dentro de + time**.

Dentro de tres años, me **habré graduado**.
*In three years, **I will have graduated**.*

▶ You also use the **future perfect** to speculate about something that may have happened in the past.

—Todavía no han llegado tus primos.
Your cousins haven't arrived yet.

—Se **habrán perdido**.
*They **probably got lost**.*

—Miguel está deprimido. ¿Qué le **habrá pasado**?
*Miguel is depressed. What **could have happened** to him?*

—No sé. No le **habrán dado** el puesto que quería.
*I don't know. **Perhaps they didn't give him** the job he wanted.*

Practice: Actividades
12 13 14 15

Más práctica
cuaderno p. 112
Para hispanohablantes
cuaderno pp. 109–110

 Online Workbook
CLASSZONE.COM

12 ¡Pobre Carlos!

Escuchar/*Escribir* Carlos tuvo un día malísimo. Dos de sus amigos comentan sobre lo que le pasó o lo que le habrá pasado. Primero, copia la siguiente tabla. Luego escucha la conversación. Si la persona sabe lo que le pasó a Carlos, marca «sabe». Si la persona está especulando sobre lo que le habrá pasado, marca «no sabe».

	sabe	no sabe
1.		
2.		
3.		
4.		
5.		
6.		

También se dice

Para referirse a los distintos tipos de industrias, no siempre se usan las mismas expresiones. Por ejemplo, se habla de **la industria ganadera** o de **la ganadería**; de **la industria petrolera** o de **la industria del petróleo**; de **la industria agrícola** o de **la agricultura**. Además, las fábricas también se conocen como **factorías** y la bolsa de valores se nombra familiarmente como **la bolsa**.

13 La economía chilena

Hablar/*Escribir* Los chilenos miran con optimismo el futuro económico de Chile. ¿Qué habrá pasado antes del año 2015?

modelo

economía chilena / florecer (to flourish)

Antes del año 2015, la economía chilena habrá florecido.

1. las fábricas / aumentar en número
2. las compañías multinacionales / incorporar el uso de las telecomunicaciones
3. la industria del turismo / aumentar dramáticamente
4. la ganadería / crecer
5. los productos / exportarse en cantidades más grandes
6. los precios / bajar
7. la industria del petróleo / desarrollarse
8. la inflación / controlarse

Vocabulario

Tipos de compañías

la bolsa de valores *stock exchange*

la fábrica *factory*

el laboratorio *laboratory*

la multinacional *multinational business*

la sociedad anónima (S.A.) *corporation (Inc.)*

▶ ¿Puedes usar estas palabras para describir empresas locales?

Activity 14:
Narrate in the future

14 ¿Qué habrá pasado?

Hablar/*Escribir* Imagina qué habrá pasado en el mundo económico y profesional al final del día. Contesta las siguientes preguntas.

modelo

¿Cómo se comunicaron?

No sé. ¿Se habrán comunicado por Internet?

1. ¿Qué pasó hoy con la bolsa de valores?
2. ¿Descubrieron algo los científicos?
3. ¿Cuántos boletos vendió el agente de viajes?
4. ¿Quién tradujo las conversaciones diplomáticas?
5. ¿Qué manejaron los banqueros?
6. ¿A quiénes ayudaron los trabajadores sociales?
7. ¿Qué inventaron en el laboratorio?
8. ¿Qué hicieron en la fábrica?
9. ¿Qué tradujo el traductor del periódico?
10. ¿Qué vendió la compañía multinacional?

15 El año 2025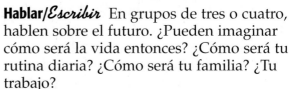

Hablar/*Escribir* En grupos de tres o cuatro, hablen sobre el futuro. ¿Pueden imaginar cómo será la vida entonces? ¿Cómo será tu rutina diaria? ¿Cómo será tu familia? ¿Tu trabajo?

modelo

Tú: *Para el año 2025, habremos construido casas en el planeta Marte.*

Compañero(a) 1: *No, yo no lo creo. Para el año 2025, habremos curado todas las enfermedades.*

Compañero(a) 2: *No, yo no lo creo. Lo que yo creo es que para el año 2025, habrán inventado carros que pueden volar.*

Amigo(a) 3: *No, yo no lo creo. Lo que yo creo es que para el año 2025…*

Nota cultural

En muchos países de Latinoamérica, la gente tiene una forma especial de ahorrar dinero. Van a una casa de cambio o a un banco y compran dólares estadounidenses. Cuando necesitan usar el dinero, cambian los dólares nuevamente.

Activity **16** brings together all concepts presented.

16 Los bosques

Leer/Escribir Lee la tabla sobre los bosques de Latinoamérica (Iberoamérica) y contesta las preguntas.

1. ¿Qué porcentaje de la tierra paraguaya está cubierta de bosques?

2. ¿Cuál es el área de tu país?

3. Tienes un amigo panameño. ¿Cuál es la tasa anual de cambio de su país?

4. ¿Cuántas especies de árboles tiene Paraguay que son comercialmente explotables? ¿Cuántas de ésas se exportan?

5. PNB quiere decir «Producto Nacional Bruto».¿Sabes cómo se dice eso en inglés?

Fuente de divisas

Un tercio de la tierra paraguaya es boscosa. Hasta 1959, la madera fue el principal producto de exportación del país, y aunque actualmente su contribución al PNB es pequeña, representa una fuente importante de divisas. Unas cuarenta y cinco especies de árboles de los bosques de Paraguay son comercialmente explotables y siete se exportan.

BOSQUES DE IBEROAMÉRICA

PAÍS	ÁREA (1000 Ha)	TASA ANUAL DE CAMBIO
Costa Rica	1,456	2.44
El Salvador	127	1.85
Guatemala	4,253	1.58
Honduras	4,608	1.94
México	48,695	1.21
Nicaragua	6,027	1.69
Panamá	3,123	1.70
Cuba	1,960	0.19
Rep. Dominicana	1,084	2.43
Argentina	34,436	0.57
Chile	8,033	0.07
Uruguay	813	0.12
Bolivia	49,345	1.12
Brasil	566,007	0.58
Colombia	54,190	0.62
Ecuador	12,007	1.65
Paraguay	12,868	2.38
Perú	68,090	0.37
Venezuela	45,943	1.13

More Practice: Más comunicación *p. R14*

Online Workbook CLASSZONE.COM

Refrán

Promete poco y haz mucho.

¿Qué quiere decir el refrán? En tu opinión, ¿por qué es mejor decir poco y dejar que tus acciones muestren tus intenciones?

«Prometo que lo hago más tarde.»

«¡Gracias! ¡Qué sorpresa magnífica!»

Read short stories

AUDIO

En voces
LECTURA

PARA LEER • **STRATEGY: READING**

Speculate about the author From your reading, what do you think was the age and professional status of Isabel Allende during her career? What other qualities does she reveal? Do you think it is better to read a piece of literature with or without knowledge about the author? Explain your answer.

EL TRABAJO

a cargo de tener responsabilidad por algo
asomar tras un vidrio verse por un cristal
el canal la compañía de televisión
el guión las palabras de un programa
las orejas una manera de decir "personas"
la pantalla por donde se ve la televisión
puntual a tiempo
el vacío donde no hay nadie

Sobre la autora

Isabel Allende, novelista chilena, nació en Lima, Perú, en 1942. Su familia tuvo que exiliarse de Chile cuando su tío Salvador Allende, el presidente del país, fue vencido por una junta militar en 1973. Isabel Allende empezó a escribir a la edad de diecisiete años y escribió su primera novela, *La casa de los espíritus*, en 1982. También ha trabajado como periodista y en la televisión.

Introducción

Allende comenzó a escribir su libro autobiográfico *Paula* mientras su hija estaba muy enferma. Es una historia que ofrece mucha información y varias anécdotas sobre la familia de Allende y sobre la historia y la política de Chile. En la selección que vas a leer, Allende le habla a su hija sobre su trabajo en Chile.

314 trescientos catorce
Unidad 4

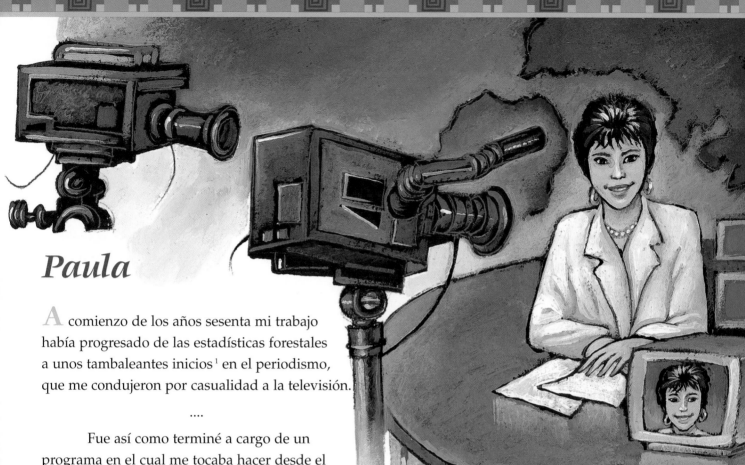

Paula

A comienzo de los años sesenta mi trabajo
había progresado de las estadísticas forestales
a unos tambaleantes inicios[1] en el periodismo,
que me condujeron por casualidad a la televisión.

....

Fue así como terminé a cargo de un
programa en el cual me tocaba hacer desde el
guión hasta los dibujos de los créditos. El trabajo
en el Canal consistía en llegar puntual, sentarme
ante una luz roja y hablar al vacío; nunca tomé
conciencia de que al otro lado de la luz un millón
de orejas esperaban mis palabras y de ojos
juzgaban mi peinado[2], de ahí mi sorpresa[3]
cuando desconocidos[4] me saludaban por la calle.
La primera vez que me viste aparecer en la
pantalla, Paula, tenías un año y medio y el susto[5]
de ver la cabeza decapitada de tu mamá
asomando tras un vidrio, te dejó un buen rato[6]
en estado catatónico… Me convertí en la persona
más conspicua del barrio, los vecinos me
saludaban con respeto y los niños me señalaban[7]
con el dedo… (Michael y yo) conseguimos un
par de becas[8], partimos a Europa y llegamos a
Suiza contigo de la mano, tenías casi dos años y
eras una mujer en miniatura.

---

[1]shaky beginning   [2]judged my hairdo   [3]surprise
[4]strangers   [5]shock, fright   [6]quite a while
[7]gestured to me   [8]scholarships

¿Comprendiste?

1. ¿En qué campos trabajaba Isabel Allende?
2. ¿En qué consistía su trabajo en la televisión?
3. ¿Por qué se fue la escritora de Chile?
4. ¿De qué se trata el libro *Paula*?

¿Qué piensas?

1. ¿Cómo se explica la reacción de Paula al ver
   a su madre en la televisión?
2. ¿Por qué crees que Isabel Allende comenzó
   a escribir su autobiografía en 1992 a la edad
   de 50 años?

Hazlo tú

¿Te parece interesante trabajar en la televisión?
Si pudieras trabajar en la televisión, ¿qué harías
— noticias, pronóstico del tiempo, telenovelas,
o programas para niños? Explica tu preferencia.

Acquire knowledge and new information from comprehensive, authentic texts when reading

En colores
CULTURA Y COMPARACIONES

México
NÁHUATL

Algunas palabras del náhuatl son:
aguacate
cacahuete
chocolate
nopal

Guatemala
El Salvador
Honduras
Nicaragua

PARA CONOCERNOS
STRATEGY: CONNECTING CULTURES

Observe how language reflects culture Each language reveals the background of the people who speak it. For example, arithmetic is derived from Latin and mathematics from Greek. There is not one English language but several, including Australian, Canadian, British, and American versions. Think about these examples and conjecture what events and experiences cause language to evolve. Organize your ideas in a chart.

Cosas que cambian un idioma
1.
2.
3.

Which of your ideas are represented in S*e hablan… ¡muchos idomas!*?

Se hablan... ¡muchos idiomas!

Galicia
GALLEGO

País Vasco
VASCUENCE

Cataluña
CATALÁN

ESPAÑA

El español o castellano es el idioma oficial de los países hispanohablantes, pero también se hablan otros idiomas. ¡A ver cuáles son!

España

El castellano, que también se conoce como español, se originó en España. En el este de España también se habla el **catalán** y en el noroeste, el **gallego**. El **euskera**, o **vascuence**, se habla en el País Vasco desde antes que llegaran los romanos a España en 202 antes de Cristo[1].

---
[1] before the Christian era

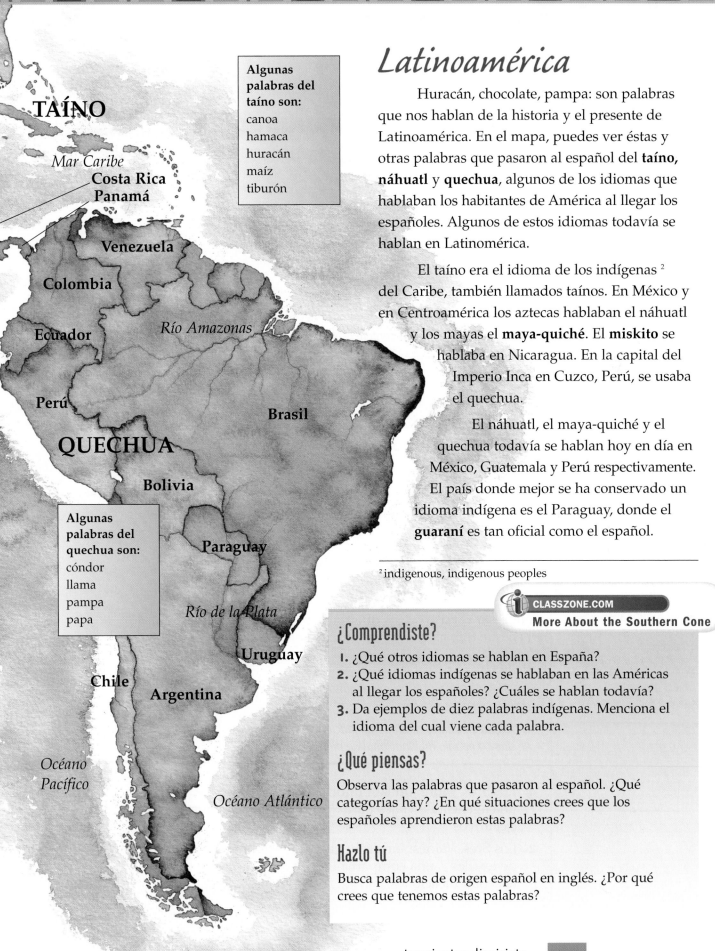

Latinoamérica

Huracán, chocolate, pampa: son palabras que nos hablan de la historia y el presente de Latinoamérica. En el mapa, puedes ver éstas y otras palabras que pasaron al español del **taíno, náhuatl** y **quechua**, algunos de los idiomas que hablaban los habitantes de América al llegar los españoles. Algunos de estos idiomas todavía se hablan en Latinomérica.

El taíno era el idioma de los indígenas[2] del Caribe, también llamados taínos. En México y en Centroamérica los aztecas hablaban el náhuatl y los mayas el **maya-quiché**. El **miskito** se hablaba en Nicaragua. En la capital del Imperio Inca en Cuzco, Perú, se usaba el quechua.

El náhuatl, el maya-quiché y el quechua todavía se hablan hoy en día en México, Guatemala y Perú respectivamente. El país donde mejor se ha conservado un idioma indígena es el Paraguay, donde el **guaraní** es tan oficial como el español.

---

[2] indigenous, indigenous peoples

TAÍNO

Mar Caribe

Costa Rica
Panamá

Venezuela

Colombia

Ecuador

Río Amazonas

Perú

Brasil

QUECHUA

Bolivia

Paraguay

Río de la Plata

Uruguay

Chile

Argentina

Océano
Pacífico

Océano Atlántico

Algunas palabras del taíno son:
canoa
hamaca
huracán
maíz
tiburón

Algunas palabras del quechua son:
cóndor
llama
pampa
papa

CLASSZONE.COM
More About the Southern Cone

¿Comprendiste?

1. ¿Qué otros idiomas se hablan en España?
2. ¿Qué idiomas indígenas se hablaban en las Américas al llegar los españoles? ¿Cuáles se hablan todavía?
3. Da ejemplos de diez palabras indígenas. Menciona el idioma del cual viene cada palabra.

¿Qué piensas?

Observa las palabras que pasaron al español. ¿Qué categorías hay? ¿En qué situaciones crees que los españoles aprendieron estas palabras?

Hazlo tú

Busca palabras de origen español en inglés. ¿Por qué crees que tenemos estas palabras?

Activity 3: Generally use appropriate behavior in social situations

En uso
REPASO Y MÁS COMUNICACIÓN

OBJECTIVES

• Learn about Latin American economics
• Clarify possession
• Express possession
• Express past probability

Now you can...

• discuss Latin American economics.

1 La población

Estás creando una encuesta para buscar unas estadísticas demográficas. Primero escribe preguntas y luego contéstalas con la información indicada.

modelo

porcentaje de la población (¿habla español?): $\frac{1}{5}$

Compañero(a): ¿Qué porcentaje de la población habla español?

Tú: Un quinto de la población habla español.

1. porcentaje de la población (¿de habla hispana?): $\frac{1}{3}$
2. edad promedio: 25
3. porcentaje de la población (¿vivir en la ciudad?): 50%
4. porcentaje de la población (¿vivir en el campo?): 50%
5. parte de la población (¿graduarse de la universidad?): la mayoría
6. parte de la población (¿trabajar en la ganadería?): la menor parte

Now you can...

• avoid redundancy.

To review

• subject and stressed object pronouns see p. 306.

2 ¿Él o ella?

Conoces a varias parejas que trabajan en industrias diferentes. Di en qué trabaja él y en qué trabaja ella.

modelo

Los Sres. Mendoza: una compañía multinacional de turismo / una compañía multinacional de petróleo

Él trabaja en una compañía multinacional que se dedica al turismo. Ella trabaja en una compañía multinacional de petróleo.

1. Los Sres. Moré: un laboratorio / una agencia de viajes para ejecutivos
2. Los Sres. Valdés: una fábrica de textiles / una compañía de telecomunicaciones
3. Los Sres. Puente: una compañía de exportaciones / un banco
4. Los Sres. Colón: un taller de artesanías / la bolsa de valores
5. Los Sres. Prado: un laboratorio/ una refinería de petróleo

Now you can...

• express possession.
• clarify possession.

To review

• possessive pronouns see p. 308.

3 **¡No!**

Estás en una fiesta y ahora tú y tus amigos se están despidiendo de la anfitriona. Ella trata de devolverte cosas que no son tuyas. También trata de devolverles cosas a tus amigos que no son suyas. ¿Cómo le respondes?

modelo

tu paraguas: *negro*

Anfitriona: *Ten, aquí está tu paraguas.*

Tú: *No, ése no. El mío es negro.*

1. tu abrigo: azul
2. la mochila de Hernán: verde
3. la bolsa de Mariluz: amarilla
4. el sombrero de Juan: rojo
5. los platos de Minerva: nuevos
6. los zapatos de tenis de Arnoldo: viejo
7. tu chaqueta: de cuero

Now you can...

• express past probability.

To review

• the future perfect see p. 310.

4 **Para ese entonces**

Tu abuelo(a) está pensando en el futuro de su familia. ¿Qué cree que va a pasar en veinte años? Sigue el modelo.

modelo

(tú) comprar una casa

Para ese entonces, habrás comprado una casa.

1. (nosotros) viajar a Argentina
2. (tú) empezar tu carrera en la industria del petróleo
3. (Enrique y Elena) casarse
4. (Anilú) graduarse de la universidad
5. (Rudi y Luisa) empezar una familia
6. (Felipe) hacerse banquero
7. (ustedes) ahorrar mucho dinero
8. (tú) realizar tus sueños

Activity 5–6: Tend to become less accurate as the task or message becomes more complex, and some patterns of error may interfere with meaning

5 ¿Dónde está Gerardo?

STRATEGY: SPEAKING

Speculate about the past When the unexpected occurs, it is natural to express opinions about what may have happened. Your conjecture about Gerardo's absence can be humorous, pleasant, logical or illogical: **¿Por qué no vino Gerardo? Se habrá perdido en el parque zoológico.** Be inventive!

Hablar/*Escribir* Gerardo prometió que iba a venir a la reunión del consejo estudiantil. ¡Pero no llegó! Todos tienen ideas de por qué no vino. Dramaticen esta situación.

modelo

Tú: *¿Pero dónde está Gerardo? Dijo que iba a venir.*

Compañero(a) 1: *Se le habrá olvidado.*

Compañero(a) 2: *Se habrá acostado muy tarde y no se despertó a tiempo.*

Compañero(a) 3: *No, no es eso. Yo creo que…*

6 El mío es de...

En grupos de dos o tres, conversen sobre las cosas que tengan y de qué tienda son.

modelo

Tú: *Yo compré mi chaqueta de piel en Ropafina.*

Compañero(a) 1: *¿Ah, sí? La mía es de Ropafina también.*

Compañero(a) 2: *Yo no tengo una chaqueta de piel, pero mi collar de oro es de la tienda en la plaza.*

Tú: *El mío es de la misma tienda.*

7 En tu propia voz

ESCRITURA Escribe un informe sobre el perfil económico de tu estado. Destaca el producto de más importancia. Las siguientes categorías pueden ayudarte a comenzar. Si quieres, incluye fotos en tu informe.

Mi estado:

Capital:

Unidad monetaria:

Perfil económico:

Productos de exportación:

La agricultura:

La ganadería:

El petróleo:

Los productos forestales:

La minería:

El turismo:

La industria pesquera:

Las telecomunicaciones:

Los textiles:

CONEXIONES

Los estudios sociales ¿Qué sabes de la ONU (Organización de las Naciones Unidas)? ¿Has oído alguna vez de la OEA (Organización de los Estados Americanos)? ¿Cuál es el propósito de estas dos organizaciones internacionales? ¿Quiénes son los miembros? Busca la información por Internet o en la biblioteca y escribe un reporte. Comparte tu reporte con la clase.

En resumen
REPASO DE VOCABULARIO

LEARN ABOUT LATIN AMERICAN ECONOMICS

Careers in Spanish

el (la) académico(a)	academic
el (la) agente de ventas	sales agent
el (la) banquero(a)	banker
el (la) bibliotecario(a)	librarian
el (la) corresponsal	correspondent
el (la) diplomático(a)	diplomat
el (la) financiero(a)	financial expert
el (la) intérprete	interpreter
el (la) trabajador(a) social	social worker
el (la) traductor(a)	translator

Industries

la agricultura	agriculture
los cereales	grains
el cobre	copper
la exportación	export
exportar	to export
la ganadería	livestock industry
el ganado	livestock
el hierro	iron
la importación	import
la industria	industry
la industria pesquera	fishing industry
el maíz	corn
la minería	mining
el perfil económico	economic profile
el petróleo	petroleum
principal	principal
los productos forestales	forestry products
la refinería	refinery
las telecomunicaciones	telecommunications
los textiles	textiles
el trigo	wheat
el turismo	tourism
la unidad monetaria	currency

Statistics

comparar	to compare
las estadísticas	statistics
mil millones	billion
un millón de millones	trillion
la mitad de	one half of
por ciento	percent
el porcentaje	percentage
el promedio	average
el quinto	one fifth
sumar	to add
el tercio	one third

 Ya sabes

un cuarto	one fourth
un décimo	one tenth
la mayoría	majority
medio(a)	half

Types of companies

la bolsa de valores	stock exchange
la fábrica	factory
el laboratorio	laboratory
multinacional	multinational
la sociedad anónima (S.A.)	corporation (Inc.)

AVOID REDUNDANCY

 Ya sabes

a mí	to me
a ti	to you
yo	I
tú	you (fam.)
usted	you (for.)
él	he
ella	she
nosotros(as)	we
vosotros(as)	you (fam. pl.)
ustedes	you (for. pl.)
ellos	they
ellas	they (fem.)

EXPRESS POSSESSION

Ya sabes

mi/mío(a)	my/mine
tu/tuyo(a)	your (fam.)/yours (fam.)
su/suyo(a)	your (for.), his, her/ yours (for.), his, hers, its
nuestro(a)	our/ours
vuestro(a)	your (pl. fam.)/yours (pl. fam.)
su/suyo(a)	your (pl.), their/ yours (pl.), theirs

EXPRESS PAST PROBABILITY

The future perfect tense

No sé dónde está Élmer. Fue a la oficina. **Habrá encontrado** más trabajo allí.

Juego

¿Cuál es tu profesión?

¿Puedes encontrar en el dibujo dos profesiones cuyos nombres empiecen con la letra **a**?

Create paragraphs
when writing

En tu propia voz
ESCRITURA

Una carrera: ¿Dónde empezar?

Una empresa local busca internos para su programa de entrenamiento. El conocimiento del español es esencial y también una familiaridad con administración de empresas, economía, humanidades o matemáticas. Tu carta adjunta (*cover letter*) debe resumir tus experiencias escolares.

Función: Describirse a sí mismo
Contexto: Informar al Jefe de personal

Contenido: Relación entre tu educación, experiencia y habilidad
Tipo de texto: Carta adjunta

PARA ESCRIBIR • STRATEGY: WRITING

Use cause and effect to demonstrate ability A good cover letter highlights the relationship between your education and experiences (cause) and your ability to do the job (effect). You must impress the potential employer and show that you can handle the position by applying your knowledge to the work.

Modelo del estudiante

A salutation in a business letter is formal, using **Estimado(a)** and the person's title, and ending with a colon.

Estimado Licenciado Ramírez:

El motivo de la presente es solicitar el puesto de interno en su compañía. Actualmente estoy tomando cursos en mercadeo y economía en mi escuela secundaria. También estoy estudiando español y pienso participar en un programa de estudios en el extranjero el año que viene. Tengo buenas notas en estos cursos y por eso creo que tengo la educación y las habilidades necesarias para contribuir al éxito de su distinguida compañía.

The phrase **por eso** indicates the connection between the writer's coursework and her ability to contribute to the company.

El verano pasado trabajé en el departamento de ventas y mercadeo de una compañía multinacional. Estaba trabajando directamente con el gerente del departamento, así que entiendo bien las responsabilidades de un interno internacional. El gerente me escribió una carta de recomendación diciendo que siempre desempeñé todos mis cargos de una manera excelente.

The expression **así que** points out the relationship between the writer's previous experience and her understanding of the needs for the current position.

Adjunto mi currículum vitae. Espero que me encuentre bien capacitada para servirle.

Atentamente,

Karen Willis

Karen Willis

Typical closings to business letters include **Atentamente** and **Le saluda muy atentamente.**

**Language Arts
Writing Standard 2.3a**
Write fictional, autobiographical, or biographical
narratives. Narrate a sequence of events and
communicate their significance to the audience.

Writing Center
CLASSZONE.COM

Estrategias para escribir

Antes de escribir...

Piensa en las calificaciones que necesita un(a)
candidato(a) para un trabajo que conoces. Considera
la educación, experiencia y las habilidades que se
requieren. Inventa un(a) candidato(a) perfecto(a) y
crea su perfil. Crea una tabla como la de la derecha
para analizar la calificación del (de la) candidato(a).

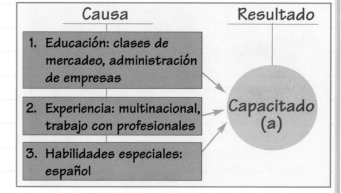

Causa	Resultado
1. Educación: clases de mercadeo, administración de empresas	
2. Experiencia: multinacional, trabajo con profesionales	Capacitado (a)
3. Habilidades especiales: español	

Revisiones

Después de escribir el primer borrador de la carta adjunta, trabajen
en grupos de cuatro para intercambiar las cartas y leerlas en voz alta.
Decidan qué aspectos de cada carta son más efectivos. Revisen cada
carta en grupo para incorporar las técnicas más eficientes y convincentes.

- *¿Qué expresiones usaron para demostrar la conexión entre las
  calificaciones y la capacidad de hacer el trabajo?*
- *¿Qué datos mencionaron para indicar lo que ha hecho el (la) candidato(a)
  y lo que está haciendo para desarrollar sus habilidades?*

La versión final

Para completar tu carta, léela de nuevo y repasa los
siguientes puntos:

- *¿Usé bien el **presente continuo** (present progressive)
  o el **presente perfecto** (present perfect)?*

Haz lo siguiente: Subraya los verbos en estos tiempos.
¿Usaste la forma correcta de **estar**? ¿Del presente
participio?

- *¿Usé bien el **potencial compuesto** (conditional perfect)
  o el **pluscuamperfecto de subjuntivo** (past perfect
  subjunctive)?*

Haz lo siguiente: Repasa las conjugaciones de **haber** en
estos tiempos. Haz un círculo alrededor de estos verbos.
¿Está conjugado correctamente **haber** y está seguido por
un participio pasado correcto?

> Sr. Gerente:
>
> Le escribo para solicitar un puesto de
> trabajo en su compañía editorial. Acabo
> de terminar mis estudios de periodismo
> y quisiera trabajar para usted. Además,
> estoy tomando clases de francés e
> italiano. Habría estudiando alemán si
> hubiera
> no habría tenido que viajar a Europa a
> hacer una entrevista. He trabajado en
> algunos periódicos y revistas y quisiera
> tener una entrevista con usted para
> ofrecerle mis servicios.

5

STANDARDS

Communication
- Discussing and describing art forms and crafts
- Identifying and specifying
- Requesting clarification
- Expressing relationships
- Referring to people and objects
- Making generalizations
- Talking about literature
- Talking about film
- Avoiding redundancy

Cultures
- Art forms in Spain and the Americas
- Well-known authors, artists, architects, and filmmakers from Spain and the Americas
- Pre-Columbian civilizations

Connections
- Social Studies: Creating a time line
- Mathematics: Investigating a mathematical system

Comparisons
- Spanish-speaking authors
- Architecture in Mexico and in the U.S.
- Spanish-language movies and English-language movies

Communities
- Using Spanish in the workplace
- Using Spanish with family and friends

 INTERNET Preview

CLASSZONE.COM
- More About Spain and the Americas
- Webquest
- Self-Check Quizzes
- Flashcards
- Writing Center
- Online Workbook

• eEdition Plus Online

ARTES EN ESPAÑA Y LAS AMÉRICAS

ESPAÑA Y LATINOAMÉRICA
CHOCOLATE Y CHURROS
¿Por qué crees que esta combinación es tan popular en el mundo hispanohablante? ¡Muestra la mezcla de sabores entre España y Latinoamérica!

ALMANAQUE CULTURAL

POBLACIÓN DE ESPAÑA: 40.037.995

ALTURA: 3718 m sobre el nivel del mar (Pico del Teide, Islas Canarias)

TEMPERATURA: 66°F (20°C) Sevilla (temp. promedio más alta); 57°F (25°C) Madrid (temp. promedio más baja)

COMIDA TÍPICA: paella, zarzuela de mariscos, tortilla española, cochinillo

GENTE FAMOSA: Julio Iglesias (cantante), Andrés Segovia (músico), Pablo Casals (músico), Sus Majestades Juan Carlos y Sofía (rey y reina), Adolfo Domínguez (diseñador)

 VIDEO   DVD   Mira el video para más información.

CLASSZONE.COM
More About Spain and the Americas

ESPAÑA
LA REINA ISABEL
(1451–1504) fue la esposa de Fernando V. Apoyó el viaje de Cristóbal Colón. ¿Qué argumentos crees que utilizó Colón para convencerla de que apoyara su viaje?

ARGENTINA
TEATRO COLÓN Ecléctico, histórico y super elegante: el Teatro Colón en Buenos Aires impresiona por la calidad de los espectáculos que se han presentado allí desde que se construyó en el siglo XIX. ¿Qué te sugieren los detalles del edificio sobre el intercambio de la cultura europea y latinoamericana?

COLOMBIA
FERNANDO BOTERO
Nació en 1932 en Medellín, Colombia y tal vez sea el artista latinoamericano de más fama internacional. Sorprende cómo utiliza la proporción de las cosas. Al observar la imagen, ¿qué ideas crees que impulsaron los cambios de forma que ves en la pintura?

ESPAÑA
SALVADOR DALÍ
Junto a Pablo Picasso y Joan Miró, Dalí es uno de los artistas españoles más importantes del siglo XX. Es uno de los mayores exponentes del surrealismo. Observa la pintura. ¿Qué crees que la hace surrealista?

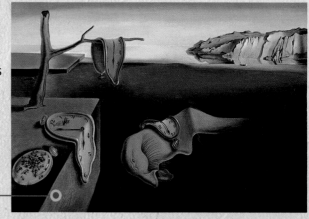

Salvador Dalí, *La persistencia de la memoria*, (1931)

325

ARTES EN ESPAÑA Y LAS AMÉRICAS

- Comunicación

- Culturas

- Conexiones

- **Comparaciones**

- Comunidades

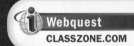
Webquest
CLASSZONE.COM

Explore comparisons in Spain and in the Americas through guided Web activities.

Comparaciones

La arquitectura de una cultura cambia a través de los años. Muchas veces, sin embargo, los edificios de diferentes épocas tienen estilos similares y es interesante compararlos. ¿En tu comunidad hay edificios modernos o antiguos? ¿Cuáles son las diferencias?

Comparaciones en acción

Compara los dos edificios que se ven en las fotos; ¿Crees que a la gente le gustan más los edificios antiguos o los modernos? ¿Cuál es la importancia de cada uno?

Culturas

Las perspectivas de las culturas hispanas se reflejan en sus artes. Describe las perspectivas reflejadas en este baile español.

Conexiones

Los mayas usaban un sistema de numeración diferente del nuestro. ¿Qué opinas de su sistema? ¿Te parece más fácil o más difícil?

NUMERACIÓN MAYA

 CERO

● UNO

●● DOS

●●● TRES

●●●● CUATRO

▬ CINCO

●
▬ SEIS

●●
▬ SIETE

●●●
▬ OCHO

●●●●
▬ NUEVE

Comunicación

En esta unidad comunicaremos información sobre las artes, incluido el cine. ¿Te gustaría ver esta película? Explica tu respuesta.

La prima Angélica (1973)

Comunidades

En esta unidad conocerás a una alumna que usa el español para ayudar a su familia y también en el trabajo. ¿Y tú? ¿Cómo usas el español fuera de la clase? ¿Hablas español con alguien de tu familia?

Fíjate

Identifica las formas de arte que ves representadas en las fotos de estas páginas.

_____ la arquitectura   _____ la escultura

_____ el baile   _____ la literatura

_____ el canto   _____ la música

_____ el cine   _____ la pintura

UNIDAD 5

ETAPA

Tradiciones españolas

OBJECTIVES

- **Identify and specify**

- **Request clarification**

- **Express relationships**

- **Discuss art forms**

¿Qué ves?

Mira la foto. Contesta las preguntas.

1. ¿Qué están haciendo las personas en la foto?

2. ¿Cómo puedes describir a las personas que están allí?

3. ¿Crees que es una celebración? ¿Por qué?

4. Mira el póster. ¿Crees que hay una conexión entre el baile mostrado ahí y la foto? ¿Cuál?

328

En contexto

Understand narration in the present

VOCABULARIO

España para jóvenes

MADRID: EL MUSEO DEL PRADO

¡Hola! Soy Miguel Antonio Ramírez Benavente. Bienvenidos al Museo del Prado, uno de mis lugares preferidos. Quiero ser artista y por eso paso muchas horas aquí. Este museo tiene obras de varias escuelas de pintura europea de los siglos XII al XIX.

Descubre

Lee las siguientes definiciones. ¿Puedes adivinar el significado de las palabras en azul?

1. Un **siglo** es un espacio temporal de cien años.
2. Un **cuadro** es lo mismo que una pintura.
3. Un **autorretrato** es un retrato de una persona hecho por ella misma.
4. Se dice que una imagen está al **fondo** de una pintura cuando ocupa el punto más distante de la **perspectiva**.
5. Una imagen ocupa el **primer plano** cuando ocupa el punto más cercano de la perspectiva.
6. Un **tapiz** es un cuadro grande de lana o seda, hecho algunas veces con oro y plata.
7. El **bailaor** de flamenco ejecuta el **zapateado** – unos ruidos rítmicos – cuando **da golpes** en el suelo con los zapatos rápidamente al **compás** de la música.

Diego de Velázquez (1599–1660)

Si pasas por El Prado, tienes que ver a Diego de Velázquez. Él es uno de los maestros de la perspectiva. *Las meninas* es su obra más importante. Te recomiendo que vengas temprano para ver este cuadro al óleo. ¡Siempre hay mucha gente!

Las meninas, 1656, Diego de Velázquez

Las figuras del rey Felipe IV y de la reina Mariana se ven en el espejo al fondo, como si posaran para su retrato.

Y por supuesto, el autorretrato de Velázquez es una de las cosas que más me gusta de esta pintura.

La Infanta Margarita está en primer plano con sus cortesanas, las meninas.

La familia de Carlos IV (c.1800), Francisco de Goya

Francisco de Goya es otro pintor que yo estudio mucho. Cuando Goya comenzó a pintar se especializó en diseños para tapices y luego en la decoración de iglesias con frescos. En 1799 se hizo el pintor de la corte de Carlos IV y luego de Fernando VII.

Francisco de Goya, 1746–1828

España para jóvenes

El *flamenco* y otros bailes típicos

¡Saludos! Soy María del Pilar Arriaga Méndez y me dedico al flamenco. Lo he estudiado desde niña.

En un tablao, *generalmente hay por lo menos cuatro personas en el* tablado:

el guitarrista
el cantaor
el bailaor *o la* bailaora
el vestido tradicional de lunares
las que dan palmadas

El flamenco es una expresión artística. Aunque se interpreta por toda España, este baile se asocia con Andalucía, que está en el sur. Hay muchos estilos de cante, el tipo de canción con que se acompaña el flamenco. El cante es una parte integral del flamenco.

El cante jondo es el cante más serio y más apasionado. Tiene un compás muy marcado. Otro estilo de cante, la saeta, es el único que no tiene compás fijo. Los bailaores siguen el ritmo de la guitarra y del momento. La coreografía es espontánea. La guitarra, las palmas y el zapateado del bailaor crean el compás del flamenco.

Otros bailes típicos

La jota es un baile de Aragón, pero también se interpreta en otras regiones españolas.

La sardana es el baile tradicional de Cataluña.

¿Comprendiste?

1. ¿Te gustan los museos? ¿Te gusta admirar y analizar las pinturas en los museos? ¿Por qué sí o por qué no?
2. ¿Qué piensas de la pintura *Las meninas* de Velázquez? ¿Por qué crees que el artista se incluyó en el retrato?
3. ¿Alguna vez has visto un espectáculo de flamenco? ¿Qué pensaste? Si nunca lo has visto, ¿crees que te gustaría? ¿Por qué?

trescientos treinta y uno
España y las Américas Etapa 1
331

Listen to
audio texts

AUDIO

En vivo
SITUACIONES

Un paseo por El Prado

Estás en el Museo del Prado en Madrid,
España. Como vas a ir por el museo en
grupo, primero miras el mapa del museo.
Luego escuchas a la guía turística mientras
habla del museo y de las obras de arte que
se encuentran allí.

❶ Mirar

Antes de entrar al Museo del Prado, estudia el
mapa de la planta baja del museo. ¿Qué tipo
de arte vas a ver?

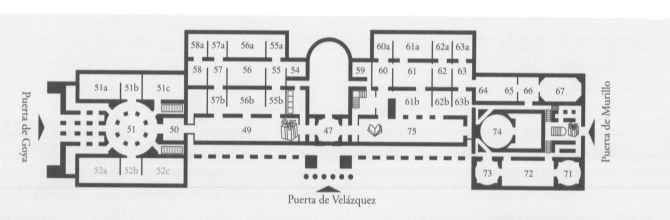

PLANTA BAJA

ESCUELA ESPAÑOLA: Pintura Medieval
Salas 49, 50 y 51C.
ESCUELA ESPAÑOLA: Siglo XVI
Salas 55, 55B, 56, 56B, 57 y 57B.
EL GRECO
Salas 64, 65 y 66.

ESCUELA FLAMENCA: Siglos XV y XVI
Salas 55A, 56, 56A, 57A, 58 y 58A.
ESCUELA ALEMANA
Sala 54.
ESCUELA HOLANDESA
Sala 59.

ESCUELA FLAMENCA: Siglo XVII
Salas 60, 60A, 61, 61A, 62, 62A, 63A y 75.
PINTURAS NEGRAS DE GOYA
Salas 51A y 51B.
ESCULTURA CLÁSICA
Salas 47, 67, 71, 72, 73 y 74.
ESCUELA ITALIANA: Pintura Veneciana Siglo XVI
Salas 61B, 62B, 63 y 63B.

MUSEO DEL PRADO

② Escuchar 🎧

La guía española les da una breve introducción al museo y las obras que se encuentran allí. Escucha su narración y decide si las oraciones que siguen son ciertas o falsas. Si la oración es falsa, cámbiala para que sea cierta.

1. El Museo del Prado es el museo más importante de México.

2. Carlos III mandó a construir el museo como Museo de Ciencias Naturales.

3. Napoleón inauguró el museo en el año 1819.

4. Las obras de arte del museo eran de las colecciones de los reyes y reinas de España de los tres siglos anteriores.

5. Las escuelas de arte representadas en el Museo del Prado son la norteamericana, la argentina y la chilena.

6. El Prado posee la colección más grande de pintura española que existe.

7. De las treinta mil pinturas, casi una quinta parte es de la escuela española.

8. Francisco de Goya expresó mucha alegría en las obras de los últimos años de su vida.

9. El Greco es de origen español.

10. Las obras de El Greco tratan de temas religiosos y místicos.

MUSEO DEL PRADO
Precio de entrada
2€ : Estudiante

③ Hablar 👥

En grupos de dos o tres, conversen sobre la obra de Velázquez, *Las meninas*. ¿Qué ven en la obra? Nombren todo lo que puedan detalladamente. Vayan a la biblioteca y busquen un libro sobre algún (alguna) artista que les interese. Cada uno debe escoger una obra favorita y compartir con sus compañeros tres observaciones sobre la obra.

trescientos treinta y tres
España y las Américas Etapa 1
333

En acción

PARTE A — Práctica del vocabulario

Objectives for Activities 1-3
• Identify and specify • Request clarification

1 Las meninas

Escribir Acabas de estudiar la pintura más famosa de Velázquez. Completa las oraciones con estas palabras.

retrato
escuela
la lista
figura
al óleo
perspectiva
siglo
en primer plano
fondo

1. Diego de Velázquez es un pintor de la _____ española.

2. Las pinturas de Velázquez son del _____ XVII.

3. *Las meninas* es una pintura _____.

4. En *Las meninas*, Velázquez pintó a la Infanta Margarita _____.

5. El rey y la reina se pueden ver reflejados en el espejo al _____ de la pintura.

6. La _____ de Velázquez también se puede ver en el cuadro.

7. La _____ de la pintura parece ser del rey y la reina, donde posan para su _____.

2 El flamenco

Leer/Escribir Querías saber más sobre el flamenco y le pediste a Isabel, tu amiga andaluza, que te lo explicara. Ella te escribió una carta para contestar tus preguntas. Completa su carta con las palabras de la lista.

cantaor   golpes   interpreta   tablao

bailaores   palmadas   zapateado   cante jondo

Querido(a) amigo(a):

Para contestarte, el flamenco es el baile tradicional de Andalucía que se __1__ de muchas maneras. Hoy en día puedes ver el espectáculo de flamenco en un __2__ flamenco en muchas partes de España, incluso en Madrid, donde lo bailan unos de los mejores __3__ del mundo.

El __4__ tiene que cantar con una voz dura y vibrante. El __5__ es el cante más serio y más apasionado de todos los cantes flamencos.

Los bailaores, al darle __6__ al tablado con los zapatos, ejecutan lo que se llama el __7__. Las personas que dan __8__ también contribuyen al compás del zapateado. ¡Tienes que venir a España a disfrutar de este espectáculo magnífico!

Abrazos,

Isabel

3 En el museo

Hablar Pasaste el día en El Prado donde viste la obra a la derecha. Al otro día le explicas a tu compañero(a) por qué te gustó.

modelo

Compañero(a): *¿Hay figuras en la pintura?*

Tú: *Sí, hay seis figuras en la pintura.*

- ¿De qué siglo es el cuadro?
- ¿Dónde están las figuras? ¿En el fondo o en primer plano?
- ¿Qué hacen las figuras?
- ¿Qué tipo de cuadro es?
- ¿De quién es la perspectiva?
- ¿Qué pasa en el cuadro?
- ¿…?

La defensa de Cádiz contra los ingleses, Francisco de Zurbarán, 1634

Vocabulario

La pintura

el cuadro histórico *historical painting*   el paisaje *landscape*

la naturaleza muerta *still-life*

 Ya sabes

el bote	el ejército	luchar contra
el océano	el opresor	la orilla

▶ ¿Cuál es tu estilo de pintura preferido?

También se dice

Cuando se habla de flamenco, se usa la palabra **bailaor** o **bailaora** en vez de **bailador** o **bailadora.** Esto se debe a un modo de hablar en Andalucía, España, donde el flamenco tiene su origen. Cuando pronuncian una palabra que contiene una **d** entre dos vocales, quitan la **d;** entonces, **bailador** se convierte en **bailaor.**

Objectives for Activities 4–13
• Identify and specify • Request clarification • Express relationships • Discuss art forms

REPASO **Demonstrative Adjectives and Pronouns**

You use **demonstrative adjectives** to point out specific things and to show the distance between the speaker and the item.

Demonstrative pronouns are used in place of the adjective and noun. Their forms are the same as demonstrative adjectives, but they have an **accent** over the first **e**.

	masculine singular	masculine plural	feminine singular	feminine plural
this, these *near the speaker*	este	estos	esta	estas *adjectives*
	éste	éstos	ésta	éstas *pronouns*
that, those *near the person spoken to*	ese	esos	esa	esas
	ése	ésos	ésa	ésas
that, those *not associated with either the speaker or the person spoken to*	aquel	aquellos	aquella	aquellas
	aquél	aquéllos	aquélla	aquéllas

Remember that **demonstrative** adjectives and pronouns agree in gender and number with the nouns to which they refer.

agrees

Regina: **Esta** novela me gustó mucho.
*I liked **this** novel very much.*

agrees

Carolina: Sí, **ese** libro es excelente.
*Yes, **that** book is excellent.*

Remember that there are also **demonstrative pronouns** that refer to ideas or unidentified things that do not have a specific gender.

esto   eso   aquello

Don't put an **accent** mark on these words.

—Marcos faltó otra vez.
Marcos was absent again.

—**Esto** me preocupa.
***This** worries me.*

Practice: Actividades

Más práctica
cuaderno pp. 117–118
Para hispanohablantes
cuaderno pp. 115–116

Online Workbook
CLASSZONE.COM

---

4 **Los turistas**

Escuchar/*Escribir* En el museo, oyes si los cuadros están próximos, cerca o lejos. Escribe **este** si está próximo; **ese** si está cerca o **aquel** si está lejos.

modelo
«¿Ves el cuadro de Goya? ¡Qué belleza!»
 aquel cuadro

1. el cuadro de Velázquez
2. el cuadro que quiero ver
3. el cuadro detrás de la gente
4. el cuadro de El Greco
5. el cuadro sin detalle

Vista de Toledo, El Greco

5 Los instrumentos musicales

Hablar/Escribir Tú y tu amigo(a) están en una tienda de música. Le dices a tu amigo(a) qué instrumento te gustaría comprar. Usa **este** si está próximo, **ese** si está cerca y **aquel** si está lejos. Luego cambien de papel.

modelo

Tú: *Creo que voy a comprar este violín.*

Compañero(a): *Yo prefiero aquel violín.*

Vocabulario

La música

el arpa (fem.) *harp*	**las maracas** *maracas*	**el tambor** *drum*
las castañuelas *castanets*	**la pandereta** *tambourine*	**la trompeta** *trumpet*
		el violín *violin*

▶ ¿Sabes tocar alguno de estos instrumentos?

trescientos treinta y siete
España y las Américas Etapa 1
337

Activity 6: Converse and listen during face-to-face social interactions

6 **¿Qué te parece?**

Hablar/Escribir Tú y tu compañero tienen opiniones sobre varias obras de arte. Conversen sobre sus opiniones.

modelo

Tú: *¿Qué te parece aquel cuadro histórico?*

Compañero(a): *¿Aquél? Me parece que es demasiado serio y le falta luz.*

1. el autorretrato
2. el paisaje
3. los cuadros al óleo
4. las figuras
5. la naturaleza muerta

GRAMÁTICA ¿Qué? vs. ¿Cuál?

▶ Both **qué** and **cuál** can be used to express *what* in English. **Cuál** is also used to express *which*.

¿Qué quieres ver en el museo?
***What** do you want to see in the museum?*

¿Cuál de los cuadros te interesa más?
***Which** of the paintings are you more interested in?*

....

▶ You use **qué** to ask someone to **define** or **describe** something. Use **cuál** if you are asking someone to **select** or **make a choice**, and to **identify** or **name** something.

¿Qué es un fresco?
***What** is a fresco?*

¿Cuál es el nombre de la obra que vamos a ver?
***What is the name** of the play that we are going to see?*

Practice: **Actividades** 7 8 9 10

Más práctica *cuaderno p. 119*
Para hispanohablantes *cuaderno p. 117*

Online Workbook
CLASSZONE.COM

7 ¡Viva la música!

Hablar/Escribir A tu compañero(a) le fascina la música contemporánea y ustedes hablan mucho de ella. Él (Ella) te hace una pregunta con **¿qué?** Tú contestas la pregunta y le haces otra pregunta con **¿cuál?** Sigue el modelo.

> **modelo**
>
> *la guitarra / el tambor*
>
> **Compañero(a):** *¿Qué instrumento prefieres, la guitarra o el tambor?*
>
> **Tú:** *Pues, yo prefiero el tambor. Y a ti, ¿cuál te gusta más?*

1. la letra de una canción de rock / la letra de una canción de hip-hop
2. un concierto de rock / un recital de música clásica
3. el repertorio de un conjunto de rock / el repertorio de una orquesta
4. los conjuntos de salsa / los conjuntos de rock
5. el son del piano / el son de la trompeta
6. el ritmo de la música rock / el ritmo de la rap
7. una canción en inglés / una canción en español
8. las cantantes mujeres / los cantantes hombres

Vocabulario

La música

la letra *lyrics*	**el recital** *recital*	**el son** *sound*
la melodía *melody*	**el repertorio** *repertoire*	

▶ ¿Qué te importa más, la melodía o la letra? ¿Por qué?

8 Las respuestas ♻

Hablar/Escribir Arturo estudia para el examen de geografía. Tú le haces preguntas para ayudarlo a estudiar. Usa **¿qué?** y **¿cuál?**

> **modelo**
>
> *La Pampa es una región de Argentina donde hay mucha agricultura.*
>
> **Tú:** *¿Qué es La Pampa?*

1. La industria más importante para la economía de Argentina es la industria ganadera.
2. 9,7% de las exportaciones de Argentina van destino a Estados Unidos.
3. El sector más rico de la economía chilena es el minero.
4. La unidad monetaria de Paraguay es el guaraní.
5. En Chile se producen muchas frutas como las guayabas, uvas, manzanas, peras y papayas.
6. Paraguay tiene más bosques que los otros países sudamericanos.
7. Los uruguayos toman mucho mate.
8. La Paz y Sucre son las capitales de Bolivia.

Activity 9:
Read articles

9 Cinco siglos

Leer/Hablar/*Escribir* Estás en Valencia y ves este artículo en una revista. Contesta las preguntas.

Aquellos maravillosos años

CINCO SIGLOS DE PINTURA VALENCIANA

El Museo de Bellas Artes de Valencia ha organizado una de sus muestras más ambiciosas de los últimos meses. Su objetivo: ofrecer al visitante un pausado recorrido por más de Cinco Siglos de Pintura Valenciana a través de las obras más representativas de los artistas que han configurado el panorama cultural de esta región entre los siglos XIV y XIX. Un total de 70 pinturas y 30 dibujos ilustran esta brillante exposición antológica, que arranca en el Gótico y concluye en el XIX, con el panorama cultural de grandes artistas como Sorolla, Pinazo, Degrain y Agrasot. *Valencia: Museo de Bellas Artes. Hasta el 30 de septiembre.*

Museo de Bellas Artes de Valencia

1. ¿Qué institución está al centro del artículo?
2. ¿Qué se presenta en ese museo?
3. ¿Qué obras se presentan?
4. ¿De qué siglos son las pinturas?
5. ¿Cuántas pinturas se muestran en esta exposición?
6. ¿Cuántos dibujos se muestran?
7. ¿Qué artistas se mencionan en el artículo?
8. ¿En qué fecha termina la exposición?

More Practice: **Más comunicación** *p. R15*

10 En el Prado

STRATEGY: SPEAKING
Discuss a painting For a discussion about a painting, you can talk about **lo que veo, lo que siento, pienso que significa…** and artistic elements that support your ideas: **color, luz y sombra, figuras, perspectiva,** etc.

Hablar/*Escribir* Usa la siguiente lista y escribe seis preguntas para el guía turístico del Museo del Prado. Usa **¿qué?** o **¿cuál(es)?** Luego, discute las preguntas con un(a) compañero(a).

modelo

aspecto de las figuras de El Greco

Tú: *¿Qué aspectos te gustan más de la obra de El Greco?*

Compañero(a): *Me gustan sus figuras largas y también sus colores intensos.*

- **diferencia entre un tapiz y un cuadro**
- **escuela de arte más reconocida**
- **la definición de naturaleza muerta**
- **la figura del artista en** *Las meninas* **(¿hacer?)**
- **la definición de autorretrato**
- **la obra más famosa de Goya**

GRAMÁTICA — Relative Pronouns

¿RECUERDAS? *p. 136* When you learned the subjunctive, you combined two sentences using the relative pronoun **que**.

Relative pronouns are used to link information found in different parts of a sentence. The **relative clause** provides additional information about the person or thing mentioned in the first part of the sentence.

relative pronoun   *relative clause*

Quiero ir al **museo que está cerca del centro**.
*I want to go to the museum **that** is near the center of town.*

You introduce a **relative clause** with a **relative pronoun.** The most common **relative pronoun** in Spanish is **que.** You can use it to refer to both people and things.

refers to a person, the artist

el **artista que pintó este cuadro**
*the artist **who** painted this painting*

refers to a thing, a play

una **obra** teatral **que divierte a todo el mundo**
*a theater play **that** entertains everyone*

Quien (and the plural **quienes**) is the **relative pronoun** that is used to refer to people. It is usually used after a **preposition.**

la **cantante con quien** hablé
*the singer **with whom** I spoke*

los **pintores de quienes** te hablé
*the painters **of whom** I spoke*

> There is no accent mark on **quien/quienes** when you use them as **relative pronouns.**

Practice:
Actividades
11 **12** **13**

Más práctica
cuaderno p. 120
Para hispanohablantes
cuaderno p. 118

Online Workbook
CLASSZONE.COM

---

11 Conversaciones

Escuchar/*Escribir* Estás en un museo y te sientas a descansar. Oyes varias conversaciones. Escribe de nuevo lo que dicen las personas en una oración que use **que.**

modelo

Escuchas: *Ese cuadro lo pintó mi vecino, el artista.*

Escribes: *El artista que pintó ese cuadro es mi vecino.*

1. El escultor _____.
2. Las pinturas _____.
3. El profesor _____.
4. No me gustan las obras _____.
5. El cantaor _____.
6. Las pinturas _____.

Nota cultural

El Museo del Prado tiene un origen curioso. Cuando el rey Fernando VII decidió cambiar las decoraciones del Palacio Real en 1818, puso papel tapiz (*wallpaper*) francés, que no hacía juego con los cuadros de Velázquez y Tiziano que había en las paredes. Entonces el rey integró estos cuadros con los de las colecciones de Carlos III, Carlos V, Felipe II, Felipe IV y Felipe V y creó, en un edificio aparte lo que hoy es el Museo del Prado.

trescientos cuarenta y uno
España y las Américas Etapa 1
341

Activity 13: Clarify, ask for and comprehend clarification

12 ¿Qué es eso?

Hablar/Escribir Le preguntas a tu amigo(a) español(a) sobre varias cosas y personas. Cuando trate de una persona, usa **quien** y no **que**. ¿Cómo te responde?

modelo

escritora (estudié)

Tú: *¿Quién es ella?*

Compañero(a): *Es la escritora con quien estudié.*

1. cuento (leí en la clase de literatura)
2. actor (hablé después de la producción)
3. cuadro (pinté en la clase de arte)
4. obra de teatro (estamos leyendo en la clase de drama)
5. pintor (más admiro)
6. novela (compré el otro día)
7. autobiografía (quiero leer)
8. biografía (te hablé el otro día)
9. poesía (escribió mi padre)
10. actor (trabaja en la producción de la obra de teatro)

Vocabulario

La literatura

la autobiografía *autobiography*	la ficción *fiction*
la biografía *biography*	la poesía *poetry*
el cuento *short story*	la producción *production*
el drama *drama*	

♻ Ya sabes: el ensayo la novela la obra de teatro el poema

▶ ¿Qué tipo de literatura disfrutas más?

13 Los artistas ♻

Hablar/Escribir En grupos de tres o cuatro, escojan una identidad artística: uno(a) puede ser un(a) pintor(a), otro(a) un(a) bailador(a), un(a) escritor(a), un(a) actor (actriz), etc. Imaginen que están en una fiesta hablando de su trabajo, sus obras o las obras de los demás.

modelo

escritor

Tú: *¿Quién es ese señor?*

Compañero(a): 1: *¿Ése? Es el escritor que escribió El mundo del artista.*

Compañero(a): 2: *¿No te acuerdas de esa novela? Es la novela de la que te hablé el otro día.*

Compañero(a): 3: …

1. escritor(a)
2. pintor(a)
3. bailador(a)
4. actor (actriz)
5. novela
6. drama
7. cuento
8. producción

Activity 14 brings together all concepts presented.

14 Mi colección privada

Hablar En grupos de tres o cuatro, conversen sobre la colección que se ofrece en este anuncio. Hablen sobre las pinturas que les gustan o que no les gustan, por qué, si les interesa la oferta, etc.

Los más bellos cuadros del arte tradicional de España

Ahora las obras de los grandes pintores españoles están reunidas en una colección única: «*Obras Maestras de los Grandes Pintores Españoles*». Admirará retratos, paisajes, naturalezas muertas… todos en reproducciones de alta calidad.

modelo

Tú: *¿Cuál de estos pinturas te gusta más?*

Compañero(a): 1: *¿A mí? Me gusta la de El Greco.*

Compañero(a): 2: *Yo prefiero la de Goya. No hay nadie que pinte como él.*

More Practice: Más comunicación *p. R15*

 **Online Workbook**
CLASSZONE.COM

Refrán

De músico, poeta y loco, todos tenemos un poco.

¿Qué quiere decir el refrán? ¿Crees que es verdad? ¿Cuál de las características que se mencionan te parece más importante? ¿Por qué?

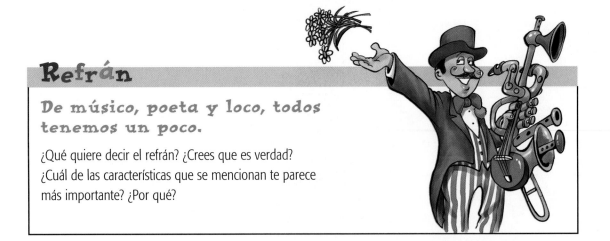

trescientos cuarenta y tres
España y las Américas Etapa 1
343

En voces

PARA LEER

STRATEGY: READING

Compare famous authors Unamuno and Matute share a love of Spain. How else are they alike or different? Use this chart to make your comparisons:

	Unamuno	Matute
fechas importantes		
eventos influyentes		
preocupaciones		
¿?		

What emotions does each express in the excerpts you read?

LA IDENTIDAD ESPAÑOLA

la guerra	lucha entre dos fuerzas
derechista	conservador en extremo
posguerra	después de la guerra
vencer	conquistar

Miguel de Unamuno

Miguel de Unamuno, escritor español, nació en Bilbao en 1864 y fue uno de los autores más importantes del siglo XIX. Tuvo una gran conciencia social y se preocupó mucho por la identidad y el futuro de España, como otros autores de su generación, que se llama «La Generación del '98».

Unamuno además pensaba mucho sobre cuestiones filosóficas de la vida y de la inmortalidad. Cultivó todos los géneros literarios, pero sus ensayos y novelas son los más famosos.

Unamuno escribió sobre su amor a España y los problemas políticos del país durante un período de cambio violento. De él es la cita conocida:

> **❝** ¡Me duele España! **❞**

Durante esa época, una guerra civil empezó en España. Unamuno se opuso a Francisco Franco, el general derechista. Por esto, las fuerzas de Franco mantuvieron a Unamuno bajo arresto a domicilio. Unamuno murió el último día de 1936–el primer año de la guerra–sin cambiar su opinión política. Dijo:

> **❝** Venceréis pero no convenceréis. **❞**

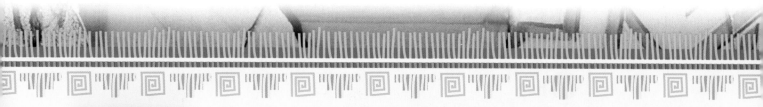

Ana María Matute

Ana María Matute, novelista española contemporánea, nació en Barcelona en 1926. Tenía diez años cuando comenzó la guerra civil y su familia se mudaba de Madrid a Barcelona para escaparse de la violencia. El punto de vista triste de sus novelas refleja[1] la desilusión que ella sintió durante la guerra civil y los años de represión que la siguieron. Ella dice:

> " Todo era injusto e incomprensible. El mundo no era tal y como nos lo habían explicado. Yo creo que nuestra generación dio tantos grandes escritores porque fuimos víctimas de un trauma muy fuerte. No se podía hacer ni decir nada. De ahí nació un sentimiento de rebeldía que creo aún mantengo. "

A pesar de sus experiencias traumáticas de niña, Matute ha tenido un éxito extraordinario con su producción literaria. Además de ser una de los autores más importantes de la narrativa de posguerra española, es la única mujer en la Real

---

[1] reflects

Academia de España. Entre sus novelas más conocidas están *Los Abel, Los hijos muertos, La trampa, El río* y *Olvidado Rey Gudú.*

Ella reconoce la dificultad de escribir novelas. Dice:

> " Quien diga lo contrario o miente[2], ¡o es un genio[3] o es un desastre! "

---

[2] is lying
[3] genius

¿Comprendiste?

1. ¿A qué generación literaria pertenece Unamuno?
2. ¿Cuál es la actitud de Unamuno hacia Francisco Franco y sus fuerzas?
3. ¿Cómo influyó la guerra civil de España en el pensamiento y la vida de Matute?
4. Según Matute, ¿por qué hay tantos escritores buenos en su generación?

¿Qué piensas?

1. ¿Qué quiere decir el comentario famoso de Unamuno: «¡Me duele España!»?
2. En tu opinión, ¿qué significa el comentario de Matute sobre la dificultad de escribir novelas?

Hazlo tú

Haz una investigación sobre la guerra civil española y escribe un ensayo breve sobre ella. Explica sus causas, quiénes participaron, etc. O escoge un cuento, una película o un libro que tenga una guerra como tema y escribe un ensayo sobre éste.

trescientos cuarenta y cinco
España y las Américas Etapa 1
345

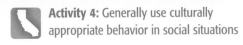

Activity 4: Generally use culturally appropriate behavior in social situations

En uso
REPASO Y MÁS COMUNICACIÓN

OBJECTIVES

- Identify and specify
- Request clarification
- Express relationships
- Discuss art forms

Now you can...

- discuss art forms.
- identify and specify.

To review

- demonstrative adjectives and pronouns see p. 336.

1 Hoy

Completa varias conversaciones que escuchaste hoy con las formas correctas de las palabras entre paréntesis.

Conversación No. 1 (ese)

José: ¿Ves __1__ guitarra?

María: ¿__2__ guitarra en la mesa?

José: Sí, __3__ ¡La acabo de comprar!

Conversación No. 2 (este)

Ana: __4__ cuadro es muy interesante.

Irma: ¿__5__? No me gusta.

Ana: ¿De veras? Yo pienso que __6__ cuadro es el mejor en el museo.

Conversación No. 3 (aquel)

Andrés: __7__ muchachos son artistas.

Beto: ¿Estás seguro? ¿__8__?

Andrés: Sí. El muchacho de la camisa azul pintó __9__ cuadro.

Now you can...

- request clarification.

To review

- ¿qué? vs. ¿cuál? see p. 338.

2 La clase de arte

Estás en la clase de arte. Durante la clase, tu compañero(a) te pregunta varias cosas. Primero, escribe preguntas para cada tema y luego contéstalas.

> **modelo**
>
> *autorretrato*
> *¿Qué es un autorretrato?*
> *Un autorretrato es un retrato de una persona hecho por ella misma.*
>
> *tapiz o pintura al óleo*
> *¿Cuál te gusta más: el tapiz o la pintura al óleo?*
> *A mí me gusta más el tapiz.*

1. naturaleza muerta o paisaje

2. tapiz

3. escuela española o escuela flamenca

4. naturaleza muerta

5. los cuadros de Goya o los cuadros de Velázquez

6. el primer plano

Now you can...

• request clarification.

To review

• ¿qué? vs. ¿cuál?
see p. 338.

3 La profesora de música

Tomas una clase de música y tu profesora quiere saber más sobre tus intereses. Completa sus preguntas con **¿qué?** o **¿cuál(es)?**

modelo

¿ <u>Qué</u> instrumento quieres aprender a tocar?

1. ¿_____ crees es más fácil de aprender, la guitarra o el violín?
2. ¿_____ clase de música quieres aprender?
3. ¿_____ de las canciones quieres aprender primero?
4. ¿_____ son tus grupos musicales favoritos?
5. ¿_____ estudiaste en el colegio?
6. ¿_____ entrenamiento musical has tenido?

Now you can...

• express relationships.

To review

• relative pronouns see p. 341.

4 La reunión

Estás en una reunión estudiantil en Madrid con tu amigo(a) madrileño(a). Pregúntale a tu compañero(a) quiénes son las personas que ves, basándote en los dibujos y las siguientes palabras.

modelo

¿él?

Tú: ¿Quién es él?

Compañero(a): ¿Él? Es el estudiante que toca la trompeta.

pintar   tocar   bailar jota   escribir   marcar el compás

1. ¿ella? 2. ¿él? 3. ¿ella? 4. ¿ella? 5. ¿él? 6. ¿él?

5 Investigación

STRATEGY: SPEAKING

Organize ideas for research Good research begins with good questions, but first categorize the areas to investigate. Where do you want to begin? Use the following chart to ask questions that will direct your research.

Francisco de Goya
¿Cuándo empezó a pintar?
¿Dónde vivía?
¿Cuántos años vivió?
¿Cuántos cuadros pintó?
¿A quiénes conocía?

En grupos de dos o tres, escriban seis preguntas sobre los grandes pintores españoles. Busquen las respuestas en una enciclopedia o en Internet.

modelo

Tú: *Yo quiero saber cuándo y por qué empezó a pintar Francisco de Goya.*

Compañero(a) 1: *Yo quiero saber cuántas obras pintó Goya durante su vida y cuántos años vivió.*

Compañero(a) 2: …

6 *En tu propia voz*

ESCRITURA Trabajas como crítico(a) de arte para el periódico estudiantil. Usa la pintura aquí o busca otra pintura de uno de los grandes pintores españoles y escribe un breve ensayo. Incluye observaciones sobre lo siguiente:

Pablo Picasso, Las meninas, 1957

- cómo te hace sentir
- colores
- figuras
- escuela
- tipo de pintura
- algún detalle de la vida del pintor
- ¿…?

modelo

Usando Las meninas *de Velázquez como modelo, Picasso rehace el cuadro en su propio estilo, usando técnicas modernas de composición y color.*

CONEXIONES

Los estudios sociales La guerra civil de España fue un evento muy importante en la vida de la gente española. Esta tragedia también tuvo una gran influencia sobre algunos de los artistas y autores más conocidos de España. ¿Por qué empezó la guerra civil? ¿Cómo terminó? Crea una cronología *(timeline)* de la época, incluyendo por lo menos diez eventos importantes de la guerra. Describe en una o dos oraciones la importancia de cada evento.

General Primo de Rivera establece una dictadura militar.		Elecciones municipales
1923		**1931**

En resumen
REPASO DE VOCABULARIO

Flashcards
CLASSZONE.COM

DISCUSS ART FORMS

Art and paintings

al óleo	oil (painting)
el autorretrato	self-portrait
el cuadro	painting
el cuadro histórico	historical painting
la escuela	school (of art)
la figura	figure
el fondo	background
el fresco	fresco
la naturaleza muerta	still life
el paisaje	landscape
la perspectiva	perspective
el primer plano	foreground
el retrato	portrait
el siglo	century
el tapiz	tapestry

 Ya sabes

el bote	boat
el ejército	army
luchar contra	to fight against
el océano	ocean
el (la) opresor(a)	oppressor
la orilla	shore

Dance

el (la) bailaor(a)	flamenco dancer
el (la) cantaor(a)	flamenco singer
el cante	flamenco song
el cante jondo	tragic flamenco song
dar golpes	to stamp
dar palmadas	to clap hands
el flamenco	flamenco dancing
interpretar	to interpret
la jota	Aragonese folk dance
la saeta	Andalusian song
la sardana	Catalan folk dance
el tablado	stage floor
el tablao	flamenco group
el zapateado	rhythmic heel tapping

Literature

la autobiografía	autobiography
la biografía	biography
el cuento	short story
el drama	drama
la ficción	fiction
la poesía	poetry
la producción	production

Music

el compás	rhythm, beat
la letra	lyrics
la melodía	melody
el recital	recital
el repertorio	repertoire
el ritmo	rhythm
el son	sound

Musical instruments

el arpa (fem.)	harp
las castañuelas	castanets
las maracas	maracas
la pandereta	tambourine
el tambor	drum
la trompeta	trumpet
el violín	violin

 Ya sabes

el ensayo	essay
la novela	novel
la obra teatral	theatrical work, play
el poema	poem

IDENTIFY AND SPECIFY

Ya sabes

este, esta, estos, estas
éste, ésta, éstos, éstas
ese, esa, esos, esas
ése, ésa, ésos, ésas
aquel, aquella, aquellos, aquellas
aquél, aquélla, aquéllos, aquéllas
esto, eso, aquello

REQUEST CLARIFICATION

¿Qué? vs. ¿cuál?

¿Qué es el cante jondo?
¿Cuál prefieres, el flamenco o la jota?

EXPRESS RELATIONSHIPS

Relative pronouns

Ésta es la bailaora que vimos anoche.
Aquél es el cantaor con quien hablamos.

Juego

Sopa de letras
¡Nunca tuve vida pero ahora estoy muerta! ¿Qué soy?

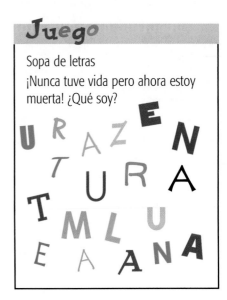

trescientos cuarenta y nueve
España y las Américas Etapa I
349

ETAPA
2

El Nuevo Mundo

OBJECTIVES

• Refer to people and objects

• Express relationships

• Make generalizations

• Describe arts and crafts

¿Qué ves?

Mira la foto. Contesta las preguntas.

1. ¿En qué lugar de América Latina se encuentran los estudiantes?

2. ¿Cómo puedes describir la escena?

3. ¿Qué crees que existía en este lugar? ¿Por qué?

En contexto

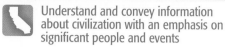

Understand and convey information
about civilization with an emphasis on
significant people and events

VOCABULARIO

Las civilizaciones precolombinas

Descubre

En español, como en inglés, hay verbos
que tienen la misma raíz que el sustantivo
que les corresponde. Si sabes qué quiere
decir el verbo, puedes adivinar qué quiere
decir el sustantivo. Primero decide cuál es
el significado del verbo, y luego da el
significado del sustantivo.

construir → construcción

to construct → construction

1. civilizar → **civilización**
2. creer → **creencia**
3. descender → **descendencia**
4. descifrar → **cifra**
5. reflejar → reflejo

Se encuentran dos mundos

El año es 1518 en el calendario europeo. Barcos misteriosos
llegan a la costa atlántica de México. Un indio sale para dar las
noticias a su emperador, Moctezuma. Él gobierna al imperio
azteca, una de las civilizaciones precolombinas de América.

Ha llegado a México Hernán Cortés
(1485–1547), un conquistador español.
Él sigue el ejemplo de Cristóbal Colón,
quien abrió el paso entre Europa y
América en 1492. Bernal Díaz del
Castillo (1492–1584), un conquistador
joven, viene con Cortés. Para Bernal y
los demás europeos, están en un Nuevo
Mundo. Éste es el nombre que los
europeos le dan a América.

Hernán Cortés (centro)

Tenochtitlán

Bernal, que también es cronista, escribe:
«*Que por una parte había grandes ciudades…
Y en la laguna otras muchas, y veíamoslo todo
lleno de canoas… Y por delante estaba la gran
Ciudad de México*».
Los españoles quedaron impresionados al ver la
ciudad azteca de Tenochtitlán. Cuando Moctezuma
y los aztecas recibieron a los españoles, los llevaron
al Templo Mayor. El Templo tenía varias pirámides

grandes. Pero la amistad entre los españoles y los aztecas no duró mucho.
Pronto hubo una guerra y los españoles conquistaron a los aztecas. Después
de la conquista española, Tenochtitlán desapareció debajo de los edificios
europeos. En 1978, se encontraron las ruinas del Templo Mayor, que hoy
se pueden ver en la Ciudad de México.

Los libros mayas

Al mismo tiempo, en la península de Yucatán y por Centroamérica, los hombres religiosos de los mayas escribieron los libros *Chilam Balam* y *Popol Vuh*. Utilizaron jeroglíficos, un sistema complicado de escritura. Con los jeroglíficos narraron la historia de casi tres mil años del pueblo maya y también sus creencias. Hoy, los arqueólogos continúan descifrando la escritura jeroglífica con sus cifras misteriosas.

Las tejedoras de Los Altos de Chiapas

Los textiles son una tradición indígena. Las mujeres de descendencia maya han mantenido viva esta tradición. Ellas aprenden las técnicas para hacer los textiles de sus madres y las enseñan a sus hijas. Las imágenes en sus tejidos reflejan diseños mayas de casi 1200 años.

Los incas

Los incas, otra civilización indígena avanzada, establecieron un imperio en la costa del Pacífico de América del Sur. En la década de 1530, Atahualpa era el Inca, el título del gobernante que significa «hijo del sol». Francisco Pizarro (c.1478–1541) y otros conquistadores llegaron hasta Cuzco. Ellos lucharon por el oro y objetos prehispánicos preciosos, que los incas hicieron antes de la llegada de los españoles. Los incas construyeron ciudades como Machu Picchu, una obra maestra de arquitectura precolombina.

¿Comprendiste?

1. ¿Te interesan la historia y el arte precolombino? ¿Qué te interesa más, la arquitectura, el arte o los textiles? ¿Por qué?
2. ¿Qué piensas de las ruinas de Tenochtitlán?
3. La escritura jeroglífica de los mayas no se ha descifrado por completo. ¿Qué te parece la invención de una forma de escribir tan complicada?
4. Las pirámides en México y en Perú son obras de arquitectura que no se pueden reproducir hoy. ¿Cómo crees que se construyeron estos edificios enormes sin la ayuda de máquinas?

trescientos cincuenta y tres
España y las Américas Etapa 2
353

Listen to
audio texts

AUDIO

En vivo

SITUACIONES

PARA ESCUCHAR

STRATEGIES: LISTENING

Pre-listening Most people can remember about seven items briefly. Read the phrases under **Escuchar** three times. Then close your eyes and say as many as you can. How many can you recall?

Improve your auditory memory To help you remember what is mentioned in the conversation, (a) re-read the list, (b) close your eyes to shut out distractions, (c) when you hear one of the phrases, say it and check it on the list. Did your auditory memory improve?

Una visita virtual

Estás en la biblioteca buscando información para tu curso sobre el arte y la historia precolombina de las Américas. Mientras lees, oyes cuatro estudiantes españoles conversando. Por casualidad, hablan del mismo tema — las contribuciones artísticas de las civilizaciones indígenas precolombinas.

❶ Leer

Has encontrado una página-web que ofrece «una visita virtual» al mundo del arte y de la historia precolombina de América Latina.

| Regresar | Adelantar | Inicio | Recargar | Imágenes | Abrir | Imprimir | Buscar | Finalizar |

Dirección: http://www.arteprecolombino.com

UNA VISITA VIRTUAL AL ARTE PRECOLOMBINO

GUATEMALA
El Templo del Jaguar en Tikal en la región del Petén

MÉXICO
Teotihuacán, «la ciudad de los Dioses» cerca de la Ciudad de México, la Pirámide del Sol y la Pirámide de la Luna

Un ejemplo de los textiles mayas tejidos por las tejedoras de los Altos de Chiapas

PERÚ
Machu Picchu, las ruinas de la antigua ciudad inca, a unos 112 kilómetros de Cuzco

PANAMÁ
La mola es el arte tradicional de los indígenas Kuna Yala de Panamá.

COLOMBIA
El arte «quimbaya» se distingue por el detalle de la ornamentación de las piezas, como se puede ver en esta máscara de oro.

❷ Escuchar

Mientras estudias, escuchas la conversación de cuatro estudiantes españoles que están sentados en la mesa junto a la tuya. Decide si mencionan cada objeto, tema o concepto en la siguiente lista.

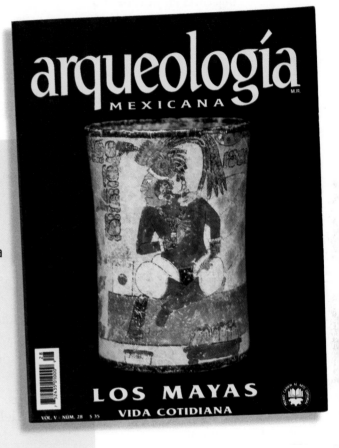

1. la escritura jeroglífica
   ❑ La mencionan.
   ❑ No la mencionan.

2. los chibchas
   ❑ Los mencionan.
   ❑ No los mencionan.

3. las creencias mayas sobre el cosmos
   ❑ Las mencionan.
   ❑ No las mencionan.

4. las tejedoras mayas
   ❑ Las mencionan.
   ❑ No las mencionan.

5. Tenochtitlán
   ❑ La mencionan.
   ❑ No la mencionan.

6. las pirámides de Teotihuacán
   ❑ Las mencionan.
   ❑ No las mencionan.

7. la joyería de oro precolombina
   ❑ La mencionan.
   ❑ No la mencionan.

8. el descubrimiento del Nuevo Mundo
   ❑ Lo mencionan.
   ❑ No lo mencionan.

9. las molas
   ❑ Las mencionan.
   ❑ No las mencionan.

10. los incas
    ❑ Los mencionan.
    ❑ No los mencionan.

❸ Hablar/Escribir

Escoge una de las culturas precolombinas mencionadas en esta lección y haz una investigación sobre la arquitectura, el arte o los textiles de esa cultura. Escribe tres cosas que no sabías antes de hacer la investigación. Trae tus observaciones a la clase. En grupos de tres o cuatro, comparen los resultados de sus investigaciones. Hagan una lista de los aspectos de las culturas que son similares y los aspectos que son diferentes.

trescientos cincuenta y cinco
España y las Américas Etapa 2
355

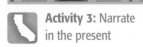

En acción

Práctica del vocabulario

Objectives for Activities 1-3
• Refer to people and objects • Express relationships

1 El examen

Escribir Mario tuvo un examen sobre la historia precolombina de las Américas. Completa las oraciones que él tuvo que completar para su examen.

- **a.** descendencia
- **b.** ruinas
- **c.** pirámides
- **d.** civilización
- **e.** precolombinos
- **f.** cronista
- **g.** escritura jeroglífica

1. En la Ciudad de México, encontraron las _____ de una _____ muy avanzada.
2. Bernal Díaz de Castillo era un conquistador y _____ español.
3. Las tejedoras de los Altos de Chiapas son de _____ maya.
4. En el centro de Tikal se encuentran dos _____; la primera se llama El Templo del Gran Jaguar.
5. La _____ de los mayas narra la historia del pueblo.
6. En las excavaciones del Templo Mayor, se encontraron objetos _____ de mucho valor.

2 Comentarios

> **STRATEGY: SPEAKING**
>
> **Maintain a discussion** In a serious discussion, words may be used that are not known to everyone. Ask for a definition or be prepared to give one. Example: **¿Qué es precolombina? — La cultura que existía en América antes de los viajes de Colón.**

Hablar/*Escribir* Tú y tu compañero(a) comentan sobre las civilizaciones precolombinas de las Américas. Conversen sobre los siguientes temas.

modelo

el sistema de escritura de los mayas: escritura jeroglífica

Tú: *El sistema de escritura de los mayas era muy complicado.*

Compañero(a): *Sí, la escritura jeroglífica de los mayas es tan complicada que los arqueólogos siguen tratando de descifrarla.*

> los cronistas españoles: crónicas
>
> las técnicas para construir pirámides: olvidarse
>
> los mayas: extenso conocimiento de ingeniería y matemáticas
>
> los tejidos: expresión artística
>
> las tradiciones de los mayas: pasar siglo tras siglo
>
> las civilizaciones precolombinas: avanzadas
>
> las pirámides: obras de arquitectura

Nota cultural

Cuando los españoles llegaron al Nuevo Mundo adoptaron muchas palabras indígenas, como las siguientes.

• **huracán** (taíno) • **tomate** (náhuatl) • **papa** (quechua)

3 Teotihuacán

Hablar/*Escribir* Elena y Enrique fueron a Teotihuacán, «la ciudad de los dioses». Imagina y escribe la conversación que tienen los dos compañeros. Luego, léela en voz alta.

1.

2.

3.

4.

5.

6.

Objectives for Activities 4-14
• Refer to people and objects • Express relationships • Make generalizations • Describe arts and crafts

REPASO   Direct Object Pronouns

You use **direct object pronouns** in Spanish to refer to items or people that have already been mentioned.

Direct Object Pronouns

me	nos
te	os
lo/la	los/las

becomes

—¿Has visto **las ruinas** en Chichén Itzá?
*Have you seen **the ruins** at Chichén Itzá?*

—Sí, **las** vi el verano pasado.
*Yes, I saw **them** last summer.*

Third-person direct object pronouns (**lo, la, los, las**) refer to **usted** and **ustedes** as well as to **él, ella, ellos,** and **ellas.**

Perdón señora. No **la** vi.
*I'm sorry ma'am. I didn't see **you.***

Direct object pronouns go before the conjugated verb except in **affirmative commands**, where you attach them.

Alfredo y Marta no saben que Uds. van a la exhibición de máscaras. **Invíten**los. *attaches*
*Alfredo y Marta don't know that you are going to the mask exhibit. **Invite them.***

Direct object pronouns come before **conjugated verbs** or attached to **infinitives** and **-ndo forms**.

—Esta novela es muy buena. ¿Quieres **leer**la? *attaches*
*This novel is very good. Do you want to read **it**?*

¿**La quieres** leer?
*Do you want to read **it**?*

—Mira. Estoy **leyéndo**la. *attaches*
*Look. I'm reading **it**.*

La estoy leyendo.
*I'm reading **it**.*

Practice:
Actividades
4 5 6

Más práctica
cuaderno p. 125
Para hispanohablantes
cuaderno p. 123

 Online Workbook
CLASSZONE.COM

---

4 ¿Lo conoces?

Escuchar/Escribir Escucha las conversaciones de varias personas. Di si la persona que contesta conoce o no las cosas mencionadas.

modelo
No las conoce.

1. _____
2. _____
3. _____
4. _____
5. _____
6. _____

Figura precolombina de un hombre. Cultura quimbayá.

---

5 **Los bailes**

Hablar/*Escribir* Tú y tu compañero(a) quieren saber si el otro (la otra) sabe bailar varios bailes. Pregúntale a tu compañero(a) si sabe bailarlos y luego, él (ella) te pregunta a ti.

modelo

Tú: *¿Sabes bailar la sardana?*

Compañero(a): *Sí, (No, no) sé bailarla.*

1. el mambo	**5.** el tango
2. la cumbia	**6.** el jarabe tapatío
3. el merengue	**7.** la bamba
4. la habanera	**8.** ¿…?

Vocabulario

Bailes típicos

el jarabe tapatío

el tango

el baile folklórico *folk dance*

la bamba *Mexican dance from Veracruz*

la cumbia *cumbia*

la danza *dance*

la habanera *habanera*

el mambo *mambo*

el merengue *merengue*

▶ ¿Conoces a algunos músicos que toquen este tipo de música?

6 **Mi primo(a) panameño(a)**

Hablar/*Escribir* Tu primo(a) de Panamá vino a visitarte y quieres hacerle muchas preguntas. Primero tu compañero(a) hace el papel del (de la) primo(a). Luego, cambien de papel. Usen las ideas de la lista o inventen otras.

modelo

Tú: *¿Viste la nueva película de Benicio del Toro?*

Compañero(a): *No, no la he visto.*

More Practice: **Más comunicación** *p. R16*

ver (la película nueva de…)
leer (la última novela de…)
escribir (la tarjeta postal para tu…)
escuchar (el nuevo disco compacto de…)
limpiar (tu cuarto)
hacer (los quehaceres)
visitar (a tu familia)
mandar (la carta por Internet)

Nota cultural

Parece que cada región de América Latina tiene su propio baile típico. Estos bailes muestran una mezcla de tradiciones precolombinas, africanas y europeas. Además de los mencionados a la izquierda, aquí hay otros muy conocidos:

- **el candombe** (Uruguay)
- **la cueca** (Chile)
- **la marinera** (Perú)
- **la bachata** (República Dominicana)

trescientos cincuenta y nueve
España y las Américas Etapa 2
359

Activity 8: Express and understand opinions

REPASO — Indirect Object Pronouns

You use indirect object pronouns in Spanish to refer to the person or thing that is **receiving the action** of the verb.

Indirect Object Pronouns

me	nos
te	os
le	les

Mandé las fotos a María.
I sent the photos to María.

Le mandé las fotos.
*I sent **her** the pictures.*

Indirect object pronouns, like the direct object pronouns, precede the **conjugated verbs**.

—¿Qué le regalaste?
*What (gift) did you give **her**?*

—Le regalé una pulsera de jade.
*I gave **her** a jade bracelet.*

Remember that sometimes you use **a + person** to clarify to whom the indirect object pronouns le and les are referring.

—¿Les escribes a tus amigos?
Do you write to your friends?

—A Magdalena le escribo mucho.
I write to Magdalena a lot.

You attach indirect object pronouns to **affirmative commands** just like you do with direct object pronouns.

Préstame tu libro de arquitectura.
Lend me your architecture book.

You can attach indirect object pronouns to **infinitives** and **progressive tenses** or you can put them before the **conjugated verbs**.

¿Puedes prestarle tu libro a José también?
Can you lend your book to José also?

↔

¿Le puedes prestar tu libro a José también?

¿Estás prestándole tu libro a Luisa?
Are you lending your book to Luisa?

↔

¿Le estás prestando tu libro a Luisa?

Practice: Actividades
7 8 9

Más práctica
cuaderno p. 126
Para hispanohablantes
cuaderno p. 124

Online Workbook
CLASSZONE.COM

---

7 Después de la entrevista

Hablar/*Escribir* Tu compañero(a) hace el papel de supervisor(a) y tú le hablas sobre una entrevista que hiciste. Luego, cambien de papel.

modelo
pedir los datos
Tú: *¿Le pediste los datos?*
Compañero(a): *Sí, señor(a), le pedí los datos.*

1. pedir la solicitud
2. explicar el puesto
3. explicar los beneficios
4. contestar sus preguntas
5. informar del sueldo
6. pedir tres recomendaciones

Nota cultural

El Inca Garcilaso de la Vega, hijo del español Garcilaso de la Vega y la princesa india Isabel Chimpu Ocllo, escribió sus *Comentarios reales del Perú* dando los detalles de la vida diaria de los incas. Se considera uno de los textos claves para entender la sociedad inca precolombina, ya que describe la conquista desde la perspectiva indígena.

8 El viaje (primera parte)

Hablar/*Escribir* Vas de viaje a México y Guatemala con tu familia. Piensan visitar las ruinas de algunas civilizaciones precolombinas. Conversa con tu compañero(a) sobre lo que quieres ver. Usa ideas de la lista si es necesario.

modelo

Compañero(a): *¿Te interesan las ruinas de Tenochtitlán?*

Tú: *Sí, hombre, ¡me fascinan! Me gustaría pasar más tiempo estudiando las civilizaciones precolombinas.*

interesar
fascinar
gustar
parecer
explicar
¿...?

la escritura jeroglífica
los tejidos de las mujeres mayas
las ruinas de Tenochtitlán
las pirámides de Tikal
el Templo Mayor
los objetos precolombinos de jade
las blusas bordadas de Guatemala
las tradiciones de los mayas
las creencias sobre el cosmos
los objetos de piedra labrada
¿...?

Vocabulario

Las artesanías

bordado(a)

el mural

tallado(a)

decorado(a) *decorated*
el jade *jade*

labrado(a) *worked, cut*
el (la) muralista *muralist*

▶ ¿Comprarías algún objeto así? ¿Por qué?

9 El viaje (segunda parte)

Hablar Sigues hablando con tu compañero(a) sobre tu viaje a México y Guatemala. Él (Ella) te hace muchas preguntas sobre lo que compraste, lo que viste, lo que le preguntaste al guía turístico, etc. Usen los verbos de la lista si quieren.

modelo

Compañero(a): *¿Qué le compraste a tu mamá?*

Tú: *Le compré una blusa bordada muy bonita de Guatemala.*

Compañero(a): *¿Qué le preguntaste al guía turístico?*

Tú: *Le pregunté en qué año se habían descubierto las pirámides de Teotihuacán.*

comprar	leer
contestar	llevar
dar	mandar
decir	pedir
escribir	preguntar
prestar	traer
hablar	servir

trescientos sesenta y uno
España y las Américas Etapa 2
361

Activity 12: Use strings of related sentences when speaking

GRAMÁTICA   More on Relative Pronouns

 ¿RECUERDAS? *p. 341* You have already learned to use the **relative pronouns** que or quien to provide additional information.

You can also use the **relative pronouns,** el que and el cual, in the same way, but to show a stronger relationship or provide greater emphasis.

	singular		plural	
masculine	el cual	el que	los cuales	los que
feminine	la cual	la que	las cuales	las que

You will often use el que, el cual, etc. after prepositions where you wish to show a stronger relationship or greater emphasis than que or quien would provide.

agrees

un templo en el que hay jeroglíficos
a temple in which there are hieroglyphics

agrees

la guitarrista sin la cual no podemos tocar
the guitarist without whom we can't play

> Notice that these forms agree in gender and number with the noun to which they refer.

Note that in formal uses, such as writing, it is more common to use el cual instead of el que.

informal

Hablé con el que llegó tarde.
I spoke with the one who arrived late.

formal

Cristóbal Colón fue un explorador famoso, el cual llegó a las Américas.
Christopher Columbus was a famous explorer who arrived in the Americas.

The **relative pronoun** cuyo(a) means *whose*. Remember to make it agree in gender and number with the noun that follows:

agrees

el pintor cuyos cuadros nos gustaron
the painter whose paintings we liked

Practice:
🔟 ⓫ ⓬

Más práctica
cuaderno p. 127
Para hispanohablantes
cuaderno p. 125

 **Online Workbook**
CLASSZONE.COM

🔟 México

Escribir Tu amigo(a) te escribe y te cuenta todo lo que hizo y vio en sus viajes a México. Completa sus oraciones con la forma correcta de **el que**.

1. Vimos un templo en _____ hay jeroglíficos.

2. Entramos a las pirámides en _____ enterraban (*buried*) a los reyes mayas.

3. Visitamos los templos en _____ encontraron los objetos de arte precolombino.

4. Leí la crónica de Colón en _____ escribió sobre el descubrimiento de las Indias.

5. Caminamos por el centro en _____ encontraron las ruinas del Templo Mayor.

6. Vi unas piedras en _____ están escritas las historias de los reyes mayas.

7. Hicimos una excursión en _____ vimos muchas ruinas.

8. Compré un libro en _____ explican el significado de palabras mayas.

9. Entramos a una tienda en _____ había artesanías muy bellas.

10. Observamos unos lugares en _____ se hacían ceremonias.

11 Revisiones

Hablar/Escribir Escribiste un ensayo pero hay unas oraciones que quieres cambiar para decirlas en clase. Sigue el modelo.

modelo

La escritura jeroglífica es un sistema de escritura pictórica. Todavía no se ha descifrado por completo.

La escritura jeroglífica, la cual todavía no se ha descifrado por completo, es un sistema de escritura pictórica.

I. Las composiciones jeroglíficas también contenían sus creencias sobre el cosmos. Narraban la historia del pueblo.

2. Bernal Díaz del Castillo escribió una descripción de Tenochtitlán en su crónica. Era cronista español.

3. Cristóbal Colón murió sin saber qué había descubierto. Abrió el paso entre Europa y las Américas.

4. Los mayas construyeron grandes pirámides. Tenían un conocimiento de ingeniería extenso.

5. La cultura mexica no desapareció. Dejó su influencia en el mundo del arte y de la arquitectura.

12 El Templo Mayor

Leer/Hablar Lee la información de un folleto del Templo Mayor. Hazle tres preguntas a tu compañero(a). Luego cambien de papel.

modelo

Tú: *¿Qué se encuentra en 1790?*

Compañero(a): *En 1790 se encuentran la Coatlicue y la Piedra del Sol, las cuales actualmente pueden verse en el Museo Nacional.*

Principales excavaciones realizadas en el Templo Mayor

1790 El 13 de agosto y el 17 de diciembre se encuentran la Coatlicue y la Piedra del Sol, respectivamente. Actualmente pueden verse en el Museo Nacional de Antropología.

1901 Se encuentran escalinatas, una gran cabeza de serpiente y el *ocelotl-cuauxicalli*, una enorme escultura que representa un jaguar que actualmente está a la entrada de la sala Mexica del Museo Nacional de Antropología.

1913– 1914 Don Manuel Gamio excava y encuentra restos de la esquina suroeste del Templo Mayor, así como una de las cabezas de serpiente del extremo sur de la escalinata de Huitzilopochtli.

1964 Eduardo Matos Moctezuma realiza el rescate de un adoratorio decorado con mascarones del dios Tlaloc, al norte de la calle de Justo Sierra.

1978– 1982 Desde el 20 de marzo de 1978 hasta noviembre de 1982, un equipo de especialistas realizan el Proyecto Templo Mayor, que da por resultado el descubrimiento del principal templo de los mexicas y de edificios aledaños al mismo.

trescientos sesenta y tres
España y las Américas Etapa 2
363

Activity 13: Understand most spoken language when the message is deliberately and carefully conveyed by a speaker accustomed to dealing with learners when listening

GRAMÁTICA **Lo que**

▶ The relative phrase **lo que** means *what* or *that which*. You use it when there is no direct person, place or thing in the main clause to which you are referring. It refers to a more generalized idea or concept.

No comprendo **lo que** quieres decir.
*I don't understand **what** you mean.*

¿Por qué no me dices **lo que** piensas?
*Why don't you tell me **what** you think?*

▶ When you use **lo que** after **todo** it means *all that, everything that.*

Tienes que decirme **todo** lo que sabes.
*You have to tell me **everything that** you know.*

Nos dieron **todo** lo que tenían.
*They gave us **everything that** they had.*

Practice: Actividades 13 14

Más práctica
cuaderno p. 128
Para hispanohablantes
cuaderno p. 126

Online Workbook
CLASSZONE.COM

---

13 Efraín

Escuchar/Escribir Efraín hizo un viaje educativo a Guatemala para estudiar la historia maya del período clásico. Escucha lo que dice de su viaje y escribe una oración que haga un resumen de cada cosa mencionada. Sigue el modelo.

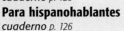

> *modelo*
>
> el período clásico de los mayas
> *Lo que aprendió fue sobre el período clásico de los mayas.*

1. las pirámides de Tikal
2. la historia maya
3. la escritura jeroglífica
4. los objetos de arte
5. la historia maya
6. unos tejidos

Nota cultural

Uno de los lugares más importantes para visitar son las ruinas de Tikal, en Guatemala. En ese sitio se levantaba la antigua ciudad maya del mismo nombre. Tikal está considerado como uno de los sitios arqueológicos más importantes de Mesoamérica. Esta imagen nos muestra el Templo del Jaguar, animal que representaba la sabiduría maya.

14 Lo que quieren

Hablar/*Escribir* Estás de viaje con tu clase de español. Todo el mundo quiere hacer algo diferente. Hablas con tu compañero(a) sobre lo que quieren hacer todos.

modelo

Ernesto prefiere ver el Templo Mayor.

Lo que Ernesto prefiere es ver el Templo Mayor.

1. Javier necesita más tiempo para ver las ruinas.
2. Elena quiere comprar una camisa bordada.
3. El Sr. Quintana quiere descifrar los jeroglíficos.
4. A Daniela le gustaría subir las pirámides.
5. Juan Felipe prefiere ir a Teotihuacán.
6. Amalia quiere ver los murales en el museo.

15 Un viaje ideal

Hablar En grupos de tres o cuatro, conversen sobre un viaje imaginario que hicieron. ¿Qué vieron? ¿Qué les gustó? Mencionen algunos elementos que aparecen en las fotos y otros que ustedes escojan.

modelo

Tú: *Cuéntenme sobre su viaje a Mesoamérica.*

Compañero(a) 1: *Vimos muchas artesanías, las cuales estaban hechas por los indígenas.*

Compañero(a) 2: *Sí, y les mandamos fotos de las pirámides a nuestros amigos.*

More Practice: **Más comunicación** *p. R16*

Online Workbook
CLASSZONE.COM

Refrán

Lo que fue y no es ya, menos es que lo que será.

¿Qué quiere decir el refrán? ¿Estás de acuerdo o no? ¿Por qué? ¿Piensas que el futuro siempre es mejor que el pasado? ¿Crees que nuestras civilizaciones van mejorándose?

Acquire knowledge and new information from comprehensive, authentic texts when reading

En colores

CULTURA Y COMPARACIONES

UN ARQUITECTO Y SUS OBRAS

PARA CONOCERNOS

STRATEGY: CONNECTING CULTURES

Use architecture as a cultural text Look at important buildings in your community. Are there any that reflect an earlier time or culture? Are there any that reflect the present or represent the future? Categorize them in the chart below. Be sure to name the buildings.

Edificios del pasado	Edificios del presente	Edificios del futuro
1.	1.	1.
2.	2.	2.
3.	3.	3.

After reading and learning more about Barragán and Legorreta, which one would you choose to design a major building for your town? Why?

Instituto Salk, Luis Barragán
(La Jolla, California)

La vegetación y el agua son componentes esenciales de la arquitectura de Barragán. La presencia del agua es especialmente expresiva en La Jolla, California. Allí, la contribución de Barragán al Instituto Salk de Louis Kahn fue la plaza creada entre las dos grandes alas[3] de hormigón[4] y separada por un canal de agua, que simboliza el límite entre el continente y el océano.

Un estilo cultural: Ricardo Legorreta

¿Qué espacios ves cuando piensas en los edificios de México? ¿Pirámides? ¿Plazas? ¿Espacios de colores vivos? El arquitecto Ricardo Legorreta usa elementos arquitectónicos[1] de éstos para crear estructuras como las que ves a la derecha.

Legorreta fue alumno de Luis Barragán (1902–1988), sin duda el arquitecto mexicano más importante del siglo XX. Barragán buscó las huellas[2] de su propia cultura para dar lugar a una obra muy personal en la que la tradición combina con la modernidad.

[1] architectural [3] wings
[2] footprints, impressions [4] concrete

Ricardo Legorreta es el arquitecto más importante de México hoy en día. Su padre apoyó el interés que sintió desde niño por los pueblos y las ciudades de México. En oposición a Barragán, Legorreta da énfasis a la belleza de las cosas ordinarias, que tras el proceso de la arquitectura se convierten en extraordinarias.

AUTOMEX, Ricardo Legorreta
(Toluca, México)
El primer gran proyecto de Legorreta fue la Fábrica de Automóviles Automex. «Cuando diseñé Automex fue como un gran grito[7]: ¡Viva México! ¡Viva Automex!»

Hotel Camino Real, Ricardo Legorreta
(Cancún, México)
En los hoteles que ha construido Legorreta, el agua parece fluir[5] a través de estos edificios y la luz solar proyecta sombras espectaculares sobre las baldosas[6]. Los tonos azulados, rosas y el amarillo tostado nos recuerdan al sol, al cielo y a las buganvillas, invitándonos a disfrutar y relajarnos.

CLASSZONE.COM
More About Spain and the Americas

¿Comprendiste?

1. ¿Qué distingue la obra de Luis Barragán?
2. ¿Qué elementos son esenciales en la arquitectura de Barragán? Da un ejemplo de un edificio donde los utiliza.
3. ¿Quén es el arquitecto más importante de México hoy en día?
4. ¿Qué sugieren los colores de Legorreta?

¿Qué piensas?

¿Qué elementos crees que comparten los edificios de Barragán y Legorreta? ¿Cuáles crees que son distintos?

Hazlo tú

¡Diseña tu casa ideal! Al pensar en el diseño, piensa en el lugar donde la construirás, si utilizarás colores vivos u oscuros, cuánta luz quieres, la forma de las habitaciones, cuántas tendrá, y si pondrás árboles y agua en los jardines. Puedes dibujar la casa o describirla. Usa estas imágenes o revistas para inspirarte.

---
[5] to flow
[6] paving stones, tiles
[7] shout

trescientos sesenta y siete
España y las Américas Etapa 2
367

ETAPA 2

Activity 4: Generally choose appropriate vocabulary for familiar topics, but as the complexity of the message increases, there is evidence of hesitation and groping for words, as well as patterns of mispronunciation and intonation

En uso
REPASO Y MÁS COMUNICACIÓN

OBJECTIVES

- Refer to people and objects
- Express relationships
- Make generalizations
- Describe arts and crafts

Now you can...

- describe arts and crafts.

To review

- direct object pronouns see p. 358.

1 ¿Qué viste?

Acabas de regresar de un viaje a México donde viste cosas muy interesantes. Tu compañero(a) quiere saber qué viste. Contéstale.

modelo

las ruinas

Compañero(a): *¿Viste las ruinas?*

Tú: *Sí, sí las vi.*   **o** *No, no las vi.*

1. el Templo Mayor
2. los jeroglíficos en el Templo de las Inscripciones
3. la Pirámide del Sol
4. un mural de Diego Rivera
5. los tejidos de las mujeres mayas
6. las ollas prehispánicas en el Museo de Antropología

Now you can...

- refer to people and objects.

To review

- indirect object pronouns see p. 360.

2 Mi amiga española

Tu compañero(a) fue a España y se hizo amigo(a) de una joven española. Tú le haces muchas preguntas. Sigue el modelo.

modelo

¿regalar / a ella?

Tú: *¿Qué le regalaste?*

Compañero(a): *Le regalé una pulsera de jade.*

1. ¿traer de España / a mí?
2. ¿mandar / a ustedes?
3. ¿pedir / a ella?
4. ¿llevar / a ella?
5. ¿dar / a ti?
6. ¿prestar / a ti?

Now you can...

• express relationships.

To review

• relative pronouns see p. 362.

❸ Natalia

Tu amiga mexicana Natalia te escribió esta carta. Complétala con las formas correctas de **el que**, **el (la) cual** o **cuyo** para aprender más sobre los mayas.

> Hola,
>
> Como me fascina la historia maya, es el tema sobre __1__ escribí para la clase de cultura. Te voy a contar algunas de las cosas que aprendí. La civilización maya, __2__ era muy avanzada, tenía un calendario muy preciso. También entendían el concepto del cero, __3__ nos sorprende porque en esos tiempos el cero todavía no se usaba en el sistema europeo. Muchas de las técnicas de los mayas se han perdido, pero la tradición de los textiles es una de __4__ no ha desaparecido. Linda Schiele, una arqueóloga __5__ estudios de los mayas son mundialmente reconocidos, ha descubierto muchas cosas fascinantes sobre esta cultura precolombina. Sin embargo, los arqueólogos no saben por completo las condiciones bajo __6__ despareció el pueblo de los mayas. Te escribo después.
>
> Abrazos,
>
> Natalia

Now you can...

• make generalizations.

To review

• uses of **lo que** see p. 364.

❹ Lo que...

Tienes un(a) amigo(a) que acaba de volver de Centroamérica. Le preguntas lo que hizo. Tu amigo(a) te contesta con muchos detalles.

modelo

¿hacer?

Tú: *¿Hiciste lo que querías hacer?*

Tú amigo(a): *Sí. Hice muchas excursiones y visité templos mayas.*

1. ¿ver?
2. ¿comprar?
3. ¿aprender?
4. ¿visitar?
5. ¿traer?
6. ¿llevar?

trescientos sesenta y nueve
España y las Américas Etapa 2
369

Activity 5: Tend to become less accurate as the task or message becomes more complex, and some patterns of error may interfere with meaning

5 Los bailes latinoamericanos

STRATEGY: SPEAKING

Discuss Latin American dance Dance is an important part of Latin culture. But how do you discuss and demonstrate musical and physical events? You can tell the place of origin (**origen**), musical terms (**compás, ritmo, melodía**), steps (**pasos**). If you can dance it, teach it to others.

En grupos de tres o cuatro, escojan el baile latinoamericano que más les interesa. Busquen información sobre el baile por Internet o en la biblioteca. Traigan un casete o un CD de música y traten de aprender a bailarlo. Cuando sepan más, hagan una presentación para la clase. Expliquen qué van a presentar y por qué lo escogieron.

modelo

Vamos a hacer una presentación sobre el merengue. Escucharemos una canción de Juan Luis Guerra. Buscamos el CD en la tienda de música. El merengue es la música de la República Dominicana.

6 En tu propia voz

ESCRITURA Escoge una civilización precolombina que te interese (los mayas, los incas, los aztecas u otra). Prepara una gráfica para organizar tus ideas en una hoja grande de papel. Escribe el nombre de la civilización en el centro. Luego añade palabras que se asocian con la primera. Estudia el modelo antes de empezar. Después de acabar con la gráfica, escribe un párrafo breve que describa la civilización que investigaste.

modelo

LOS MAYAS
1. Período Clásico: 300 - 900 d.C.
2. escritura
3. textiles
4. pirámides
5. ¿...?

CONEXIONES

Las matemáticas Los mayas tenían un sistema de matemática diferente al nuestro. Busca información sobre su sistema en la biblioteca o por Internet. Crea un póster con los «números» entre 0 y 20. Incluye también algunos números más grandes como tu fecha de nacimiento y cualquier otro número que tenga importancia para ti.

NUMERACIÓN MAYA

- CERO
- UNO
- DOS
- TRES
- CUATRO
- CINCO
- SEIS
- SIETE
- OCHO
- NUEVE

370 trescientos setenta
Unidad 5

En resumen
REPASO DE VOCABULARIO

DESCRIBE ARTS AND CRAFTS

Crafts

bordado(a)	embroidered
decorado(a)	decorated
el jade	jade
labrado(a)	worked, cut
el mural	mural
el (la) muralista	muralist
precioso(a)	precious, valuable
tallado(a)	carved
el tejido	weaving

Dances

el baile folklórico	folk dance
la bamba	dance from Veracruz
la cumbia	cumbia
la danza	dance
la habanera	habanera
el jarabe tapatío	dance from Guadalajara
el mambo	mambo
el merengue	merengue
el tango	tango

The New World

abrir el paso	to open the way
avanzado(a)	advanced
la cifra	statistic, code
la civilización	civilization
el conquistador	conqueror
la creencia	belief
el cronista	chronicler
la descendencia	descent
descifrar	to decipher
los jeroglíficos	hieroglyphics
el Nuevo Mundo	the New World
la pirámide	pyramid
precolombino(a)	pre-Columbian
reflejar	to reflect
las ruinas	ruins
la técnica	technique
el templo	temple
la tradición	tradition

REFER TO PEOPLE AND OBJECTS

Direct objects

—¿Viste el Templo Mayor cuando fuiste a México?
—Sí, lo vi.

Indirect objects

—¿Qué te puedo traer de Chiapas?
—¿Me puedes traer un tejido tradicional?

EXPRESS RELATIONSHIPS

Relative pronouns

Ésta es la Pirámide del Sol, la que visitamos cuando fuimos a México.

MAKE GENERALIZATIONS

Lo que

Para las culturas precolombinas, lo que consideramos el «descubrimiento» del Nuevo Mundo no fue un descubrimiento verdadero.

Juego

Los jeroglíficos

Usa la siguiente clave (key) para descifrar este mensaje de cuatro palabras que está escrito en jeroglíficos de una civilización desconocida.

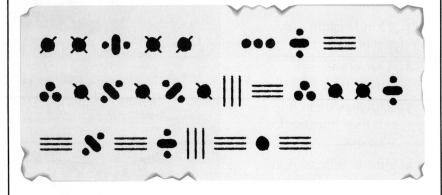

trescientos setenta y uno
España y las Américas Etapa 2
371

ETAPA
3

Lo mejor de dos mundos

OBJECTIVES

- Talk about literature

- Talk about film

- Avoid redundancy

¿Qué ves?

Mira la foto. Contesta las preguntas.

1. ¿Dónde están estas personas?

2. ¿Qué cosas están mirando?

3. ¿Quiénes de la foto se ven más interesados? ¿Por qué?

4. Mira el folleto. ¿Qué anuncia? ¿Has visto folletos así en tu ciudad?

mayo

Trayendo a los chicos y chicas contemporáneos lo mejor de su mundo

Un talento deslumbrante por Pepe A. Álvarez Gómez

LECTORES DE LA ONDA: En esta edición, les traigo una entrevista con Josefina Teresa Almodóvar Pérez, la joven **poeta, cuentista** y más recientemente **novelista** de 17 años que ha causado una sensación increíble entre los **críticos** literarios al publicar su novela *Bajo otro cielo*.

Encontré a Josefina en el balcón de su casa, descansando después del trabajo que ha **culminado** en un éxito absoluto. Josefina se ha hecho famosa de la noche a la mañana; **sin embargo,** se mantiene amistosa y no se olvida de sus compañeros de clase.

Descubre

A. Los cognados De todas las palabras en azul en el artículo, haz una lista de cognados y sus equivalentes en inglés.

B. Palabras en contexto Ahora lee el artículo. Trata de decidir el significado de las palabras que no son cognados según el contexto.

1. trama	a. praise
2. personaje	b. dazzling
3. cibernético	c. character
4. elogian	d. plot
5. dentro del alcance	e. relating to cyberspace
6. deslumbrante	f. predictable
7. amenazadores	g. within reach
8. predecible	h. threatening
9. sin embargo	i. blind
10. ciega	j. nevertheless

También se dice

Un dos por tres Josefina dice que los protagonistas no se enamoraron en **un dos por tres.** Es una manera de decir que no se enamoraron muy rápido.

La Onda: Josefina, es un placer conocerte. ¿Te gusta ser famosa?

Josefina: Pues Pepe, es **emocionante,** pero también estoy muy ocupada con las ventas de mi novela y pensando en la próxima.

La Onda: ¡En la próxima! ¿No es demasiado pronto?

Josefina: (ríe) Parece **irónico,** ¿no? Cuando escribo algo, siempre estoy pensando en otra cosa. Aunque *Bajo otro cielo* ha tenido mucho éxito y sé que los críticos la **elogian** en la prensa, estoy lista para intentar otros **géneros** y **estilos.** Además quiero seguir mejorando mi **prosa.**

La Onda: ¿Qué puedes decirnos sobre *Bajo otro cielo*?

Josefina: Es una novela **contemporánea** de **personajes** que luchan por sus vidas respectivas en lugares distintos del planeta. Margarita Buscasueños y Joaquín Esperanzado—los **protagonistas**— viven en Europa y en Estados Unidos. Parte de la **trama** se desarrolla en aeropuertos, cuando Joaquín se lleva por equivocación las maletas de Margarita y regresa a España. Al abrir las maletas, encuentra el libro de direcciones de Margarita y le escribe por correo electrónico…

La Onda: Eso de las maletas me parece muy interesante. ¿Hay algún **simbolismo** en esa confusión?

Josefina: Puede ser. Hay objetos en la vida de uno que tienen un significado especial, que llegan a ser un símbolo.

La Onda: Volvamos a Margarita, Joaquín y a los aeropuertos. ¿Le das una interpretación **innovadora** a eso de que el amor es **ciego**?

Josefina: No, Pepe. Eso sería bastante **predecible.** No, el encuentro de Joaquín y Margarita es hasta cierto punto una **sátira** de las novelas de amor populares durante el **romanticismo,** y también populares ahora, donde un chico y una chica se conocen y en un dos por tres se enamoran. Esto es un poco más complicado. Cuando Joaquín comienza a escribirle a Margarita, ella lo encuentra **amenazador;** él, al ver la reacción que ella tiene, cree que ella es paranoica.

La Onda: El **clímax** de la novela es mucho más emocionante. ¿Cómo es que Joaquín y Margarita descubren a los terroristas?

Josefina: Me gusta mucho esa parte también, pero mejor es que los lectores la lean.

La Onda: *Bajo otro cielo* es una historia que sin duda está **dentro del alcance** de muchos gustos. ¿Tienes un título para tu próxima novela?

Josefina: No sé cómo se llamará, pero tratará algún tema **cibernético** siguiendo el estilo del **realismo mágico,** que mezcla la realidad y la fantasía.

La Onda: Te acabas de ganar el Premio del Escritor Joven del Año. ¿Quisieras ganarte el **Premio Nóbel** alguna vez?

Bajo otro cielo

Josefina Teresa Almodóvar Pérez

Josefina: ¡Qué pregunta, Pepe! Pero no escribo porque quiero ganarme premios. Escribo porque me gusta, porque no me imagino la vida sin escribir.

La Onda: Creo que nuestra entrevista está llegando a su final. ¿Tienes algún consejo para nuestros lectores?

Josefina: Bueno, que disfruten la vida y que lean, claro.

La Onda: Te deseamos mucho éxito y será hasta la próxima.

Online Workbook
CLASSZONE.COM

¿Comprendiste?

1. ¿Has pensado alguna vez en hacerte famoso(a) haciendo algo creativo? ¿Qué has querido hacer?
2. ¿Cuál es el (la) cuentista, novelista o poeta que más te gusta? ¿Por qué?
3. ¿Qué preguntas le harías tú a un(a) escritor(a) famoso(a)?
4. ¿Qué clase de libros te gusta leer? ¿Cuál es tu estilo preferido?
5. ¿Crees que los temas cibernéticos van a ser más o menos importantes en el futuro? ¿Por qué?

trescientos setenta y cinco
España y las Américas Etapa 3
375

Listen to
audio texts

El club de cine

Estás en el club de cine de la escuela, donde un grupo de estudiantes se reúne para discutir ideas para la próxima película que van a hacer.

❶ Leer

En esta reunión, hablan de hacer una película sobre la novela de Josefina Teresa Almodóvar Pérez. Lee la cubierta (*cover*) del libro.

Margarita Buscasueños es una muchacha luchadora y Joaquín Esperanzado vive para un mundo mejor. Su amistad virtual comienza a través de una equivocación, cuando Joaquín se lleva las maletas de Margarita. Al principio ella cree que él es un tipo amenazador, pero luego llega a saber la verdad. Lo que comienza como conflicto termina en una gran amistad *Bajo otro cielo*.

«Drámatico. Alucinante. No me pude acostar hasta que leí la última página de *Bajo otro cielo*». Arturo Costas, *Nuestro País*

«¡Qué delicia! Por fin, una novela sin trama predecible. Cada página es una sorpresa y una lección inolvidable sobre la psicología de los ciber-fanáticos». Alma Reyes, *Ser es Leer*

Josefina Teresa Almodóvar Pérez

«¿Qué pasa cuando el realismo mágico se une con el romanticismo? *Bajo otro cielo*, ¡por supuesto! Si lee sólo un libro este año, tiene que ser éste». Rosalinda Salinas, *Opiniones Literarias*

ISBN 0-385-47137-8

9 780385 471374

Bajo otro cielo

Josefina Teresa Almodóvar Pé...

② Escuchar

Escucha la discusión del club de cine. Decide cuál(es) de estas palabras se usa(n) para describir el concepto en **negrilla**. ¡OJO! A veces se menciona más de una palabra.

1. **la protagonista:** ❏ inocente ❏ inteligente ❏ cómica

2. **el simbolismo:** ❏ profundo ❏ emocionante ❏ absurdo

3. **la sátira:** ❏ deslumbrante ❏ cómica ❏ emocionante

4. **la autora:** ❏ talento ❏ Premio Nóbel ❏ innovadora

5. **el estilo:** ❏ complicado ❏ predecible ❏ claro

6. **el género:** ❏ simbolismo ❏ romanticismo ❏ realismo mágico

7. **el final:** ❏ irónico ❏ profundo ❏ emocionante

③ Hablar/Escribir

En grupos de tres o cuatro, hagan una lista de cinco novelas favoritas del grupo. Hagan una tabla con los cinco títulos de las novelas. Bajo cada título, escriban el nombre del autor y el género. Luego, para cada novela, escriban una oración sobre una de las siguientes cosas.

- la trama
- el protagonista
- el simbolismo
- la sátira
- el estilo
- el final

Hablen entre sí sobre sus opiniones antes de escribir las oraciones finales. Compartan sus tablas con la clase.

trescientos setenta y siete
España y las Américas Etapa 3
377

En acción

PARTE A

Práctica del vocabulario

Objectives for Activities 1-4
• Talk about literature

1 La librería

Hablar/*Escribir* Vas a la librería a comprar varios libros. Le pides ayuda al dependiente. ¿Qué le dices?

modelo

Balance total *por Aída Estrada*

¡La mejor novela del año!

Busco una novela que se titula Balance total. *La novelista se llama Aída Estrada.*

1. *Avenida nueve de julio* por Marcos Ybarra

¡Una colección de cuentos inolvidable!

2. *El jaguar en mi corazón* por Sonia Cisneros

Una colección de poemas para el romántico

3. *Siete años en Costa Rica* por Andrés Gutiérrez

Con esta colección de ensayos, conozca Costa Rica por dentro y por fuera.

4. *Una vida artística: La vida de Pablo Pérez* por Amalia de la Rosa

La biografía de un artista sin límites

2 Los críticos

Escribir Cada semana lees las opiniones de los críticos en el periódico. Escoge la palabra de la lista que mejor define o explica lo que dice cada crítico.

modelo

«Es un tema que trata de la vida moderna.»

f. *contemporáneo*

a. estilo
b. romanticismo
c. realismo mágico
d. sátira
e. clímax
f. contemporáneo
g. personaje

1. «La trama se concentra en un romance misterioso.»

2. «Hay escenas que convierten la realidad en algo mágico.»

3. «La protagonista es una persona muy inteligente.»

4. «La novela pone en ridículo a los cibernéticos.»

5. «La autora escribe oraciones muy claras y sencillas.»

6. «La novela culmina en una escena muy explosiva.»

3 El grupo de lectores

Hablar/*Escribir* Tú y tu compañero(a) son miembros de un grupo de lectores que analiza una novela cada mes. Háganse preguntas y expresen sus opiniones sobre alguna novela que hayan leído recientemente.

modelo

Tú: *¿Qué pensaste de la trama de Mi doble vida?*

Compañero(a): *Pensé que la trama era predecible.*

trama (complicada, innovadora, formulista…)

estilo del autor (claro, expresivo, creativo…)

final (emocionante, irónico, predecible…)

protagonista (inocente, inteligente, valiente…)

novela (dramática, deslumbrante, contemporánea, inolvidable, impresionante…)

Vocabulario

La literatura

creativo(a) *creative*

derivado(a) *derivative, unoriginal*

dramático(a) *dramatic*

expresivo(a) *expressive*

formulista *formulaic*

impresionante *impressive*

original *original*

simbólico(a) *symbolic*

titularse *to be titled*

el título *title*

tratarse de (¿De qué se trata?) *to be about*

▶ ¿Puedes usar algunas de estas palabras para describir algo que has leído recientemente?

4 La clase de literatura

STRATEGY: SPEAKING

Discuss a novel Sharing and discussing a book can be as rewarding as reading it. Here are some topics for the discussion: **lo que me gustó más, lo que me molestó, lo que no comprendí, el personaje más interesante, el mensaje del autor, cómo se compara con otras novelas del mismo autor, el valor de la novela.** Encourage others to talk by asking their opinions.

Hablar En la clase de literatura, tú y tu compañero(a) tienen que analizar su novela favorita. Escriban cinco opiniones sobre varios aspectos de la novela: el tema, la trama, el protagonista, el clímax y el estilo.

modelo

Tú: *¿Cuál es tu novela favorita?*

Compañero(a): *Pues, me gustan muchas, pero creo que la que más me gusta es* The Great Gatsby, *de F. Scott Fitzgerald.*

Tú: *¿De qué se trata?*

Compañero(a): *Pues, el protagonista, Jay Gatsby, es un hombre muy rico pero misterioso…*

Objectives for Activities 5-14
• Talk about literature • Talk about film • Avoid redundancy

REPASO **Double Object Pronouns**

 ¿RECUERDAS? *pp. 358, 360* You have already learned to use **direct** and **indirect object pronouns** to avoid redundancy. You can also use these two kinds of object pronouns together. When you do, you put the **indirect** object **before** the **direct** object.

Ya tengo el libro de cuentos. Me lo prestó mi amiga Julia.
*I already have the book of short stories. My friend Julia lent **it** to **me**.*

Comprendemos los poemas porque nos los explicó
el profesor.
*We understand the poems because the teacher explained **them** to **us**.*

There is a special rule for verbs with two pronouns when both are **third person:** change the **indirect** object pronoun to se.

le + lo = se lo

¿Le mostraste el dibujo a Carlos?   Sí, se lo mostré.
Did you show the drawing to Carlos?   *Yes, I showed **it** to **him**.*

¿Les diste el guión a las actrices?   Sí, se lo di.
Did you give the script to the actresses?   *Yes, I gave **it** to **them**.*

Don't forget to put **object pronouns** before all **conjugated verbs** except **affirmative commands**, where you attach them. When you attach them, put an accent mark on the verb.

Mauricio quiere usar nuestro carro.   **Préstenselo**.
Mauricio wants to use our car.   *Lend **it** to **him**.*

Remember that when you use **object pronouns** with **infinitives** and the **-ndo forms**, you can put the pronouns either before or after the verb.

Me gusta esa computadora.   Me gusta esa computadora.
Me la **quiero** comprar.   Quiero **comprármela**.
*I like that computer. I want to buy **it***   *I like that computer. I want to buy **it***
*for **myself**.*   *for **myself**.*

Mamá nos los **está**   Mamá está
preparando.   **preparándonoslos**.
*Mom is preparing **them** for **us**.*   *Mom is preparing **them** for **us**.*

Practice: **Más práctica**   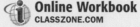 **Online Workbook**
Actividades   *cuaderno* pp. 133–134   CLASSZONE.COM
5 6 7   **Para hispanohablantes**
                     cuaderno pp. 131–132

También se dice
Hay muchas maneras de decir que algo no es interesante o que… ¡es aburrido!

• **¡Qué lata!**
  (la mayoría de los países hispanohablantes)

• **¡Qué plomo!**
  (España, Argentina)

• **¡Qué denso!**
  (Argentina)

• **¡Qué barro! / ¡Qué charro!**
  (Colombia)

• **¡Qué pejiguera! /
  ¡Qué hartero!**
  (Puerto Rico)

• **¡Qué palo! / ¡Qué rollo!**
  (España)

5 El (La) presidente(a)

Hablar/Escribir Tú eres el (la) secretario(a) del club literario. El (la) presidente(a) te pregunta si has hecho varias cosas que te había pedido. ¿Cómo le respondes?

modelo

¿Compraste el libro de poemas para Marta?

Sí, yo se lo compré. **o** *No, no se lo compré.*

1. ¿Prestaste la colección de ensayos a Miguel?
2. ¿Regalaste la biografía de Pablo Neruda a Carlos?
3. ¿Devolviste el guión a Anilú?
4. ¿Diste el libro de cuentos a Marcelo?
5. ¿Recomendaste esa novela a Juan?
6. ¿Mandaste la colección de obras teatrales a Lisa?

Nota cultural

Muchas de las películas y programas de televisión producidos en Estados Unidos son populares en los países de habla española. En algunas ocasiones, los títulos en inglés pueden ser traducidos literalmente al español. Pero en otras ocasiones las traducciones al español no son tan exactas, y a veces completamente diferentes. Por ejemplo, la película americana *It Could Happen to You*, con Nicolas Cage, Bridget Fonda y Rosie Pérez, en español se llama «La lotería del amor». Otros títulos diferentes son:

* *Ghost*: «La sombra del amor»
* *Star Trek:* «Viaje a las estrellas»
* *A River Runs Through It*: «Nada es para siempre»

6 En la cafetería

Hablar/Escribir Estás hablando con tu mejor amigo(a) en la cafetería. Usando elementos de las tres columnas, hazle preguntas sobre varias acciones. Sigue el modelo.

modelo

Tú: *¿Le diste el disco compacto a Mireya?*

Compañero(a): *Sí, se lo di ayer.*

Tú: *¿Le comentaste el simbolismo del poema a Martín?*

Compañero(a): *Sí, se lo comenté anoche.*

Acción	Objetos/ Conceptos	Persona
dar	novela(s)	¿a quién?
mostrar	colección de poemas	
prestar	disco(s) compacto(s)	
comentar	juego(s) electrónico(s)	
mandar	película(s)	
recomendar	video(s)	
hablar de	carta(s)	
devolver	tarjeta(s)	
	postal(es)	
	poema	
	trama	
	sátira	
	simbolismo	
	...	

Activity 9: Understand most spoken language when the message is deliberately and carefully conveyed by a speaker accustomed to dealing with learners when listening

7 **El cumpleaños**

Hablar Muchos de tus amigos y parientes cumplen años este mes. Conversa con tu compañero(a) sobre tus ideas para varios regalos. Menciona cinco regalos para cinco personas. Luego, cambien de papel.

modelo

Tú: *Quiero comprarle una novela latinoamericana a mi novio(a) para su cumpleaños.*

Compañero(a): *Pues, ¡cómprasela!*

Apoyo para estudiar

Negative command

Remember that in a negative command, object pronouns precede the verb. So you can advise against an action (¡No se la compre!), but you should give a reason why (porque ella prefiere la poesía).

GRAMÁTICA **Nominalization**

If you want to avoid repeating the same word over again in a sentence, you can use **nominalization.**

You can **drop the noun** if it's used with an **adjective** and use just the **adjective** and **article** instead.

drop the noun   *becomes the noun*

el libro nuevo → **el nuevo**
the new book   *the new one*

las películas cómicas → **las cómicas**
the comic films   *the comic ones*

Me gusta **la camisa amarilla**. No me pondría **la roja**.
*I like the **yellow shirt**. I wouldn't wear the **red one**.*

> You know that the **red one** is referring to the **red shirt**.

This structure also works with **indefinite articles, demonstrative adjectives,** and **numbers.**

No quiero un carro viejo. Quiero **uno nuevo**.
*I don't want an old car. I want **a new one**.*

Esas novelas realistas no son muy buenas. **Estas simbólicas** son mejores.
*Those realist novels aren't very good. **These symbolic ones** are better.*

Necesito cajas para enviar regalos. Quiero **tres cuadradas** y **dos redondas**.
*I need boxes to send gifts. I want **three square ones** and **two round ones**.*

Practice: Actividades
8 **9** **10** **11**

Más práctica
cuaderno p. 135

Para hispanohablantes
cuaderno p. 133

 Online Workbook
CLASSZONE.COM

8 Los actores

Hablar/*Escribir* Paco y Elena son dos actores que se están vistiendo para una obra. Eligen ropa distinta para sus personajes. ¿Qué dice cada uno sobre las piezas de ropa?

> **modelo**
>
> *Elena (jeans azules); yo (jeans negros)*
>
> *Elena se puso los jeans azules.*
>
> *Yo me puse los negros.*

1. Elena (zapatos marrones); yo (zapatos blancos)

2. Paco (bufanda a cuadros); yo (bufanda verde)

3. Paco (chaqueta anaranjada); yo (chaqueta amarilla)

4. Elena (blusa púrpura); yo (camisa roja)

5. Paco (chaleco de piel); yo (chaleco blanco)

6. Elena (sandalias cafés); yo (sandalias negras)

9 Ana y Manuel

Escuchar/*Escribir* Ana y Manuel han sido novios por muy poco tiempo, por eso no conocen los gustos del otro muy bien todavía. Escucha su conversación y di cuál de las cosas mencionadas prefiere cada uno.

> **modelo**
>
> *Ana: Prefiere la romántica.*
>
> *Manuel: Prefiere la cómica.*

1. Ana:
   Manuel:
2. Ana:
   Manuel:
3. Ana:
   Manuel:
4. Ana:
   Manuel:
5. Ana:
   Manuel:

Activity 11:
Read articles

⑩ Los gustos

Hablar/Escribir Acabas de conocer a un(a) amigo(a) nuevo(a). Quieres saber más sobre sus gustos. Hazle cinco preguntas a tu nuevo(a) amigo(a), y él (ella) te hace cinco preguntas a ti. Usa ideas de las dos columnas. Estudia los modelos.

modelo

Tú: *¿Cuáles botas te gustan más?*

Compañero(a): *Me gustan las negras.*

Tú: *¿Cuál película te gustó más, la de Argentina o la de España?*

Compañero(a): *Me gustó más la española.*

> ropa
> accesorios
> persona
> objeto
> obra de arte
> color
> nacionalidad
> número

⑪ Eva Luna

Leer/Hablar/Escribir Con tu compañero(a), lee el artículo sobre la novela *Eva Luna* de Isabel Allende. Juntos escriban cinco preguntas sobre el director, el cineasta, la novela, la novelista, la protagonista, el guión o el guionista. Usando sus preguntas, conversen sobre el artículo.

modelo

Tú: *¿Quién va a ser el director de* Eva Luna*?*

Compañero(a): *El británico Michael Radford.*

Tú: *¿Qué otras películas ha dirigido?*

Compañero(a): …

Radford adapta novela de Allende

EFE. Santiago de Chile.

El británico Michael Radford, director de *Il Postino*, basado en una novela del escritor chileno Antonio Skármeta, se encuentra en Chile trabajando en la adaptación de una novela de Isabel Allende al cine.

Eva Luna, la obra de la chilena Isabel Allende que relata la vida de una huérfana que de la miseria pasa a la fama y fortuna, es el proyecto del director.

«Originalmente el guión lo escribiría Laura Esquivel, pero ella no estaba disponible y yo sugerí que lo hiciera Antonio».

La cinta sobre *Eva Luna* será en inglés, «con un presupuesto de Hollywood»

y tendrá ciertas licencias sobre el escrito original a fin de «rejuvenecerlo» y «rescatar la riqueza de sus personajes distintivamente sudamericanos».

Radford sostuvo que la adaptación de la obra implica «un enorme reto por la envergadura de la historia» y anticipó que su filmación no será en Chile, debido a requerimientos del libro, que exigen paisajes selváticos que no se encuentran en el país.

More Practice: **Más comunicación** *p. R17*

Vocabulario

Las películas

el (la) **cineasta** *filmmaker*

el (la) **cinematógrafo(a)** *cinematographer*

el (la) **director(a)** *director*

dirigir *to direct*

el **guión** *script*

el (la) **guionista** *scriptwriter*

hacer el papel *to play the role*

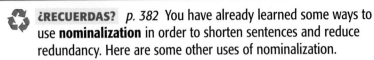

GRAMÁTICA More on Nominalization

♻ **¿RECUERDAS?** *p. 382* You have already learned some ways to use **nominalization** in order to shorten sentences and reduce redundancy. Here are some other uses of nominalization.

▶ With possessives:

la fiesta **de Ana María** ➡ **la** de Ana María
los cuentos **de este autor** ➡ **los** de este autor

Me interesan las películas de Saura, pero me gustan más **las** de **Almodóvar**.
*I'm interested in Saura's films, but I like **Almodóvar's** better.*

El bordado de tu mamá es más bonito que **el** de la señora Vélez.
*Your mother's embroidery is prettier than **Mrs. Vélez's**.*

▶ With other phrases beginning with **de**:

Hay muchos abrigos en esta tienda. **Los de cuero** son más caros que **los de lana**.
*There are many coats in this store. The leather **ones** are more expensive than the woolen **ones**.*

▶ To shorten clauses:

la fiesta **que vimos** ➡ **la** que vimos
el programa **que mencioné** ➡ **el** que mencioné

Vimos dos películas la semana pasada. **La que vimos** el lunes era mucho mejor que **la que vimos** el martes.
*We saw two movies last week. **The one** we saw on Monday was much better than **the one** we saw on Tuesday.*

▶ **Lo de** is a phrase that doesn't refer to any specific noun. You use it to mean "the matter of," "the news about," etc.

Lo del Premio Nóbel es muy interesante.
That business about the Nobel Prize is very interesting.

Practice: Actividades
(12)(13)(14)

Más práctica
cuaderno pp. 135-136
Para hispanohablantes
cuaderno pp. 133-134

 **Online Workbook**
CLASSZONE.COM

(12) Clarificaciones ♻ 👥

Hablar/*Escribir* Tu compañero(a) acaba de regresar de un viaje a México. Tú le haces muchas preguntas sobre su viaje y él (ella) te pide clarificaciones. Sigue el modelo.

modelo
subir (pirámide: está al sur / está al norte)

Tú: *¿Subiste la pirámide?*

Compañero(a): *¿Cuál? ¿La que está al sur o la que está al norte?*

1. ver (objeto de arte precolombino: está hecho de jade / está hecho de oro)
2. gustar (mural: pintó Diego Rivera / pintó David Alfaro Siqueiros)
3. visitar (ruinas: están en el centro de la ciudad / están más lejos del centro)
4. interesar (tradiciones: de los mayas / de los mexica)
5. gustar (blusa: está bordada / tiene diseños mayas)
6. gustar (CD: de música mariachi / de salsa)

trescientos ochenta y cinco
España y las Américas Etapa 3
385

Activity 14: Express and understand opinions

13 Lo contemporáneo

Escuchar/Escribir Estás en una clase en la cual van a estudiar la literatura y el cine contemporáneo de España y Latinoamérica. Escucha y escribe lo que pregunta el profesor a los estudiantes.

modelo

El profesor les pregunta a los estudiantes si prefieren estudiar los cuentos de Borges o los de Allende *.*

1. El profesor les pregunta a los estudiantes si prefieren estudiar ____.

2. El profesor les pregunta a los estudiantes si prefieren estudiar ____.

3. El profesor les pregunta a los estudiantes si prefieren estudiar ____.

4. El profesor les pregunta a los estudiantes si prefieren estudiar ____.

5. El profesor les pregunta a los estudiantes si prefieren estudiar ____.

6. El profesor les pregunta a los estudiantes si prefieren estudiar ____.

14 ¿Cuál?

Hablar/Escribir Pregúntale a tu compañero(a) sobre varios libros que ha leído y sobre algunas películas que ha visto. Luego, cambien de papel. Traten de expresar de una forma clara y precisa por qué les gustó alguna obra más que otra.

modelo

Tú: *¿Cuál te gustó más, la novela de García Márquez o la de Esquivel?*

Compañero(a): *Me gustó más la de Esquivel.*

Tú: *¿Por qué?*

Compañero(a): ...

Ideas: ✔

1	la colección de cuentos de ...	
2	la colección de poemas de ...	
3	la colección de ensayos de ...	
4	la novela de ...	
5	la autobiografía de ...	
6	la biografía de ...	
7	la película de ...	

Nota cultural

La casa de la laguna (1997) es una de las obras más famosas de la escritora puertorriqueña Rosario Ferré (1942). En su obra, Rosario Ferré observa el impacto que han tenido las relaciones de Puerto Rico con España y con Estados Unidos. Como otros escritores latinoamericanos, ella utiliza la historia y las tradiciones de Puerto Rico para darles un contexto a sus personajes.

ROSARIO FERRÉ
La casa de la laguna
GRANDES NOVELISTAS EMECÉ

Activity 15 brings together all concepts presented.

15 Comentarios literarios

Hablar/*Escribir* En grupos de tres o cuatro, lean la información que viene de la cubierta de la novela *Como agua para chocolate*. Luego, conversen sobre la autora, la novela o la película según lo que aprendieron.

modelo

Tú: *¿Leíste la novela Como agua para chocolate?*

Compañero(a) 1: *No, pero vi la película.*

Compañero(a) 2: *¿Quién es la protagonista?*

Compañero(a) 3: *Tita, la tía de la narradora.*

Tú: *¿Qué sabes de Laura Esquivel, la autora?*

More Practice:

Más comunicación *p. R17*

Online Workbook
CLASSZONE.COM

Como agua para chocolate

Como agua para chocolate, el libro de mayor venta en México en 1990, es una novela romántica e intensa, condimentada con momentos dulces y agrios. Parecida en su estructura a *How To Make An American Quilt*, a «Tampopo» en su celebración de la comida, y a «Heartburn» en su ironía y agudeza, *Como agua para chocolate* es una historia animada y divertida sobre la vida familiar en México a principios de siglo.

Tita, la tía de la narradora, es la más joven de las hijas de Mamá Elena, la tiránica dueña del rancho de la familia De la Garza. Al crecer, Tita se convirtió en una cocinera extraordinaria. Cada capítulo comienza con una receta de Tita y sus esmeradas instrucciones de preparación.

En ciertas familias mexicanas la tradición determina que la hija menor no puede casarse ya que debe permanecer en el hogar al cuidado de su madre. Tita se enamora de Pedro, pero Mamá Elena decide respetar la tradición y arregla que Pedro se case con la hermana mayor.

La voz de Laura Esquivel es directa, simple y fascinante. Ella ha escrito una novela fresca e innovadora, brindando su inimitable talento a una clásica historia de amor.

Laura Esquivel se inició como guionista. Su guión *Chido One* fue nominado al premio «Ariel» que otorga la Academia Mexicana de Ciencias y Artes Cinematográficas. Este año, la versión fílmica de *Como agua para chocolate*, arrasó con los premios ganando un total de diez «Arieles,» incluyendo el de mejor guión para Laura Esquivel, quien vive en la Ciudad de México acompañada de su esposo e hija.

ISBN 0-385-47137-8

9 780385 471374

Refrán

La imaginación hace cuerpo de lo que es visión.

¿Qué quiere decir el refrán? En tu opinión, si tienes una idea, ¿qué más necesitas para realizarla? ¿Cuál es más importante – la visión original o la imaginación para darle vida?

En voces

AUDIO

LECTURA

PARA LEER

STRATEGY: READING

Interpret a drama Reading a play requires
the interpretation of characters, their
motivations, even their movements and
gestures. In a novel those elements are often
described. Imagine yourself as the director
of «La casa de Bernarda Alba». First read
the entire scene; then read the lines of each
character separately, ignoring all others.
How would you advise each actress to play
her role?

Cómo hacer el papel de...

*   Magdalena
*   Martirio
*   Amelia

EL AMOR Y EL MATRIMONIO

buen mozo	guapo
emisario	representante
pretender	venir en busca de una novia
pretendiente	el que busca una novia
rondar la casa	visitar frecuentemente
soltero	no se ha casado
tener buenas condiciones	ser rico

Sobre el autor

Federico García Lorca nació en
Granada en 1898. Vivió durante la
Guerra Civil en España, período
turbulento, y murió a manos del
ejército del General Francisco
Franco. García Lorca es famoso por su poesía lírica y sus
obras teatrales. Tal vez sea mejor conocido por su gran
trilogía de dramas: *Bodas de sangre*, *Yerma* y *La casa de
Bernarda Alba*.

Federico García Lorca

Introducción

La casa de Bernarda Alba, «Drama de mujeres en los pueblos de España», tiene tres actos. Vas a leer unas líneas de una escena del primer acto. Tres hijas de Bernarda Alba discuten sobre el pretendiente de su hermana Angustias. Hablan Magdalena, que tiene 30 años, Martirio, de 24 años y Amelia, de 27 años.

La casa de Bernarda Alba

MAGDALENA ¡Ah! Ya se comenta por el pueblo. Pepe el Romano viene a casarse con Angustias. Anoche estuvo rondando la casa y creo que pronto va a mandar un emisario.

MARTIRIO Yo me alegro. Es buen mozo.

AMELIA Yo también. Angustias tiene buenas condiciones.

MAGDALENA Ninguna de las dos os alegráis.

MARTIRIO ¡Magdalena! ¡Mujer!

MAGDALENA Si viniera por el tipo de Angustias, por Angustias como mujer, yo me alegraría; pero viene por el dinero. Aunque Angustias es nuestra hermana, aquí estamos en familia y reconocemos que está vieja, enfermiza[1], y que siempre ha sido la que ha tenido menos méritos de todas nosotras. Porque si con veinte años parecía un palo[2] vestido, ¡qué será ahora que tiene cuarenta!

MARTIRIO No hables así. La suerte viene a quien menos la aguarda[3].

AMELIA ¡Después de todo dice la verdad! ¡Angustias tiene todo el dinero de su padre, es la única rica de la casa y por eso ahora que nuestro padre ha muerto y ya se harán particiones[4] viene por ella!

MAGDALENA Pepe el Romano tiene veinticinco años y es el mejor tipo de todos estos contornos[5]. Lo natural sería que te pretendiera a ti, Amelia, o a nuestra Adela, que tiene veinte años, pero no que venga a buscar lo más oscuro[6] de esta casa, a una mujer que, como su padre, habla con las narices.

MARTIRIO ¡Puede que a él le guste!

MAGDALENA ¡Nunca he podido resistir[7] tu hipocresía!

---

[1] sickly
[2] stick
[3] luck comes to he who least expects it
[4] the inheritance will be divided up
[5] surrounding area
[6] the least desirable
[7] to stand, put up with

¿Comprendiste?

1. ¿Cuál es la actitud de cada una de las hermanas en cuanto al novio de Angustias?
2. ¿Quién es Adela? ¿Cuántos años tiene? ¿Quién es Pepe el Romano?
3. ¿Cómo es Angustias, según Magdalena?

¿Qué piensas?

1. ¿A quién se refiere Magdalena cuando habla de la hipocresía?
2. ¿Crees que las hermanas Alba tienen motivos que no se expresan? ¿Qué podrán ser?

Hazlo tú

1. Representen la escena para la clase.
2. Imagina los motivos de Pepe el Romano en pretender a Angustias y escribe una escena entre Pepe el Romano y su mejor amigo en la cual hablan de estos motivos.

trescientos ochenta y nueve
España y las Américas Etapa 3
389

En colores
CULTURA Y COMPARACIONES

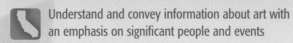

TRES DIRECTORES

PARA CONOCERNOS
STRATEGY: CONNECTING CULTURES
Reflect on the international appeal of movies

In many ways the director is the "author" of a movie using the skill and art of writers, actors, and camera person to form a finished work. What film directors can you name? Here you will read about directors of international fame. What aspects of a film help or hinder its international appeal?

Influencias	Para un interés internacional
Lengua	
Aspectos visuales	
Guión -original -basado en novela	
Renombre -de actores -de director	
¿?	

¡Luces, cámara, acción! ¿Alguna vez has querido actuar o dirigir una película? ¿Y una película en español? Entonces, te presentamos a este grupo de directores. Carlos Saura, Fina Torres y María Luisa Bemberg son tres de los directores más famosos del mundo hispanohablante. Si te gusta el cine, ¡estás en buena compañía!

Carlos Saura

Carlos Saura es el director clásico del cine español moderno. Una de sus primeras películas, *La caza* (1965), se considera el modelo del «nuevo cine español», un período entre 1960 y 1975. Otras películas importantes de Saura son *La prima Angélica* (1973), *Cría cuervos* (1975) y *Mamá cumple cien años* (1979).

La prima Angélica (1973)

Fina Torres

Fina Torres, directora venezolana de cine, ha vivido en París desde los años '70. *Oriana* (1985), su primera película, ganó el premio la Cámara de Oro en el Festival de Cannes. *Mecánicas celestes* (1996), su segunda película, es una versión moderna y romántica del cuento de la Cenicienta[1]. Fue presentada en el Festival de Sundance en 1996 y premiada como mejor película en el Festival de Cine Venezolano.

---

[1] Cinderella

Mecánicas celestes (1996)

María Luisa Bemberg

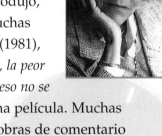

María Luisa Bemberg, directora argentina, escribió obras de teatro y guiones de película. También produjo, escribió y dirigió muchas películas: *Momentos* (1981), *Miss Mary* (1986), *Yo, la peor de todas* (1990), y *De eso no se habla* (1993), su última película. Muchas películas suyas son obras de comentario político y social, como su película más famosa, *Camila* (1984), que fue nominada al Oscar.

Miss Mary (1986)

CLASSZONE.COM
More About Spain and the Americas

¿Comprendiste?

1. ¿Qué importancia tiene *La caza* de Carlos Saura?
2. ¿Cuándo pasó el cine español por un período de transición? ¿Qué director se asocia con esta época?
3. ¿Qué tema trata la película *Mecánicas celestes* de Fina Torres?
4. ¿Qué fama tiene *Camila*? ¿Quién la dirigió?

¿Qué piensas?

1. ¿Crees que las películas deben tratar temas realistas o románticas?¿Por qué?
2. ¿Cuál de estos directores te interesa más? ¿Y de las películas?

Hazlo tú

Escribe un ensayo sobre tu director(a) favorito(a). Di qué películas ha hecho, qué premios ha ganado, cuál película te gusta más y cómo es su estilo.

 Activity 4: Generally choose appropriate vocabulary for familiar topics, but as the complexity of the message increases, there is evidence of hesitation and groping for words, as well as patterns of mispronunciation and intonation

En uso

REPASO Y MÁS COMUNICACIÓN

OBJECTIVES
- Talk about literature
- Talk about film
- Avoid redundancy

Now you can...
- talk about literature.

To review
- double object pronouns see p. 380.

1 Los regalos

Tu compañero(a) quiere saber qué les regalaste a varias personas y cuál fue la ocasión. Sigan el modelo.

modelo

Elvira (una colección de poemas / las Navidades)
Compañero(a): *¿Qué le regalaste a Elvira?*
Tú: *Le regalé una colección de poemas.*
Compañero(a): *¿Cuándo se la regalaste?*
Tú: *Se la regalé en las Navidades.*

1. Diana (un libro de cuentos / su cumpleaños)
2. Antonio (unos libros de poemas / las Navidades)
3. Marta (una novela latinoamericana / su cumpleaños)
4. Tomás (una colección de ensayos / el día de la Amistad)
5. Marcos (una biografía / la Hanuka)
6. Inés (una autobiografía / su cumpleaños)

Now you can...
- avoid redundancy.

To review
- double object pronouns see p. 380.

2 El profesor de literatura

Joaquín describe su clase de literatura. Completa las oraciones para saber qué piensa Joaquín de la clase y de su profesor.

modelo

«El profesor nos explicó el simbolismo. Nos lo explicó de una manera muy original».

1. «Nos recomendó la última novela de Laura Esquivel. _____ _____ recomendó enfáticamente».
2. «Nos mostró la cubierta. _____ _____ mostró durante la clase».
3. «Nos devolvió los exámenes. _____ _____ devolvió ayer».
4. «Me dio una nota muy buena. _____ _____ dio porque contesté todas las preguntas correctamente».
5. «Me prestó su diccionario. _____ _____ prestó porque no sabía algunas palabras».

Now you can...

• talk about literature.

• talk about film.

To review

• nominalization

• see pp. 382, 385.

③ Preferencias

Tú y tu amigo(a) tienen preferencias diferentes. Explica.

modelo

la novela romántica / la novela cómica

Yo prefiero la novela romántica. Mi amigo(a) prefiere la cómica.

1. la película romántica / la película cómica

2. la pintura realista / la pintura abstracta

3. el último libro de Esquivel / el primer libro de Esquivel

4. los poemas largos de Neruda / los poemas cortos de Neruda

5. los dos murales de Rivera / los dos murales de Orozco

6. el guión dramático / el guión cómico

Now you can...

• avoid redundancy.

To review

• nominalization

• see pp. 382, 385.

④ ¿Cuál compraste?

Fuiste de compras con tu compañero(a), pero tuviste que irte antes de que él (ella) terminara sus compras. Lo (La) ves después y quieres saber qué decidió comprar y por qué. Sigan el modelo.

modelo

¿color claro o color oscuro?

Tú: *¿Compraste el chaleco de color claro o el de color oscuro?*

Compañero(a): *Compré el de color oscuro.*

Tú: *¿Por qué lo compraste?*

Compañero(a): *Lo compré porque es ideal para la fiesta de Susana.*

1. ¿verde o azul?

2. ¿estampado(a) o a rayas?

3. ¿de tacón alto o tacón bajo?

4. ¿azul o gris?

5. ¿oro o plata?

6. ¿de lunares o de un solo color?

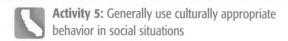

Activity 5: Generally use culturally appropriate behavior in social situations

5 Mi película favorita

STRATEGY: SPEAKING

Critique a film In what ways can you share your enthusiasm about a movie with others who may or may not have seen it? Give as much information as possible about the plot, director, actors, and script. Tell what impressed you, and tell just enough to build interest and curiosity. If everyone has seen your favorite, different opinions may be expressed.

En grupos de dos o tres, conversen sobre sus películas favoritas. Hablen sobre el director, la trama, los protagonistas, el clímax, los actores, el guión, etc. Expliquen por qué la película es su favorita.

modelo

Tú: *Mi película favorita es…*

Compañero(a): *¿Quién fue el director?*

6 En tu propia voz

ESCRITURA ¡Tú eres crítico(a) para una revista! Escribe una crítica de un libro o de una película. Antes de empezar a escribir, organiza tus ideas en un esquema. Da por lo menos tres opiniones para cada categoría.

**Nombre de la película
(o del libro)**

El (La) director(a) (o el (la) autor(a))

Los actores

Los protagonistas

La trama

El final

¿…?

TÚ EN LA COMUNIDAD

Laura es alumna en Maryland. Ayuda a su hermano, quien está aprendiendo a hablar español. Laura trabaja en un restaurante y ayuda a los clientes que están aprendiendo inglés a pedir la comida. Laura también traduce para su madre. Su madre es enfermera y frecuentemente trata con hispanohablantes en su trabajo.

En resumen
REPASO DE VOCABULARIO

TALK ABOUT LITERATURE

Literature

el clímax	*climax*
el (la) cuentista	*short story writer*
el estilo	*style*
el final	*ending*
el género	*genre*
el (la) novelista	*novelist*
el personaje	*character*
el (la) poeta	*poet*
la prosa	*prose*
el (la) protagonista	*protagonist*
la sátira	*satire*
titularse	*to be titled*
el título	*title*
la trama	*plot*
tratarse de	*to be about*

Literary criticism

amenazador(a)	*threatening*
cibernético(a)	*relating to cyberspace*
ciego(a)	*blind*
contemporáneo(a)	*contemporary*
creativo(a)	*creative*
el (la) crítico(a)	*critic*
culminar	*to end, to culminate*
dentro del alcance	*within reach*
derivado(a)	*derivative, unoriginal*
deslumbrante	*dazzling*
dramático(a)	*dramatic*
elogiar	*to praise*
emocionante	*exciting*
expresivo(a)	*expressive*
formulista	*formulaic*
impresionante	*impressive*
innovador(a)	*innovative*
irónico(a)	*ironic*
original	*original*
predecible	*predictable*
el Premio Nóbel	*the Nobel Prize*
el realismo mágico	*magical realism*
el romanticismo	*romanticism*
simbólico(a)	*symbolic*
el simbolismo	*symbolism*
sin embargo	*nevertheless*

TALK ABOUT FILM

Films

el (la) cineasta	*filmmaker*
el (la) cinematógrafo(a)	*cinematographer*
el (la) director(a)	*director*
dirigir	*to direct*
el guión	*script*
el (la) guionista	*scriptwriter*
hacer el papel	*to play the role*

AVOID REDUNDANCY

Double object pronouns

—¿Tienes la revista de arte para Marisol?
—No, ya **se la** di.

Nominalization

—¿Cuál de los libros prefieres, **el de Matute** o **el de Lorca**?
—Prefiero **el de Lorca**.
—¿Y entre las novelas contemporáneas y **las tradicionales**?
—Me gustan más **las contemporáneas**.

Juego

La creación literaria

¿Cuál de estas palabras no se relaciona con el dibujo?

a. el autor **b.** el crítico **c.** la protagonista **d.** el título

LA SELVA OSCURA

trescientos noventa y cinco
España y las Américas Etapa 3
395

Write essays
Create paragraphs when writing

En tu propia voz

ESCRITURA

Mitos, leyendas, ficciones

La Nación, un periódico español, ha organizado un concurso literario para jóvenes estadounidenses. Tienen que escribir reseñas breves de sus novelas favoritas. Las mejores reseñas se publicarán en una edición especial. Piensa en una obra que te guste. Luego escribe un resumen breve con tu propio punto de vista.

Función: Resumir una novela

Contexto: Informar a los lectores jóvenes

Contenido: Descripción de una novela

Tipo de texto: Una reseña

PARA ESCRIBIR · STRATEGY: WRITING

Support an opinion with facts and examples A good review begins with a thesis statement that clearly gives your opinion. Support that opinion with facts and examples from the work reviewed, and supply a brief plot summary for readers who have not read the work.

Modelo del estudiante

> The writer includes a **clear thesis statement**, identified as personal opinion.

La Casa en Mango Street por Sandra Cisneros

En mi opinión, la historia de Esperanza Cordero es una de las historias inmigrantes que forman parte de la narrativa nacional de Estados Unidos. La experiencia de Esperanza en un barrio hispano de Chicago es particular, también es una con la cual muchos lectores pueden identificarse.

> The writer includes a **brief summary** of the book's content.

La familia de Esperanza llega a vivir a Mango Street cuando Esperanza tiene doce años. Desde ese momento, Esperanza se ve obligada a analizar sus sueños en un ambiente nuevo. Se siente decepcionada cuando sus padres la llevan a la casa. «La casa de Mango Street no es de ningún modo como ellos la contaron».

Este tono de desilusión continúa durante la primera parte del libro. Pero, poco a poco, Esperanza descubre que la realidad que la rodea es tan interesante como su mundo interior. Observa de cerca a su familia, a sus amigos y a sus vecinos y escribe las historias que los unen a todos.

> The writer **supports his opinion** with a scene from the book.

Sus descripciones tienen un lenguaje detallado y colorido. El retrato de su casa ideal es un ejemplo: «Nuestra casa sería blanca, rodeada de árboles, un jardín enorme»... Además de las descripciones detalladas, Esperanza escribe experiencias, como la vez que se disgusta con su mejor amiga.

Para Esperanza, la casa en Mango Street es un lugar desagradable al principio. Al final, pasa a ser el lugar donde una familia hace su vida y trabaja para tener una vida mejor.

Language Arts
Writing Standard 2.2a
Write responses to literature.
Demonstrate a comprehensive understanding
of the significant ideas in works or passages.

Estrategias para escribir

Antes de escribir...

Usa la tabla de la derecha para ayudarte a
escoger una tésis. Puedes analizar la novela
que escogiste e identificar las cosas que te
gustaron (P: positivo), las que no te gustaron
(N: negativo) y las interesantes (I: interesante).
Así puedes organizar tus reacciones e
identificar una opinión que te puede servir de
tésis. Luego busca ejemplos y datos específicos
del libro que apoyen esta opinión.

Libro: *La Casa en Mango Street* por Sandra Cisneros	
P (+)	• La personalidad de Esperanza
	• La amistad de Esperanza y su mejor amiga
	• La trama
N (-)	• Los episodios a veces no dan muchos detalles
	• Todo se ve a través de los ojos de un personaje
I (¿?)	• La experiencia de cambiar de hogar
	• Una protagonista de dos culturas
	• El plan que tiene Esperanza para lograr su casa ideal
	• Las diferencias y las cosas en común que tiene la familia de Esperanza con otras familias en Estados Unidos

Revisiones

Después de escribir el primer borrador de la reseña,
pídele a un(a) compañero(a) que la lea y comente sobre
la tésis y los ejemplos que la apoyan. Tienes que seleccionar
y resumir sólo los ejemplos importantes para apoyar tu
perspectiva. Pregúntale:

- *¿Cuál es la tesis? ¿Fue fácil o difícil identificarla?*
- *¿Cuáles son los ejemplos de libro que apoyan la tésis?*
- *¿Cómo la apoyan? ¿Hay otros ejemplos que serían mejores? ¿Cuáles?*

La versión final

Para completar tu reseña, léela de nuevo y repasa los
siguientes puntos:

- *¿Usé **la nominalización** para hablar de cosas e ideas que mencioné?*

Haz lo siguiente: Subraya todos los ejemplos de la
nominalización. Luego busca la idea o cosa a la que se
refiere. ¿Usaste correctamente los artículos o pronombres?

- *¿Usé **lo** + adjetivo o **lo que** para referir a ideas o conceptos abstractos?*

Haz lo siguiente: Repasa el uso de **lo** y **lo que**. Haz un círculo
alrededor de la palabra **lo**. ¿Has usado **lo** de una manera apropiada?

> La caso en Mango St. es un
> cuento que narra la historia de
> Esperanza Cordero. Lo que me
> impresionó es que el nombre de
> Esperanza tiene mucho significado.
> ~~La chica llamada~~ Esperanza vive
> con la ilusión de vivir en una casa
> grande. La que alquilan es pequeña
>   ^
>   la familia

trescientos noventa y siete
España y las Américas Etapa 3
397

UNIDAD 6

¡YA LLEGÓ EL FUTURO!

 ## STANDARDS

Communication
- Narrating in the past
- Expressing doubt and certainty
- Reporting what others say
- Talking about television
- Talking about technology
- Stating locations
- Making contrasts
- Describing unplanned events
- Comparing and evaluating
- Expressing precise relationships
- Navigating cyberspace

Cultures
- The history and culture of Venezuela, Colombia, Ecuador, Peru, and Bolivia
- Television in the Spanish-speaking world
- Technology in the Spanish-speaking world

Connections
- Art: Designing an ad for an electronic device
- Technology: Choosing a computer system

Comparisons
- Television programming in the Spanish-speaking world and in the U.S.
- Technology in the Spanish-speaking world and in the U.S.

Communities
- Using Spanish in the workplace
- Using Spanish to help others

INTERNET Preview
CLASSZONE.COM
- More About Colombia, Venezuela, and the Andean Countries
- Webquest
- Self-Check Quizzes
- Flashcards
- Writing Center
- Online Workbook
- eEdition Plus Online

VENEZUELA, COLOMBIA Y ECUADOR
PLÁTANOS FRITOS Éste es un plato típico que sirve para acompañar casi todas las comidas de estos países. Es muy sabroso y se destaca por la combinación del sabor dulce de los plátanos con el sabor característico del aceite. ¿Qué otras comidas como ésta conoces?

ECUADOR
EL TELÉFONO CELULAR En la Etapa 2 vas a ver que el teléfono celular tiene una importancia especial para las áreas remotas de América Latina. ¿Puedes adivinar qué tipo de importancia?

POBLACIÓN: Bolivia: 8.300.463, Colombia: 40.349.388, Ecuador: 13.183.978, Perú: 27.483.864, Venezuela: 23.916.810

ALTURA: 6882 m sobre el nivel del mar (Cerro Illimani, Bolivia)

CLIMA: 84°F (29°C) Maracaibo, Venezuela, 46°F (10°C) La Paz, Bolivia

COMIDA TÍPICA: locro, llapingachos, ocopa, carapulcra, chuao, arepa, hallaca, sancocho de sábalo, rendón, asado de llama, chuñia, salteños

GENTE FAMOSA: Gabriel García Márquez (escritor), Jaime Freire (escritor), Simón Bolívar (político), Fina Torres (directora), María Reiche (estudiosa de los misterios de Nazca, Perú)

VIDEO DVD Mira el video para más información.

CLASSZONE.COM
More About Colombia, Venezuela, and the Andean Countries

COLOMBIA
PARQUE DE CIENCIA Y TECNOLOGÍA MALOKA
¿Un parque de atracciones dedicado a la ciencia y la tecnología? Lo puedes encontrar en Colombia; es el único parque de este tipo en América Latina. ¿Qué clase de atracciones crees que tiene?

VENEZUELA
SIMÓN BOLÍVAR (1783-1830) nació en Venezuela y se llama «El Libertador de América» por sus esfuerzos para la independencia de América Latina. Se compara con figuras históricas de EE.UU. ¿Quiénes serán?

PERÚ
MACHU PICCHU
Los incas usaron tecnologías nuevas para construir sus templos y ciudades. ¿Puedes pensar en otras tecnologías del pasado que cambiaron la forma de hacer las cosas?

VENEZUELA
ARMANDO REVERÓN Este artista usa la realidad y lo moderno como inspiración para sus obras de arte. En tu opinión, ¿qué comentario hace esta escultura sobre la tecnología?

6

- Comunicación

- Culturas

- Conexiones

- Comparaciones

- **Comunidades**

¡YA LLEGÓ EL FUTURO!

Comunidades

Existen muchas oportunidades de usar el español en nuestras comunidades. Piensa en las diferentes situaciones en que podrías usar el español para ayudar a otras personas.

Comunidades en acción **Explica cómo usarías el español en estas situaciones:**
- **con pacientes hispanohablantes en un hospital**
- **con turistas hispanohablantes en una tienda**
- **con niños hispanohablantes después de la escuela**

Conexiones

¿Es importante la tecnología en tu vida? ¿Cómo la usas para mejorar tu español y para aprender más de las culturas hispanas?

Comparaciones

¿Te gusta mirar la televisión? ¿Ves muchas telenovelas? Las telenovelas latinoamericanas son muy populares. Pero hay una gran diferencia entre las telenovelas latinoamericanas y las estadounidenses. ¿Sabes cuál es?

Comunicación

La comunicación es un elemento esencial de la vida diaria. Frecuentemente usamos Internet para comunicarnos con otras personas y para buscar información nueva. ¿Cómo lo usas tú?

Fíjate

Imagina tu vida en el futuro. ¿Cómo usarás el español? Piensa en tu carrera profesional y en tus pasatiempos. Haz una lista, incorporando ideas de estas páginas.

Culturas

El teléfono celular es un medio de comunicación usado en todo el mundo. ¿Crees que se usa mucho o poco en los países andinos? ¿Por qué?

¿Qué quieres ver?

OBJECTIVES

- Narrate in the past

- Express doubt and certainty

- Report what others say

- Talk about television

¿Qué ves?

Mira la foto. Contesta las preguntas.

1. ¿Qué cosas en la foto te dicen dónde están los actores?

2. ¿Cuál es la actitud del hombre? ¿Y de la mujer?

3. Mira la revista. ¿De qué crees que se trata? ¿Cómo lo sabes?

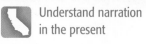
Recomendaciones para la semana

TELE-GUÍA: PROGRAMACIÓN TV-CABLE UNIVISA

Descubre

A. Las películas Las películas se clasifican en categorías como Acción, Comedia, Documental, Misterio, etc. Haz una lista de todas las categorías en la sección **En contexto**. Luego, da dos ejemplos de películas para cada categoría.

B. Índice de audiencia Las películas siempre tienen un índice de audiencia *(ratings)*. Este índice les indica a los padres si es apropiado que sus hijos vean esa película o no. Adivina qué quieren decir los siguientes índices.

1. apto para toda la familia

2. se recomienda discreción

3. prohibido para menores
  a. PG–13
  b. R–rated
  c. G–rated

DOCUMENTAL
Las Islas Galápagos: Paraíso biológico
CANAL 5   **DOMINGO 14.00h**
Viajen con un científico y una bióloga a las islas que siguen fascinando al mundo científico. **Apto para toda la familia.**

CIENCIA FICCIÓN
CINE: Los robots humanos
CANAL 6   **JUEVES 20.00h**
Unos robots extraterrestres eligen un pueblo venezolano como su próxima base. **Se recomienda discreción.**

COMEDIA
TELESERIE Los líos de Olivia
CANAL 3   **MARTES 18.00h**
En este **episodio**, Olivia hace el papel de Cupido y le presenta su mejor amiga a su agente. El instante que se conocen, ¡se odian!, pero Olivia no lo quiere dejar así. Apto para toda la familia.

MISTERIO
CINE: Mi tío, el ladrón
CANAL 3   **DOMINGO 16.00h**
El tío del narrador tiene una falta pequeña: por las noches es ladrón. Pero no sabe lo que hace –todo ocurre mientras camina dormido. Apto para toda la familia.

MISTERIO
CINE: La vida secreta del gobernador
CANAL 5   **VIERNES 22.30h**
El gobernador tiene un secreto que sólo su **guardaespaldas** sabe. ¿Qué hace el empleado leal con la información explosiva? **Prohibido para menores**.

HORROR
CINE: Vino del lago

CANAL 6   **DOMINGO 16.30h**

Un animal acuático misterioso
sale del lago para aterrorizar un
pueblo ecuatoriano. Se recomienda
discreción.

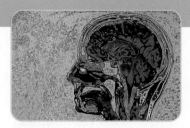

DRAMA
TELEDRAMA: El pasado perdido

CANAL 3   **DOMINGO 15.00h**

El protagonista pierde la memoria
en un accidente automovilístico.
Cuando sale del coma, no conoce
ni a su familia ni a sus amistades.

AVENTURA
Héroe sin hogar

CANAL 5   **DOMINGO 16.30h**

Un agente del gobierno descubre
prácticas ilegales. En vez de ser
elogiado, huye por su vida. Se
recomienda discreción.

ACCIÓN
Alarma nuclear

CANAL 6   **DOMINGO 14.00h**

Dos pilotos norteamericanos tienen
que desarmar unos misiles
nucleares. Prohibido para menores.

DIBUJOS ANIMADOS
Thumbelina

CANAL 5   **SÁBADO 10.00h**

Este clásico infantil es la historia
de una niña que se despierta y
encuentra que es tan pequeña como
un dedo. Apto para toda la familia.

CONCURSO
Inteligencia

CANAL 3   **LUNES 9.00h**

Tendrás la oportunidad de mostrar
tu inteligencia mientras ganas
premios valiosos.

PROGRAMA DE ENTREVISTAS
ROSIE O'GORMAN

CANAL 3   **LUNES 11.00h**

En vivo y directo desde Nueva York,
Rosie habla con los miembros de la
obra teatral «Titanic», la actriz
puertorriqueña Rita Moreno y el
cantante Luis Miguel.

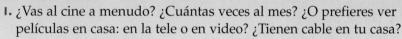
Online Workbook
CLASSZONE.COM

¿Comprendiste?

1. ¿Vas al cine a menudo? ¿Cuántas veces al mes? ¿O prefieres ver películas en casa: en la tele o en video? ¿Tienen cable en tu casa?
2. ¿Qué clase de película o programa te interesa más? ¿Por qué? ¿Hay algunos actores que se asocien con ese tipo de película o programa? Nombra dos o tres.
3. ¿Cómo se decide en tu casa qué programa van a poner? ¿Es fácil o difícil que los miembros de tu familia se pongan de acuerdo?
4. Escoge una película o teleserie de la tele-guía en esta sección y convence a tus compañeros que es el programa que deben ver. Explica por qué te parece interesante.

Listen to
audio texts

AUDIO

En vivo
SITUACIONES

PARA ESCUCHAR
STRATEGY: LISTENING

Keep up with what is said and agreed Reporting a lively conversation requires accurate listening. What does the Domínguez family finally agree to watch on television? To keep everything straight, focus only on facts:

1. ¿Qué programas discuten?
2. ¿A qué hora empiezan? ¿Hay conflictos?
3. Después de la discusión, ¿qué programa(s) escogen?

¿Qué vamos a ver?

Estás en Quito, Ecuador. Pasas un domingo con los Domínguez, una familia que conoces desde chico(a). Todos quieren ver la tele y miran la Tele-Guía para ver qué programas hay. Luego hablan de lo que quieren ver.

1 Mirar

Estudia la siguiente programación que salió en la edición del domingo en la TeleGuía.

TELE– GUÍA domingo   **C11**

Programas en la tarde   domingo, 13 de marzo

Acción
Alarma nuclear

14.00h CANAL 6

Dos pilotos norteamericanos tienen que desarmar unos misiles nucleares. Prohibido para menores.

Teledrama
El pasado perdido

15.00h CANAL 3

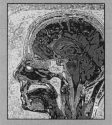

El protagonista pierde la memoria en un accidente automovilístico. Cuando sale del coma, no conoce ni a su familia ni a sus amistades.

Documental
Las Islas Galápagos: Paraíso biológico

14.00h CANAL 5

Viajen con un científico y una bióloga a las islas que siguen fascinando al mundo científico. Apto para toda la familia.

Horror
Vino del lago

16.30h CANAL 6

Un animal acuático misterioso sale del lago para aterrorizar un pueblo ecuatoriano. Se recomienda discreción.

Misterio
Mí tío, el ladrón

16.00h CANAL 3

El tío del narrador tiene una falta pequeña: por las noches es ladrón. Pero no sabe lo que hace —todo ocurre mientras camina dormido. Apto para toda la familia.

② Escuchar

Escucha la conversación entre los Domínguez. Copia la siguiente tabla y marca **sí** junto a los programas que deciden ver, y **no** junto a los que no deciden ver. Luego, escribe la hora de los programas que van a ver este domingo.

	Sí	No	Hora
acción			
ciencia ficción			
comedia			
dibujos animados			
documental			
teledrama			
horror			
misterio			

③ Hablar

En grupos de tres o cuatro, comenten sobre las decisiones de los Domínguez. Traigan una teleguía de su periódico a la clase. Comparen los programas de su periódico con los de los Domínguez. Escojan un día de programación y decidan qué verían si estuvieran en la misma situación que los Domínguez.

407

En acción

PARTE A

Práctica del vocabulario

Objectives for Activities 1–4
• Narrate in the past • Talk about television

1 ¿Por qué no...?

Hablar/Escribir Estás en Bogotá con tu compañero(a) y quieren ir al cine. Tú le sugieres una película a tu compañero(a) y él (ella) te dice si la película es buena o mala.

¡Escape!
☆☆☆

modelo

Tú: *¿Por qué no vemos esa película de acción?*

Compañero(a): *Recibió tres estrellas. Quiere decir que es una película buena.*

1/2☆ MALA	☆☆☆	BUENA
☆ REGULAR	☆☆☆☆	MUY BUENA
☆☆ INTERESANTE	☆☆☆☆☆	EXCELENTE

1.

¡Qué confusión!
☆☆☆☆

2.

Los mayas
☆☆☆

3.

La familia
☆☆

4.
LA MANO
☆

2 Quiero ver...

Hablar/Escribir Tus padres salieron y tú tienes que cuidar a tu hermanito(a) de ocho años. Él (Ella) quiere ver ciertas películas. Tú le tienes que decir si puede verlas, según los índices de audiencia.

modelo

Blancanieves (G)

Compañero(a): *Quiero ver la película Blancanieves.*

Tú: *Está bien. Puedes verla. Es «apta para toda la familia».*

1. *Aladín (G)*

2. *Los robots humanos (PG-13)*

3. *Alarma nuclear (R)*

También se dice

En español se usa la palabra **televisor** para referirse al aparato que transmite programas. La palabra **televisión** se refiere a la programación de los varios canales. Muchos hispanohablantes usan la palabra **tele** para hablar de la televisión: «Voy a ver la tele». o «Vamos a mirar la televisión». Los dos verbos, **mirar** y **ver,** pueden usarse igualmente para hablar de esta actividad.

3 La tele-guía

Hablar/*Escribir* Tu mejor amigo(a) ha venido a tu casa a pasar unas horas hablando y viendo la tele contigo. Tienen que decidir qué van a ver. Conversen sobre las selecciones en la tele-guía.

modelo

Tú: *¿Qué hay en la tele?*

Compañero(a): *A ver, hay un documental sobre los misterios del universo en el canal 2 a las 9 de la noche.*

Tú: *No quiero ver un documental. ¿Qué más hay?*

Compañero(a): *…*

4 ¡No! ¡Mejor ésta!

Hablar Con la tele-guía del periódico o la de «En contexto» elige un programa y convence a tu compañero(a) de que deben verlo.

modelo

Tú: *Esta película se ve interesante y recibió cuatro estrellas.*

Compañero(a): *¿A qué hora es?*

Tú: *A las tres de la tarde.*

Compañero(a): *¡Ay, no! Yo quería ver ésta…*

8.05

La historia del béisbol
¡desde los comienzos de este deporte excitante!

9 Muzzik

9.50 Itzhak Perlman: En casa del violinista **10.50** Mozart, por Natalie Dessay **12.30** E. Ansermet dirige la O.S.R.: Overtura op. 115 de Beethoven **12.45** E. Ansermet dirige la O.S.R.: El vals de Ravel **13.25** Cierre

50 Documanía

8.00 Microdocus: Misterios del universo **9.05** Explorer: Rastreadores de tiburones **9.50** Nova: Hawai, nacida del fuego **10.45** Explorer: Carreras de submarinos **11.30** Agencia Capa: Los marinos de Cronstadt/ La ciudad de la alegría **12.25** Explorer: Pirañas **13.20** El siglo del cine: El cine ruso según…**14.15** Explorer: Frailecillo, un pájaro viajero **15.00** Nova: Rescatando crías de ballena **16.20** Explorer: Fotógrafos de acción **17.05** Agencia Capa: Barcelona por Javier Mariscal **20.05** Explorer: Inteligencia animal **21.00** Microdocus: Misterios del universo **21.05** National Geographic **2.10** Cierre

60 Estaciones

13.15 Miles de palomas en el País Vasco 2/2 **13.55** Las crónicas de Walker's Cay: Lo mejor de Walker's Cay **14.15** Pesca mayor: Guadalupe y sus marlines azules **15.35** Terra animae: Arrecifes de coral **16.25** La gran enciclopedia de la caza: Caza y gastronomía **16.50** Temporada no 25 **17.55** El mercado del hurón **18.55** Cazas del mundo: Gansos silvestres de Patagonia **20.20** Cuando despiertan las marmotas **20.50** Serie Brasil: Pesca del tucunare Açu **21.35** La afición de la pesca: El gave se rebela **22.00** Truchas arco iris del lago Washington **22.30** Pez vela de Costa Rica **1.20** Cierre

Práctica: gramática y vocabulario

Objectives for Activities 5–16
• Narrate in the past • Express doubt and certainty • Report what others say • Talk about television

REPASO   Preterite vs. Imperfect

▶ Remember there are two different tenses to speak about the past, the **preterite** and the **imperfect**.

Use the **preterite:**   • to describe a past action with a **specific beginning and ending.**

Encendí la tele, **vi** las noticias y luego la **apagué.**
*I **turned on** the TV, **saw** the news, and then **turned** it **off.***

Use the **imperfect:**   • to talk about past actions **without saying when they began or ended.**

Eran las once de la mañana y **llovía.**
*It **was** eleven o'clock in the morning and it **was raining.***

> You use the **imperfect** to talk about time and describe weather in the past.

• to describe **habitual** or **repeated**, or two **simultaneous** actions in the past.

Cuando **éramos** chicos, **mirábamos** los dibujos animados.
*When we **were** children, we **used to watch** cartoons.*

▶ When you use the imperfect and preterite together, use the **imperfect** to tell what was **going on** in the background and the **preterite** to express **what happened.**

Cuando **vi** que **transmitían** un programa que no me **interesaba, cambié** de canal.
*When I **saw** that they **were broadcasting** a program that **didn't interest** me, I **changed** channels.*

▶ Some verbs have different meanings in the **preterite** and **imperfect.**

	preterite	imperfect
saber	supe *I found out*	sabía *I knew*
conocer	conocí *I met*	conocía *I knew, used to know (a person/place)*
querer	quise *I tried to*	quería *I wanted*
no querer	no quise *I refused to*	no quería *I didn't want*
poder	pude *I could (and did)*	podía *I was able to (but didn't necessarily do it)*
tener	tuve *I got*	tenía *I had*

Practice: **Actividades**   **5 6 7**   **Más práctica** *cuaderno p. 141*
Para hispanohablantes *cuaderno p. 139*

Online Workbook
CLASSZONE.COM

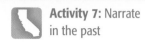
5 La escritora

Escuchar/Escribir Estás viendo un documental sobre escritores famosos. Escucha la entrevista con la escritora. Decide si ella siempre hacía las cosas indicadas, o si las hizo sólo una vez.

modelo

ir al cine los domingos
Siempre lo hacía.

1. ver una comedia nueva
2. enfermarse
3. escribir un guión
4. escribir por horas
5. decirles a sus padres que iba a ser famosa
6. darse cuenta de que iba a ser escritora

6 Cuando era niño(a)

Escribir Quieres comparar tus hábitos relacionados a la tele cuando eras pequeño(a) con tus hábitos de hoy. Escribe dos oraciones: en la primera, emplea el imperfecto y en la segunda emplea el pretérito. Sigue el modelo.

modelo

Cuando era niño(a), veía los dibujos animados el sábado por la mañana. Este sábado por la mañana vi videos musicales.

ver los dibujos animados

ir al cine a ver películas de horror

pasar horas enfrente de la tele

hacer la tarea antes de encender la tele

participar en los concursos en la tele

pelearse por el control remoto

7 El (La) guionista

Escribir Eres el (la) guionista de un documental sobre una persona que admiras mucho. ¿Sobre quién es el documental? En la primera escena cuenta la historia de su vida. ¿Cómo era cuando era joven? ¿Qué pasó para cambiar su vida? Escribe la narración de la primera escena. Usa tu imaginación.

modelo

Adela Quiñones era una muchacha sencilla, de una familia muy unida. Cuando tenía diez años, pensaba que el mundo era como una película. Pensaba que los finales siempre serían felices. Pero entonces, un día, todo cambió…

More Practice: **Más comunicación** *p. R18*

Activity 9: Express and understand opinions

| REPASO | Indicative vs. Subjunctive |

You can use the indicative or the subjunctive after some conjunctions or verbs, depending on what you want to express.

Use the indicative after these phrases:

• when the outcome of the action is **certain**.

Use the subjunctive after these phrases:

• when the outcome of the action is **uncertain** or after a **command**.

cuando   tan pronto como   en cuanto

después (de) que   hasta que

Esperaron hasta que terminó la fiesta.

They waited until the party ended.

Te **llamaré** cuando vengan nuestros amigos.

I'll call you when (as soon as) our friends come.

• after verbs and phrases that indicate certainty or opinion.

No dudo que quieren venir.

I don't doubt that they want to come.

• after verbs and phrases that indicate doubt or disbelief.

Es dudoso que puedan venir.

It's doubtful they'll be able to come.

You use the subjunctive after **aunque,** *although,* if you are **not sure** about whether the action of the subordinate clause is happening or not.

indicative: **I know it's bad.**

Tengo que salir **aunque** hace mal tiempo.

I have to go out even though the weather is bad.

subjunctive: **I'm not sure if it's bad.**

Tengo que salir **aunque** haga mal tiempo.

I have to go out even though the weather may be bad.

Practice:

Actividades 8 9 10

Más práctica *cuaderno p. 142*
Para hispanohablantes *cuaderno p. 140*

Online Workbook
CLASSZONE.COM

Nota cultural

Las telenovelas son muy populares en Latinoamérica. México, Venezuela y Argentina son los productores principales de este tipo de programación, que además de verse en su país de origen, se exporta a otros países. Las de mayor éxito pueden llegar a los canales hispanos de Estados Unidos, como por ejemplo "La venganza", Vale todo" y "Las vías del amor". Al contrario de las telenovelas americanas, la mayoría de las telenovelas en español tienen un comienzo, un desarrollo y un final, ¡generalmente feliz!

8 Mis amigos venezolanos

Hablar/*Escribir* Estás en casa de unos amigos venezolanos. Oyes comentarios sobre varias cosas. Completa las oraciones con el indicativo o el subjuntivo según el contexto.

modelo

yo: *tener el dinero*

Compraré una videocasetera tan pronto como tenga el dinero.

1. nosotros: terminar la cena
   Pondré la tele tan pronto como _____.
2. yo: encontrar el video
   Grabaré el programa tan pronto como _____.
3. nosotros: comprar el televisor
   Tendremos que pedir el servicio de cable después que _____.
4. ellos: encontrar un programa de dibujos animados
   Los niños no estarán felices hasta que _____.
5. sus padres: llegar a casa
   Apagaron el televisor cuando _____.
6. el programa: empezar
   Cambiaré de canal en cuanto _____.

Vocabulario

La televisión

la antena parabólica *satellite dish*

cambiar de canal *to change channels*

grabar *to record*

la televisión por cable *cable television*

la televisión por satélite *satellite television*

la videocasetera *videocassette recorder*

el control remoto *remote control*

► ¿Cuáles de estos aparatos tienes en casa?

9 ¡Es dudoso!

Hablar/*Escribir* Tú y tu compañero(a) nunca están de acuerdo en sus opiniones sobre las novelas que han leído o las películas que han visto. Sigan el modelo.

modelo

ser buena novelista

Tú: *Es buena novelista.*

Compañero(a): *Dudo que sea buena novelista. Sus novelas son aburridas.*

1. final ser predecible
2. tener un estilo moderno
3. escribir prosa buena
4. ser su primera película como director
5. tener que filmar esa escena otra vez
6. preparar un guión corto
7. llamar a buenos actores
8. usar música clásica
9. hacer una película de su novela
10. ¿...?

Activity 11: Understand most spoken language when the message is deliberately and carefully conveyed by a speaker accustomed to dealing with learners when listening

10 Los hermanos Saldívar

STRATEGY: SPEAKING

Negotiate Negotiation is the art of reaching an agreement in which everyone wins. In order to reach an agreement, set conditions for doing something later that both want: **Te daré el control remoto tan pronto como empiece nuestro programa favorito.**

Hablar Los hermanos Saldívar se pelean siempre porque cada uno quiere quedarse con el control remoto. Con un compañero(a), haz el papel de los hermanos. ¡Sean creativos!

modelo

Tú: *Dame el control remoto.*

Compañero(a): *Te daré el control remoto en cuanto termines la tarea.*

REPASO Reported Speech

You have learned two ways to indicate what someone is saying:

Direct quote: Carlos: «No salgo».
Carlos: "I'm not going out."

Reported Speech: Carlos **dice que** no sale.
Carlos says he's not going out.

You use the **indicative** to summarize what someone said.

- When you report what someone said (*dijo*), you use one of the **past tenses** or the **conditional**.

  Carlos dijo que **no salía.**
  Carlos said he wasn't going out.

  Carlos dijo que **no saldría.**
  Carlos said he wouldn't go out.

- When you report what someone says or is saying (*dice*), you use the **present tense**, **future tense** or **ir + a + infinitive**.

  Carlos dice que **no saldrá.**
  Carlos says he won't go out.

  Carlos dice que **no va a salir.**
  Carlos says he is not going out.

Remember that if you are using **decir** to indicate what someone tells another person to do, you use the **subjunctive** to express that idea.

Carlos **dice que** **no salgas.**
Carlos is telling you not to go out.

Practice: Actividades 12 13

Más práctica cuaderno p. 143
Para hispanohablantes cuaderno p. 141

 **Online Workbook** CLASSZONE.COM

11 Jorge

Escuchar/Escribir Escuchas la conversación de Jorge. Mientras escuchas, tu hermanita te pregunta qué dice Jorge. Más tarde, tu hermanito te pregunta qué dijo Jorge.

modelo

¿Qué dice? Dice que irá al cine esta tarde.

¿Qué dijo? Dijo que iría al cine esta tarde.

1. ¿Qué dice? / ¿Qué dijo?
2. ¿Qué dice? / ¿Qué dijo?
3. ¿Qué dice? / ¿Qué dijo?
4. ¿Qué dice? / ¿Qué dijo?

12 Tu papá dijo...

Hablar/*Escribir* Estás cuidando a tu primito(a) de nueve años y le dices las instrucciones que le dijo su papá.

modelo

Tu papá me dijo que te acostaras a las nueve.

> hacer la tarea
>
> cepillarte los dientes
>
> (no) hablar por teléfono con tus amigos antes de cenar
>
> (no) navegar por Internet
>
> (no) ver esa película prohibida para menores esta noche
>
> obedecerme todo el tiempo

13 Mi novio(a)

Hablar Hablas por teléfono con tu novio(a). Tu mejor amigo(a) quiere saber lo que te está diciendo. Cuéntale a tu amigo(a) qué te dice tu novio(a).

modelo

Compañero(a): ¿Qué dice? o ¿Qué dijo?

Tú: *Dice que vendrá a verme esta tarde. o Dijo que vendría a verme esta tarde.*

comprarme   invitarme a...

ir al (a la) ...   conseguir boletos para ...

¿...?

llevarme al (a la)...

GRAMÁTICA   **Sequence of Tenses**

You have already learned a number of indicative and subjunctive tenses in Spanish. Here's a guide to how they work together.

Indicative verb in the main clause: Subjunctive verb in the subordinate clause:

With **present** or **present perfect**	**Quiero** que **cambies** de canal. **He dicho** que no **mires** ese programa. **Me alegro de** que **hayas grabado** el programa.	use **present subjunctive** or **present perfect subjunctive**
With **preterite** **imperfect** **past perfect** **conditional**	Te **dije** que **cambiaras** de canal. **Quería** que **miraras** ese programa. Te **había dicho** que no **miraras** ese programa. **Preferiría** que no **hubieras grabado** el programa.	use **imperfect subjunctive** or **past perfect subjunctive**

Practice: **Actividades** 14 15 16   **Más práctica** *cuaderno* p. 144   **Para hispanohablantes** *cuaderno* p. 142    **Online Workbook** CLASSZONE.COM

Activity 17: Converse and listen
during face-to-face social interactions

14 Abuelo

Hablar/Escribir Tu abuelo viene de visita.
Cree que no oyes lo que te pide. ¿Qué te dice?

modelo

¡Cambia de canal!

He dicho que cambies de canal.

1. ¡Apaga la tele!
2. ¡No veas ese programa sensacionalista!
3. ¡Dame el control remoto!
4. ¡Pide el servicio de televisión por satélite!
5. ¡No manipules a tus hermanos!
6. ¡No discutas conmigo!
7. ¡Pon otro programa!
8. ¡No le hagas caso a la crítica!
9. ¡Baja el volumen!
10. ¡Tráeme palomitas de maíz!

Vocabulario

La crítica

controlar *to control*

entretenido(a) *entertaining*

influir *to influence*

manipular *to manipulate*

la percepción *perception*

el público *audience*

la reacción crítica *critical response*

sensacionalista *sensationalized*

▶ ¿Crees que la televisión nos influye mucho?
¿Por qué?

15 El (La) director(a)

Hablar/Escribir Eres crítico(a) de cine para
una revista boliviana. Le haces una entrevista
a un(a) director(a) que acaba de salir con una
película muy bien recibida. Dramatiza la
situación con tu compañero(a).

modelo

Tú: *La película le gustó al público.*

Compañero(a): *Me alegro de que la película le haya
gustado al público.*

1. La reacción de los críticos ha sido
muy positiva.
2. El público ha dicho que la película
es entretenida.
3. El público ha comparado su trabajo al
de Bergman.
4. Por fin usted ha completado su obra
maestra.
5. A usted lo han invitado al festival de
cine en Cannes.
6. La película ha recibido diez nominaciones
para el premio Óscar.

Nota cultural

Cuando le haces una invitación a otra persona, debes
usar la frase «Te invito». Por ejemplo, si haces una fiesta
en tu casa, puedes decir «Te invito a mi fiesta».
Y si quieres invitar a alguien a comer, debes decir,
«Te invito a comer…»
Pero… ¡cuidado! En los
países de habla
hispana, si «invitas» a
alguien a comer o a
tomar un refresco, tú
debes pagar.

16 El documental

Hablar/Leer Están reunidos para estudiar juntos y ven que hay un documental en la televisión. En grupos de tres o cuatro lean la descripción del documental. Hablen de por qué quisieran verlo o no usando frases como **pienso que, dudo que, tan pronto como, en cuanto, hasta que** y **después de que.**

modelo

Tú: *Pienso que si viera el documental de los osos polares, no podría terminar mi trabajo a tiempo. No lo quiero ver.*

Compañero(a) 1: *Yo lo voy a ver en cuanto termine de leer este capítulo.*

Compañero(a) 2: *No creo que puedas terminar tu trabajo sin ver el documental…*

Documentales 3

Osos polares, los señores del Ártico

El mayor carnívoro del mundo tiene su reducto en el Ártico, en las regiones polares del norte del planeta. Durante más de 100 mil años los osos polares han sido los señores supremos de esta tierra de hielo y nieve, pero su señorío acaba donde comienza la aldea de Churchill, en Canadá.

Viernes, 20.00h Canal 3

17 Mis reacciones

Hablar En grupos de tres o cuatro, hablen sobre alguna película o teleserie que hayan visto recientemente. Hablen sobre los temas en la lista y los que les interesen.

modelo

Tú: *¿Viste la película Mi tío, el ladrón?*

Compañero(a) 1: *Sí, claro, ¿quién no la ha visto?*

Compañero(a) 2: *Es una comedia muy buena.*

Compañero(a) 3: *¿Qué pensaste de…es más divertida?*

Compañero(a) 4: *¿Creo que la película…?*

el género (acción, drama, etc.)
la reacción crítica (¿cuántas estrellas?)
la percepción del público
los actores
¿…?

More Practice:
Más comunicación *p. R18*

Online Workbook
CLASSZONE.COM

Refrán

Lo futuro aún no ha llegado y lo presente es casi pasado.

¿Qué quiere decir el refrán? En tu opinión, ¿qué comentario hace sobre el tiempo? ¿Crees que el tiempo pasa rápido para personas de todas edades? ¿Pasa más rápido cuando uno es joven o mayor (o adulto)?

Read
articles

En voces

LECTURA

PARA LEER • STRATEGY: READING

Distinguish facts from interpretations Magazine articles offer both factual information and the author's interpretations. Based on your reading of *Brillo afuera, oscuridad en casa*, decide which of these statements are fact and which are interpretation.

1. El autor representa una perspectiva venezolana.
2. «Amor mío» es más popular en EE.UU. que en Venezuela.
3. La población hispana en EE.UU. ha aumentado. Por eso, la popularidad de programas hispanos ha aumentado también.
4. Tres programas venezolanos están entre los diez primeros espacios en popularidad.
5. Era inevitable que «Amor mío» tuviera éxito internacional.
6. Es mejor tener éxito internacional que éxito local.

Use the text of the article to justify your choices.

PROGRAMAS DE TELEVISIÓN

actual	en este momento, presente
brillar	iluminar
el brillo	una luz brillante
los creadores del dramático	escritores de guion
creciente	va aumentando
emanar	tener origen en
idolatrados	admirados, populares

Farándula es una revista sobre la programación de televisión en Venezuela. Esta revista también ofrece artículos sobre actores populares de América Latina, Estados Unidos y Europa.

Introducción

Vas a leer un artículo sobre una telenovela que se llama «Amor mío», uno de los programas de origen venezolano más populares en Estados Unidos.

Brillo afuera, oscuridad en casa

«Amor mío»: Si bien en nuestra tierra esta telenovela pasó por debajo de la mesa[1], en los Estados Unidos brilla como un sol, al estar en el primer lugar de los veinte programas más vistos de ese país. Astrid Gruber y Julio Pereira (sus protagonistas) consiguieron fuera el éxito que nunca encontraron en casa....

Los programas de habla hispana son cada vez más populares en Estados Unidos. Al parecer, la creciente población latina de ese país es la razón principal. Pero lo importante de todo esto es que en esta área le llevamos ventaja a muchas naciones, pues nuestra televisión es una de las más vistas.

Es así como en los actuales momentos, según el *ranking* que incluyen los veinte primeros espacios[2] de la televisión emanados de la Nielsen Hispanic Television, tenemos tres muy buenas posiciones. «Sábado sensacional», por su parte, ocupa el puesto número siete, «Maite» está en el puesto tres y, como gran victoria, encontramos a la novela «Amor mío» en el primer lugar.

Sus protagonistas, Astrid Gruber y Julio Pereira, son idolatrados (como nunca aquí) en los Estados Unidos al igual que los creadores del dramático que son Isamar Hernández, Ricardo García y Manuel Manzano.

Ahora sí pueden cantar victoria y no pasar por debajo de la mesa. Por tanto, demostramos que, una vez más, existe brillo afuera y oscuridad en casa.

---

[1] went unnoticed
[2] television programs

¿Comprendiste?

1. ¿Por qué son cada vez más populares en Estados Unidos los programas en español?
2. ¿Qué programas venezolanos son más populares entre los televidentes hispanos? ¿Cómo lo sabes?
3. Según el artículo, el éxito de «Amor mío» demuestra que «existe brillo afuera y oscuridad en casa». ¿Qué quiere decir?

¿Qué piensas?

1. ¿Qué clase de programa será «Sábado sensacional»? ¿Y «Maite»?
2. ¿Por qué es «Amor mío» más popular afuera de Venezuela?

Hazlo tú

Tú eres creador(a) de una telenovela. Escribe una escena de la telenovela. Escribe cómo se titula, de qué se trata, cómo se llaman los actores y los personajes, cuál es el tema y dónde y cuándo sucede la acción.

Activity 2: Converse in simple transactions on the phone

En uso
REPASO Y MÁS COMUNICACIÓN

OBJECTIVES

- Narrate in the past
- Express doubt and certainty
- Report what others say
- Talk about television

Now you can...

- narrate in the past.

To review

- preterite vs. imperfect, see p. 410.

1 El viaje inolvidable

Viste una película venezolana en la tele titulada *El viaje inolvidable*. Completa la narración de la película para saber cómo se sintió la protagonista de la película. Usa el pretérito o el imperfecto de los verbos entre paréntesis.

Cuando __1__ (ser) chica, mis padres __2__ (decidir) que __3__ (ir) a vivir en Venezuela. ¡Yo __4__ (estar) aterrorizada! No __5__ (saber) hablar español y tampoco __6__ (querer) dejar atrás mis amiguitos de clase.

__7__ (Ser) un día muy bonito cuando nos __8__ (despedir) de los vecinos y de la familia. Yo __9__ (estar) muy triste. No __10__ (poder) imaginar cómo __11__ (ir) a ser mi nueva vida.

Pero en fin, no __12__ (tener) de qué preocuparme. Me __13__ (encantar) Venezuela y ahora como adulta no puedo imaginar cómo hubiera sido mi vida sin ese viaje inolvidable.

Now you can...

- express doubt and certainty.

To review

- indicative vs. subjunctive, see p. 412.

2 La conversación telefónica

Hablas con un(a) amigo(a) colombiano(a) por teléfono. Completa sus oraciones para saber qué te dijo. Usa el indicativo o el subjuntivo de los verbos entre paréntesis.

modelo

«Te llamé tan pronto como supe que iban a dar esa película que querías ver. ¿Por qué no hacemos planes para ir a verla?» (yo: saber)

1. «¡Hola! Mis primos van a venir tan pronto como _____ la tarea. ¿Quieres venir?». (terminar)

2. «¡Espera! Primero tenemos que comer así que iremos al cine después de que _____ el almuerzo. ¿Luego te llamo?». (tomar)

3. «¡Hola, Julio! Invita a tus hermanos. No dudo que _____ venir». (querer)

4. «¡Buenas tardes, Sra. Díaz! Nos quedaremos en el centro comercial hasta que (usted) _____ por nosotros». (venir)

5. «Mamá nos llevará a la discoteca después de que nos _____. ¿Te vea a las?». (reunir)

6. «¡Hola, Luis! La semana pasada, fuimos a comer después de que _____ de la discoteca. ¿Quieres ir de nuevo.». (nosotros: salir)

Now you can...

• talk about
  television.

To review

• sequence of tenses,
  see p. 415.

③ Entre hermanos

Mario y Miguel son hermanos. ¡Siempre se pelean! Completa su diálogo con la forma correcta de los verbos entre paréntesis para ver quién se quedó con el control remoto.

Mario: Quiero que __1__ de canal, por favor. (cambiar)

Miguel: Pero yo quiero ver este programa.

Mario: ¡Te dije que __2__ de canal! (cambiar)

Miguel: ¡Yo tengo el control remoto y digo que no!

Mario: He dicho que me __3__ el control remoto. (dar)

Miguel: Preferiría que te __4__ a tu cuarto. (ir)

Mario: Voy a llamar a papá.

Miguel: Llámalo. He dicho que te __5__ a tu cuarto. (ir)

Papá: Bien. Denme el control remoto. No va a haber televisión por tres días.

Mario: ¡Te dije que __6__ caso! (hacerme)

Now you can...

• report what others
  say.

To review

• reported speech,
  see p. 414.

④ El (La) hermano(a) mayor

Los padres de Antonio salieron a cenar con unos amigos y lo dejaron a él en casa. Ahora él te cuenta todo lo que dijo su mamá.

modelo

«Mamá dijo que no trabajara en la computadora».

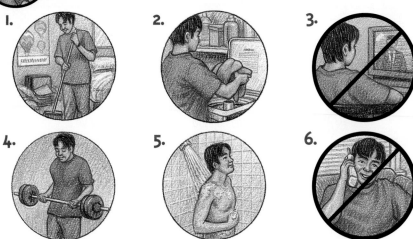

Activity 5: Generally choose appropriate vocabulary for familiar topics, but as the complexity of the message increases, there is evidence of hesitation and groping for words, as well as patterns of mispronunciation and intonation

5 Programas de ayer y hoy

STRATEGY: SPEAKING

Retell memories Reminiscences are memories of the past relived in the present. Tell about your childhood preferences and experiences relating to television. **(Me encantaban los dibujos animados, pero mis padres me decían que no podría verlos antes de acostarme.)** Then contrast those memories with your current tastes, experiences, and what others say to you about them. **(Anoche los vi otra vez. No me gustaron tanto.)**

En grupos de dos o tres, conversen sobre los tipos de programas que les gustaban cuando eran chicos. Compárenlos con los que les gustan ahora.

modelo

Tú: *Cuando era chico(a), me encantaban los dibujos animados.*

Compañero(a) 1: *Y, ¿ahora?*

Tú: *Ahora prefiero las películas de acción. El otro día vi una película que se titulaba…*

6 En tu propia voz

ESCRITURA Vas a inventar tu propia película o teleserie. Primero decide las siguientes cosas:

- ¿género? (aventura, acción, etc.)
- ¿título?
- ¿trama?

Ahora escribe una descripción breve de la película o teleserie. Cuando estés satisfecho(a) con tu descripción, haz un póster de tu programa. Usa fotos o dibujos para ilustrar tu concepto. Si quieres, puedes escribir un lema *(slogan)* para el programa o película.

TÚ EN LA COMUNIDAD

Lucille es alumna en Florida. Habla español con su familia. También hablaba español en su trabajo de voluntaria en un hospital. Cuando no había enfermeras que hablaran español, ella traducía. Lucille siempre ofrece su ayuda cuando está en una tienda y ve a turistas hispanohablantes que no hablan inglés.

En resumen
REPASO DE VOCABULARIO

Equipment

la antena parabólica	satellite dish
cambiar de canal	to change channels
el control remoto	remote control
grabar	to record
la televisión por cable	cable television
la televisión por satélite	satellite television
la videocasetera	video cassette recorder

Programs

los dibujos animados	cartoons
el documental	documentary
en vivo y directo	live programming
el episodio	episode
el (la) guardaespaldas	bodyguard
el programa de acción	action program
el programa de ciencia ficción	science fiction program
el programa de concurso	game show
el programa de entrevistas	talk show
el programa de horror	horror program
el programa de misterio	mystery program
la tele-guía	television guide
el teledrama	mini-series
la teleserie	TV series

Reactions

apto(a) para toda la familia	G-rated
controlar	to control
entretenido(a)	entertaining
influir	to influence
manipular	to manipulate
la percepción	perception
prohibido(a) para menores	R-rated
el público	audience
la reacción crítica	critical response
se recomienda discreción	PG-13 rated
sensacionalista	sensationalized

Preterite vs. imperfect

Eran las seis de la tarde cuando **nos sentamos** para ver el documental.

Subjunctive vs. indicative

Veremos la teleserie en cuanto **lleguen** los abuelos.
Vimos los dibujos animados tan pronto como **llegaron** los primos.

Reported speech

Mamá **nos dijo que el documental empezaba** a las diez.
Mamá **nos dijo que no miráramos** la película prohibida para menores.

Juego

¿Qué vamos a ver?

Completa las frases con las palabras apropiadas. Luego pon en orden las letras de los círculos para saber qué es lo que todo el mundo quiere tener cuando mira la tele.

A Marisol le gustan mucho las sorpresas y lo desconocido. Por eso, siempre mira los ___ __○—○

___ ___ ○○ ___ ○ ○ ___

A Jorge le gusta obtener nueva información. Así que él prefiere ver ○○ __ ___ ○ ___ ___

○○○ ___ ○ ___ ___.

Respuesta: __ __ __ __ __ __ __

__ __ __ __ __ __

ETAPA

2

Aquí tienes mi número...

OBJECTIVES

- Talk about technology

- State locations

- Make contrasts

- Describe unplanned events

¿Qué ves?

Mira la foto. Contesta las preguntas.

1. ¿En qué lugares puedes encontrar esta clase de tienda?

2. ¿Qué cosas comprarías?

3. ¿Cómo crees que se sienten estos amigos?

4. Mira el anuncio. ¿Conoces otras tiendas que vendan lo mismo?

425

En contexto

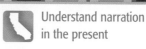

Understand narration in the present

VOCABULARIO

Descubre

En ElectroMundo Estás en el almacén ElectroMundo y escuchas estas conversaciones. ¿Puedes adivinar qué quieren decir las palabras en azul?

1. Tengo un radio **portátil** y una computadora **portátil**. Los puedo llevar adondequiera.

2. Es difícil decidir qué **marca** debo comprar. ¡Hay tantas **marcas** de todos los productos!

3. Dice el anuncio que están ofreciendo un **descuento** de 10 por ciento. ¡Puedo ahorrar mucho dinero!

4. Quiero un teléfono **inalámbrico** porque quiero hacer mis llamadas de cualquier sitio.

5. Compré este modelo de fax por su **durabilidad**. Tengo un amigo que ha tenido el mismo fax por cinco años.

6. Esta computadora tiene una **garantía** de dos años. Si le pasa algo dentro de dos años, me la componen gratis.

   a. *cordless*
   b. *brand*
   c. *portable*
   d. *guarantee*
   e. *discount*
   f. *durability*

ELECTROMUNDO

Donde la tecnología existe solamente para ti. ¿Quieres tener a la mano la última tecnología personal? ¡Ven a ElectroMundo! Puedes contar con la durabilidad de las marcas más conocidas y los descuentos más bajos. Ofrecemos garantías de un año para todos los productos electrónicos. ¡Ven hoy a ElectroMundo! ¡Hay una tienda accesible cerca de ti!

Radio portátil

14.000 B
Ahora 12.790 B

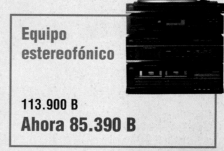

Equipo estereofónico

113.900 B
Ahora 85.390 B

Walkman™ con audífonos

25.600 B
Ahora 20.500 B

Altoparlantes

102.500 B
Ahora 82.000 B

¡REBAJA!

Videocámara

284.800 B
Ahora 242.100 B

Televisor portátil

56.900 B
Ahora 48.400 B

Teléfono celular

19.900 B
Ahora 17.900 B

Contestadora automática

(Espacio para dos horas de **telemensajes**)
69.000 B
Ahora 48.400 B

Teléfono con identificador de llamadas

153.800 B
Ahora 115.300 B

Teléfono inalámbrico

22.700 B
Ahora 15.900 B

Pilas/Baterías

2.800-8.500 B
Ahora 2.200-6.800 B

Asistente electrónico

512.600 B
Ahora 435.800 B

Beeper

42.700 B
(servicio17.000 B/mes)
Ahora 36.300 B
(servicio 14.500 B/mes)

Fax multi–funcional

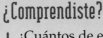

(fax, impresora y fotocopiadora)
125.000 B
Ahora 112.800 B

Computadora portátil

1.196.000 B
Ahora 957.012 B

¡Todos los sábados, descuentos del 10% al 30% para ciertos modelos! Precios razonables todos los días.

**Centro Comercial Chacaíto,
Chuao, Caracas
tel. 02/959-3824**

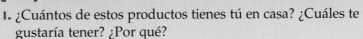

¿Comprendiste?

1. ¿Cuántos de estos productos tienes tú en casa? ¿Cuáles te gustaría tener? ¿Por qué?
2. ¿Tienes una computadora portátil? Si tuvieras una, ¿adónde la llevarías?
3. ¿Qué tres productos electrónicos han cambiado tu vida? Di cómo era tu vida antes de haber comprado cada producto y cómo cambió después de haberlo comprado.
4. ¿Crees que es importante siempre comprar la marca más conocida de un producto? ¿Por qué?
5. Algunos creen que la vida se ha complicado mucho con la tecnología personal. Hay otros que piensan que se ha hecho más sencilla y más fácil. ¿Qué opinas tú? Explica.

AUDIO

En vivo

SITUACIONES

PARA ESCUCHAR

STRATEGY: LISTENING

Analyze the appeal in radio ads Commercials contain a double appeal: one is monetary, the other psychological. First listen to and identify what is being sold and at what discount. Then listen again to determine in what other ways the ad appeals to potential customers. Check which elements you believe are present:

	sí	no
Comunicación con otros		
Conveniencia personal		
Necesidad personal		
Personas famosas lo usan		
Reputación del producto		

Which ad appeals to you most? Why?

¡Grandes rebajas!

Estás en Caracas, Venezuela y ves este anuncio en el periódico. Luego, escuchas un anuncio de radio para saber más sobre los descuentos que se ofrecen.

❶ Leer

Quieres comprar algunos productos electrónicos, pero primero quieres saber qué descuentos se ofrecen. Lee el anuncio de ElectroMundo y haz una lista de todos los productos en el anuncio.

ELECTROMUNDO

¡Venga el sábado para ahorrar con nuestros súper descuentos! Ofrecemos descuentos sensacionales de ciertos productos populares.

¡Escuche WVZA-109 FM a las diez de la mañana para saber cuáles son y cuánto va a ahorrar!

❷ Escuchar

Pones la radio a las diez de la mañana y escuchas el anuncio
del almacén ElectroMundo. Si el anuncio ofrece un descuento
para el producto, escribe **sí** y el porcentaje del descuento en
la lista que hiciste en la Actividad 1. Si el anuncio no ofrece
un descuento para ese producto, escribe **no** junto
a ese producto.

❸ Hablar

En grupos de tres o cuatro, comenten sobre los productos
electrónicos que usan todos los días. ¿Cuánto les costaron?
¿Dónde los compraron? ¿Recibieron un descuento? ¿Hay
algunos que siempre han tenido? Inventen un producto
electrónico que no existe todavía pero que tendría usos
importantes para el estudiante de hoy.

En acción

Práctica del vocabulario

Objectives for Activities 1–3
• Talk about technology

1 La tecnología personal

Escribir Estás en Colombia en una fiesta con tus amigos. Todos comentan las ventajas de sus productos electrónicos. Completa sus comentarios.

a. baterías
b. teléfono celular
c. computadora portátil
d. altoparlantes
e. Walkman™
f. grabadora
g. contestadora automática

1. «Viajo mucho y necesito trabajar en el avión. Por eso compré una _____.»

2. «Cuando hago ejercicio, me gusta oír música. Siempre llevo mi _____ al gimnasio.»

3. «El radio no funciona. Creo que tengo que comprar unas _____.»

4. «Tengo un equipo estereofónico pero me faltan los _____.»

5. «Para mi proyecto quiero entrevistar a los miembros de mi familia. ¿Me prestas tu _____?»

6. «Viajo mucho por la noche en el coche. Por eso quiero comprar un _____, en caso de emergencia.»

2 ¡Voy a ElectroMundo!

Hablar/*Escribir* Quieres ir a ElectroMundo para aprovechar (*take advantage of*) los descuentos. Invita a tu compañero(a). Dile qué vas a comprar y por qué necesitas ese producto. Luego, cambien de papel.

modelo

Tú: *Voy a ElectroMundo. ¿Quieres ir conmigo?*

Compañero(a): *Sí, claro. ¿Qué vas a comprar?*

Tú: *Me gustaría comprar una videocámara. Quiero grabar mi fiesta de cumpleaños.*

1.

2.

3.

4.

5.

6.

3 El (La) vendedor(a)

Hablar Estás en un almacén que vende productos electrónicos. Quieres comprar un identificador de llamadas. Tu compañero(a) hace el papel del (de la) vendedor(a) y trata de convencerte de que compres El Óptimo. Luego, cambien de papel.

modelo

Tú: *Quiero comprar un identificador de llamadas.*

Compañero(a): *Déjeme mostrarle El Óptimo de ComSinc.*

Tú: *¿Cuáles son las ventajas de Él Óptimo?*

Compañero(a): …

Vocabulario

De compras

la confiabilidad *reliability, dependability*

convencer *to convince*

devolver *to return something*

distinguir entre *to distinguish between*

equivocarse *to make a mistake*

estar descompuesto(a)/roto(a) *to be broken*

(no) funcionar *to (not) work*

fijarse en *to notice*

inigualable *unequalled*

la nitidez *clarity, sharpness*

respaldado(a) *supported (by); backed (by)*

tomar en cuenta *to take into account*

> ¿Cuáles de estas palabras puedes usar para decir lo que más te importa al momento de comprar?

ComSinc anuncia
El Óptimo

¿Quieres saber quién te llama cuando no estás en casa? ¿O saber quién es antes de contestar el teléfono? Con el servicio Identificador de Llamadas ComSinc tendrás la seguridad de siempre saber quién llama.

CAPACIDAD

- Registra el número de quien llama, su nombre, la fecha y la hora
- Guarda hasta 60 números
- Puedes borrar los números en la memoria de uno en uno o todos a la vez
- Se usa con cualquier sistema telefónico

TECNOLOGÍA

- Tecnología FLEX.
- La pantalla electrónica con mayor nitidez.
- Y a un PRECIO accesible.

Todas estas ventajas, respaldadas por el servicio, confiabilidad y profesionalismo que sólo ComSinc le puede ofrecer.

Ventas: • D.F.: 221-7569 • Guadalajara: 667-0568
• Monterrey: 310-0700 • Llamada sin costo: 01-800-92-304 00
ó consulte a su Distribuidor Autorizado.

ComSinc®

Práctica: gramática y vocabulario

REPASO   Conjunctions

 ¿RECUERDAS? *pp. 162, 190, 234* You have already learned some **conjunctions** you can use to relate events in time or to express cause-and-effect relationships.

▶ Always use the subjunctive after these **conjunctions** listed in the vocabulary box at the right.

▶ You can use either the subjunctive or the indicative after the following **conjunctions**:

**cuando   en cuanto   hasta que
tan pronto como**

• Use the subjunctive if the outcome of the action is **uncertain** or after a **command**.

Lo veré cuando entre.
*I'll see him when **he comes in.***

Dime cuando entre.
*Tell me whenever **he comes in.***

• Use the indicative if the outcome of the action is **certain**.

Lo vi cuando entró.
*I saw him when **he came in.***

▶ You use the indicative after **aunque** *(although, even though, even if)* if the action of the subordinate clause is a **fact**. Use the subjunctive if it is just a **possibility**.

It is a possibility that it may be expensive.

Voy a comprar esa computadora **aunque** sea cara.
*I'm going to buy that computer **even if it's expensive.***

It is a fact that it's expensive.

Voy a comprar esa computadora **aunque** es cara.
*I'm going to buy that computer **although it's expensive.***

Vocabulario

 Ya sabes

a menos que
antes (de) que
con tal (de) que
en caso (de) que
para que
sin que

**Practice:
Actividades**
4 5 6

Más práctica
cuaderno p. 149
Para hispanohablantes
cuaderno p. 147

 Online Workbook
CLASSZONE.COM

---

4 ¿Para qué?

Hablar/Escribir Tu hermanito(a) siempre quiere saber para qué quieres ciertas cosas. Dile tus razones. Primero tu compañero(a) hará el papel de tu hermanito(a). Luego, cambien de papel.

modelo

videocámara: grabar la boda de nuestro primo

Tú: *¿Para qué me das la videocámara?*

Compañero(a): *Para que grabes la boda de nuestro primo.*

1. Walkman™: escuchar música mientras hacer ejercicio

2. teléfono celular: llamar del carro si hay una emergencia

3. contestadora automática: recibir mensajes

4. identificador de llamadas: saber quién llama sin contestar

5. radio portátil: oír música en la playa

6. computadora portátil: trabajar en el avión

7. baterías: escuchar la radio en el parque

8. teléfono: invitar a Diana al cine

9. discos compactos: poner música alegre

10. disquetes: guardar el archivo electrónico

5 Aunque

Hablar/Escribir Estás seguro(a) de que quieres comprar ciertos productos aunque existan razones para no comprarlos. Di que vas a comprar cuatro productos y menciona la razón contraria. Usa la conjunción **aunque** en tus oraciones.

modelo

Voy a comprar un teléfono celular aunque no lo necesite.

no necesitarlo(la)	no ofrecer descuento
ser caro(a)	no tener garantía
ya tener uno (una)	no ser la marca que
no gustarme	quiero
	¿...?

6 ¿Lo vas a comprar?

Hablar Estás de compras en una tienda de equipo electrónico con un(a) amigo(a). Tu amigo(a) no ha decidido qué va a comprar. Conversen sobre el producto y usen las frases de la lista.

modelo

Tú: *¿Vas a comprar la computadora portátil?*

Compañero(a): *Sí, la voy a comprar a menos que cueste demasiado.*

a menos (de) que   con tal (de) que

antes (de) que   sin que   aunque

para que   en caso (de) que

More Practice: **Más comunicación** *p. R19*

REPASO **Prepositions and Adverbs of Location**

You can indicate one object's relation to another by using **adverbs** and **prepositions** of location.

Use **de** only when the phrase is followed by a specific location.

> *Not clear where, just inside.*

Adverb: Mi hermano salía de la casa mientras yo todavía estaba **dentro**.
*My brother was leaving the house while I was still **inside**.*

> *Exact: Inside the house.*

Preposition: El televisor está **dentro de** la casa.
*The television set is **inside** the house.*

Practice: Actividades
7 8 9

Más práctica
cuaderno p. 150
Para hispanohablantes
cuaderno p. 148

Online Workbook
CLASSZONE.COM

Vocabulario

Adverbs/Prepositions of location

afuera *outside*
al lado (de) *next to*
atrás *in back, behind*
detrás (de) *behind*
enfrente (de) *in front of*

♻ Ya sabes

abajo	encima (de)
debajo (de)	frente (a)
delante (de)	fuera (de)
dentro (de)	junto (a)

cuatrocientos treinta y tres
Colombia, Venezuela y los países andinos Etapa 2
433

Activity 10: Understand most spoken language when the message is deliberately and carefully conveyed by a speaker accustomed to dealing with learners when listening

7 En mi cuarto

Hablar/Escribir Le vas a prestar varios objetos a tu mejor amigo(a). Le tienes que decir dónde están esos objetos porque va a ir a recogerlos cuando tú no estás. ¿Qué le dices?

modelo

La video cámara está delante del televisor.

8 ¿Dónde está?

Hablar/Escribir Tú y tu compañero(a) quieren saber dónde están varias personas, cosas y mascotas. Háganse preguntas y contesten lógicamente.

modelo

Tú: *¿Dónde está tu gato?*

Compañero(a): *Está afuera.*

Tú: *Y tu perro, ¿dónde está?*

Compañero(a): *Está dentro de la casa.*

¿adentro?   ¿atrás?   ¿afuera?

¿abajo?   ¿...?   ¿al lado?

9 Perdido

Hablar Nunca puedes encontrar tus cosas. Le preguntas a tu hermano(a) si ha visto ciertas cosas. Tu compañero(a) hace el papel de tu hermano(a). Pregúntale dónde están cuatro objetos perdidos y luego cambien de papel.

modelo

Tú: *He buscado por dondequiera y no puedo encontrar mi asistente electrónico. ¿Lo has visto?*

Compañero(a): *Sí, claro, lo vi por la mañana.*

Tú: *¿Dónde lo viste?*

Compañero(a): *Estaba debajo del periódico en la mesa del comedor.*

GRAMÁTICA Pero vs. Sino

▶ You know that the word **pero** is usually the equivalent of the English conjunction *but*. However, there is another word in Spanish, **sino**, that also means *but*. It is used in situations where the idea being conveyed is *not this,* **but** *rather that.*

No vamos a comer carne, **sino** pescado.
*We're not going to eat meat, **but** (rather) fish.*

No debes vender la computadora vieja, **sino** repararla.
*You shouldn't sell the old computer, **but** (rather) fix it.*

▶ You can also use **sino** with:

no sólo… **sino también**…
*not only… **but** also…*

Compró **no sólo** una computadora portátil, **sino también** un teléfono inalámbrico.
*He bought **not only** a laptop, **but** a cordless phone **too.***

▶ When there is a **conjugated verb** in the second part of the sentence you use **sino que** instead of **sino.**

No sólo escribe cartas **sino que también manda** correo electrónico.
*He doesn't only write letters **but he** also **sends** e-mail.*

No vendí la computadora vieja, **sino que** la **reparé**.
*I didn't sell the old computer, **but** (instead) **I fixed** it.*

Practice:
Actividades
(10) (11) (12)

Más práctica
cuaderno p. 151
Para hispanohablantes
cuaderno p. 149

Online Workbook
CLASSZONE.COM

---

10 Alma

Escuchar/*Escribir* Alma salió con Juan anoche. Escucha su descripción de lo que pasó y decide si la palabra que le falta a las oraciones es **pero** o **sino**.

modelo

Alma iba a salir con Julio **pero** *cambió de opinión y salió con Juan.*

1. Alma y Juan no fueron al cine _____ al restaurante.

2. Alma no comió pescado _____ carne.

3. Alma pidió un café _____ le trajeron un refresco.

4. Alma pidió un postre _____ no le gustó.

5. Juan no pagó con tarjeta de crédito _____ en efectivo.

6. Juan no dejó propina _____ le dio las gracias al camarero.

También se dice
Hay varias maneras de hablar de las computadoras en los países hispanohablantes.
- **computador** (Latinoamérica)
- **computadora** (Latinoamérica)
- **ordenador** (España)

Activity 12: Express and understand opinions

11 ¿Qué vieron por fin? ♻ 👥

Hablar/*Escribir* Tu amigo(a) te pregunta qué viste anoche en la tele. Dramaticen la situación y cambien de papel.

modelo

acción

Compañero(a): *¿Viste el programa de acción?*

Tú: *Sí, lo vi. No, no vi el programa de acción sino el de misterio.*

1. misterio / ¿?
2. drama / ¿?
3. comedia / ¿?
4. documental / ¿?
5. ciencia ficción / ¿?
6. concurso / ¿?
7. programa de entrevistas / ¿?
8. dibujos animados / ¿?
9. la telenovela / ¿?
10. ¿…?

12 ¿Qué regalo compraré?

Escribir Estás tratando de decidir si debes comprarte un teléfono celular u otro producto electrónico. Lee el anuncio y decide si lo vas a comprar o no. Escribe cuatro oraciones explicando por qué no te lo debes comprar.

modelo

Quiero comprarme un teléfono celular, pero de veras no lo necesito.

No quiero comprarme un teléfono celular sino…

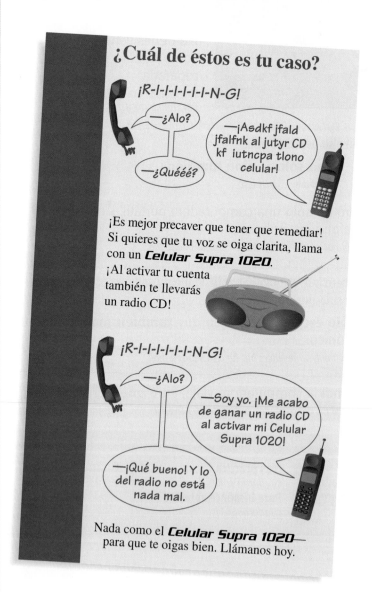

Se for Unplanned Occurrences

You can use a special construction with **se** to indicate that an action was unplanned or unexpected.

Se le cayeron los libros.
She (He, You form.) dropped the books.

Se me rompieron los anteojos.
I (accidentally) broke my glasses.

Se nos acaba la leche.
We're running out of milk.

- Notice that the **verb** is always in either the third-person singular or third-person plural.

- You use an **indirect object pronoun** to say to whom the action occurred.

- To emphasize this relationship, you can also add a phrase consisting of **a + the person** (noun or pronoun).

  A mí se me perdieron los audífonos y a Luisa **se le** perdieron las pilas.
  I lost my earphones and Luisa lost her batteries.

Vocabulario

Unplanned events

acabársele (a uno) *to run out of*

caérsele (a uno) *to drop*

descomponérsele (a uno) *to break down, to malfunction*

ocurrírsele (a uno) *to dawn on, to occur to*

olvidársele (a uno) *to forget*

perdérsele (a uno) *to lose*

quedársele (a uno) *to leave something behind*

rompérsele (a uno) *to break*

▶ ¿Te ha pasado alguna de estas cosas recientemente?

Practice:

Actividades
13 **14** **15**

Más práctica *cuaderno p. 152*
Para hispanohablantes *cuaderno p. 150*

Online Workbook
CLASSZONE.COM

Nota cultural

En Latinoamérica y España, los programas de concursos son muy populares y cuentan con miles de televidentes. Algunos de los programas más populares son «Sorpresa y media» (Argentina), «La noche del domingo» (Argentina) y «¿Quiere cacao?» (Colombia; «cacao» quiere decir «ayuda».) Generalmente los participantes tienen que cantar, responder a preguntas o participar en diversos concursos para ganar los premios

cuatrocientos treinta y siete
Colombia, Venezuela y los países andinos Etapa 2
437

Activity 15: Use strings of related sentences when reading

13 Las excusas

Escuchar/*Escribir* Hubo una fiesta y todos explicaron luego por qué no pudieron ir. Escribe las excusas de cada persona.

modelo

Marcela no pudo ir a la fiesta porque se le descompuso el coche.

1. Joaquín no fue a la fiesta porque _____.
2. Norma nunca llegó a la fiesta porque _____.
3. Arturo llegó tarde porque _____.
4. Sandra no fue a la fiesta porque _____.
5. Ileana nunca llegó porque _____.

14 Lo inesperado

Hablar/*Escribir* A menudo hay ocasiones en las cuales ocurren cosas inesperadas y uno tiene que explicar por qué no ha ocurrido lo que todo el mundo esperaba. Ahora tú estás en esa situación. Explica qué pasó.

modelo

¡Ay! ¡No te compré un regalo porque se me olvidó que hoy era tu cumpleaños!

perder	acabar
• las llaves	• la leche
• los libros	• la gasolina
descomponer	**olvidar**
• la computadora portátil	• comprar un regalo
• el equipo estereofónico	• mandar una tarjeta

15 ¡Perdona!

STRATEGY: SPEAKING

Make excuses When making an apology, you have the choice of giving a short answer: **Se me olvidó**. But consider telling the whole story that led up to your having forgotten: **Todo empezó cuando fui al centro.**

Hablar A menudo tenemos que pedir perdón cuando no cumplimos nuestras promesas. Con tu compañero(a), inventa cuatro situaciones en las cuales tienes que pedir perdón.

modelo

Tú: *¿Por qué no me llamaste anoche?*

Compañero(a): *¡Se me acabaron las baterías del teléfono celular! Y cuando por fin llegué a casa, ya era muy tarde.*

Activity **16** brings together all concepts presented.

16 El fax multifuncional

Hablar/*Escribir* En grupos de tres o cuatro, conversen sobre los modelos de fax diferentes que se ofrecen en el anuncio. Escojan el fax que comprarían si pudieran. Hagan una lista de las ventajas y desventajas.

modelo

Tú: *Yo creo que debemos comprar el fax que tiene copiadora a color.*

Compañero(a) 1: *¡Eso no importa! Lo que importa es que tenga un centro de mensajes.*

Compañero(a) 2: *En la tienda de enfrente venden muchos.*

Es un tren...
Es un avión...
¡Es... FAX ZIP!

El fax más rápido del mundo

FAX MULTIFUNCIONAL 3X2Y
• Función computarizada de teléfono

Fax Zip cuenta con todos los avances tecnológicos que harán de su fax un placer

▶ Impresora de alta resolución–ino más cartas ilegibles!

▶ Centro de mensajes–iningún recado perdido!

▶ Escáner–iimágenes claras!

▶ Función computarizada de teléfono–iuna conexión asegurada y rápida!

FAX MULTIFUNCIONAL 7Z9T
• Impresora/copiadora a colores
• Centro de mensajes

FAX CONTESTADOR 9P3S
• Impresora/copiadora a colores
• Centro de mensajes
• Función computarizada de teléfono

FAX MULTIFUNCIONAL 4J6P
• Impresora/copiadora a colores
• Centro de mensajes

Llame a su distribuidor más cercano para conseguir un catálogo.

More Practice: **Más comunicación** *p. R19*

Online Workbook
CLASSZONE.COM

Refrán

No hay cosa de más saber que a sí mismo conocer.

¿Qué quiere decir el refrán? En tu opinión, ¿qué es más importante, saber muchos datos o conocerse a sí mismo? ¿Por qué?

cuatrocientos treinta y nueve
Colombia, Venezuela y los países andinos Etapa 2
439

En colores
CULTURA Y COMPARACIONES

PARA CONOCERNOS

STRATEGY: CONNECTING CULTURES

Survey technology in daily life How common is the use of personal technology in your life? Do an informal survey in your class or school to determine what percentage of those you survey have and use a cellular phone. Document the range of specific uses among those you interview. Show the proportion of users and non-users in a pie chart and rank order the uses from the most common to the least common. How do these uses compare with those mentioned in *¿Un aparato democrático?* Which of the uses would have the most appeal in a commercial ad?

¿Un aparato democrático?

A todas horas y en todo lugar se puede contar con el timbre[1] del más moderno de los aparatos telefónicos— el teléfono celular. En muchos países de Sudamérica, este aparato está convirtiéndose en un elemento esencial de la comunicación… y no hay duda de que el uso del celular aumentará porque los precios están bajando, la competencia está aumentando y los servicios que se ofrecen están multiplicándose.

---

[1]ring

Este aparato tiene mucha importancia en Sudamérica, sobre todo en los países andinos, porque el terreno [2] de estas regiones remotas ha dificultado mucho la expansión del sistema tradicional de teléfonos. El servicio telefónico en casa es caro, a veces no muy bueno y difícil de conseguir.

El teléfono celular transmite mensajes por medio de las ondas [3] de radio. Los abonados [4] usan aparatos móviles que les dan la oportunidad de comunicarse con cualquier otro usuario [5] desde cualquier lugar. Así que es posible que el teléfono celular pueda lograr [6] lo que hasta ahora nadie ha podido hacer— ¡democratizar las comunicaciones y unir zonas remotas del continente!

El teléfono celular ofrece muchas posibilidades la participación en la red mundial [7] de comunicaciones. No sorprende que en América Latina el teléfono celular sea un servicio que todo el mundo desea tener.

---

[2] terrain, landscape
[3] waves
[4] subscribers
[5] users
[6] to achieve
[7] World Wide Web

CLASSZONE.COM
More About Colombia, Venezuela, and the Andean countries

¿Comprendiste?

1. ¿Por qué está aumentando el uso del teléfono celular en América Latina?
2. ¿Qué ventajas tiene el teléfono celular sobre el teléfono tradicional?
3. ¿Qué es posible que pueda lograr el teléfono celular? ¿Por qué?

¿Qué piensas?

1. ¿Puedes pensar en otros tipos de tecnología que sirven para unir regiones remotas?
2. ¿Cuáles son otros obstáculos físicos que la tecnología tal vez pueda superar (*overcome*)?

Hazlo tú

Busca información sobre el uso del teléfono celular en Estados Unidos y haz una presentación de lo que descubras.

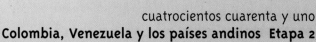

cuatrocientos cuarenta y uno
Colombia, Venezuela y los países andinos Etapa 2
441

Activity 4: Tend to become less accurate as the task or message becomes more complex, and some patterns of error may interfere with meaning

En uso
REPASO Y MÁS COMUNICACIÓN

Now you can...

• talk about technology.

To review

• conjunctions, see p. 432.

OBJECTIVES

• Talk about technology
• State locations
• Make contrasts
• Describe unplanned events

1 ¡Sí, claro!

Tu compañero(a) quiere saber si vas a comprar varios productos electrónicos. Tú le dices que sí, bajo ciertas condiciones. ¿Qué le dices?

modelo

computadora portátil (en cuanto yo tener el dinero)

Compañero(a): *¿Vas a comprar una computadora portátil?*

Tú: *Sí, claro, en cuanto tenga el dinero.*

1. contestadora automática (cuando tener mi propio teléfono)
2. beeper (para que tú poder encontrarme a cualquier hora)
3. asistente electrónico (para que yo no olvidar tu número de teléfono)
4. teléfono celular (en caso de que haber una emergencia)

Now you can...

• state location.

To review

• prepositions and adverbs of location, see p. 433.

2 El cuarto de Gloria

El cuarto de Gloria es un desastre. Su mamá entra a su cuarto y no puede creer lo que ve. ¿Qué le dice a Gloria?

modelo

las camisetas

¡Ay, hija! ¡Las camisetas están encima de la computadora!

1. los zapatos
2. los periódicos
3. los discos compactos
4. la raqueta de tenis
5. las pelotas de tenis
6. la guitarra

Now you can...

• make contrasts.

To review

• **pero** vs. **sino**, see p. 435.

③ CompuVisión

Ricardo fue con su mamá a CompuVisión para hacer unas compras.
Completa sus oraciones con **sino**, **pero** o **sino que** para saber cómo les fue.

modelo

No fuimos a ElectroMundo __sino__ a CompuVisión.

1. No fuimos por la mañana _____ por la tarde.
2. Yo iba a comprar una computadora portátil _____ no lo hice.
3. Mamá quería un televisor portátil _____ no tenían la marca que quería.
4. No compramos el teléfono celular _____ el inalámbrico.
5. Papá no fue con nosotros _____ se quedó en casa.
6. Yo quería comprar muchas cosas _____ no tenía suficiente dinero.

Now you can...

• describe unplanned events.

To review

• **se** for unplanned occurrences, see p. 437.

④ El aeropuerto

La familia Núñez acaba de llegar al aeropuerto para empezar sus
vacaciones en Perú. ¡Pero a todos se les olvidó algo! ¿Qué se les olvidó?
Luego, di qué va a pasar.

modelo

señor Núñez

Al señor Núñez se le olvidó el pasaporte. No va a poder viajar.

1. señores Núñez

2. AEROLÍNEA señor Núñez

3. Miguelito

4. Arturo y Alejandra

5. señora Núñez

6.  Miguelito

⑤ El almacén

STRATEGY: SPEAKING

Consider the factors for or against an electronic purchase For some purchases, it is useful to plan what you want to know before talking with the salesperson. Think about your personal needs and finances, and ask for information about these: **las mejores marcas, diferencias entre marcas, descuentos, rebajas, garantía, durabilidad, ventajas y desventajas de uso.**

En grupos de tres o cuatro, dramaticen la siguiente situación. Dos o tres amigos van al almacén a comprar varios productos electrónicos. Tienen muchas preguntas sobre los varios modelos y marcas. El (La) dependiente contesta sus preguntas y trata de convencerlos que compren un producto bastante caro.

modelo

Vendedor(a): *¿En qué puedo servirles?*

Compañero(a) 1: *Pues, yo buscaba un televisor.*

Vendedor(a): *Muy bien. Aquí tenemos el modelo Q2010…*

Compañero(a) 2: …

⑥ *En tu propia voz*

ESCRITURA Escoge un producto electrónico o inventa uno. Luego, escribe un anuncio para el periódico que convenza al público que es el mejor modelo de ese tipo que se vende hoy día. Asegúrate que contestaste las siguientes preguntas en tu anuncio.

• ¿Qué hace el producto?

• ¿Para quién es el producto?

• ¿Por qué es el mejor?

• ¿Cuáles son las ventajas de este modelo?

• ¿Es el precio razonable?

• ¿Se ofrece un descuento?

CONEXIONES

El arte Usando la información que sacaste de la Actividad 6, diseña un anuncio para un producto electrónico. Toma en cuenta que un anuncio debe tener mucho impacto y no puede tener demasiado texto. ¿Cuáles son los puntos más importantes que quieres comunicar? Tal vez tendrás que cortar el texto que escribiste. ¿Necesitas un lema publicitario? ¿Qué tipo de foto o dibujo necesitas para acompañar el texto? Haz un póster o diseña el anuncio en la computadora si prefieres. Piensa bien en la importancia de los colores y el diseño que escoges. Explícale a la clase por qué elegiste los elementos en tu anuncio.

Surfea la red por medio de la **Onda Cibernética**

El mejor proveedor de Internet en Caracas

En resumen
REPASO DE VOCABULARIO

TALK ABOUT TECHNOLOGY

Equipment

el altoparlante	speaker
el asistente electrónico	electronic assistant
los audífonos	headphones
la batería	battery
el beeper	beeper
la computadora portátil	laptop computer
la contestadora automática	answering machine
el equipo estereofónico	stereo equipment
el fax multifuncional	multifunctional fax machine
el identificador de llamadas	caller identification
la pila	battery
el radio portátil	portable radio
el teléfono celular/ inalámbrico	cellular / cordless telephone
el telemensaje	voice mail
el televisor portátil	portable television
la videocámara	videocamera
el Walkman™	walkman™

Shopping

accesible	available, accessible
la confiabilidad	reliability,
convencer (zo)	to convince
el descuento	discount
devolver (ue)	to return
distinguir entre	to distinguish
la durabilidad	durability
equivocarse	to make a mistake
estar descompuesto(a)	to be broken
estar roto(a)	to be broken
fijarse en	to notice
la garantía	guarantee
inigualable	unequalled
la marca	brand name
la nitidez	clarity, sharpness
(no) funcionar	to (not) work
respaldado(a)	backed (by)
tomar en cuenta	to take into account

MAKE CONTRASTS

 Ya sabes

a menos que	unless
antes (de) que	before
con tal (de) que	as long as
cuando	when
en caso (de) que	in case (of)
en cuanto	as soon as
hasta que	until
para que	so that
tan pronto como	as soon as

STATE LOCATIONS

Adverbs/prepositions of location

afuera	outside
al lado (de)	beside, next to
atrás	in back, behind
detrás (de)	behind
enfrente (de)	facing

Ya sabes

abajo	below
alrededor (de)	around
debajo (de)	underneath
delante (de)	in front of
dentro (de)	inside
encima (de)	on top of
frente (a)	facing
fuera (de)	outside of
junto (a)	next to

DESCRIBE UNPLANNED EVENTS

Accidents

acabársele (a uno)	to run out of
caérsele (a uno)	to drop
descomponérsele (a uno)	to break down, to malfunction
ocurrírsele (a uno)	to dawn on, to occur to
olvidársele (a uno)	to forget
perdérsele (a uno)	to lose
quedársele (a uno)	to leave behind
rompérsele (a uno)	to break

Juego

Sopa de letras
¡Tengo mucha energía pero nunca hago ejercicio! ¿Qué soy?

A U N A B
E N Í
T A A R

ETAPA
3

¡Un viaje al ciberespacio!

OBJECTIVES

- Compare and evaluate

- Express precise relationships

- Navigate cyberspace

¿Qué ves?

Mira la foto. Contesta las preguntas.

1. ¿Qué tipo de lugar es este?

2. ¿Qué clase de productos crees que tienen?

3. ¿Qué cosas crees que podrá hacer el chico de la mochila roja?

4. ¿Qué te dice la página-web de los servicios que ofrecen en este sitio?

CYBER CAFE

Monkey on line
Internet- phone

HAPPY HOUR
1/2 Price 9 to 10 am
1 hour 10,000

E-Mail Gratis

Tfno USA
1000/min

Premio sorpresa
Si traes un AMIGO

Happy Hour 5 to 6 pm

Foofix and the Monkey join together
Watch the World Cup at Monkey on...
Happy Hour: 9 to 10 am

Regresar	Adelantar	Inicio		Recargar	Imágenes		Abrir		Imprimir		Buscar		Finalizar

Dirección: http://www.altesa.net/monkeyonline

¿Novedades?	¿Interesante?	Búsqueda	Directorio	Software

CYBER CAFE
MONKEY on line

INFORMACIÓN IMPORTANTE

Artesanía

Espectáculos

"El mono" será tu guía.

Venga a la selva cibernética
del mono y descubra un
mundo lleno de aventura.

Agencias de viaje

español

Aprenda español

Agencias de viaje   Aprenda español   Artesanía   Espectáculos

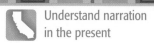
■————— J O V E N E T

Descubre

A. **Las relaciones entre palabras**
Adivina el significado de las palabras
en **azul** basándote en el significado
de los verbos.

   1. usar: **usuario**
   2. ampliar: **ampliable**
   3. configurar: **configuración**
   4. calcular: **hoja de cálculo**
     a. spreadsheet
     b. user
     c. expandable
     d. configuration

B. **Cognados falsos** En español hay
palabras que se parecen mucho a
palabras en inglés, pero que no
tienen el mismo significado. ¿Puedes
adivinar qué significa **la red
mundial**? ¡No es un color!

Perfil

de nuestros lectores #11: Jimena Villarroel

Hablamos con Jimena, una joven estudiante peruana, sobre su computadora personal.

¡Hola! Yo soy Jimena Villarroel, una estudiante de colegio en Lima, Perú. Me fascinan las computadoras— ¡mi cuarto es un verdadero laboratorio de computación! Tengo todo ordenado según mi propia configuración. También tengo mi propia página–web en la red mundial. Si quieren, ¡escríbanme para decirme qué piensan de mi sistema! Mi dirección electrónica es jimenav@colcen.edu.

Creo que nosotros, los usuarios, debemos tener los últimos y mejores programas de software disponibles para aprovechar los sistemas. Yo tengo:
• Programas anti-virus
• Procesador de texto
• Hojas de cálculo
• Base de datos

HARDWARE

• Doce discos
• 256 MB de memoria, ampliable a 1 GB
• Conexión a Internet de alta velocidad
• Tarjeta de sonido
• Tarjeta gráfica
• Micrófono Multimedia

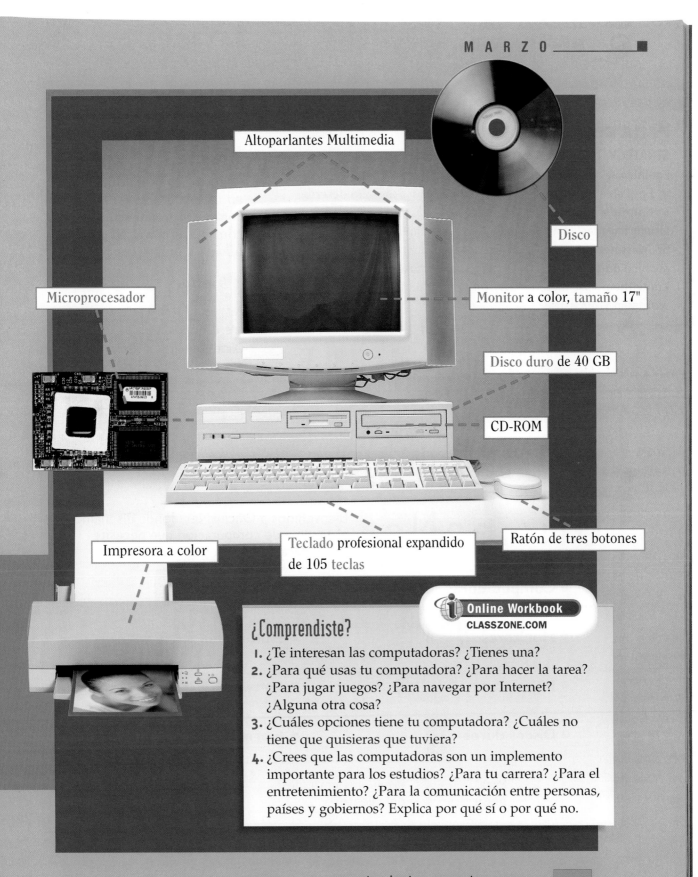

Altoparlantes Multimedia

Disco

Microprocesador

Monitor a color, tamaño 17"

Disco duro de 40 GB

CD-ROM

Impresora a color

Teclado profesional expandido de 105 teclas

Ratón de tres botones

¿Comprendiste?

1. ¿Te interesan las computadoras? ¿Tienes una?
2. ¿Para qué usas tu computadora? ¿Para hacer la tarea? ¿Para jugar juegos? ¿Para navegar por Internet? ¿Alguna otra cosa?
3. ¿Cuáles opciones tiene tu computadora? ¿Cuáles no tiene que quisieras que tuviera?
4. ¿Crees que las computadoras son un implemento importante para los estudios? ¿Para tu carrera? ¿Para el entretenimiento? ¿Para la comunicación entre personas, países y gobiernos? Explica por qué sí o por qué no.

cuatrocientos cuarenta y nueve
Colombia, Venezuela y los países andinos Etapa 3
449

Listen to
audio texts

AUDIO

En vivo
SITUACIONES

PARA ESCUCHAR
STRATEGY: LISTENING

Pre-listening Check your computer vocabulary by making a list in English and Spanish of terms you already know to describe a computer system.

Identify important computer vocabulary Listen and note the items that Sr. Martínez mentions. (See **Escuchar**.) Then verify whether they are also on your list. If not, add them because you will need to know them. Use a dictionary to find any remaining English words on your list.

El mejor sistema

Eres el (la) asistente del señor Martínez, un hombre de negocios con una oficina en La Paz, Bolivia. Él quiere comprar un sistema de computación nuevo para la oficina y quiere tus consejos.

❶ Leer

Para poder darle buenos consejos a tu jefe, decides visitar la página-web de OficinaNet, una compañía que se especializa en sistemas de computación.

| Regresar | Adelantar | Inicio | Recargar | Imágenes | Abrir | Imprimir | Buscar | Finalizar |

Dirección: http://www.OficinaNet.com

| ¿Novedades? | ¿Interesante? | Búsqueda | Directorio | Software |

OficinaNet

Bienvenidos a OficinaNet, donde diseñamos el sistema de computación que necesite usted para su oficina.

- **Computadoras**
- **Monitores**
- **Impresoras**
- **Discos duros**
- **Módem/fax**
- **Multimedia**
- **Software**
- **Juegos**

Para pedir · ¿?Preguntas comunes del cliente · Servicios para el cliente

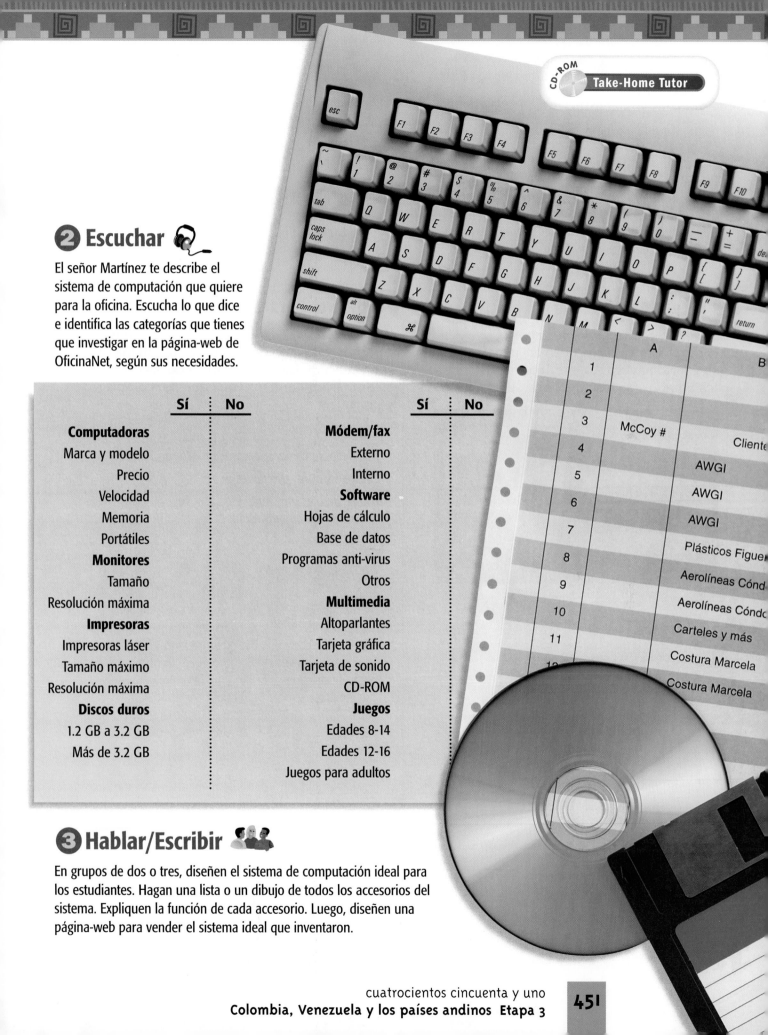

❷ Escuchar 🎧

El señor Martínez te describe el sistema de computación que quiere para la oficina. Escucha lo que dice e identifica las categorías que tienes que investigar en la página-web de OficinaNet, según sus necesidades.

	Sí	No		Sí	No
Computadoras			**Módem/fax**		
Marca y modelo			Externo		
Precio			Interno		
Velocidad			**Software**		
Memoria			Hojas de cálculo		
Portátiles			Base de datos		
Monitores			Programas anti-virus		
Tamaño			Otros		
Resolución máxima			**Multimedia**		
Impresoras			Altoparlantes		
Impresoras láser			Tarjeta gráfica		
Tamaño máximo			Tarjeta de sonido		
Resolución máxima			CD-ROM		
Discos duros			**Juegos**		
1.2 GB a 3.2 GB			Edades 8-14		
Más de 3.2 GB			Edades 12-16		
			Juegos para adultos		

❸ Hablar/Escribir 👥

En grupos de dos o tres, diseñen el sistema de computación ideal para los estudiantes. Hagan una lista o un dibujo de todos los accesorios del sistema. Expliquen la función de cada accesorio. Luego, diseñen una página-web para vender el sistema ideal que inventaron.

cuatrocientos cincuenta y uno
Colombia, Venezuela y los países andinos Etapa 3
451

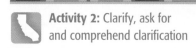

Activity 2: Clarify, ask for and comprehend clarification

En acción

Práctica del vocabulario

Objectives for Activities 2–3
• Compare and evaluate

1 El cuarto de Esteban

Escribir A Esteban le encanta su computadora y siempre lo encuentras en su cuarto navegando por Internet. Escribe los nombres de los objetos de su escritorio.

También se dice

Hay varias palabras que se usan en vez de **altoparlante**, entre las cuales están:

• **bocina** (Colombia)

• **altavoz** (Venezuela)

• **caja acústica** (España)

• **gabinete acústico** (México)

• **bafle** (Argentina)

• **parlante** (general)

2 Quiero comprar...

Hablar/*Escribir* Tú y tu compañero(a) quieren comprar varios accesorios para su sistema de computación. Usando palabras de las dos columnas, pregúntense qué quieren comprar y por qué.

modelo

Tú: *Necesito comprar un módem externo.*

Compañero(a): *¿No tienes uno?*

Tú: *Sí, pero es muy lento y para navegar por Internet necesito uno más rápido.*

módem externo
monitor
computadora
  portátil
disco duro
programa anti-virus
hoja de cálculo
juego interactivo
CD-ROM
tarjeta de sonido
tarjeta gráfica
¿...?

muy lento
pantalla más grande
viajar
más memoria
proteger mis programas
para las matemáticas
para divertirse
usar un programa
  de multimedia
¿...?

3 **La nueva computadora**

Hablar Estás pensando que te quieres comprar la nueva computadora de Xilo. Tu compañero(a) quiere saber por qué. Conversen sobre las ventajas de la computadora según el anuncio.

modelo

Tú: *Me interesa mucho la nueva Xilo.*

Compañero(a): *¿Ah, sí? ¿Por qué?*

Tú: *Pues, mira el anuncio. Es un equipo totalmente multimedia.*

Compañero(a): …

Con la NUEVA
computadora

Conecta
a tu familia

$13,999.00*

Porque la nueva computadora de **XILO** es la computadora para el hogar, que ha logrado integrar todo lo que cada miembro de tu familia estaba esperando.

La Computadora de **XILO** es mucho más que una computadora. Es un equipo totalmente multimedia, capaz de reconocer la voz, tomar llamadas y recados con alta fidelidad y consultar información por Internet. Con el microprocesador podrás jugar en tercera dimensión, consultar alguna de sus múltiples enciclopedias o trabajar en su procesador de textos, su hoja de cálculo o cualquiera de sus herramientas de productividad. No le falta nada.

XILO creó esta computadora porque piensa en tu familia.

XILO
Computadoras

*Precio en pesos más I.V.A. Cambios sujetos a la fluctuación del dólar.

Nota cultural

Hace veinte años, solamente las grandes instituciones de gobierno en Latinoamérica tenían computadoras. Éstas se encontraban en grandes cuartos con aire acondicionado y eran del tamaño de máquinas industriales. La revolución de la computadora personal cambió esto. Ahora, no sólo las oficinas de gobierno sino también las empresas privadas, y sobre todo las escuelas, pueden tener una o varias computadoras. Poco a poco se han creado cursos de computación en las escuelas privadas, públicas y más recientemente, en las escuelas primarias.

cuatrocientos cincuenta y tres
Colombia, Venezuela y los países andinos Etapa 3
453

Objectives for Activities 4–14
• Compare and evaluate • Express precise relationships • Navigate cyberspace

---

REPASO **Comparatives and Superlatives**

▶ You use the words **más, menos,** and **tan** to make **comparisons.**

- **más** + **adjective** or **adverb** + **que**
  more… than

  > Este programa es **más interesante** que el otro.
  > *This program is **more interesting than** the other one.*

- **menos** + **adjective** or **adverb** + **que**
  less… than

  > Este programa es **menos interesante** que el otro.
  > *This program is **less interesting than** the other one.*

- **tan** + **adjective** or **adverb** + **como**
  as… as

  > Este programa es **tan interesante** como el otro.
  > *This program is **as interesting as** the other one.*

....

▶ To form the **superlative** in Spanish (*the biggest, the greatest,* etc.), you use:

- **definite article** + **más/menos** + **adjective**

  > Este programa es **el** más **interesante.**
  > *This program is **the most interesting** one.*

  > Este programa es **el** menos **interesante.**
  > *This program is **the least interesting** one.*

....

▶ The adjectives **bueno** and **malo** have irregular **comparative** and **superlative** forms:

	comparative	superlative
bueno	**mejor, mejores**	**el/la… mejor, los/las… mejores**
malo	**peor, peores**	**el/la… peor, los/las… peores**

....

▶ When the **comparative word** comes before the noun, you use the same constructions except that **tan** becomes **tanto/tanta/tantos/tantas** to agree with the noun that follows.

> Yo preparé **menos documentos** que Pedro.
> *I prepared **fewer documents than** Pedro.*

matches

> Yo preparé **tantos documentos** como Pedro.
> *I prepared **as many documents as** Pedro.*

Practice: **Actividades** **4 5 6 7**

Más práctica *cuaderno pp. 157–158*
Para hispanohablantes *cuaderno pp. 155–156*

Online Workbook
CLASSZONE.COM

4 Comparaciones ♻

Hablar/*Escribir* Todos tienen diferentes opiniones sobre el objeto que es más importante o necesario para la vida moderna. Escribe tus opiniones sobre las siguientes cosas. Sigue el modelo.

modelo

«El teléfono celular es más necesario que el asistente electrónico».

(+) necesario

1. (+) importante

2. (-) útil

3. (+) necesario   **4.** (+) cara

5. (=) importantes   **6.** (-) divertido

5 Marcos ♻ 🎧

Escuchar/*Escribir* Estás en CompuVisión, un almacén de productos electrónicos, con tu amigo venezolano Marcos. Él tiene unas opiniones muy fuertes sobre varias cosas que quieres comprar. Escucha a Marcos y escribe su opinión sobre los objetos indicados.

modelo

esta computadora/ésa

Esta computadora tiene tanta memoria como ésa.

1. este juego interactivo/ése
2. este programa anti-virus
3. este módem
4. esta marca de computadora
5. esta hoja de cálculo/ésa
6. el descuento
7. el precio

Activities 6–7: Express
and understand opinions

6 Internet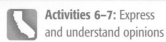

Hablar/*Escribir* Tú y tu compañero(a)
navegan por Internet y ven varias cosas
que les gustan y otras que no. Comparen
las cosas que encuentran en Internet.

modelo

Tú: *¿Te gusta esta página-web?*

Compañero(a): *Sí, es más interesante que la otra.*

Vocabulario

La red mundial

el buzón electrónico *electronic mailbox*

conectarse a/ *to connect to /*
  desconectarse de *disconnect from*

la contraseña *password*

el correo electrónico *e-mail*

en línea *online*

el enlace *link*

el grupo de conversación *chat group*

el grupo de noticias *news group*

hacer clic/doble clic *click/double click*

el icono del programa *program icon*

el Localizador Unificador de Recursos (LUR) *URL*

el servicio de búsqueda *search engine*

el sitio *site*

*correo
electrónico*

▸ ¿Cuáles de estas cosas ves con frecuencia cuando
navegas por Internet?

7 Grupo de conversación

STRATEGY: SPEAKING

Compare and evaluate films in a chat group You can
compare recent films based on their story
(**guión, personajes, trama, clímax, final**),
expertise of those involved (**cinematógrafo,
director, actores, actrices**), and emotions they
create (**emociones, sentimientos**). Can you
decide which are the best or the worst (**la
mejor, la peor película**)?

Hablar/*Escribir* Participas en un grupo de
conversación por Internet. ¿Cuál es el tema
principal del grupo? ¿Cuál es la edad
promedio de los participantes? Escribe una
conversación entre los miembros de tu grupo
de conversación.

modelo

Grupo de conversación: Películas recientes

Tú: *A mí me encantan las películas de acción.
¿Has visto la última de Bruce Willis?*

Compañero(a) 1: *No prefiero las películas menos
violentas.*

También se dice

Cuando navegas por Internet puedes ir a muchos sitios.
En algunos verás que se habla de **la Internet**, ya que la
palabra *net* significa *red* en español, una palabra
femenina (la red). En otros casos te encontrarás con **el
Internet**, porque se refiere al sistema de comunicación
(masculino). Y otras veces verás simplemente **Internet**,
sin artículo. ¡No te preocupes! Todas son correctas y
puedes usar la que más te guste.

You use **prepositions** to clarify locations and to show relationships among people, places, and things.

▶ Use the preposition **a** to express:

• **motion toward** a **place**	Vamos **al cine/a Venezuela.** *We're going **to the movies/to Venezuela.***
• how **far away** something is in space or time	Adela vive **a tres cuadras del colegio.** *Adela lives **three blocks from the school.*** Caracas está **a tres horas de mi ciudad.** *Caracas is **three hours (away) from my city.***
• a **point in time**	Se fue **a las ocho y media.** *He left **at eight-thirty.***
• units of **measurement**	Viajábamos **a 50 millas por hora.** *We were traveling **at 50 miles per hour.***

▶ Use the preposition **en** to express:

• **position** or **location**	Trabaja **en este edificio.** *He works **in this building.*** Están **en Bogotá.** *They're **in Bogotá.***
• a **period of time** *(as opposed to a specific point of time)*	Vuelvo **en una semana.** *I'll be back **in a week.*** Terminamos **en una hora.** *We'll finish **in an hour.***

▶ You have already used **de** to express **possession.** You can also use the preposition **de:**

• to form **compound nouns**	**el banco de datos** *database*	**la hoja de cálculo** *spreadsheet*
• to mark a **characteristic feature**	un niño **de tres años** *a **three-year-old** child*	

▶ Use **con** to express the idea of **accompaniment.**

Alfredo sale **con** Anita.   Tomo café **con** leche.
*Alfredo is going out **with** Anita.*   *I have coffee **with** milk.*

Practice: **Actividades**   **Más práctica** *cuaderno p. 159*    **Online Workbook**
   Para hispanohablantes *cuaderno p. 157*   CLASSZONE.COM

cuatrocientos cincuenta y siete
Colombia, Venezuela y los países andinos Etapa 3
457

Activity 11:
Narrate in the future

8 Ana

Hablar/Escribir Estás en casa de tu amiga ecuatoriana, Ana. Ella te enseña a navegar por Internet en la computadora de su padre. Para saber lo que te dice, completa sus oraciones con **a**, **con**, **de** o **en**.

modelo

«La computadora de mi padre es nueva.»

1. «La computadora vino _____ tarjeta de sonido y tarjeta gráfica.»

2. «Primero voy a hacer doble clic _____ el icono del programa.»

3. «_____ este módem puedes conectarte rápidamente a Internet.»

4. «¡Mira! Tengo seis mensajes _____ mi buzón electrónico.»

5. «Voy a copiar _____ mi amigo en ese mensaje.»

6. «Voy a tratar de responder _____ el correo electrónico de hoy.»

7. «Más tarde podemos ir al grupo _____ conversación para estudiantes de español.»

8. «Este grupo empieza _____ las ocho.»

9. «Quiero ver la página-web _____ OficinaNet.»

10. «_____ el servicio de búsqueda, es fácil encontrar lo que quieras en Internet.»

11. «Ahora me voy a desconectar _____ Internet.»

12. «Voy _____ la tienda de computación a comprar unos disquetes».

9 Entrevista en las noticias

Hablar/Escribir Ves las noticias en la tele. Completa lo que dicen la reportera y el chico con las preposiciones **a**, **con**, **de** o **en**.

Reportera:

¡Buenas tardes! Soy Ana de la Cruz y les hablo hoy desde el Café Ciberespacio __1__ el centro de Caracas. Como ya saben, Internet ha cambiado la vida __2__ todos. No podemos salir __3__ la casa sin que alguien nos hable __4__ la red mundial. Hay de todo __5__ la red: juegos, grupos de conversación y páginas-web __6__ muchas compañías e individuos. Estoy aquí __7__ Joaquín. ¿Joaquín, cuánto tiempo navegas por Internet?

Chico:

La verdad es que navego por Internet todos los días __8__ las tres. __9__ mi casa hay tres computadoras: la mía, la __10__ mi hermano y una para mis papás. Los mejores sitios son los grupos de conversación __11__ otros jóvenes.

La verdad es que navego por Internet todos los días.

¡Hola! Soy Ana de la Cruz y les hablo hoy desde el Café Ciberespacio.

10 ¿Adónde vas?

Hablar/*Escribir* Conversa con tu compañero(a) sobre lo que va a hacer después de clases. Usa las ideas de la lista.

modelo

Tú: *¿Adónde vas?*

Compañero(a): *Voy al Café Ciberespacio.*

Tú: *¿Con quién vas?*

Compañero(a): *Voy con un amigo de la clase de español.*

a (lugar)	de (persona)
a (distancia)	de (edad)
a (tiempo)	con (persona)
en (lugar)	con (objeto)
en (tiempo)	¿…?

11 Mis planes

Escribir Escribe un párrafo describiendo tus planes para el resto de la tarde. Trata de usar las preposiciones **a**, **con**, **de** y **en**.

modelo

Voy a la casa de mi amigo Arturo a las cuatro de la tarde. Vamos a navegar por Internet con unos amigos.

More Practice:

Más comunicación *p. R20*

GRAMÁTICA Verbs with Prepositions

Many verbs require the use of a certain **preposition**. With verbs of **motion** such as **entrar, ir, salir, subir, venir, volver,** you use the preposition **a** when the verb is followed by an **infinitive**. The same preposition is used with the verbs in the left column in the box below.

Vinimos a jugar.
We came to play.

Entraron al comedor.
They went into the dining room.

Other verbs like those in the right column in the box below require the use of the preposition **de** as part of their essential meaning.

Tratamos de crear una página-web.
We're trying to create a web page.

Ahora **se acuerdan de** la contraseña.
Now they remember the password.

After **insistir** you use the preposition **en**.

Insisto en que me mandes un mensaje.
I insist that you send me a message.

Remember that **tener que** means *to have to do something* and that **hay que** means *one must, you should.*

Hay que tener correo electrónico.
You should (It's a necessity to) have e-mail.

Entonces, **tengo que** comprarme un módem.
Then I have to buy a modem.

Vocabulario

 Ya sabes

aprender a	acabar de
ayudar a	acordarse de (ue)
comenzar a (ie)	dejar de
empezar a (ie)	olvidarse de
enseñar a	tener ganas de
invitar a	tratar de
prepararse a	

Practice:
Actividades
12 13 14

Más práctica
cuaderno p. 160
Para hispanohablantes
cuaderno p. 158

Online Workbook
CLASSZONE.COM

---

cuatrocientos cincuenta y nueve
Colombia, Venezuela y los países andinos Etapa 3
459

Activity 13: Understand most spoken language when the message is deliberately and carefully conveyed by a speaker accustomed to dealing with learners when listening

12 Conversación telefónica

Hablar/Escribir Alberto y Herlinda son amigos peruanos. Alberto quiere aprender más sobre cómo usar la computadora y navegar por Internet. Completa la conversación telefónica entre ellos usando los verbos de la lista y las preposiciones apropiadas.

> preparar
>
> aprender
>
> invitar
>
> tener muchas ganas
>
> ir
>
> acabar
>
> enseñar
>
> ayudar

Alberto: Quiero ___1___ usar la hoja de cálculo.

Herlinda: Yo te ___2___ usarla.

Alberto: ¿También me puedes ___3___ usar Internet?

Herlinda: ¡Sí, claro! Pero ___4___ desconectarme de Internet.

Alberto: ¡Perfecto! ___5___ aprender a navegar por Internet.

Herlinda: Mira, ___6___ salir para clases ahora, pero vuelvo a las cinco. ¿Por qué no vienes a casa a las siete?

Alberto: ¡Me parece perfecto! Te ___7___ tomar un café después de que acabemos.

Herlinda: ¡De acuerdo! ¡___8___ divertirte mucho! Internet es muy divertido.

13 La página-web

Escuchar/Escribir Escucha lo que le pasó a Eduardo cuando él y sus amigos decidieron crear una página-web. Completa las oraciones para describir su situación.

> **modelo**
>
> *Él y sus amigos trataron de crear una página-web.*

1. Ellos _____ las ocho de la mañana.
2. Eran las ocho de la noche y apenas _____ terminarla.
3. Ricardo _____ ayudarlos.
4. Para crear una página-web interesante, _____ saber muchas cosas.
5. Eduardo _____ irse.
6. Eduardo _____ la cita con su novia.
7. Eduardo _____ cenar con su novia.
8. Eduardo _____ la cita.
9. Eduardo _____ ella cuando vio su reloj.
10. Su novia _____ él sea puntual.

Nota cultural

Cuando quieres navegar por Internet y encontrar sitios en español, debes empezar tu búsqueda escribiendo lo que quieres encontrar en español. También será mucho más fácil si incluyes la palabra «español». Por ejemplo, puedes buscar «música en español» y encontrarás muchísimos sitios. Muchos de los servicios de búsqueda te dan la opción de escoger el idioma que quieres usar.

Activity 15 brings together all concepts presented.

14 ¡Tengo ganas de...!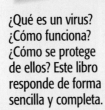

Hablar/Escribir Es el fin de semana. Tú y tu compañero(a) quieren hacer varias cosas, pero también tienen otras obligaciones. Conversen sobre lo que quieren hacer y por qué no pueden. Usen los verbos de la lista.

modelo

Tú: *Tengo ganas de navegar por Internet.*

Compañero(a): *¿No te acuerdas de la tarea para la clase de español?*

Tú: *La haré más tarde.*

> acordarse de
> aprender a
> ayudar a
> dejar de
> enseñar a
> invitar a
> olvidarse de
> tener ganas de
> tratar de
> ¿...?

15 Libros de informática

Hablar/Escribir En grupos de tres o cuatro, lean las descripciones de los libros de informática a continuación. Conversen sobre los temas y sobre los libros que les interesarían a todos.

modelo

Tú: *A mí me gustaría saber cómo proteger mi sistema de un virus.*

Compañero(a) 1: *Pues, entonces debes comprar el libro Virus Informático.*

Compañero(a) 2: *Creo que es uno de los mejores.*

Compañero(a) 3: *Hay otro más interesante. Se llama…*

¿Qué es un virus? ¿Cómo funciona? ¿Cómo se protege de ellos? Este libro responde de forma sencilla y completa.

Ayuda al usario a obtener el máximo partido del disco duro de su ordenador y resolver muchos problemas.

Explica con claridad los conceptos utilizados en el ámbito de la microinformática y otras disciplinas afines.

More Practice:
Más comunicación *p. R20*

Online Workbook
CLASSZONE.COM

Refrán

El que sabe dos lenguas vale por dos.

¿Qué quiere decir el refrán? ¿Por qué crees que es importante hablar más de una lengua? ¿Crees que esta habilidad será más o menos importante en el futuro? ¿Por qué?

cuatrocientos sesenta y uno
Colombia, Venezuela y los países andinos Etapa 3
461

En voces

AUDIO
LECTURA

PARA LEER
STRATEGY: READING

Monitor comprehension A good self-check is to restate (paraphrase) what each paragraph is about with a phrase or short sentence. If you can't, then reread, identify what seems unclear, and ask questions. Jot down a brief paraphrase of each paragraph as you read. You can then use those notes to write a brief summary about **el autor**, **su vida y su obra**. This will help you start:

1. La lectura empieza con una cita de una novela de García Márquez.
2.
3.

LA LITERATURA

los acontecimientos	eventos
cotidiano(a)	diario(a)
entretejer	combinar

Nota cultural

Macondo es el lugar imaginario donde suceden muchas de las obras de García Márquez. Los personajes allí viven en un ambiente misterioso y mágico.
Este pueblo apareció por primera vez en *La hojarasca*, luego en *Los funerales de la Mamá Grande*, y también en la famosa novela *Cien años de soledad*.

Gabriel García Márquez

❝ Muchos años después, ante el pelotón de fusilamiento[1], el coronel Aureliano Buendía había de recordar aquella tarde remota en que su padre lo llevó a conocer el hielo[2] ❞.

Así comienza *Cien años de soledad* (1967), la novela más conocida de Gabriel García Márquez, uno de los escritores principales de las Américas en el siglo XX. Se ha dicho que dentro de la literatura latinoamericana, *Cien años de soledad* tiene tanta importancia como *El Quijote* de Cervantes.

Como en muchas de las obras de García Márquez, esta novela entreteje las historias de individuos con la historia de su pueblo. En *Cien años de soledad*, el pueblo es Macondo y los individuos son los miembros de la familia Buendía. García Márquez narra los acontecimientos de la historia del pueblo y de la familia Buendía utilizando un estilo conocido como realismo mágico. Este estilo combina la realidad con elementos fantásticos.

[1] firing squad   [2] ice

El escritor insiste en que el realismo mágico no es una combinación de elementos reales y fantásticos, sino que es la manera en que sucede la vida cotidiana de Colombia. García Márquez nació en ese país en 1928 y se crió [3] con sus abuelos hasta la edad de ocho años. Los cuentos que le hacía su abuela influyeron sus escritos.

Cuando era joven, García Márquez se dedicó al periodismo. Publicó su primera novela, *La hojarasca*, en 1955. En *El coronel no tiene quien le escriba* (1957), García Márquez siguió utilizando la historia de Colombia como marco de referencia para sus protagonistas.

Su fama aumentó con *El otoño del patriarca* (1975) y *Crónica de una muerte anunciada* (1981). En 1982, García Márquez ganó el Premio Nóbel de Literatura. Continúa escribiendo novelas y guiones para películas.

Sin duda, García Márquez es una de las voces más potentes de América Latina. Su visión sugiere que el individuo es protagonista de dos historias: la de su vida y la de su pueblo. Al leer la obra de García Márquez, vemos que a veces no es fácil saber dónde comienza una historia y termina la otra. Nuestra inquietud [4] no es confusión, sin embargo: es el comienzo de la búsqueda [5] intrépida de quiénes somos y seremos en el lugar donde nos ha tocado vivir [6], sea en Macondo o Main Street.

---

[3] was raised   [5] search
[4] uncertainty   [6] where it is our fate to live

Online Workbook CLASSZONE.COM

¿Comprendiste?

1. ¿Quiénes son los protagonistas de *Cien años de soledad*? ¿Dónde sucede?
2. ¿Qué es el realismo mágico? ¿Qué combina?
3. ¿Qué influencia de la niñez de García Márquez aparece en sus obras?
4. ¿Qué papel juega la historia de un pueblo en la obra de García Márquez?

¿Qué piensas?

¿Por qué crees que un escritor como García Márquez puede afectar cómo vemos la historia?

Hazlo tú

Elige un lugar en tu estado o ciudad e investiga algún evento que te llame la atención. Luego, inventa un nombre ficticio para el lugar e imagina qué personajes (tanto reales como ficticios) participaron en ese evento. Si quieres, escribe un cuento corto o dibuja tus personajes y el lugar.

En colores
CULTURA Y COMPARACIONES

B o l i v i a

PARA CONOCERNOS
STRATEGY: CONNECTING CULTURES
Evaluate the Internet as a means of developing cultural knowledge and understanding You have learned about Latin cultures through many media (print, video, conversations, Internet). These media cover social and business interactions, traditions, history, geography, contributions in literature, fine arts, social, economic, and political institutions. Out of these general topics, think of three personal cultural goals. Then evaluate whether the Internet is high, medium, or low as a means of helping you reach those goals. Give reasons. Use **"Bolivia en la red"** plus your own knowledge to guide your thinking.

Metas	Evaluación	Comentario
1.		
2.		
3.		

¡Conoce Bolivia por Internet! Un viaje virtual puede darte mucha información y ayudarte a planificar un viaje verdadero.

Tú, como usuario(a) de Internet, puedes acceder a la página-web de Bolnet, el sitio oficial de Bolivia, y escoger entre las posibilidades del menú. Puedes explorar varios aspectos de la sociedad boliviana (educación, gobierno, el mundo de los negocios, etc.) usando los enlaces de la página. Puedes leer la prensa[1] del país en tu pantalla, conversar con abonados[2] de Bolnet o escuchar un cuento.

[1] press   [2] subscribers

464

en la red

Otro servicio boliviano de Internet, Bolivianet, ofrece a sus miembros una guía de correo electrónico, que es una lista de las direcciones electrónicas de casi dos mil personas, bolivianas en su mayoría. Esta guía existe con el fin[3] de unir a los bolivianos a través del mundo. ¡Es una guía telefónica para la época de Internet!

[3] exists with the goal

¿Quisieras hacer un viaje a Bolivia? Toda la información que necesites está disponible[4] en la red mundial. Con una página como ésta puedes conocer las regiones del país, leer una descripción de cada una, ver fotos de los lugares de interés turístico y obtener una lista de hoteles. ¡Incluso puedes reservar tu habitación de hotel por Internet! ¿No ves? Un viaje a Bolivia puede empezar con un viaje por el ciberespacio.

[4] available

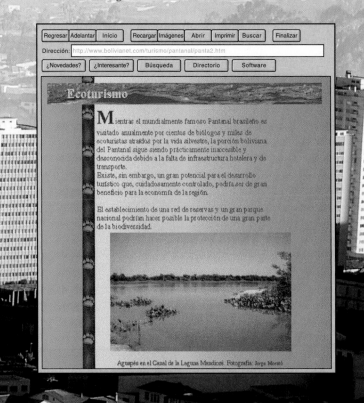

| Regresar | Adelantar | Inicio | Recargar | Imágenes | Abrir | Imprimir | Buscar | Finalizar |

Dirección: http://www.bolivianet.com/turismo/pantanal/panta2.htm

¿Novedades? ¿Interesante? Búsqueda Directorio Software

Ecoturismo

Mientras el mundialmente famoso Pantanal brasileño es visitado anualmente por cientos de biólogos y miles de ecoturistas atraídos por la vida silvestre, la porción boliviana del Pantanal sigue siendo prácticamente inaccesible y desconocida debido a la falta de infraestructura hotelera y de transporte.

Existe, sin embargo, un gran potencial para el desarrollo turístico que, cuidadosamente controlado, podría ser de gran beneficio para la economía de la región.

El establecimiento de una red de reservas y un gran parque nacional podrían hacer posible la protección de una gran parte de la biodiversidad.

Aguapés en el Canal de la Laguna Mandioré. Fotografía: Jorge Moraté.

¿Comprendiste?

1. ¿Qué es Bolnet?
2. ¿Cómo nos ayuda Internet a conocer Bolivia?
3. ¿Qué es la guía de correo electrónico? ¿Para qué sirve?
4. ¿Qué utilidad tiene Internet para el (la) turista que quiere viajar por Bolivia?

¿Qué piensas?

1. Da un ejemplo de cómo Internet facilita el contacto entre personas.
2. ¿Qué aspectos de Internet te parecen más útiles para conocer los países hispanoamericanos?

Hazlo tú

Usa un servicio de búsqueda para encontrar otras páginas sobre Bolivia o los otros países de esta unidad. Explora los enlaces de las páginas que visites y prepara un resumen de lo que se puede aprender sobre el país usando Internet.

cuatrocientos sesenta y cinco
Colombia, Venezuela y los países andinos Etapa 3
465

En uso

REPASO Y MÁS **COMUNICACIÓN**

OBJECTIVES

• Compare and evaluate
• Express precise relationships
• Navigate cyberspace

Now you can...

• navigate cyberspace.

To review

• superlatives, see p. 454.

❶ El mejor

Tu amigo(a) ecuatoriano(a) pregunta qué piensas sobre cosas que encuentras en Internet. Escríbele para contestar sus preguntas.

> **modelo**
>
> *¿Qué piensas de esta página-web? (interesante)*
>
> *Esta página-web es la más (la menos) interesante de todas.*

1. ¿Qué piensas de este grupo de conversación? (divertido)
2. ¿Qué piensas de este grupo de noticias? (útil)
3. ¿Qué piensas de este servicio de búsqueda? (mejor/peor)
4. ¿Qué piensas de este sitio? (mejor/peor)
5. ¿Qué piensas de este enlace? (mejor/peor)

Now you can...

• compare and evaluate.

To review

• comparatives, see p. 454.

❷ Opiniones

Tú y dos compañeros están en CompuVisión y comparan varios productos. Los tres tienen opiniones diferentes. ¿Qué dicen?

> **modelo**
>
> *complicado*
>
> **Tú:** *Este programa de software es más complicado que el otro.*
>
> **Compañero(a) 1:** *No, no, éste es menos complicado.*
>
> **Compañero(a) 2:** *Pienso que ése es tan complicado como el otro.*

3 **La carta**

Elena le escribió una carta a su mejor amiga, María. Completa la carta con las preposiciones **a**, **de**, **con** y **en** para saber qué le pasó el otro día.

Querida María,

El otro día, navegué por Internet __1__ mi hermano __2__ quince años. Como él quería comprar una computadora, fuimos __3__ la página-web __4__ OficinaNet. En la página-web vimos que la compañía está __5__ el centro de Caracas. Está __6__ tres cuadras de mi colegio. __7__ las dos de la tarde fuimos __8__ la oficina para ver los sistemas de computación. Hablamos con la señora Villarreal, la dueña __9__ la compañía. Ella convenció a mi hermano que la computadora que le gustaba era la computadora __10__ sus sueños. Antes de que decidiera qué iba a hacer, yo le dije a la señora Villarreal «Volvemos __11__ una semana». Saqué a mi hermano y le dije «¿Qué estabas pensando? No tienes el dinero para comprar esa computadora!».

Abrazos,

Elena

Now you can...
• express precise relationships.

To review
• prepositions, see p. 457.

4 **El diario**

Lee la entrada del 11 de mayo en el diario de Abigail. Completa la entrada con las preposiciones apropiadas.

Now you can...
• express precise relationships.

To review
• verbs with prepositions, see p. 459.

11 de mayo

Por fin aprendí __1__ navegar por Internet. Empecé __2__ conectarme cuando me di cuenta de que no podía acordarme __3__ mi contraseña. En ese momento, mi hermanita entró __4__ ver qué estaba haciendo. En vez de ayudarme __5__ recordar mi contraseña, empezó __6__ reírse. «Deja __7__ reírte de mí», le dije, enojada. «Estoy tratando __8__ aprender algo nuevo y tú, ¡burlándote de mí!» Trató __9__ pedirme perdón pero yo no tenía ganas __10__ perdonarla. «Insisto __11__ que salgas de aquí inmediatamente», le grité. «Vas __12__ quedarte tranquila», me respondió y se fue __13__ decirles a mis padres que ¡yo la había insultado a ella! ¡Imagínate!

5 CompuVisión

STRATEGY: SPEAKING

Compare and evaluate computer configurations
Judging the value of technology involves comparing performance in relation to cost. Securing the knowledgeable opinions of others also helps. Consider: **hardware (tamaño, color, memoria, velocidad, resolución); software (usos, resultados); necesidades y preferencias personales; precio, descuento, garantía**

Hablar/*Escribir* Están en CompuVisión. Examinen todas las configuraciones y hablen sobre ellas. Conversen sobre las ventajas y desventajas de todas las computadoras que ven. Dramaticen esta situación en grupos de dos o tres.

modelo

Tú: *¿Qué te parece esta computadora?*

Compañero(a) 1: *No sé. ¿Cuánta memoria tiene?*

Compañero(a) 2: *…*

6 En tu propia voz

ESCRITURA ¡Diseña tu propia página-web! Primero decide qué quieres incluir en tu página. Luego haz dibujos y escribe narraciones que pondrías en tu página–web.

- Nombre
- Mis actividades
- Mis películas
- Mis deportes
- Mis clases
- ¿…?
- Mis viajes
- Mi familia
- Mis planes para el futuro

CONEXIONES

La tecnología La escuela piensa comprar una computadora nueva para el salón de tu clase de español. Te toca a ti escribir los requisitos. Habla con el (la) profesor(a) y los otros alumnos y averigua para qué piensan usar la computadora. Tomando en cuenta esta información, decide qué tipo de sistema necesitan. Puedes consultar con el (la) profesor(a) de informática para ver si tienes toda la información que necesitas.

En resumen
REPASO DE VOCABULARIO

NAVIGATE CYBERSPACE

Computer equipment

ampliable	expandable
la configuración	configuration
el disco	disk
el disco duro	hard drive
disponible	available
externo(a)	external
el fax	fax
el hardware	hardware
interno(a)	internal
la memoria	memory
el micrófono	microphone
el microprocesador	microprocessor
el módem	modem
el monitor	monitor
el tamaño	size
la tarjeta de sonido	sound card
la tarjeta gráfica	graphics card
la tecla	key
el teclado	keyboard
el (la) usuario(a)	user

Cyberspace

el buzón electrónico	electronic mailbox
conectarse a	to connect to
la contraseña	password
el correo electrónico	e-mail
desconectarse de	to disconnect from
la dirección electrónica	e-mail address
en línea	on-line
el enlace	link
el grupo de conversación	chat group
el grupo de noticias	news group
hacer (doble) clic	to (double) click
el icono del programa	program icon

Cyberspace (continued)

el Localizador Unificador de Recursos (LUR)	URL
la página-web	web page
la red mundial	World Wide Web
el servicio de búsqueda	search engine
el sitio	site

Software

la base de datos	database
la hoja de cálculo	spreadsheet
el juego interactivo	interactive game
el programa anti-virus	anti-virus program
el software	software

COMPARE AND EVALUATE

Comparatives and superlatives

Este juego es **más** interesante **que** el otro.
Aquel programa es **el menos** interesante.
Yo preparé **tantos** documentos **como** Pedro.

EXPRESS PRECISE RELATIONSHIPS

Prepositions

Voy **a** comprar un sistema **de** computación.
Quiero uno **con** base de datos.
Pienso comprarlo **en** una semana.

♻ Ya sabes

acabar de	to have just
acordarse de (ue)	to remember
aprender a	to learn to
ayudar a	to help to
comenzar a (ie)	to begin to
dejar de	to stop doing something
empezar a (ie)	to begin
enseñar a	to teach to
invitar a	to invite
olvidarse de	to forget to
prepararse a	to prepare to
tener ganas de	to feel like to
tratar de	to try to

Juego

¿Quién ganó la computadora?

El premio del concurso de ElectroMundo fue una computadora. Mira esta información para saber quién ganó.

- Roberto tiene una tarjeta de sonido.
- El (La) ganador(a) es miembro de un grupo de conversación en la red mundial.
- Carmen tiene varios juegos interactivos para su computadora.
- A Carmen le gusta hablar con otras personas usando Internet.
- Roberto quiere comprarse un fax/módem para navegar en Internet.

cuatrocientos sesenta y nueve
Colombia, Venezuela y los países andinos Etapa 3
469

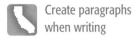

Create paragraphs when writing

En tu propia voz

ESCRITURA

La tecnología del mundo de hoy

Tú eres reportero(a) del periódico de tu escuela. Tienes que escribir un artículo para la próxima edición sobre algún evento escolar relacionado con la tecnología. Recuerda que tus lectores son estudiantes como tú, así que escoge un tema que les interese. En tu artículo, escribe un resumen dando detalles del evento y citas de los participantes.

Función: Presentar un resumen **Contenido:** Visita de estudiantes colombianos

Contexto: Informar a los estudiantes **Tipo de texto:** Artículo periodístico

---

PARA ESCRIBIR • STRATEGY: WRITING

Prioritize information in order of importance When writing a journalistic article, identify the most important facts and summarize them first. (These basic facts should answer the "who, what, when, where, why" questions of the reader.) Then present additional facts of secondary importance. Next, give information that is least important to a general understanding of the event, but that provides color or background.

Modelo del estudiante

Visita de estudiantes colombianos

> The first paragraph gives a **complete summary** of the most important aspects of the event.

El martes pasado un grupo de quince estudiantes colombianos vinieron a nuestra escuela para visitar el nuevo laboratorio de computadoras y conocer la nueva tecnología que ofrece, ya que es la más avanzada de la región.

Primero disfrutaron una recepción en su honor en el auditorio. Después, el maestro Stan Smith, especialista en computadoras y tecnología, hizo una presentación breve sobre el hardware y software que hay en el laboratorio, sus capacidades y los programas que utilizamos más frecuentemente.

> The **secondary facts** provide more details about the information in the first paragraph and break it down into smaller pieces.

Después de la presentación todos los participantes fueron al laboratorio para que los estudiantes colombianos tuvieran la oportunidad de ver y usar nuestro equipo.

Exploramos juntos en Internet y nos divertimos mucho con los programas de realidad virtual y los videojuegos.

Cuando los estudiantes tuvieron que salir, el maestro Smith propuso una visita de estudiantes de nuestra escuela a la suya, diciendo, «Ojalá que puedan regresar algún día o, aún mejor, que nosotros podamos ir a Colombia y visitarlos a Uds. la próxima vez».

> The next level of information includes a **quotation** with a hope about a future related event.

La visita terminó con sonrisas por todas partes. Los estudiantes colombianos no estaban acostumbrados al tiempo de aquí, que es mucho más frío que el de Colombia. Pero todos parecían muy entusiasmados con la visita. Guillermo Díaz, un estudiante de Bogotá, comentó, «Nos alegramos de estar aquí... ¡a pesar del frío increíble que hace!» y todos nos reímos.

> The final level of information provides some **colorful information** about the event, focusing on a funny quotation.

Language Arts
Writing Standard 2.1a
Write fictional, biographical, or autobiographical narratives.
Narrate a sequence of events and communicate their
significance to the audience.

Writing Center
CLASSZONE.COM

Estrategias para escribir

Antes de escribir ...

Piensa en un suceso reciente que quieras
describir. Para organizar la información según el
nivel (level) de importancia, usa una gráfica como
la de la derecha. La información más importante va
primero; luego la información que da más detalles. Al
final escribes datos que presenten información adicional.

el martes pasado, estudiantes colombianos, visita
a la escuela, conocer la nueva tecnología

recepción en el auditorio, presentación
sobre la tecnología, visita al
laboratorio

despedida, sugerencia del
maestro Smith sobre una
visita a su escuela

todos muy
animados, quejas
sobre el frío
de aquí

Revisiones

Después de escribir el primer borrador de tu artículo,
compártelo con un compañero de clase. Pregúntale:

- *¿Cuál es la información más importante del artículo?*
- *¿Dónde está la información que amplifica la idea central?*
  ¿Cuál es?
- *¿Qué efecto tienen los comentarios de los participantes sobre
  el tono del artículo?*

La versión final

Antes de crear tu versión final, léela de
nuevo y repasa los siguientes puntos:

- *¿Usé bien el **pretérito** y el **imperfecto**?*

Haz lo siguiente: Subraya todos los verbos
y determina si se debe usar el imperfecto o
el pretérito. Repasa el uso del pretérito y
del imperfecto.

- *¿Usé **formas comparativas y
  superlativas** para añadir más
  información a los datos del artículo?*

Haz lo siguiente: Haz un círculo alrededor de las expresiones
comparativas o superlativas. ¿Concuerda el artículo definido
con el sustantivo relacionado? ¿Concuerdan los adjetivos con
los sustantivos?

El mes pasado, hice un proyecto de
investigación por Internet. ~~Buscaba~~ Busqué
artículos periodísticos y escribí informes
sobre ciudades latinoamericanas. ~~Aprendía~~ Aprendí a
usar los servicios de búsqueda. ~~Escogía~~ Escogí
estudiar la ciudad de Bogotá porque es la
ciudad (más poblado) de Colombia. a

cuatrocientos setenta y uno
Colombia, Venezuela y los países andinos Etapa 3
471

RECURSOS

Más comunicación   R2
To help in the management of these paired activities,
the Student A cards have been printed upside-down.

Juegos – respuestas   R21

Gramática – resumen   R22

Glosario
español–inglés   R37
inglés–español   R61

Índice   R85

Créditos   R93

1 Etapa preliminar p. 13
¿Qué hizo ayer?

8:00	
10:00	Planché la ropa.
12:00	
2:00	Comí con mi familia.
4:00	
6:00	Vi una película cómica.
8:00	
10:00	Me acosté.

modelo

Estudiante A: ¿Qué hizo a las ocho de la mañana?

Estudiante B: A las ocho se despertó en su cuarto.

Estudiante A Marcela escribió todo lo que hizo ayer en su diario. Tú tienes parte de la información y tu compañero(a) tiene la otra parte. Juntos(as) describan las actividades que hizo Marcela ayer y traten de imaginar dónde las hizo.

Estudiante B Marcela escribió todo lo que hizo ayer en su diario. Tú tienes parte de la información y tu compañero(a) tiene la otra parte. Juntos(as) describan las actividades que hizo Marcela ayer y traten de imaginar dónde las hizo.

modelo

Estudiante A: ¿Qué hizo a las ocho de la mañana?

Estudiante B: A las ocho se despertó en su cuarto.

8:00	Me desperté.
10:00	
12:00	Jugué al fútbol.
2:00	
4:00	Compré una blusa nueva.
6:00	
8:00	Visité a un amigo enfermo.
10:00	

---

2 Etapa preliminar p. 25
¡Qué cita!

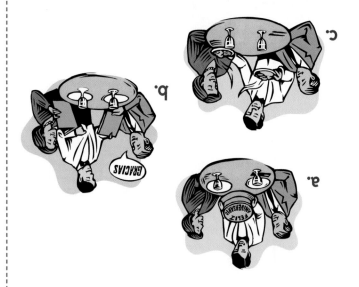

a. b. c.

Estudiante A Antonio hizo seis dibujos de lo que le pasó ayer. Tú tienes la mitad de los dibujos y tu compañero(a) tiene los demás. Usa el pretérito para poner los dibujos en orden sin mirar los de tu compañero(a).

Estudiante B Antonio hizo seis dibujos de lo que le pasó ayer. Tú tienes la mitad de los dibujos y tu compañero(a) tiene los demás. Usa el pretérito para poner los dibujos en orden sin mirar los de tu compañero(a).

d.

e.

f.

R2 RECURSOS
Más comunicación

3 Unidad 1 Etapa 1 p. 41
Cita a ciegas

Estudiante A Tu compañero(a) te quiere arreglar una cita a ciegas (blind date). Hazle preguntas para saber más de la cita. Luego cambien de papel y contesta las preguntas de tu compañero(a) con la siguiente información.

modelo

Estudiante A: ¿De dónde es?

Estudiante B: Es de Seattle.

sociable y atrevido(a)	una fiesta en casa
moreno y lacio	de Carlota
grandes y negros	el 14 de septiembre
Miami	a la una de la tarde

Origen: _____ Personalidad: _____

Cabello: _____ Ojos: _____

Fecha y hora de la cita: _____

Lugar de la cita: _____

Estudiante B Tu compañero(a) te pregunta sobre la persona que le buscaste para una cita a ciegas (blind date). Contéstale con la siguiente información. Luego cambien de papel.

modelo

Estudiante A: ¿De dónde es?

Estudiante B: Es de Seattle.

verdes	Seattle
el cine Estrella	rojizo y ondulado
considerado(a)	el 15 de septiembre
y cómico(a)	a las 7 de la tarde

Origen: _____ Personalidad: _____

Cabello: _____ Ojos: _____

Fecha y hora de la cita: _____

Lugar de la cita: _____

---

4 Unidad 1 Etapa 1 p. 45
Antes y ayer

Estudiante A Tú y tu compañero(a) quieren saber qué hicieron estas personas ayer y en su niñez (antes). Trabajen juntos(as) para completar la tabla.

modelo

Estudiante B: ¿Michalín compartió sus cosas ayer?

Estudiante A: No, no compartió sus cosas ayer. ¿Compartía sus cosas antes?

Nombre	Actividad	Antes	Ayer
Michalín	compartir sus cosas		no
Joaquín	respetar a sus maestros		no
Tú	discutir con tu hermano(a)		
Tu compañero(a)	hacerles caso a sus padres		

Estudiante B Tú y tu compañero(a) quieren saber qué hicieron estas personas ayer y en su niñez (antes). Trabajen juntos(as) para completar la tabla.

modelo

Estudiante B: ¿Michalín compartió sus cosas ayer?

Estudiante A: No, no compartió sus cosas ayer. ¿Compartía sus cosas antes?

Nombre	Actividad	Antes	Ayer
Michalín	compartir sus cosas	no	
Joaquín	respetar a sus maestros	no	
Tú	discutir con tu hermano(a)		
Tu compañero(a)	hacerles caso a sus padres		

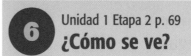

5 Unidad 1 Etapa 2 p. 67
¿Qué harán?

6 Unidad 1 Etapa 2 p. 69
¿Cómo se ve?

(contenido invertido — Estudiante A, actividad 5)

- navegar en tabla de vela
- no describirte sus talentos
- poder hacer todo lo que quiere
- coleccionar tarjetas de los jugadores

4. A Francisca le interesan las computadoras.
3. Chalo es atrevido.
2. Cristina es descarada.
1. A Norberto le encantan las montañas.

Estudiante B: *No pescará en alta mar.*

Estudiante A: *Olivia detesta andar en bote.*

modelo

Luego cambien de papel y contéstale a tu compañero(a), usando las frases de abajo.
Entonces tu compañero(a) te dice lo que cada persona hará según sus gustos y personalidad.
Estudiante A Descríbele a las siguientes personas a tu compañero(a), usando las oraciones que aparecen a continuación.

(contenido invertido — Estudiante A, actividad 6)

4. a acampar
3. la playa
2. una boda
1. el gimnasio

Estudiante B: *Llevará su blusa de lunares.*

Estudiante A: *Marta va a un concierto.*

modelo

Pedro. Sigan el modelo.
Luego cambien de papel y dile qué llevará
Estudiante A: Dile a tu compañero(a) adónde va Marta. Él (Ella) te va a decir qué ropa llevará.

Estudiante B Tu compañero(a) va a describir a varias personas. ¿Qué crees que harán las personas mencionadas según sus gustos y personalidad? Escoge actividades lógicas de la lista. Luego cambien de papel y usa las oraciones de abajo para describirle a las personas indicadas a tu compañero(a).

modelo

Estudiante A: *Olivia detesta andar en bote.*
Estudiante B: *No pescará en alta mar.*

- no hacerles caso a sus padres
- navegar por Internet
- volar en planeador
- hacer alpinismo

5. Cecilia es aficionada al béisbol.
6. Virginia es modesta.
7. A Andrés le fascinan los deportes de agua.
8. Jaime es mimado.

Estudiante B: Tu compañero(a) te dice adónde va Marta. Dile qué ropa llevará. Luego cambien de papel y dile adónde va Pedro. Sigan el modelo.

modelo

Estudiante A: *Marta va a un concierto.*
Estudiante B: *Llevará su blusa de lunares.*

a.

c.

b.

d.

5. la escuela
7. una competencia de ciclismo

6. un restaurante elegante
8. una fiesta de los setenta

7 Unidad 1 Etapa 3 p. 88
¡Adivinen!

8 Unidad 1 Etapa 3 p. 91
¿Cuáles son las diferencias?

(Estudiante A — texto invertido)

Estudiante A Da pistas (clues) para que tu compañero(a) adivine (guess) el verbo en negrita de cada oración. ¡Ojo! No puedes usar ese verbo en tu descripción. Luego, cambien de papel y trata de adivinar los verbos de tu compañero(a).

modelo

Martina y Bárbara **se conocen** bien.

Estudiante A: Martina y Bárbara se hablan de todo…
Hace muchos años que son amigas…
Frecuentemente Martina sabe lo que
Bárbara va a decir…

Estudiante B: ¿Se llevan bien?… ¿Se saludan?…
¡Se conocen bien!

1. David y su primo se **quieren**.
2. Elvira y su vecina **se telefonean**.
3. Carlos y Anita se **pelean**.
4. Luci y Ana **se ayudan** con la tarea.

(Estudiante A — texto invertido, actividad 8)

Estudiante A Con tu compañero(a), háganse preguntas sobre los diferentes objetos de sus dibujos para identificar siete diferencias entre los dos dibujos. ¡Buena suerte!

modelo

Estudiante A: Veo una bandera de Estados Unidos…

Estudiante B: Aquí se ven dos banderas, una de Estados Unidos y otra de México… ¿A qué hora cierra el…?

Estudiante B Trata de adivinar las actividades que hacen varias personas según las pistas (clues) de tu compañero(a). Luego, cambien de papel y da pistas para que tu compañero(a) adivine (guess) el verbo en negrita de cada oración de abajo. ¡Ojo! No puedes usar ese verbo en tu descripción.

modelo

Martina y Bárbara **se conocen** bien.

Estudiante A: Martina y Bárbara se hablan de todo…
Hace muchos años que son amigas…
Frecuentemente Martina sabe lo que
Bárbara va a decir…

Estudiante B: ¿Se llevan bien?… ¿Se saludan?…
¡Se conocen bien!

5. David y Lola **se cuentan** chismes.
6. Jovita y su abuela **se escriben**.
7. Los padres de César **se perdonan**.
8. Los estudiantes **se quejan**.

Estudiante B Con tu compañero(a), háganse preguntas sobre los diferentes objetos de sus dibujos para identificar siete diferencias entre los dos dibujos. ¡Buena suerte!

modelo

Estudiante A: Veo una bandera de Estados Unidos…

Estudiante B: Aquí se ven dos banderas, una de Estados Unidos y otra de México…¿A qué hora cierra el…?

9 Unidad 2 Etapa 1 p. 118
¿Me puedes ayudar?

3. **2.** **1.**

a mi mamá.
Estudiante B: *Sí, con mucho gusto. ¡Sembremos árboles! o Si pudiera, lo haría, pero tengo que ayudar*

árboles hoy.
Estudiante A: *¿Podrías darme una mano? Voy a sembrar*

sembrar árboles

modelo

Luego, cambien de papel. Sigan el modelo.
El (Ella) te puede ayudar o te dará una excusa.
compañero(a) que te ayude con las actividades.
Estudiante A Mira los dibujos y pide a tu

Estudiante B: Tu compañero(a) te pide ayuda con las actividades que se muestran en los dibujos. Dile que sí o dale una excusa. Luego, cambien de papel. Sigan el modelo.

modelo

sembrar árboles

Estudiante A: *¿Podrías darme una mano? Voy a sembrar árboles hoy.*

Estudiante B: *Sí, con mucho gusto. ¡Sembremos árboles!*
o Si pudiera, lo haría, pero tengo que ayudar a mi mamá.

4. **5.** **6.**

---

10 Unidad 2 Etapa 1 p. 121
¿Qué harías?

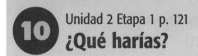

6. visitarlos / ayudarlos a ir de compras
5. buscar una escuela bilingüe … / escoger una con un buen programa de deportes
4. darle dinero / ser voluntario(a) en…
Para responder
3. buscar trabajo después de la graduación
2. conservar el agua
1. mejorar tu escuela
Para preguntar

Estudiante B: *Educaría al público.*

Estudiante A: *¿Cómo lucharías contra el racismo?*

luchar contra el racismo

modelo

preguntar y responder. Sigan el modelo.
situación. Pueden usar las siguientes ideas para
compañero(a) sobre lo que harían en cada
Estudiante A Conversa con un(a)

Estudiante B Conversa con un(a) compañero(a) sobre lo que harían en cada situación. Pueden usar las siguientes ideas para preguntar y responder. Sigan el modelo.

modelo

luchar contra el racismo

Estudiante A: *¿Cómo lucharías contra el racismo?*

Estudiante B: *Educaría al público.*

Para responder

1. ser voluntario en…

2. reciclar / educar al público sobre…

3. buscar un trabajo mejor pagado / buscar algo interesante

Para preguntar

4. ayudar a la gente sin hogar

5. escoger una universidad

6. ayudar a los ancianos

11 — Unidad 2 Etapa 2 p. 140 — Treinta segundos

Estudiante A Menciona uno de estos lugares a tu compañero(a). Él (Ella) tiene 30 segundos para hacer todas las oraciones que pueda. Tiene que incluir el lugar, las expresiones impersonales y los verbos de su lista. Luego, cambien de papel.

modelo

Estudiante A: *Estás en la clase de español.*

Estudiante B: *No es bueno que la maestra les dé mucha tarea a los estudiantes de la clase de español. Es importante que tú no salgas temprano de la clase de español…*

entender	vestirse	pensar		
ser	pedir	decir	hacer	
estar	empezar	ir	cerrar	repetir

1. el centro de la comunidad
2. la escuela
3. el parque
4. el restaurante

Estudiante B Tu compañero(a) te menciona un lugar. ¿Cuántas oraciones puedes hacer en 30 segundos? Incluye el lugar, las expresiones impersonales y los verbos de la lista en tu respuesta. Luego, cambien de papel.

modelo

Estudiante A: *Estás en la clase de español.*

Estudiante B: *No es bueno que la maestra les dé mucha tarea a los estudiantes de la clase de español. Es importante que tú no salgas temprano de la clase de español…*

dar	mentir	jugar	saber	volver	seguir
ver	servir	conocer	dormir	reír	salir

5. el cine
6. en casa
7. en la clase
8. en una tienda

12 — Unidad 2 Etapa 2 p. 143 — ¿Bueno o malo?

Estudiante A Lee las situaciones en la primera lista. Imagina quién las hizo y dónde ocurrieron. Cuéntasela a tu compañero(a) para saber su opinión. Sigue el modelo. Luego, cambien de papel y opina sobre los comentarios de tu compañero(a) usando las opciones de la segunda lista.

modelo

ir a España

Estudiante A: *Mi maestra de historia fue a España.*

Estudiante B: *Es bueno que tu maestra haya visitado ese país.*

• ayudar a proteger la capa de ozono
• no reciclar
• trabajar de voluntario(a)
• juntar fondos para una buena causa

1. haber una sequía (¿dónde?)
2. echar químicos en el aire

Estudiante B Tu compañero(a) te cuenta unas situaciones. ¿Qué opinas? Usa las ideas de la primera lista para responder. Sigue el modelo. Luego, cambien de papel. Imagina cómo pasaron las situaciones en la segunda lista. ¿Quién las hizo y dónde ocurrieron? Cuéntaselas a tu compañero(a).

modelo

ir a España

Estudiante A: *Mi maestra de historia fue a España.*

Estudiante B: *Es bueno que tu maestra haya visitado ese país.*

• no llover bastante
• embellecer la tierra
• visitar ese país
• contaminar

3. echar botellas y latas en el basurero
4. no usar aerosoles

13 Unidad 2 Etapa 3 p. 159
Chisme

14 Unidad 2 Etapa 3 p. 165
¡Vacaciones ideales!

Estudiante A Dile a tu compañero(a) algo extraordinario que sepas o puedas inventar acerca de las siguientes personas. Él (Ella) te va a dar su opinión. Luego cambien de papel. Tu compañero(a) te va a hacer un comentario. ¿Qué opinas? Responde con una expresión de emoción de la segunda.

modelo

el jugador de béisbol

Estudiante A: ¿Sabes que Barry Bonds tiene el récord de jonrones?

Estudiante B: ¡Me alegro de que él disfrute de ese honor!

1. el (la) presidente de...
2. el actor...
3. la cantante...
4. el (la) autor(a)...

• Espero que...
• Me alegro de que...
• Siento que...
• Es triste que...
• Ojalá que...

Estudiante A Pregúntale a tu compañero(a) si podrás hacer estas actividades durante tus vacaciones. Él (Ella) te responde después de consultar el folleto turístico. Sigan el modelo. Luego, cambien de papel.

modelo

andar en bicicleta

Estudiante A: ¿Voy a andar en bicicleta?

Estudiante B: Dudo que andes en bicicleta allí. o Creo que vas a andar en bicicleta allí.

1. alquilar videos
2. hacer montañismo
3. navegar en tabla de vela
4. esquiar en el agua

Estudiante B Tu compañero(a) te va a contar ciertas cosas extraordinarias de varias personas. ¿Qué opinas? Escoge una expresión de emoción de la primera lista para responder. Luego cambien de papel y dile a tu compañero(a) algo extraordinario que sepas o puedas inventar acerca de las personas de la segunda lista. Él (Ella) te va a dar su opinión.

modelo

Me alegro de que…

Estudiante A: ¿Sabes que Barry Bonds tiene el récord de jonrones?

Estudiante B: ¡Me alegro de que él disfrute de ese honor!

5. el (la) cantante…
6. uno(a) de los profesores…
7. la actriz…
8. el (la) deportista…

• Es ridículo que…
• Ojalá que…
• Espero que…
• Es una lástima que…
• Me alegro de que…

Estudiante B Tienes el folleto turístico de las vacaciones de tu compañero(a). Mira el folleto y contesta sus preguntas sobre las actividades que podrá hacer. Sigan el modelo. Luego, cambien de papel.

modelo

andar en bicicleta

Estudiante A: ¿Voy a andar en bicicleta?

Estudiante B: Dudo que andes en bicicleta allí. o Creo que vas a andar en bicicleta allí.

5. pescar en alta mar
6. esquiar
7. escalar montañas
8. volar en planeador

15 — Unidad 3 Etapa 1 p. 191 — Según ciertas condiciones

Estudiante A Pregunta a tu compañero(a) sobre las siguientes actividades. Luego contesta sus preguntas escogiendo una de las opciones de la segunda lista. Sigan el modelo.

modelo

ir al teatro

Estudiante A: ¿Vas al teatro?

Estudiante B: Voy con tal de que vea una comedia. **o**
Voy con tal de que vea un musical.

1. hablar en clase
2. ir a fiestas
3. cantar
4. esquiar

- con tal de que (ser por correo electrónico / ¿?)
- con tal de que (ser interesante / ¿?)
- antes de que (salir de casa / ¿?)
- a menos que (ser) deportes individuales / ¿?)

Estudiante B Tu compañero(a) te va a hacer preguntas. Contéstalas con oraciones escogiendo una de las opciones de la primera lista. Luego cambien de papel.

modelo

con tal de que (un musical / ¿?)

Estudiante A: ¿Vas al teatro?

Estudiante B: Voy con tal de que vea un musical. **o**
Voy con tal de que vea una comedia.

- a menos que (estar cansado(a) / ¿?)
- con tal de que (ir mis amigos / ¿?)
- con tal de que (hacer buen tiempo / ¿?)
- para que (todos saber mis opiniones / ¿?)

5. trabajar
6. practicar deportes
7. hacer la tarea
8. escribir cartas

16 — Unidad 3 Etapa 1 p. 195 — ¿Qué querían?

Estudiante A Las personas que aparecen a continuación querían que otras personas hicieran algo. Pregúntale a tu compañero(a) qué querían. Luego cambien de papel y contesta las preguntas de tu compañero(a), usando la segunda lista.

modelo

los músicos

Estudiante A: ¿Qué querían los músicos?

Estudiante B: Deseaban que la gente bailara la bomba.

1. la madre de Hernán • la profesora no dar un examen difícil
2. el graduando • las fotos salir bien
3. la mesera • a la graduanda gustarle su regalo
4. el presidente de la compañía • los estudiantes entrenarse todos los días

Estudiante B Contesta las preguntas de tu compañero(a) usando la información de la primera lista. Luego cambien de papel y pregunta a tu compañero(a) qué querían estas personas. Sigan el modelo.

modelo

la gente bailar la bomba

Estudiante A: ¿Qué querían los músicos?

Estudiante B: Deseaban que la gente bailara la bomba.

- los clientes dejar una buena propina
- su hijo hacer la limpieza
- los empleados trabajar más rápidamente
- todos sus parientes y amigos venir a su fiesta

5. los padrinos de la graduanda
6. la profesora de educación física
7. los estudiantes de español
8. el fotógrafo

17 Unidad 3 Etapa 2 p. 213
Anuncios personales

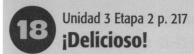

HOMBRE
Trabajo de voluntario después de la escuela. Los fines de semana juego al fútbol. Si te gusta hacer esas actividades, llama al 2-53-56-52.

MUJER
Soy una persona positiva y me río mucho. Me gusta bailar y salir con amigos. ¿Tienes interés? Mi número: 1-74-80-34.

HOMBRE
Soy una persona muy honesta y fiel. Si buscas a una persona así, llámame al 5-35-98-80.

MUJER
En mi tiempo libre hago esculturas. También escribo poemas. Si tenemos algo en común, llámame al 7-74-12-38.

1. Manolo / intelectual
2. Perla / activo
3. Javier / cómica
4. Cristóbal / sociable

Mensaje: «Soy una mujer atleta interesada en una persona a quien le gusta entrenarse».

Estudiante B: Recomiendo que llame al número 3–34–61–57.

Estudiante A: Cristina busca a alguien que sea atleta.

Cristina/atleta

modelo

Estudiante A Busca un(a) amigo(a) para estas personas. Lee las características que busca y tu compañero(a) leerá los anuncios. ¿A qué número debe llamar? Escriban un mensaje para cada una.

Estudiante B Tu compañero(a) te lee las características de personas que buscan amigos. Decide a qué número debe llamar la persona. Escriban un mensaje para cada una.

modelo

Cristina / atleta

Estudiante A: *Cristina busca a alguien que sea atleta.*

Estudiante B: *Recomiendo que llame al número 3–34–61–57.*

Mensaje: «*Soy una mujer atleta interesada en una persona a quien le gusta entrenarse*».

5. Julieta / activo
6. Víctor / divertida
7. Nadia / sincero
8. Jesús / creativa

MUJER.
Me encanta conocer a otra gente. Voy a fiestas y bailes con frecuencia. Si te gusta charlar, me puedes llamar al 5-22-47-34.

MUJER.
Paso mucho tiempo leyendo. Me fascinan las matemáticas y algún día quiero descubrir una fórmula nueva. Si quieres conocerme, llama al 3-23-40-68.

MUJER.
Conmigo, todo el mundo se ríe. Dicen que cuento los mejores chistes...pero siempre son de buen gusto. Llámame al 6-77-71-52.

HOMBRE.
Mis pasatiempos incluyen volar en planeador y escalar montañas. Si te gusta la aventura, soy la persona que buscas. Marca el 8-98-14-15.

---

18 Unidad 3 Etapa 2 p. 217
¡Delicioso!

1. pedir / tener mucha hambre
2. servir / celebrar la fiesta de Navidad
3. tomar / comer pastel de chocolate
4. comer / ser las 10:00 de la noche

Estudiante B: Comería mucho queso y helado.

Estudiante A: ¿Qué comerías si quisieras ser más grande?

comer / querer ser más grande

modelo

Estudiante A Pregunta a tu compañero(a) qué comería en estas ocasiones. Luego, cambien de papel y responde a sus preguntas basándote en el dibujo.

Estudiante B Tu compañero(a) quiere saber qué comerías en estas ocasiones. Usa el dibujo para responder a sus preguntas. Luego, cambien de papel.

modelo

comer / querer ser más grande

Estudiante A: *¿Qué comerías si quisieras ser más grande?*

Estudiante B: *Comería mucho queso y helado.*

5. pedir / querer ser más delgado(a)
6. beber / tener mucha sed
7. servir / celebrar una fiesta de graduación
8. comer / estar en la casa de mis abuelos

19 Unidad 3 Etapa 3 p. 237
¡A jugar!

Estudiante A Juega a esto con tu compañero(a). El objeto es que tu compañero(a) adivine quiénes son las personas que tú describes. ¡Ojo! Sólo puedes describir los deseos de la persona. Sigue el modelo. ¿Cuántas personas puede adivinar en cuatro minutos? Luego, cambien de papel y trata de adivinar quiénes son las personas que describe tu compañero(a).

modelo

Estudiante A: Insiste en que los escritores terminen su trabajo a tiempo. Quiere que no haya errores en el libro… Desea que muchas personas compren el libro…

Estudiante B: ¿El maestro?… ¡La editora!

policía	agente de viajes
peatón(a)	piloto
reina	novio(a)
actor	estudiante

Estudiante B Adivina quiénes son las personas que describe tu compañero(a). Tienes cuatro minutos para adivinarlas. Luego cambien de papel y describe a las personas a continuación. ¡Ojo! Sólo puedes contar a tu compañero(a) los deseos de la persona. Sigue el modelo.

modelo

Estudiante A: Insiste en que los escritores terminen su trabajo a tiempo. Quiere que no haya errores en el libro… Desea que muchas personas compren el libro…

Estudiante B: ¿El maestro?… ¡La editora!

doctor(a)	**bebé**
presidente	maestro(a)
pintor(a)	comediante
mesero(a)	ladrón

20 Unidad 3 Etapa 3 p. 239
¿Verdad o mentira?

Estudiante A Di algo cierto o inventa algo falso sobre las siguientes personas. ¿Lo cree tu compañero(a)? Si no te cree, te va a responder con una expresión de duda. Si te cree, responderá con una expresión positiva. ¿Puedes engañarlo(la)? Luego, cambien de papel.

modelo

tus tíos / verdad

Estudiante A: Mis tíos tienen nueve hijos.

Estudiante B: Dudo que tus tíos tengan nueve hijos. **o:** Pienso que tus tíos tienen nueve hijos.

Estudiante A: ¡Tienes razón! Es verdad que tienen nueve hijos. **o:** ¡Te engañé!

1. tu madre / verdad
2. tu mejor amigo(a) / verdad
3. tu primo(a) / mentira
4. un(a) maestro(a) / mentira

Estudiante B Tu compañero(a) te va a decir algo cierto o inventar algo falso. Si le crees, responde con una expresión positiva. Si no, responde con una expresión de duda. ¿Puede engañarte? Luego, cambien de papel y usa la siguiente lista.

modelo

tus tíos / verdad

Estudiante A: Mis tíos tienen nueve hijos.

Estudiante B: Dudo que tus tíos tengan nueve hijos. **o:** Pienso que tus tíos tienen nueve hijos.

Estudiante A: ¡Tienes razón! Es verdad que tienen nueve hijos. **o:** ¡Te engañé!

5. tu(s) hermano(s) / mentira
6. un actor (una actriz) / mentira
7. tus padres / verdad
8. tu abuelo(a) / verdad

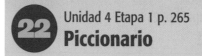

21 — Unidad 4 Etapa 1 p. 263
Una solicitud

Nombre: _____
Fecha de solicitud: 25-4-05 _____
Teléfono: 9-76-42-98 _____
Fecha de nacimiento: _____
Ciudadanía _____
Estado civil: soltera _____
Educación: licenciatura _____
Campo de estudio: _____
Sueldo corriente: _____
Recomendación: José Cruz _____

modelo

Estudiante A: ¿Cómo se llama?

Estudiante B: …

Estudiante A Con tu compañero(a), completa la información en la solicitud. Usa palabras interrogativas para obtener la información necesaria.

Estudiante B Con tu compañero(a), completa la información en la solicitud. Usa palabras interrogativas para obtener la información necesaria.

modelo

Estudiante A: ¿Cómo se llama?

Estudiante B: Se llama Juana Aiken.

Nombre: Juana Aiken _____
Fecha de solicitud: _____
Teléfono: _____
Fecha de nacimiento: 21-3-78 _____
Ciudadanía argentina _____
Estado civil: _____
Educación: _____
Campo de estudio: informática _____
Sueldo: $30.000 _____
Recomendación: _____

22 — Unidad 4 Etapa 1 p. 265
Piccionario

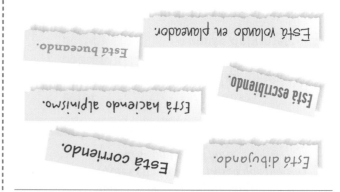

Está buscando.
Está volando en planeador.
Está haciendo alpinismo.
Está escribiendo.
Está corriendo.
Está dibujando.

Estudiante A: (Dibuja a una persona cantando.)

Estudiante B: Está hablando… Está gritando… ¡Está cantando!

modelo

(Está cantando.)

Estudiante A Juega a esto con tu compañero(a). Haz dibujos para comunicar las siguientes acciones y tu compañero(a) va a adivinarlas. Luego cambien de papel y adivina las acciones que dibuja tu compañero(a).

Estudiante B Juega a esto con tu compañero(a). Él (Ella) va a hacer dibujos para representar a varias acciones. Adivina lo que está pasando. Luego cambien de papel y haz dibujos para comunicar las siguientes acciones a tu compañero(a).

modelo

(Está cantando.)

Estudiante A: (Dibuja a una persona cantando.)

Estudiante B: Está hablando… Está gritando… ¡Está cantando!

Está comiendo.
Está bailando.
Está durmiendo.
Está estudiando.
Está lloviendo.
Está graduándose.

23 ¿Qué hay?
Unidad 4 Etapa 2 p. 285

Estudiante A ¿Son iguales los dibujos que tienen tú y tu compañero(a)? Hazle preguntas a tu compañero(a) para saber más de su dibujo. Después contesta sus preguntas sobre la escena que ves. Sigan el modelo.

modelo

Estudiante A: ¿Hay algún anuncio publicitario en la puerta?

Estudiante B: No hay ningún anuncio publicitario en la puerta.

Estudiante B ¿Son iguales los dibujos que tienen tú y tu compañero(a)? Contesta las preguntas de tu compañero(a) sobre la escena que ves. Luego hazle preguntas para saber más sobre su dibujo. Sigan el modelo.

modelo

Estudiante A: ¿Hay algún anuncio publicitario en la puerta?

Estudiante B: No hay ningún anuncio publicitario en la puerta.

---

24 ¿Ideas diferentes?
Unidad 4 Etapa 2 p. 288

1. Bárbara / maestra   2. Oscar / mecánico
3. Lola / abogada   4. Miguel / ingeniero

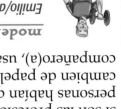

modelo

Emilio/artista

Estudiante A: Emilio es artista.

Estudiante B: Sus padres querían que hubiera sido deportista.

Estudiante A Aquí ves las profesiones de varias personas. Pregúntale a tu compañero(a) si son las profesiones que los padres de las personas habían querido para sus hijos. Luego cambien de papel y contesta las preguntas de tu compañero(a), usando la información de abajo.

Estudiante B Tu compañero(a) te va a decir las profesiones que escogieron varias personas. ¿Es lo que querían sus padres? Usa la siguiente información para contestar. Luego cambien de papel y describe a tu compañero(a) las profesiones de las personas de abajo.

modelo

Emilio/artista

Estudiante A: Emilio es artista.

Estudiante B: Sus padres querían que hubiera sido deportista.

5. Chela / arquitecta   7. Félix / juez
6. Paco / intérprete   8. Diana / taxista

1.   2.   3.   4.

25 Unidad 4 Etapa 3 p. 308
Estadísticas

```
100%
 75%
 50%
 25%
  0%
```
trabajadores sociales
banqueros
traductores
diplomáticos

4. financieros   2. agentes de viajes
3. corresponsales   1. académicos

El veinte por ciento de ellos son intérpretes.

Estudiante B: Un quinto de ellos son intérpretes. o

Estudiante A: ¿Cuántos son intérpretes?

intérprete

modelo

de abajo.

compañero(a), usando los datos de la gráfica
papel y contesta las preguntas de tu
para las siguientes carreras. Luego cambien de
de la universidad. Pregunta cuáles son los datos
estudiantes bilingües que acaban de graduarse
estadísticas sobre las carreras de varios

Estudiante A Tu compañero(a) tiene

Estudiante B Tu compañero(a) busca
estadísticas sobre las carreras de estudiantes
bilingües que acaban de graduarse de la
universidad. Contesta sus preguntas basándote
en la información de la siguiente gráfica. Luego
cambien de papel y pregunta cuáles son los
datos para las siguientes carreras.

modelo

intérprete

Estudiante A: ¿Cuántos son intérpretes?

Estudiante B: Un quinto de ellos son intérpretes. o
El veinte por ciento de ellos son intérpretes.

5. diplomáticos
6. trabajadores
   sociales
7. traductores
8. banqueros

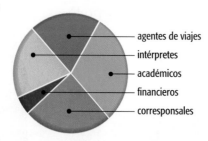

agentes de viajes
intérpretes
académicos
financieros
corresponsales

---

26 Unidad 4 Etapa 3 p. 313
¿Cuándo?

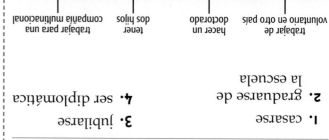

OCTAVIO

| 5 años | 10 años | 15 años | 20 años |

trabajar de voluntario en otro país | hacer un doctorado | tener dos hijos | trabajar para una compañía multinacional

2. graduarse de la escuela   4. ser diplomática
1. casarse   3. jubilarse

¿Y tú?...

Estudiante B: Habrá hecho eso dentro de cinco años.

países?

Estudiante A: ¿Cuándo habrá viajado Estela a muchos

viajar a muchos países

modelo

de Octavio.

preguntas de tu compañero(a) según las metas
(ella)? Luego, cambien de papel y contesta las
cuándo Estela habrá hecho estas cosas. ¿Y él

Estudiante A Pregunta a tu compañero(a)

Estudiante B Contesta las preguntas de tu
compañero(a) sobre Estela. Luego, pregunta
cuándo Octavio habrá hecho estas cosas y
cuándo las hará tu compañero(a).

modelo

viajar a muchos países

Estudiante A: ¿Cuándo habrá viajado Estela a muchos
países?

Estudiante B: Habrá hecho eso dentro de cinco años.
¿Y tú?...

graduarme | viajar a muchos países | casarme | ser diplomática | jubilarme

| 1 año | 5 años | 10 años | 15 años | 30 años |

ESTELA

5. hacer un doctorado
6. tener dos hijos
7. trabajar para una compañía multinacional
8. trabajar de voluntario en otro país

27 Unidad 5 Etapa 1 p. 340
¿Cuántas diferencias?

Estudiante A Hay algunas diferencias entre tu dibujo y el de tu compañero(a). Hazle preguntas y contesta las preguntas de tu compañero(a) para determinar cuántas diferencias hay.

modelo

¿cuál? / nombre

Estudiante A: ¿Cuál es su nombre?

Estudiante B: No sé cuál es su nombre. ¿Cuál es el nombre de la joven?

1. ¿qué? / primer plano   5. ¿qué? / instrumentos

3. ¿cuál? / diferencia / panderetas   7. ¿?

---

28 Unidad 5 Etapa 1 p. 343
Intereses personales

Estudiante A Pregúntale a tu compañero(a) sobre sus intereses, usando la lista de la parte A. Luego contesta las preguntas de tu compañero(a), basándote en la lista de la parte B. Sigue el modelo.

modelo

literatura (cómica / ciencia ficción)

Estudiante A: ¿Qué tipo de literatura te interesa?

Estudiante B: Me interesa la literatura que sea muy cómica.

A	B
1. música / interesar	4. (incluye / no incluye) letra
2. pintores / admirar	5. darme (risa / miedo)
3. bailarines / gustar mirar	6. ser (siglo 17 / siglo 20)

Estudiante B Hay algunas diferencias entre el dibujo que tienes y el de tu compañero(a). Hazle preguntas usando la siguiente lista y contesta las preguntas de tu compañero(a) para determinar cuántas diferencias hay.

modelo

¿cuál? / nombre

Estudiante A: ¿Cuál es su nombre?

Estudiante B: No sé cuál es su nombre. ¿Cuál es el nombre de la joven?

2. ¿qué? / fondo   6. ¿qué? / tocar

4. ¿qué? / hacer   8. ¿?

Estudiante B Tú y tu compañero(a) quieren conocerse mejor. Contesta las preguntas de tu compañero(a) basándote en la información de la lista A. Sigue el modelo. Luego hazle preguntas a tu compañero(a) usando la lista B.

modelo

literatura (cómica / ciencia ficción)

Estudiante A: ¿Qué tipo de literatura te interesa?

Estudiante B: Me interesa la literatura que sea muy cómica.

A	B
1. darme (ánimo / tranquilidad)	4. música / escuchar más
2. pintar (naturalezas muertas / cuadros históricos)	5. obras teatrales / preferir ver
3. bailar (el tango / el flamenco)	6. estilo de cuadros / interesar más

29 Unidad 5 Etapa 2 p. 359
¿Lo hizo?

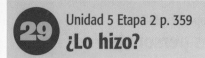

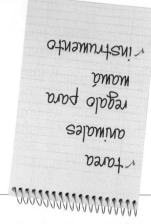

Estudiante A Tu compañero(a) escribió una lista de actividades. Pregúntale si ya las hizo, usando la siguiente información. Luego cambien de papel e imagínate que la lista en el cuaderno es tuya. Una marca indica que cumpliste la actividad. Sigue el modelo.

modelo

pintar la máscara

Estudiante A: ¿Pintaste la máscara?

Estudiante B: Sí, la pinté.

1. hacer los quehaceres
2. escuchar salsa
3. escribir el ensayo
4. navegar por Internet para buscar información

Notebook list:
✓ tarea
✓ animales
regalo para mamá
✓ instrumento

Estudiante B Imagínate que la lista en el cuaderno es tuya. Una marca indica que hiciste la actividad indicada. Contesta las preguntas de tu compañero(a) según la lista. Luego cambien de papel y pregúntale a tu compañero(a) si hizo cada una de las siguientes actividades.

modelo

Pintar la máscara

Estudiante A: ¿Pintaste la máscara?

Estudiante B: Sí, la pinté.

1. bañar los perros
2. leer la biografía
3. practicar el violín
4. comprar el tejido

Notebook list:
✓ nuevo disco compacto
tarea
✓ limpiar
✓ computadora

30 Unidad 5 Etapa 2 p. 363
¿Quién es?

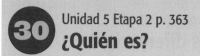

Estudiante A Juega a esto con tu compañero(a). Para cada categoría de la siguiente lista, escoge un ejemplo y escríbelo en una hoja aparte. Dale pistas (clues), usando la forma correcta de **el que** o **el cual,** para que tu compañero(a) lo adivine. Sigue el modelo. Luego, cambien de papel y trata de adivinar los ejemplos de tu compañero(a).

modelo

jueza

Estudiante A: Esta jueza es la que fue la primera mujer en la Corte Suprema de Estados Unidos.

Estudiante B: ¡Sandra Day O'Connor!

1. político
2. actriz
3. conquistador
4. artista
5. monumento
6. película

Estudiante B Juega a esto con tu compañero(a). Tu compañero(a) te va a dar pistas (clues) para ver si puedes adivinar una cosa o una persona que él (ella) ha escrito en otra hoja de papel. Sigue el modelo. Luego cambien de papel. Para cada categoría de la siguiente lista, escoge un ejemplo y escríbelo en una hoja aparte. Dale pistas a tu compañero(a), usando la forma correcta de **el que** o **el cual,** para que tu compañero(a) lo adivine.

modelo

jueza

Estudiante A: Esta jueza es la que fue la primera mujer en la Corte Suprema de Estados Unidos.

Estudiante B: ¡Sandra Day O'Connor!

7. músico o conjunto
8. explorador
9. actor
10. libro
11. país
12. deportista

MÁS COMUNICACIÓN

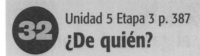

31 Unidad 5 Etapa 3 p. 384
¡No repitas!

Estudiante A Entrevista a tu compañero(a), usando las siguientes opciones. Luego cambien de papel y contesta las preguntas de tu compañero(a). Cuando contestes, no repitas el sustantivo. Sigan el modelo.

modelo

comer comidas (picantes / blandas)

Estudiante A: ¿Prefieres comer las comidas picantes o las comidas blandas?

Estudiante B: Me gustan más las picantes.

1. comprar un carro (nuevo / usado)
2. tomar una clase (fácil / interesante)
3. ver finales (feliz / irónico)
4. leer novelas (innovador / tradicional)
5. hacer deportes (individuales / en equipo)

Estudiante B Tu compañero(a) te va a entrevistar. Cuando contestes sus preguntas, no repitas el sustantivo. Sigan el modelo. Luego cambien de papel y entrevista a tu compañero(a), usando las opciones de la siguiente lista.

modelo

comer comidas (picantes / blandas)

Estudiante A: ¿Prefieres comer las comidas picantes o las comidas blandas?

Estudiante B: Me gustan más las picantes.

6. ver una película (creativo / dramático)
7. escuchar una canción (expresivo / divertido)
8. mirar a modelos (rubio / pelirrojo)
9. llevar la ropa (clásico / original)
10. hacer los papeles (romántico / cómico)

32 Unidad 5 Etapa 3 p. 387
¿De quién?

Oprah Winfrey   Tiger Woods

Elvis Presley   Laura Esquivel

Estudiante A A ti y a tus amigos siempre les fascinan las personas famosas. Dile a tu compañero(a) la actividad que hiciste y él (ella) tiene que identificar a la persona famosa relacionada con la actividad. Sigue el modelo. Luego cambien de papel.

modelo

visitar una compañía de tecnología fenomenal

Estudiante A: Visité una compañía de tecnología fenomenal.

Estudiante B: Ah, visitaste la de Bill Gates.

1. mirar una película de ciencia ficción
2. leer un libro sobre la esclavitud
3. estudiar el cubismo
4. coleccionar tarjetas de un atleta

Estudiante B A ti y a tus amigos siempre les fascinan las personas famosas. Tu compañero(a) te dice la actividad que hizo y tienes que identificar a la persona famosa relacionada con la actividad. Sigue el modelo. Luego cambien de papel.

modelo

visitar una compañía de tecnología fenomenal

Estudiante A: Visité una compañía de tecnología fenomenal.

Estudiante B: Ah, visitaste la de Bill Gates.

Michael Jordan   **George Lucas**

Pablo Picasso   **Toni Morrison**

5. visitar una casa famosa llamada «Graceland»
6. leer una novela que en cada capítulo describe cómo preparar comida
7. ver un programa de tele con una anfitriona que también es actriz
8. mirar partidos de golf increíbles

33 Unidad 6 Etapa 1 p. 411
El perro, el gato y el ratón

Estudiante A Los dibujos representan lo que pasó en un dibujo animado, pero no aparecen en el orden correcto. Hay seis dibujos en total. Junto a tu compañero(a), determinen el orden lógico de los dibujos.

modelo

Estudiante A: ¿Cuál es el primer dibujo?

Estudiante B: Pues, el gato...

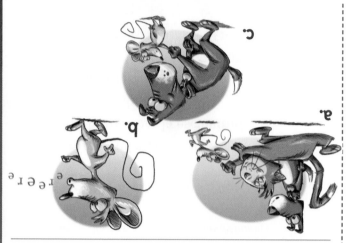

Estudiante B Los dibujos representan lo que pasó en un dibujo animado, pero no aparecen en el orden correcto. Hay seis dibujos en total. Junto a tu compañero(a), determinen el orden lógico de los dibujos.

modelo

Estudiante A: ¿Cuál es el primer dibujo?

Estudiante B: Pues, el gato...

34 Unidad 6 Etapa 1 p. 417
¿Qué dijo?

Estudiante A Pregúntale a tu compañero(a) qué dijeron las siguientes personas. Luego, cambien de papel y escoge una de las opciones de la derecha para contestar lógicamente las preguntas de tu compañero(a).

modelo

los jóvenes «Vamos a alquilar un video para el sábado.»

Estudiante A: ¿Qué dijeron los jóvenes?

Estudiante B: Dijeron que iban a alquilar un video para el sábado.

1. el actor
2. la crítica
3. los padres
4. la deportista

• «Es bastante difícil dirigir un programa en vivo y directo.»
• «Queremos mirar los dibujos animados.»
• «No miren esa película porque es prohibida para menores.»
• «Es el mejor guión que he escrito.»

Estudiante B Tu compañero(a) quiere saber qué dijeron varias personas. Escoge una de las siguientes opciones para contestar sus preguntas lógicamente. Luego, cambien de papel y pregúntale qué dijeron las personas de la lista de la derecha.

modelo

los jóvenes «Vamos a alquilar un video para el sábado.»

Estudiante A: ¿Qué dijeron los jóvenes?

Estudiante B: Dijeron que iban a alquilar un video para el sábado.

• «Todos deben ir a ver esta película de cuatro estrellas.»
• «Ojalá que alguien grabe el partido.»
• «Quiero hacer el papel de Juan Carlos.»
• «No miren la tele antes de hacer la tarea.»

5. los niños
6. el guionista
7. los padres
8. la directora

35 Unidad 6 Etapa 2 p. 433
¿Qué quieres?

Estudiante A Pregúntale a tu compañero(a) si quiere las siguientes cosas. Luego, cambien de papel y contéstale sus preguntas, usando las expresiones de la lista a la derecha.

modelo

radio portátil   en cuanto

Estudiante A: ¿Te gustaría un radio portátil?

Estudiante B: Sí, voy a comprar uno en cuanto tenga bastante dinero.

1. videocámara   • hasta que
2. contestadora   • aunque
   automática   • para que
3. asistente electrónico • con tal que
4. beeper

Estudiante B Cuando tu compañero(a) te pregunte si quieres varias cosas, usa las siguientes expresiones para contestar. Luego, cambien de papel y pregúntale a él (ella) si quiere las cosas de la lista a la derecha.

modelo

radio portátil   en cuanto

Estudiante A: ¿Te gustaría comprar un radio portátil?

Estudiante B: Sí, voy a comprar uno en cuanto tenga bastante dinero.

- a menos que   5. audífonos
- tan pronto como   6. computadora portátil
- para que   7. telemensaje
- en caso de que   8. equipo estereofónico

---

36 Unidad 6 Etapa 2 p. 434
Dibujos con acción

Estudiante A Dibuja las siguientes cosas. Tu compañero(a) describe cada objeto y su posición en la escena. Luego, cambien de papel y adivina (describe) lo que dibuja tu compañero(a). ¿Quién adivina más rápido?

1. El teléfono inalámbrico está delante de la computadora portátil.
2. El beeper está encima de la contestadora automática.
3. Los audífonos están alrededor del radio portátil.
4. La grabadora está debajo de la mesa.

Estudiante B Tu compañero(a) te va a dibujar cuatro escenas. ¿Puedes describir cada objeto y su posición en la escena? Luego, cambien de papel y dibuja las siguientes escenas para tu compañero(a). ¿Quién adivina más rápido?

5. La pila está junto a la videocámara.
6. El equipo estereofónico está abajo.
7. El teléfono celular está dentro del carro.
8. El televisor portátil está frente al altoparlante.

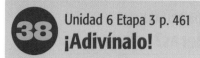

37 Unidad 6 Etapa 3 p. 459
Información

Estudiante A Cuatro de tus amigos van a viajar este verano. Trabajando junto a tu compañero(a), completa la tabla con la información que falta para saber más sobre los viajes.

modelo

Estudiante A: ¿Cuál es la hora de salida de Esteban?
Estudiante B: La hora de salida de Esteban es las 13:15.

	Esteban	Jesús	Emilia	Dani
destinación		Cali		Ponce
distancia		675 mi.		1800 mi.
hora de salida		14:30		8:45
duración del viaje		7 días		15 días

---

38 Unidad 6 Etapa 3 p. 461
¡Adivínalo!

Estudiante A Juega a esto con tu compañero(a). Da pistas para que tu compañero(a) adivine el verbo y la preposición en negrita de cada una de las siguientes oraciones. ¡Ojo! No puedes usar esas palabras en tu descripción. Luego cambien de papel y trata de adivinar las expresiones de tu compañero(a).

modelo

El niño **viene con** su mamá.

Estudiante A: El niño acompaña a su mamá… Él llega acompañado por ella… Su mamá lo trae…

Estudiante B: El niño va con su mamá… sale con su mamá… ¡**Viene con** su mamá!

1. La niña **tiene que** comprar un disco.
2. El estudiante **aprende a** crear una página-web.
3. La secretaria **se olvida de** enviar el fax.
4. El programador **acaba de** limpiar la pantalla.

Estudiante B Cuatro de tus amigos van a viajar este verano. Trabajando junto a tu compañero(a), completa la tabla con la información que falta para saber más sobre los viajes.

modelo

Estudiante A: ¿Cuál es la hora de salida de Esteban?

Estudiante B: La hora de salida de Esteban es las 13:15.

	Esteban	Jesús	Emilia	Dani
destinación		Mérida		Madrid
distancia	850 mi.		540 mi.	
hora de salida	13:15		14:30	
duración del viaje		12 días		5 días

Estudiante B Juega a esto con tu compañero(a). Trata de adivinar las expresiones según las pistas de tu compañero(a). Luego cambien de papel y da pistas para que tu compañero(a) adivine la expresión en negrita de cada una de las siguientes oraciones. ¡Ojo! No puedes usar esas palabras en tu descripción.

modelo

El niño **viene con** su mamá.

Estudiante A: El niño acompaña a su mamá… Él llega acompañado por ella… Su mamá lo trae…

Estudiante B: El niño va con su mamá… sale con su mamá… ¡**Viene con** su mamá!

5. El programa **insiste en** que uses la contraseña.
6. **Hay que** hacer doble clic.
7. La profesora **comienza a** participar en un grupo de conversación.
8. **Tengo ganas de** navegar por Internet.

Juegos—respuestas

UNIDAD 1

Etapa 1 p. 53: c. descarado

Etapa 2 p. 75: monedero

Etapa 3 p. 99: reparar, desyerbar, vaciar, descansar

UNIDAD 2

Etapa 1 p. 127: 1. c, 2. b, 3. a

Etapa 2 p. 149: la capa de ozono

Etapa 3 p. 173: mono, mariposa

UNIDAD 3

Etapa 1 p. 201: b. prohibir

Etapa 2 p. 223: pavo, pastel

Etapa 3 p. 247: faro

UNIDAD 4

Etapa 1 p. 275: 2

Etapa 2 p. 297: jubilarse, beneficios, aumentar, no trabajar

Etapa 3 p. 321: académico, agente de ventas

UNIDAD 5

Etapa 1 p. 349: una naturaleza muerta

Etapa 2 p. 371: Somos una civilización avanzada.

Etapa 3 p. 395: b. el crítico

UNIDAD 6

Etapa 1 p. 423: programas de misterio, los documentales, el control remoto

Etapa 2 p. 445: una batería

Etapa 3 p. 469: Carmen

Gramática-resumen

Grammar Terms

Adjective (p. 336): a word that describes a noun
*A Francisco le gustan las pinturas **modernas.***

Adverb (p. 433): a word that modifies a verb, an adjective, or another adverb
*Puerto Rico es **muy** bonito, **especialmente** la playa.*

Article: a word that identifies the class of a noun: masculine or feminine, singular or plural
***El** Yunque es **un** bosque tropical. **El** perfume de **las** orquídeas es muy agradable.*

Auxiliary Verb (pp. 46, 141): a secondary verb that is used with a main verb
*Nuestra visita al Prado **ha** sido increíble y ya **estamos** pensando ir otra vez.*

Command (pp. 114, 116): a verb form used to tell someone to do something
***Haga** ejercicio para bajar el estrés.*

Comparative (p. 454): a phrase that compares two different things
*El teatro español es **más divertido que** el cine.*

Conditional Tense (pp. 119, 214, 289, 414): a verb form that indicates that the action in a sentence could happen at a future time
*¿Te **gustaría** proteger el medio ambiente?*

Conditional Perfect Tense (p. 289): a verb form that expresses actions that would have been done if something else had been true
***Habríamos trabajado** en finanzas si hubiéramos estudiado economía.*

Conjugated verb (pp. 360, 380): a verb whose endings reflect person and number (as opposed to an infinitive)
*Muchas personas **quieren** prepararse para carreras en tecnología.*

Conjunction (pp. 162, 190, 234, 412, 432, 435): a word that acts as a connector between words, phrases, clauses, or sentences
*También queremos ir al Parque Nacional Volcán Póas, **pero** no pensamos que haya tiempo.*

Demonstrative (pp. 336, 382): an adjective or a pronoun that points out someone or something
*Francisco, ¿escribiste **este** poema?–¿**Éste**? Sí, lo escribí.*

Direct Object (pp. 46, 264, 306, 358, 380): a noun or pronoun that receives the action of the main verb in a sentence
*La mesera puso **la mesa** con cubiertos. **La** puso con cubiertos.*

Future Tense (pp. 65, 68, 119): a tense that indicates that the action in a sentence will happen in the future
*Francisco **escribirá** un artículo sobre la conservación de la naturaleza.*

Future Perfect Tense (p. 310): a verb form that expresses what will have happened by a certain time
*A las dos, ya **habremos salido.***

Gender: a term that categorizes a noun or pronoun as masculine or feminine
*Laura no quiere **el azúcar** en **la sopa.***

Imperfect Tense (pp. 42, 43, 268, 286, 410): a verb form that notes incomplete or repeated actions or states with reference to the past
***Juntábamos** fondos para el centro de la comunidad cuando supimos cuántos servicios **se ofrecían** allí.*

Indicative Mood (pp. 238, 412, 414, 415, 432): the mood of the verb used for statements that report what is/was and for questions (as opposed to the subjunctive mood)
***Voy** a celebrar el día de la Independencia en Puerto Rico. ¿Adónde **vas** tú?*

Indirect Object (pp. 46, 62, 360, 380): a noun or pronoun that tells to or for whom/what the action is done
*Catalina recomendó el libro a sus **amigos.** Catalina **les** recomendó el libro.*

GRAMÁTICA–RESUMEN

Infinitive: the basic form of a verb that ends in **-ar, -er, -ir.**
*Susana quiere **ser** una estrella.*

Interrogative (pp. 262, 338): a word that asks a question
*¿**Qué** te gusta hacer durante las vacaciones?*

Main Clause (pp. 46, 141, 210, 214, 232, 234, 288, 364, 415): in a sentence with two clauses, the main clause is the one that can function alone as a complete sentence
*Con tal de que haya buen tiempo, **vamos a hacer alpinismo este fin de semana.***

Negative Command (p. 114): a verb form used to tell someone not to do something
*Alejandro, **no vayas** al campamento hasta mañana.*

Nominalization (pp. 382, 385): to use another part of speech (such as an article, an adjective or a pronoun) instead of a noun
*Esta computadora es increíble, pero **la tuya** es mejor. **La suya** es **la más impresionante.***

Noun: a word that names a person, animal, place, or thing
***Frida Kahlo**, una **pintora** muy famosa, pintó **pinturas** sobre su **vida** en **México**.*

Number: a term that categorizes a noun or pronoun as singular or plural
*Donde hay **una piñata**, hay **una fiesta**. Hay **muchas fiestas** en **México**.*

Past Participle (pp. 46, 141, 286, 289, 310): a verb form that indicates past action but does not specify person or number; used in the present and past perfect tenses; also is used as an adjective
*La impresora está **rota**. La hemos **comprado** hace sólo dos semanas. Hemos **vendido** la otra.*

Past Perfect Tense (p. 46): a verb form that focuses on actions that were completed before others in the past, consisting of an auxiliary verb and the past participle
*Ya **habíamos regado** las plantas cuando empezó a llover.*

Past Progressive Tense (p. 268): a verb form that describes an action that was in progress in the past at a certain point in time
***Estábamos mirando** un espectáculo de flamenco cuando Mario tuvo que irse.*

Possessive (pp. 308, 385): an adjective or a pronoun that tells to whom the noun it describes belongs
*Don Miguel dice que son **sus** marionetas. Son las marionetas*

suyas. *Las **suyas** son muy comunes en México.*

Preposition (pp. 63, 341, 433, 457): a word that shows the relationship between its object and another word
***Por** su segundo álbum Cristian Castro ganó un premio.*

Present Participle (p. 46): the **-ando/-endo/-iendo** form of a verb
*Estamos **regando** las plantas.*

Present Perfect Tense (p. 46): a verb form that indicates that the action in a sentence has been done in the past
*¡Me **ha gustado** tanto Ecuador!*

Present Progressive Tense (pp. 264, 266): a verb form that indicates that an action is in progress at this very moment
*Alicia **está preparándose** para su entrevista.*

Present Tense (pp. 4, 8, 264): a verb form that indicates that the action is happening now, does happen, or will happen in the near future.
*Antonio Banderas **participa** en películas en inglés y español.*

Preterite Tense (pp. 12, 16, 20, 24, 43, 268, 286, 410): a verb form that indicates that the action in a sentence happened at a particular time in the past
*¿Cuándo **se inventó** la guitarra eléctrica?*

Pronoun (pp. 84, 264, 306, 308, 336, 358, 360, 362, 364, 380): a word that takes the place of a noun
*Catalina siempre compra el periódico. **Ella** siempre **lo** compra.*

Reciprocal Verb (p. 86): verb that expresses actions that people do for or to each other
*Marta y Fernanda **se escriben** todos los días y **se hablan** por teléfono una vez a la semana.*

Reflexive Verb (pp. 84, 86, 89): a verb for which the subject receives the action
*Las letras D.F. **se refieren** a la Ciudad de México.*

Relative Clause (p. 341): a subordinate clause that is introduced by a relative pronoun
*Quiero ir al museo **que está cerca del centro**.*

Relative Pronoun (pp. 341, 362, 364): a pronoun that refers to something that has been mentioned previously
*Esa celebración patriótica, **la cual** te describí antes, es famosa por todo el mundo.*

Si Clause (p. 214): a clause that expresses a hypothetical or contrary-to-fact situation
Si tenemos tiempo, vamos a ver una película de Carlos Saura.

Stem (p. 20): the part of the infinitive that remains after the **-ar, -er,** or **-ir** ending is deleted
*Para formar el futuro del verbo **hablar**, hay que añadir las terminaciones a la forma **habl-**.*

Subject: the noun or noun phrase in a sentence that tells who or what does the action
Isabel saltaba la cuerda cuando era niña.

Subjunctive Mood (pp. 136, 138, 139, 141, 158, 160, 162, 188, 190, 192, 210, 212, 214, 232, 234, 238, 286, 412, 414, 415, 432): a verb form in a dependent clause that indicates that a sentence expresses doubt, emotion, opinion or an unlikely happening (as opposed to the indicative mood)
*Francisco no cree que **vaya** a nevar.*

Subordinate Clause (pp. 46, 141, 210, 232, 415, 432, 435): in a sentence with two clauses, this clause is incomplete and cannot function as a sentence on its own
Además de ser atrevida, Manuela también es muy independiente.

Superlative (p. 454): a phrase that describes which item has the most or least of a quality
*¡Isabel y Andrea van a ser **las más elegantes** de todas las chicas en la fiesta!*

Tense: the conjugated form of a verb whose endings follow a pattern
*Soy muy creativo. **Creo** que **puedo** inventar anuncios para la tele.*

Verb: a word that expresses action or a state of being
*Don Miguel **viene** al parque con sus marionetas.*
*Está contento cuando los niños **se sonríen**.*

Nouns, Articles, and Pronouns

Nouns

Nouns identify people, animals, places, or things. Spanish nouns are either **masculine** or **feminine**. They are also either **singular** (identifying one thing) or **plural** (identifying more than one thing). **Masculine nouns** usually end in **-o** and **feminine nouns** usually end in **-a**.

To make a noun **plural,** add **-s** to a word ending in a vowel and **-es** to a word ending in a consonant. When a noun ends in **z**, change the **z** to **ces** to form the plural: *actriz, actrices.* Generally, the same syllable carries the force of pronunciation in plural and in singular forms: *lápiz, lápices; pantalón, pantalones; joven, jóvenes.*

Singular Nouns		Plural Nouns	
Masculine	**Feminine**	**Masculine**	**Feminine**
amigo	amiga	amigos	amigas
chico	chica	chicos	chicas
hombre	mujer	hombres	mujeres
suéter	blusa	suéteres	blusas
zapato	falda	zapatos	faldas

Articles

Articles identify the class of a noun: masculine or feminine, singular or plural. **Definite articles** are the equivalent of the English word *the*. **Indefinite articles** are the equivalent of *a, an*, or *some*.

Definite Articles		
	Masculine	**Feminine**
Singular	**el** amigo	**la** amiga
Plural	**los** amigos	**las** amigas

Indefinite Articles		
	Masculine	**Feminine**
Singular	**un** amigo	**una** amiga
Plural	**unos** amigos	**unas** amigas

Pronouns

A **pronoun** can take the place of a noun. The choice of pronoun is determined by how it is used in the sentence.

Subject Pronouns	
yo	nosotros(as)
tú	vosotros(as)
usted	ustedes
él, ella	ellos(as)

Pronouns Used After Prepositions	
de **mí**	de **nosotros(as)**
de **ti**	de **vosotros(as)**
de **usted**	de **ustedes**
de **él**, de **ella**	de **ellos(as)**
After **con, mí** and **ti** become **conmigo, contigo**.	

Direct Object Pronouns	
me	nos
te	os
lo, la	los, las

Indirect Object Pronouns	
me	nos
te	os
le	les

Reflexive Pronouns	
me	nos
te	os
se	se

Demonstrative Pronouns	
éste(a), esto	éstos(as)
ése(a), eso	ésos(as)
aquél(la), aquello	aquéllos(as)

Adjectives

Adjectives describe nouns. In Spanish, adjectives must match the **number** and **gender** of the nouns they describe. When an adjective describes a group with both genders, the masculine form is used. To make an adjective plural, apply the same rules that are used for making a noun plural. Most adjectives are placed after the noun.

Adjectives		
	Masculine	**Feminine**
Singular	el chico **guapo**	la chica **guapa**
	el chico **paciente**	la chica **paciente**
	el chico **fenomenal**	la chica **fenomenal**
	el chico **trabajador**	la chica **trabajadora**
Plural	los chicos guapo**s**	las chicas guapa**s**
	los chicos paciente**s**	las chicas paciente**s**
	los chicos fenomenal**es**	las chicas fenomenal**es**
	los chicos trabajador**es**	las chicas trabajadora**s**

Adjectives cont.

Sometimes adjectives are placed before the noun and **shortened. Grande** is shortened before any singular noun. Several others are shortened before a masculine singular noun.

Shortened Forms			
alguno	**algún** chico	primero	**primer** chico
bueno	**buen** chico	tercero	**tercer** chico
malo	**mal** chico	grande	**gran** chico(a)
ninguno	**ningún** chico		

Possessive adjectives identify to whom something belongs. They agree in gender and number with the possessed item, not with the person who possesses it. These forms always come before nouns.

Possessive Adjectives				
	Masculine		**Feminine**	
Singular	**mi** amigo	**nuestro** amigo	**mi** amiga	**nuestra** amiga
	tu amigo	**vuestro** amigo	**tu** amiga	**vuestra** amiga
	su amigo	**su** amigo	**su** amiga	**su** amiga
Plural	**mis** amigos	**nuestros** amigos	**mis** amigas	**nuestras** amigas
	tus amigos	**vuestros** amigos	**tus** amigas	**vuestras** amigas
	sus amigos	**sus** amigos	**sus** amigas	**sus** amigas

Demonstrative adjectives point out which noun is being referred to. Their English equivalents are *this, that, these,* and *those.*

Demonstrative Adjectives		
	Masculine	**Feminine**
Singular	**este** amigo	**esta** amiga
	ese amigo	**esa** amiga
	aquel amigo	**aquella** amiga
Plural	**estos** amigos	**estas** amigas
	esos amigos	**esas** amigas
	aquellos amigos	**aquellas** amigas

Interrogatives

Interrogatives		
¿Adónde?	¿Cuándo?	¿Por qué?
¿Cómo?	¿Cuánto(a)? ¿Cuántos(as)?	¿Qué?
¿Cuál(es)?	¿Dónde?	¿Quién(es)?

Comparatives and Superlatives

Comparatives

Comparatives are used when comparing two different things.

Comparatives		
más (+) **más** interesante **que…** Me gusta correr **más que** nadar.	menos (−) **menos** interesante **que…** Me gusta nadar **menos que** correr.	tan(to) (=) **tan** interesante **como…** Me gusta leer **tanto como** escribir.

There are a few irregular comparatives:
• When talking about people, use **mayor** and **menor.**

Age	Quality
mayor	mejor
menor	peor

• **Mejor** and **peor** are the comparative forms of **bueno(a)** and **malo(a).**

• When talking about numbers, use **de** instead of **que.**
>   **más de** cien…
>   **menos de** cien…

Superlatives

Superlatives are used to distinguish one item from a group. They describe which item has the most or least of a quality.

The ending **-ísimo(a)** can be added to an adjective to form a superlative.

Irregular comparatives are also irregular as superlatives.

Use "de" after a superlative:
El más alto de la clase

Superlatives		
	Masculine	**Feminine**
Singular *Plural*	**el** chico **más** alto **el** chico **menos** alto **los** chicos **más** altos **los** chicos **menos** altos	**la** chica **más** alta **la** chica **menos** alta **las** chicas **más** altas **las** chicas **menos** altas
Singular *Plural*	mole buen**ísimo** frijoles buen**ísimos**	pasta buen**ísima** enchiladas buen**ísimas**

Prepositions and Adverbs of Location

Prepositions and Adverbs of Location					
abajo afuera al lado (de)	alrededor (de) atrás debajo (de)	delante (de) dentro (de) detrás (de)	encima (de) enfrente (de) frente (a)	fuera (de) hacia hasta	junto (a) sobre

Affirmative and Negative Words

Affirmative	Negative
a menudo	jamás
algo	nada
alguien	nadie
algún (alguna)	ningún (ninguna)
alguno(a)	ninguno(a)
muchas veces	rara vez
o… o…	ni… ni…
siempre	nunca
también	tampoco

Adverbs

Adverbs modify a verb, an adjective, or another adverb. Many adverbs in Spanish are made by changing an existing adjective.

Adjective	→	Adverb
reciente	→	reciente**mente**
frecuente	→	frecuente**mente**
fácil	→	fácil**mente**
normal	→	normal**mente**
especial	→	especial**mente**
feliz	→	feliz**mente**
cuidadoso(a)	→	cuidadosa**mente**
rápido(a)	→	rápida**mente**
lento(a)	→	lenta**mente**
tranquilo(a)	→	tranquila**mente**

Verbs

Simple Tenses

		Indicative					Subjunctive	
		Present	**Imperfect**	**Preterite**	**Future**	**Conditional**	**Present**	**Imperative**
Infinitive *Present Participle* *Past Participle*	hablar hablando hablado	hablo hablas habla hablamos habláis hablan	hablaba hablabas hablaba hablábamos hablabais hablaban	hablé hablaste habló hablamos hablasteis hablaron	hablaré hablarás hablará hablaremos hablaréis hablarán	hablaría hablarías hablaría hablaríamos hablaríais hablarían	hable hables hable hablemos habléis hablen	habla no hables hable hablemos hablen
Infinitive *Present Participle* *Past Participle*	comer comiendo comido	como comes come comemos coméis comen	comía comías comía comíamos comíais comían	comí comiste comió comimos comisteis comieron	comeré comerás comerá comeremos comeréis comerán	comería comerías comería comeríamos comeríais comerían	coma comas coma comamos comáis coman	come no comas coma coman
Infinitive *Present Participle* *Past Participle*	vivir viviendo vivido	vivo vives vive vivimos vivís viven	vivía vivías vivía vivíamos vivíais vivían	viví viviste vivió vivimos vivisteis vivieron	viviré vivirás vivirá viviremos viviréis vivirán	viviría vivirías viviría viviríamos viviríais vivirían	viva vivas viva vivamos viváis vivan	vive no vivas viva vivan

Perfect Tenses

Perfect Tenses		
Present Perfect	**Present Perfect Subjunctive**	**Future Perfect**
he has ha hemos habéis han } hablado comido vivido	haya hayas haya hayamos hayáis hayan } hablado comido vivido	habré habrás habrá habremos habréis habrán } hablado comido vivido
Past Perfect	**Past Perfect Subjunctive**	**Conditional Perfect**
había habías había habíamos habíais habían } hablado comido vivido	hubiera hubieras hubiera hubiéramos hubierais hubieran } hablado comido vivido	habría habrías habría habríamos habríais habrían } hablado comido vivido

Progressive Tenses

Progressive Tenses			
Present Progressive	**Present Participle**	**Past Progressive**	**Present Participle**
estoy estás está estamos estáis están	hablando comiendo viviendo	estaba estabas estaba estábamos estabais estaban	hablando comiendo viviendo

Stem-Changing Verbs

Infinitive in -ar	Present Indicative	Present Subjunctive
cerrar e→ie	cierro cierras cierra cerramos cerráis cierran	cierre cierres cierre cerremos cerréis cierren
probar o→ue	pruebo pruebas prueba probamos probáis prueban	pruebe pruebes pruebe probemos probéis prueben
jugar u→ue	juego juegas juega jugamos jugáis juegan	juegue juegues juegue juguemos juguéis jueguen

like **cerrar:** comenzar, despertar(se), empezar, merendar, nevar, pensar, recomendar, regar, sembrar, sentar(se)

like **probar:** acostar(se), almorzar, contar, costar, encontrar(se), mostrar, recordar, rogar, sonar, volar

Infinitive in -er	Present Indicative	Present Subjunctive
perder e→ie	pierdo pierdes pierde perdemos perdéis pierden	pierda pierdas pierda perdamos perdáis pierdan
volver o→ue	vuelvo vuelves vuelve volvemos volvéis vuelven	vuelva vuelvas vuelva volvamos volváis vuelvan

like **perder:** atender, entender, querer
like **volver** (past participle: vuelto)**:** devolver (devuelto), doler, encender, llover, mover, poder, resolver (resuelto)

Infinitive in -ir	Indicative		Subjunctive
	Present	**Preterite**	**Present**
pedir e→i pidiendo	pido pides pide pedimos pedís piden	pedí pediste pidió pedimos pedisteis pidieron	pida pidas pida pidamos pidáis pidan
dormir o→ue, u durmiendo	duermo duermes duerme dormimos dormís duermen	dormí dormiste durmió dormimos dormisteis durmieron	duerma duermas duerma durmamos durmáis duerman
sentir e→ie, i sintiendo	siento sientes siente sentimos sentís sienten	sentí sentiste sintió sentimos sentisteis sintieron	sienta sientas sienta sintamos sintáis sientan

like **pedir:** competir, despedir(se), repetir, seguir, servir, vestir(se)
like **dormir(se):** morir (past participle: **muerto**)
like **sentir:** divertir(se), mentir, preferir, requerir, sugerir

Spell-Changing Verbs

buscar

Preterite: bus**qu**é, buscaste, buscó, buscamos, buscasteis, buscaron
Present Subjunctive: bus**que**, bus**qu**es, bus**que**, bus**que**mos, bus**qu**éis, bus**que**n

like **buscar:** marcar, pescar, sacar, secar(se), tocar

conducir

Present Indicative: condu**zc**o, conduces, conduce, conducimos, conducís, conducen
Preterite: condu**j**e, condu**j**iste, condu**j**o, condu**j**imos, condu**j**isteis, condu**j**eron
Present Subjunctive: condu**zc**a, condu**zc**as, condu**zc**a, condu**zc**amos, condu**zc**áis, condu**zc**an

like **conducir:** producir, reducir, traducir

conocer

Present Indicative: cono**zc**o, conoces, conoce, conocemos, conocéis, conocen
Present Subjunctive: cono**zc**a, cono**zc**as, cono**zc**a, cono**zc**amos, cono**zc**áis, cono**zc**an

like **conocer:** crecer, nacer, ofrecer, pertenecer

conseguir

Present Indicative: consi**g**o, consigues, consigue, conseguimos, conseguís, consiguen
Present Subjunctive: consi**g**a, consi**g**as, consi**g**a, consi**g**amos, consi**g**áis, consi**g**an

like **conseguir:** seguir

construir

Present Indicative: constru**y**o, constru**y**es, constru**y**e, construimos, construís, constru**y**en
Preterite: construí, construiste, constru**y**ó, construimos, construisteis, constru**y**eron

creer

Preterite: creí, creíste, cre**y**ó, creímos, creísteis, cre**y**eron
Present Participle: cre**y**endo
Past Participle: creído

like **creer:** leer

cruzar

Preterite: cru**c**é, cruzaste, cruzó, cruzamos, cruzasteis, cruzaron
Present Subjunctive: cru**c**e, cru**c**es, cru**c**e, cru**c**emos, cru**c**éis, cru**c**en

like **cruzar:** almorzar (o→ue), comenzar (e→ie), empezar (e→ie)

escoger

Present Indicative: esco**j**o, escoges, escoge, escogemos, escogéis, escogen
Present Subjunctive: esco**j**a, esco**j**as, esco**j**a, esco**j**amos, esco**j**áis, esco**j**an

like **escoger:** proteger

esquiar

Present Indicative: esqu**í**o, esqu**í**as, esqu**í**a, esquiamos, esquiáis, esqu**í**an
Present Subjunctive: esqu**í**e, esqu**í**es, esqu**í**e, esquiemos, esquiéis, esqu**í**en

llegar

Preterite: lle**gu**é, llegaste, llegó, llegamos, llegasteis, llegaron
Present Subjunctive: lle**gu**e, lle**gu**es, lle**gu**e, lle**gu**emos, lle**gu**éis, lle**gu**en

like **llegar:** apagar, jugar, pagar

Irregular Verbs

andar

Preterite: anduve, anduviste, anduvo, anduvimos, anduvisteis, anduvieron

caer

Present Indicative: caigo, caes, cae, caemos, caéis, caen
Preterite: caí, caíste, cayó, caímos, caísteis, cayeron
Present Subjunctive: caiga, caigas, caiga, caigamos, caigáis, caigan
Present Participle: cayendo
Past Participle: caído

dar

Present Indicative: doy, das, da, damos, dais, dan
Preterite: di, diste, dio, dimos, disteis, dieron
Present Subjunctive: dé, des, dé, demos, deis, den
Commands: da (tú), no des (neg. tú), dé (Ud.) den (Uds.)

decir

Present Indicative: digo, dices, dice, decimos, decís, dicen
Preterite: dije, dijiste, dijo, dijimos, dijisteis, dijeron
Future: diré, dirás, dirá, diremos, diréis, dirán
Conditional: diría, dirías, diría, diríamos, diríais, dirían
Present Subjunctive: diga, digas, diga, digamos, digáis, digan
Commands: di (tú), no digas (neg. tú), diga (Ud.), digan (Uds.)
Present Participle: diciendo
Past Participle: dicho

estar

Present Indicative: estoy, estás, está, estamos, estáis, están
Preterite: estuve, estuviste, estuvo, estuvimos, estuvisteis, estuvieron
Present Subjunctive: esté, estés, esté, estemos, estéis, estén

haber

Present Indicative: he, has, ha, hemos, habéis, han
Preterite: hube, hubiste, hubo, hubimos, hubisteis, hubieron
Future: habré, habrás, habrá, habremos, habréis, habrán
Conditional: habría, habrías, habría, habríamos, habríais, habrían
Present Subjunctive: haya, hayas, haya, hayamos, hayáis, hayan

hacer

Present Indicative: hago, haces, hace, hacemos, hacéis, hacen
Preterite: hice, hiciste, hizo, hicimos, hicisteis, hicieron
Future: haré, harás, hará, haremos, haréis, harán
Conditional: haría, harías, haría, haríamos, haríais, harían
Present Subjunctive: haga, hagas, haga, hagamos, hagáis, hagan
Commands: haz (tú), no hagas (neg. tú), haga (Ud.), hagan (Uds.)
Past Participle: hecho

ir

Present Indicative: voy, vas, va, vamos, vais, van
Imperfect: iba, ibas, iba, íbamos, ibais, iban
Preterite: fui, fuiste, fue, fuimos, fuisteis, fueron
Present Subjunctive: vaya, vayas, vaya, vayamos, vayáis, vayan
Commands: ve (tú), no vayas (neg. tú), vaya (Ud.), vayan (Uds.)
Present Participle: yendo

oír

Present Indicative: oigo, oyes, oye, oímos, oís, oyen
Preterite: oí, oíste, oyó, oímos, oísteis, oyeron
Present Subjunctive: oiga, oigas, oiga, oigamos, oigáis, oigan
Present Participle: oyendo
Past Participle: oído

poder

Present Indicative: puedo, puedes, puede, podemos, podéis, pueden
Preterite: pude, pudiste, pudo, pudimos, pudisteis, pudieron
Future: podré, podrás, podrá, podremos, podréis, podrán
Conditional: podría, podrías, podría, podríamos, podríais, podrían
Present Subjunctive: pueda, puedas, pueda, podamos, podáis, puedan
Present Participle: pudiendo

poner

Present Indicative: pongo, pones, pone, ponemos, ponéis, ponen
Preterite: puse, pusiste, puso, pusimos, pusisteis, pusieron
Future: pondré, pondrás, pondrá, pondremos, pondréis, pondrán
Conditional: pondría, pondrías, pondría, pondríamos, pondríais, pondrían
Present Subjunctive: ponga, pongas, ponga, pongamos, pongáis, pongan
Commands: pon (tú), no pongas (neg. tú), ponga (Ud.), pongan (Uds.)
Past Participle: puesto

like **poner:** descomponer(se), imponer, oponer(se)

querer

Present Indicative: quiero, quieres, quiere, queremos, queréis, quieren
Preterite: quise, quisiste, quiso, quisimos, quisisteis, quisieron
Future: querré, querrás, querrá, querremos, querréis, querrán
Conditional: querría, querrías, querría, querríamos, querríais, querrían
Present Subjunctive: quiera, quieras, quiera, queramos, queráis, quieran

saber

Present Indicative: sé, sabes, sabe, sabemos, sabéis, saben
Preterite: supe, supiste, supo, supimos, supisteis, supieron
Future: sabré, sabrás, sabrá, sabremos, sabréis, sabrán
Conditional: sabría, sabrías, sabría, sabríamos, sabríais, sabrían
Present Subjunctive: sepa, sepas, sepa, sepamos, sepáis, sepan
Commands: sabe (tú), no sepas (neg. tú), sepa (Ud.), sepan (Uds.)

salir

Present Indicative: salgo, sales, sale, salimos, salís, salen
Future: saldré, saldrás, saldrá, saldremos, saldréis, saldrán
Conditional: saldría, saldrías, saldría, saldríamos, saldríais, saldrían
Present Subjunctive: salga, salgas, salga, salgamos, salgáis, salgan
Commands: sal (tú), no salgas (neg. tú), salga (Ud.), salgan (Uds.)

ser

Present Indicative: soy, eres, es, somos, sois, son
Imperfect: era, eras, era, éramos, erais, eran
Preterite: fui, fuiste, fue, fuimos, fuisteis, fueron
Present Subjunctive: sea, seas, sea, seamos, seáis, sean
Commands: sé (tú), no seas (neg. tú), sea (Ud.), sean (Uds.)

tener

Present Indicative: tengo, tienes, tiene, tenemos, tenéis, tienen

Preterite: tuve, tuviste, tuvo, tuvimos, tuvisteis, tuvieron

Future: tendré, tendrás, tendrá, tendremos, tendréis, tendrán

Conditional: tendría, tendrías, tendría, tendríamos, tendríais, tendrían

Present Subjunctive: tenga, tengas, tenga, tengamos, tengáis, tengan

Commands: ten (tú), no tengas (neg. tú), tenga (Ud.), tengan (Uds.)

like **tener:** mantener, obtener

traer

Present Indicative: traigo, traes, trae, traemos, traéis, traen

Preterite: traje, trajiste, trajo, trajimos, trajisteis, trajeron

Present Subjunctive: traiga, traigas, traiga, traigamos, traigáis, traigan

Present Participle: trayendo

Past Participle: traído

valer

Present Indicative: valgo, vales, vale, valemos, valéis, valen

Preterite: valí, valiste, valió, valimos, valisteis, valieron

Future: valdré, valdrás, valdrá, valdremos, valdréis, valdrán

Conditional: valdría, valdrías, valdría, valdríamos, valdríais, valdrían

Present Subjunctive: valga, valgas, valga, valgamos, valgáis, valgan

Commands: val (tú), no valgas (neg. tú), valga (Ud.), valgas (Uds.)

venir

Present Indicative: vengo, vienes, viene, venimos, venís, vienen

Preterite: vine, viniste, vino, vinimos, vinisteis, vinieron

Future: vendré, vendrás, vendrá, vendremos, vendréis, vendrán

Conditional: vendría, vendrías, vendría, vendríamos, vendríais, vendrían

Present Subjunctive: venga, vengas, venga, vengamos, vengáis, vengan

Commands: ven (tú), no vengas (neg. tú), venga (Ud.), vengan (Uds.)

Present Participle: viniendo

ver

Present Indicative: veo, ves, ve, vemos, veis, ven

Preterite: vi, viste, vio, vimos, visteis, vieron

Imperfect: veía, veías, veía, veíamos, veíais, veían

Past Participle: visto

GLOSARIO
español–inglés

This Spanish-English glossary contains all of the active vocabulary words that appear in the text as well as passive vocabulary from readings and culture sections. Most inactive cognates have been omitted. The active words are accompanied by the number of the unit and etapa in which they are presented. For example, **la autobiografía** can be found in **5.1** (*Unidad* **5**, *Etapa* **1**). The roman numerals **I** and **II** indicate words that were taught in Levels 1 and 2.

a to, at **I**
  a la(s)… at … o'clock **I**
  a continuación next **II**
  a gusto comfortable, happy
  a la derecha (de)
    to the right (of) **I**
  a la izquierda (de)
    to the left (of) **I**
  a la venta for sale
  a lo largo de throughout
  a menos que unless **3.1**
  a menudo often **4.2**
  a pie on foot **I**
  ¿A qué hora es…?
    (At) What time is…? **I**
  a tiempo on time **II**
  A todos nos toca…
    It is up to all of us… **II**
  a veces sometimes **I, II, 4.2**
abajo down **I, II**
el abecedario alphabet
abierto(a) open **I, II**
el (la) abogado(a) lawyer **II, 4.2**
el (la) abonado(a) subscriber
abordar to board (a plane) **II**
el abrazo hug **II**
el abrelatas can opener **II, 2.3**
el abrigo coat **I**
abril April **I**
abrir to open **I**
abrir el paso to open the way **5.2**

la abuela grandmother **I**
el abuelo grandfather **I**
los abuelos grandparents **I**
aburrido(a) boring **I**
aburrir(se) to get bored **II**
acá here **I**
acabar de to have just **I**
acabársele (a uno) to run out of **6.2**
el (la) académico(a) academic **4.3**
acampar to camp **II, 1.2**
accesible available, accessible **6.2**
el aceite oil **I, II**
las aceitunas olives **I**
la acera sidewalk **II**
acertado(a) right
aconsejar to advise **II**
acordarse (ue) de to remember **6.3**
acostar (ue) to go to bed, to lie down **II**
el (la) actor/actriz actor/actress **II**
la actuación performance **II**
acudir a to attend **3.3**
acuerdo
  estar de acuerdo to agree **I, II**
el acumulador battery
adaptarse to adapt oneself **4.1**
Adiós. Good-bye. **I**
la administración de empresas business administration **4.1**
adónde (to) where **I**
los adornos decorations **II**
la aduana customs **II**
la aerolínea airline **II**
el aeropuerto airport **I**

el aerosol aerosol **2.2**
afeitarse to shave oneself **I, II**
afuera outside **6.2**
el (la) agente de ventas sales agent **4.3**
el (la) agente de viajes travel agent **II**
agosto August **I**
agotado
  Estoy agotado(a).
    I'm exhausted. **2.1**
agradecer to thank **3.1**
el (la) agricultor(a) farmer **II, 4.2**
la agricultura agriculture **4.3**
la agronomía agronomy **4.1**
el agua (fem.) water **I**
  el agua de coco coconut milk **II**
  el agua dulce fresh water **2.3**
  esquiar en el agua to water-ski **1.2**
el aguacero downpour **II, 2.3**
ahora now **I**
  ¡Ahora mismo! Right now! **I**
el aire
  al aire libre outdoors **I**
  la contaminación del aire air pollution **I, 2.2**
el aire acondicionado air conditioning **II**
el ajedrez
  jugar al ajedrez to play chess **II**
al to the **I**
  al aire libre outdoors **I**
  al contrario on the contrary **II**

al lado (de) beside, next to I
al óleo oil painting **5.1**
el ala wing
el (la) alcalde mayor **3.3**
alegrar(se) de que
to be happy that II
alegre happy I
alemán(ana) German II
la alfabetización literacy
algo something I
el algodón cotton **1.2**
alguien someone I
   conocer a alguien to know, to
     be familiar with someone I
alguno(a) some I
la alimentación nourishment II
el alimento food II
allá there I
allí there I
el alma (fem.) soul
el almirante admiral **3.3**
la almohada pillow II, **2.3**
almorzar (ue) to eat lunch I, II
el almuerzo lunch I
alquilar to rent I
   alquilar un video
     to rent a video I
alrededor around II
alto(a) tall I
el altoparlante speaker **6.2**
la altura altitude, height II, **2.2**
   la altura del terreno
     altitude of terrain
el aluminio aluminum II
amable nice II
   Muy amable.
     That's kind of you. **3.2**
el amanecer dawn **2.3**
amanecer to start the day **2.3**
amarillo(a) yellow I
la ambulancia ambulance II
amenazador(a) threatening **5.3**
el (la) amigo(a) friend I
la amistad acquaintance,
   friendship II
el amor love II
ampliable expandable **6.3**
anaranjado(a) orange I
ancho(a) wide I, II
los ancianos the elderly **2.1**
andar to walk II
   andar en bicicleta
     to ride a bike I

andar en patineta
   to skateboard I
el (la) anfitrión(a) host(ess) **3.2**
el anillo ring I
animado(a) animated II
el animal animal I
animarse to get encouraged,
   interested **1.3**
el aniversario anniversary II
anoche last night I, II
anochecer to get dark **2.3**
el anochecer nightfall **2.3**
anteayer day before yesterday I, II
la antena parabólica
   satellite dish **6.1**
los anteojos glasses **1.1**
los antepasados ancestors I, **3.3**
antes (de) before I
antiguo(a) old, ancient I, II
el anuncio commercial II
el año year I
   el año escolar the school year II
   el Año Nuevo New Year **3.2**
   el año pasado last year I
   ¿Cuántos años tiene…?
     How old is…? I
   ¡Próspero Año Nuevo!
     Happy New Year! **3.2**
   Tiene… años.
     He/She is… years old. I
   apagar la luz
     to turn off the light I
el apartamento apartment I
aparte separate
   Es aparte. Separate checks. I
el apellido last name, surname I
apenas scarcely II
apoyarse to support each other **1.3**
el apoyo support **3.1**
apreciar to appreciate **3.1**
aprender to learn I
apretado(a) tight II
aprovechar to take advantage of
apto(a) para toda la familia
   G-rated (movie) **6.1**
aquel(la) that (over there) I, **5.1**
aquél(la) that one (over there) I, **5.1**
aquello that (over there) I, **5.1**
aquí here I
el árbol tree I, **2.1**
   trepar a un árbol
     to climb a tree II
la arena sand II

el arete earring I
argentino(a) Argentine II
armar to assemble **1.3**
el armario closet I, II
el arpa (fem.) harp **5.1**
el (la) arquitecto(a)
   architect I, II, **4.2**
arquitectónico(a) architectural
la arquitectura architecture I
arreglar(se) to get ready; to get
   dressed up II
el arreglo arrangement
arriba up I, II
el arroz rice I
   el arroz con dulce dessert of
     rice, cinnamon, and coconut
     milk **3.2**
   el arroz con gandules
     rice and pigeon peas **3.2**
   el arroz con leche dessert of
     sweet rice and milk **3.2**
el arte art I
la artesanía handicraft I
el (la) artesano(a) artisan I, II, **4.2**
el artículo article II
los artículos de cuero
   leather goods I
el (la) artista artist II
el ascensor elevator II
así fue que and so it was that II
el asiento seat II
el (la) asistente assistant II, **402**
   el asistente electrónico
     electronic assistant **6.2**
asistir (a) to attend II
la aspiradora vacuum cleaner I
   pasar la aspiradora
     to vacuum I, II
la aspirina aspirin II
el asta (fem.) flagpole
asustar to frighten
   asustar(se) (de)
     to be scared of II
atardecer to get dark **2.3**
el atardecer late afternoon **2.3**
el atletismo athletics II
atrás in back, behind **6.2**
atrevido(a) daring **1.1**
el atún tuna II
los audífonos headphones **6.2**
el auditorio auditorium I
aumentar to increase **4.2**
aunque even though II

la **autobiografía** autobiography **5.1**
el **autobús** bus **I**
el (la) **autor(a)** author **II**
el **autorretrato** self-portrait **5.1**
el (la) **auxiliar de vuelo**
   flight attendant **II**
avanzado(a) advanced **5.2**
avanzar to advance **3.1**
el **ave** (fem.) bird
la **avenida** avenue **I**
las **aventuras** adventures **II**
el **avión** airplane **I**
   pilotar una avioneta to fly a
      single-engine plane **1.2**
avisar to announce
ayer yesterday **I, II**
ayudar (a) to help **I**
   ¿Cómo puedo ayudarte?
      How can I help you? **2.1**
   ¿Me ayuda a pedir?
      Could you help me order? **I**
ayudarse to help (each other) **1.3**
el **azúcar** sugar **I**
azul blue **I**

el (la) **bailaor(a)** flamenco dancer
   5.1
bailar to dance **I**
el **bailarín/la bailarina**
   dancer **II, 4.2**
el **baile folklórico** folk dance **5.1**
bajar (por) to go down,
   to descend **II**
   bajar un río en canoa to go
      down a river by canoe **II**
bajo(a) short (height) **I**
balanceado(a) balanced **II**
el **balde** bucket **II**
la **baldosa** paving stone, tile
la **ballena jorobada**
   humpback whale **2.3**
el **balón** soccer ball **1.1**
el **baloncesto** basketball **I**
la **bamba** Mexican dance from
   Veracruz **5.2**
el **banco** bank **I**
la **banda** band **3.3**
la **bandera** flag **3.3**
el (la) **banquero(a)** banker **4.3**

bañarse to take a bath **I, II**
la **bañera** bathtub **II**
el **baño** bathroom **I, II**
barato(a) cheap, inexpensive **I**
la **barba** beard **1.1**
el **barco** ship **I**
barrer to sweep **I**
   barrer el piso
      to sweep the floor **II**
la **base de datos** database **6.3**
bastante enough **II**
la **basura** trash, garbage **I, 2.1**
   sacar la basura
      to take out the trash **I**
el **basurero** trash can,
   wastebasket **II, 1.3**
el **bate** bat **I**
la **batería** battery **6.2**
el **batido** milk shake **II**
el **bebé** baby **II**
beber to drink **I**
   ¿Quieres beber…?
      Do you want to drink…? **I**
   Quiero beber…
      I want to drink… **I**
la **bebida** beverage, drink **I**
la **beca** scholarship
el **beeper** beeper **6.2**
el **béisbol** baseball **I**
las **bellas artes** fine arts **II**
la **belleza** beauty **II**
los **beneficios** benefits **II, 4.2**
el **beso** kiss **II**
la **biblioteca** library **I**
el (la) **bibliotecario(a)** librarian **4.3**
la **bicicleta** bike
   andar en bicicleta
      to ride a bike **I**
bien well **I**
   (No muy) Bien, ¿y tú/usted?
      (Not very) Well, and you? **I**
el **bienestar** well-being **II**
bienvenido(a) welcome **I**
el **bigote** mustache **1.1**
la **billetera** wallet **1.2**
la **biografía** biography **5.1**
el **birrete** cap **3.1**
el (la) **bisabuelo(a)**
   great grandfather/
   great grandmother **II**
el **bistec** steak **I**
blanco(a) white **I**
la **blusa** blouse **II**

la **boa constrictora**
   boa constrictor **2.3**
la **boca** mouth **I**
la **boda** wedding **II**
la **bola** ball **I**
el **boleto** ticket **II**
boliviano(a) Bolivian **II**
la **bolsa** bag **I**
la **bolsa de valores**
   stock exchange **4.3**
el **bolso** shoulder bag **1.2**
la **bomba**
   Afro-Caribbean dance **3.3**
el **bombero**
   firefighter, fireman **I, II, 4.2**
la **bombilla** lightbulb **1.3**
bonito(a) pretty **I**
el **bordado** embroidery **5.2**
el **borrador** eraser **I**
el **bosque** forest **I, 2.2**
las **botas** boots **I**
el **bote** boat **II**
la **botella** bottle **I, II, 2.2**
el **brazo** arm **I, II**
brindar to make a toast **3.1**
el **brindis** toast **3.1**
el **broche** fastener, clip
el **bronceador** suntan lotion **I**
bucear to scuba-dive **2.3**
el **budín** pudding **3.2**
bueno(a) good **I**
   ¡Buen provecho!
      Enjoy! (your meal) **3.2**
   Buenas noches.
      Good evening. **I, II**
   Buenas tardes.
      Good afternoon. **I, II**
   Buenos días.
      Good morning. **I, II**
   Es bueno que…
      It's good that… **II**
   Hace buen tiempo.
      It is nice outside. **I**
   lo bueno the good thing **1.1**
la **bufanda** scarf **I**
el **búho** owl **2.3**
buscar to look for, to search **I**
la **búsqueda** search
   el servicio de búsqueda
      search engine **6.3**
el **buzón** mailbox **II**
   el buzón electrónico
      electronic mailbox **6.3**

C

el **caballo** horse I
el **cabello** hair 1.1
la **cabeza** head I, II
   lavarse la cabeza
      to wash one's hair I
cada each, every I
la **cadena** chain 1.2
caer(se) (me caigo) to fall down II
   caer bien/mal
      to like/dislike II, 1.2
   caérsele (a uno) to drop 6.2
el **café** café, coffee I
la **cafetería** cafeteria, coffee shop I
la **caja registradora** cash register II
el (la) **cajero(a)** cashier II
el **cajero automático**
      ATM II
los **calamares** squid I
el **calcetín** sock I
la **calculadora** calculator I
la **calefacción** heat, heating II
la **calidad** quality I
caliente hot, warm I
¡Cállate! Be quiet! I
la **calle** street I
el **calor**
   Hace calor. It is hot. I
   tener calor to be hot I
la **caloría** calorie II
calvo(a) bald 1.1
la **cama** bed I, II
   hacer la cama to make the bed I
la **cámara** camera I, II
cambiar to change, to exchange I
   cambiar de canal
      to change channels (TV) 6.1
el **cambio** change, money
      exchange I
caminar con el perro
      to walk the dog I
el **camino** path, road I, 3.1
la **camisa** shirt I
la **camiseta** T-shirt I
el **campamento** camp II
la **campana** bell 3.2
la **campanada** tolling of the bell 3.2
la **campaña** campaign 2.1
el **campo**
      field, countryside, country I

el **campo de estudio**
      field of study 4.1
canadiense Canadian II
el **canal** channel II
la **cancha** court I
el **cangrejo** crab
cansado(a) tired I
cansarse to get tired II
el (la) **cantante** singer II
el (la) **cantaor(a)** flamenco singer
   5.1
cantar to sing I
   cantar en el coro
      to sing in the chorus II
el **cante**
      el cante jondo
         tragic flamenco song 5.1
la **capa de ozono** ozone layer II, 2.2
capacitado(a) qualified II, 4.1
la **capilla** chapel
la **cara** face I, II
el **caracol** shell II
la **careta** mask
la **carne** meat I
      la carne de res beef I, II
la **carnicería** butcher's shop I
caro(a) expensive I
      ¡Es muy caro(a)!
         It's very expensive! I
la **carrera** career II, 4.2
la **carretera** road, highway I, 4.1
el **carro** car I
la **carta** letter I
      mandar una carta
         to send a letter I
la **cartera** wallet I
el (la) **cartero(a)**
      mail carrier I, II, 4.2
el **cartón** cardboard, cardboard
      box, carton II, 2.2
la **casa** house I
casarse to get married II
el **casco** helmet I
el **casete** cassette I
casi almost II
castaño(a) brown (hair) I
las **castañuelas** castanets 5.1
la **catarata** waterfall
catorce fourteen I
la **causa** cause II
la **cebolla** onion I, II
celebrar to celebrate I
la **cena** supper, dinner I

cenar to eat dinner I, II
Cenicienta Cinderella
el **centígrado** centigrade II, 2.3
el **centro** center, downtown I
   el **centro comercial**
      shopping center I
   el **centro de la comunidad**
      community center 2.1
   el **centro de rehabilitación**
      rehabilitation center 2.1
cepillarse el pelo
      to brush one's hair II
el **cepillo** hairbrush I, II
el **cepillo (de dientes)**
      brush (toothbrush) I, II
la **cerámica** ceramics I
cerca (de) near to I
la **cerca** fence I
el **cerdo** pig I
el **cereal** cereal I, II
los **cereales** grains 4.3
la **ceremonia de graduación**
      graduation ceremony 3.1
la **cereza** cherry II
cero zero I
cerrado(a) closed I, II
cerrar (ie) to close I
el **chaleco** vest II
el **champú** shampoo I, II
Chao. Goodbye. II
la **chaqueta** jacket I
los **cheques** checks II
   los **cheques de viajero**
      travelers' checks II
chévere awesome I
      ¡Qué chévere! How awesome! I
los **chicharrones** pork rinds I
      comer chicharrones
         to eat pork rinds I
el (la) **chico(a)** boy/girl I
chileno(a) Chilean II
chino(a) Chinese II
chismoso(a) gossipy 4.2
el **chorizo** sausage I
cibernético(a) relating to
      cyberspace 5.3
el **ciclo** level of high school
      curriculum
ciego(a) blind 5.3
el **cielo** sky 2.2
cien one hundred I
la **ciencia ficción** science fiction II
las **ciencias** science I

cierto(a) certain
  Es cierto. It's certain.
  No es cierto que…
    It is not certain that… I
la cifra number, numeral 5.2
cinco five I
cincuenta fifty I
el cine movie theater I
  ir al cine to go to the movies I
el (la) cineasta filmmaker 5.3
el (la) cinematógrafo(a)
  cinematographer 5.3
el cinturón belt I
la cita appointment I
la ciudad city I
la ciudadanía citizenship II
el (la) ciudadano(a) citizen 2.1
la civilización civilization 5.2
claro
  ¡Claro que sí! Of course! I
la clase class, classroom I
el (la) cliente customer II
el clima climate II, 2.2
el clímax climax 5.3
el cobre copper 4.3
el coche car 4.2
la cocina kitchen I, II
cocinar to cook I
el codo elbow II
el cofre chest
el cohete firecracker 3.2
la cola de caballo ponytail 1.1
colaborar con
  to collaborate with 2.1
coleccionar to collect 1.2
el colegio school I
la colina hill II, 2.2
el collar necklace I
Colombia Colombia I
colombiano(a) Colombian II
el color color I
  ¿De qué color…?
    What color…? I
  el color brillante bright color 1.2
  el color claro pastel 1.2
  el color oscuro dark color 1.2
  de un solo color solid color 1.2
el columpio de mimbre
  wicker rocking chair
el combustible fuel II, 2.2
el (la) comediante
  comedian/comedienne II
el comedor dining room I, II

el comedor de beneficencia
  soup kitchen 2.1
comenzar (ie) to begin II
comer to eat I, II
  comer chicharrones
    to eat pork rinds I
  darle(s) de comer to feed I
  ¿Quieres comer…?
    Do you want to eat…? I
  Quiero comer…
    I want to eat… I
el comercio business 4.1
cómico(a) funny, comical I
la comida food, meal I
cómo how I
  ¿Cómo es?
    What is he/she like? I
  ¿Cómo está usted?
    How are you? (formal) I
  ¿Cómo estás?
    How are you? (familiar) I
  ¿Cómo me veo?
    How do I look? II
  ¡Cómo no! Of course! I
  ¿Cómo se llama?
    What is his/her name? I
  ¿Cómo se va a…?
    How do you get to…? II
  ¿Cómo te llamas?
    What is your name? I
  ¿Cómo te queda?
    How does it look on you? II
  Perdona(e), ¿cómo llego a…?
    Pardon, how do I get to…? I
cómodo(a) comfortable II, 1.2
el (la) compañero(a) companion II
la compañía company I
comparar to compare 4.3
compartir to share I, 1.1
el compás rhythm, beat 5.1
la competencia competition 3.3
competir (i) to compete II
complicado(a) complicated II, 2.2
comprar to buy I
comprender to understand I
comprensivo(a) understanding 1.1
la computación
  computer science I, II
la computadora computer I
  la computadora portátil
    laptop computer 6.2
común common II
  tener en común

to have in common 1.1
la comunidad community
con with I
  con rayas striped I
  con tal de que
    provided that, as long as 3.1
el concierto concert I
el concurso contest I, II, 3.2
conducir to drive II
el (la) conductor(a) driver II
conectar to connect 1.3
  conectarse de to connect to 6.3
la confiabilidad reliability,
  dependability 6.2
la configuración configuration 6.3
el congelador freezer I
el conjunto musical group 3.3
conmemorar to commemorate 3.3
conmigo with me I
conmovedor(a) moving
conocer to know, to be familiar
  with I, II
  conocer a alguien to know, to
    be familiar with, someone I
  conocerse bien (mal) to know
    each other (not very) well 1.3
el conocimiento knowledge 4.2
el conquistador
  conqueror, ladykiller 5.2
conseguir (i) to obtain II
el consejo advice II
conservador(a) conservative 3.3
conservar to conserve II, 2.1
considerado(a) considerate 1.1
la constitución constitution 3.3
construir to construct II
la consulta consultation II
el consultorio office (doctor's) II
consumir to consume 2.1
la contabilidad accounting 4.1
el (la) contador(a) accountant I, II,
  4.2
la contaminación pollution II
  la contaminación del aire
    air pollution I, 2.2
el contaminante pollutant 2.2
contaminar to pollute II
contar (ue) to count, to tell or
  retell I
  contar chistes to tell jokes II
  contarse chismes
    to tell each other gossip 1.3

contarse secretos
to tell each other secrets **1.3**
contemporáneo(a)
contemporary **5.3**
contento(a) content, happy, pleased **I**
la contestadora automática
answering machine **6.2**
contestar to answer **I**
contigo with you **I**
el contorno surrounding area
la contraseña password **6.3**
el contrato contract **II, 4.2**
el control remoto
remote control **6.1**
controlar to control **6.1**
convencer to convince **6.2**
la conversación
el grupo de conversación
chat group **6.3**
convivir
to live together, to get along **2.1**
el coquito eggnog **3.2**
el corazón heart **I**
el corral corral, pen **I**
el correo post office **I**
el correo electrónico e-mail **6.3**
correr to run **I**
correr riesgos to take risks **4.1**
el (la) corresponsal
correspondent **4.3**
el cortacésped lawnmower **1.3**
cortar el césped to cut the grass **II**
cortarse to cut one's self **II**
corto(a) short (length) **I**
la cosa thing **I**
costar (ue) to cost **I**
a mí me cuesta mucho
it's hard for me
¿Cuánto cuesta(n)…?
How much is (are)…? **I**
costarricense Costa Rican **II**
la costumbre custom **3.3**
crear to create **II, 2.1**
creativo(a) creative **5.3**
crecer to grow up **II**
la creencia belief **5.2**
creer to think, to believe **I, II**
Creo que sí/no.
I think so. I don't think so. **I**
¿Tú crees? Do you think so? **II**
la crema cream **I, II**
criarse to be raised

la crítica criticism **II**
el (la) crítico(a) critic **5.3**
el (la) cronista chronicler **5.2**
el cruce crossing **II**
cruzar to cross **I**
el cuaderno notebook **I**
la cuadra city block **I**
cuadrado(a) square **1.1**
el cuadro painting **5.1**
el cuadro histórico
historical painting **5.1**
cuál(es) which (ones), what **I**
¿Cuál es la fecha?
What is the date? **I**
¿Cuál es tu teléfono? What is your phone number? **I, II**
cuando when, whenever **I, 2.3**
cuando era niño(a) when I/he/she was young **II**
cuándo when **I**
cuánto how much **I**
¿A cuánto está(n)…?
How much is (are)…? **I**
¿Cuánto cuesta(n)…?
How much is (are)…? **I**
¿Cuánto es? How much is it? **I**
¿Cuánto le doy de propina?
How much do I tip? **I**
¿Cuánto tiempo hace que…?
How long is it since…? **I**
cuántos(as) how many **I**
¿Cuántos años tiene…?
How old is…? **I**
cuarenta forty **I**
el cuarto room **I**
limpiar el cuarto
to clean the room **I, II**
cuarto(a) quarter, fourth **I, II**
y cuarto quarter past **I**
cuatro four **I**
cuatrocientos four hundred **I**
cubano(a) Cuban **II**
los cubiertos utensils **II**
la cuchara spoon **I**
el cuchillo knife **I**
el cuello neck **II**
la cuenta bill, check **I, II**
la cuenta de ahorros
savings account **II**
La cuenta, por favor.
The check, please. **I**
el (la) cuentista
short-story writer **5.3**

el cuento short story **5.1**
la cuerda rope
saltar la cuerda to jump rope **II**
el cuero leather **I, 1.2**
los artículos de cuero
leather goods **I**
el cuerpo body **I, II**
cuidado
tener cuidado to be careful **I, II**
cuidadosamente carefully **I**
cuidadoso(a) careful **I**
cuidar to take care of **I, 2.1**
culminar to end, to culminate **5.3**
la cumbia cumbia (Latin music) **5.2**
el cumpleaños birthday **I**
el (la) cuñado(a)
brother-in-law/sister-in-law **II**
el currículum résumé, curriculum vitae **II, 4.1**
cursar to be enrolled in, to take

la danza dance **5.2**
dañino(a) damaging **2.2**
dar (doy) to give **I**
dar palmadas to clap hands **5.1**
dar una vuelta to take a walk, stroll, or ride **II**
dar(se) cuenta de to realize **II**
darle(s) de comer to feed **I**
los datos facts; information **II, 4.1**
de of, from **I**
de buen humor
in a good mood **I**
de cuadros plaid, checkered **I**
de la mañana in the morning **I**
de la noche at night **I**
de la tarde in the afternoon **I**
de mal humor in a bad mood **I**
de maravilla marvelous **II**
De nada. You're welcome. **I**
¿De veras? Really? **II**
de vez en cuando
once in a while **I**
debajo (de) below, underneath **I, II**
deber should, ought to **I**
decidir to decide **I**
décimo(a) tenth **I, II**
decir (digo) to say, to tell **I, II**

decisiones
   tomar decisiones
     to make decisions **4.1**
decorado(a) decorated **5.2**
dedicarse a to apply oneself to
  (something) **1.3**
los dedos fingers, toes **II**
dejar to leave (behind) **I**;
  to allow **3.1**
   dejar la propina to leave the tip **I**
   dejar un mensaje
     to leave a message **I**
   Deje un mensaje después del
     tono. Leave a message after
     the tone. **I**
   Le dejo… en…
     I'll give…to you for… **I**
   Quiero dejar un mensaje
     para… I want to leave a
     message for… **I**
dejar de to stop doing something
del from the **I**
delante de in front of **I, II**
delgado(a) thin **I**
delicioso(a) delicious **I**
demasiado(a) too much **I, II**
la democracia democracy **3.3**
democrático(a) democratic **3.3**
dentro (de) inside (of) **I**
   dentro del alcance
     within reach **5.3**
el (la) dependiente(a)
  salesperson **II**
el deporte sport
   practicar deportes
     to play sports **I**
el (la) deportista sportsman,
  sportswoman **II**; athlete **4.2**
deprimido(a) depressed **I**
la derecha right **I**
   a la derecha (de) to the right (of) **I**
derecho straight ahead **I**
el derecho law; right **3.3**
los derechos (humanos) (human)
  rights **2.1**
derivado(a) derivative,
  unoriginal **5.3**
el derrame de petróleo oil spill **2.2**
desafortunadamente
  unfortunately **II**
desagradable unpleasant **1.1**
desanimarse to get discouraged **1.3**
desarmar to take apart **1.3**

desarrollar to develop **2.2**
el desarrollo development **II, 2.1**
desayunar to have breakfast **I, II**
el desayuno breakfast **I**
descansar to rest **I**
descarado(a) insolent, shameless **1.1**
la descendencia descendants **5.2**
descifrar to decipher **5.2**
descomponérsele (a uno)
  to break down, to malfunction
  6.2
descompuesto
   estar descompuesto(a)
     to be broken **6.2**
desconectar to turn off **1.3**
   desconectarse de
     to disconnect from **6.3**
los desconocidos strangers
el descubrimiento discovery **3.3**
descubrir to discover **II, 2.2, 2.3**
el descuento discount **6.2**
desde from **I**
   desde allí from there **II**
desear to desire **II**
desempeñar un cargo
  to carry out a responsibility **4.1**
desenchufar to unplug **1.3**
el desfile parade, procession **I, 3.1**
el desierto desert **I**
deslumbrante dazzling **5.3**
el desodorante deodorant **II**
desorganizado(a) disorganized **1.3**
la despedida good-bye **3.2**
despedirse (i) to say good-bye **II**
el desperdicio waste **2.2**
el despertador alarm clock **I, II**
despertarse (ie) to wake up **I**
después (de) after, afterward **I**
destacarse to stand out **1.2**
la destrucción destruction **II, 2.2**
el desván attic **1.3**
la desventaja disadvantage **II, 4.2**
desyerbar to weed **1.3**
el detalle detail **II**
detestar to hate **1.2**
detrás (de) behind **I**
devolver (ue) to return **I, 6.2**
el día day **I**
   Buenos días. Good morning. **I**
   el Día de Acción de Gracias
     Thanksgiving Day **3.2**
   el Día de la Abolición de
     Esclavitud Abolition Day **3.3**

   el Día de la Amistad
     Valentine's Day **3.2**
   el Día de la Independencia
     Independence Day **3.2**
   el Día de la Madre/del Padre
     Mother's/Father's Day **3.2**
   el Día de la Navidad
     Christmas Day **3.2**
   el Día de la Raza
     Columbus Day **3.2**
   ¿Qué día es hoy?
     What day is today? **I**
   Tal vez otro día.
     Maybe another day. **I**
   todos los días every day **I**
diario(a) daily **II**
dibujar to draw **II**
el dibujo drawing
   el dibujo técnico
     technical drawing **4.1**
   los dibujos animados
     cartoons **6.1**
el diccionario dictionary **I**
diciembre December **I**
diecinueve nineteen **I**
dieciocho eighteen **I**
dieciséis sixteen **I**
diecisiete seventeen **I**
el diente tooth **I, II**
   lavarse los dientes
     to brush one's teeth **I, II**
la dieta diet **II**
diez ten **I**
diferencia
   a diferencia de
     as contrasted with **1.1**
difícil difficult, hard **I**
el dinero money **I**
el diploma diploma **3.1**
el (la) diplomático(a) diplomat **4.3**
la dirección address, direction **I**
   la dirección electrónica
     e-mail address **6.3**
el (la) director(a) director **5.3**
dirigir (dirijo) to direct **5.3**
el disco disk **6.3**
   el disco compacto
     compact disc **I**
   el disco duro hard drive **6.3**
la discriminación
  discrimination **2.1**
disculpar(se) to apologize **II**
   disculpe(me) excuse (me) **II**

el **discurso** speech **3.1**
discutir to discuss, to argue **1.1**
el **(la) diseñador(a)** designer **1.2**
el **diseño** design **4.1**
el **disfraz** costume
disfrutar con los amigos
　　to enjoy time with friends **II**
disponible available **6.3**
dispuesto
　　estar dispuesto(a)
　　　　to be willing to **4.1**
la **distancia** distance **II**
distinguir entre
　　to distinguish between **6.2**
diverso(a) diverse **II, 2.2**
divertido(a) enjoyable, fun,
　　entertaining **I, II**
divertirse (ie) to enjoy oneself **II**
doblar to turn **I**
doce twelve **I**
la **docena** dozen **I**
el **(la) doctor(a)** doctor **II**
el **doctorado** doctorate **4.1**
el **documental** documentary **6.1**
el **dólar** dollar **I**
doler (ue) to hurt **II**
el **dolor** pain
　　el **dolor de cabeza** headache **II**
domingo Sunday
dominicano(a) Dominican **II**
don/doña Mr./Mrs. **I, II**
donar to donate **2.1**
dónde where **I**
　　¿De dónde eres?
　　　　Where are you from? **I**
　　¿De dónde es?
　　　　Where is he/she from? **I**
　　¿Dónde tiene lugar?
　　　　Where does it take place? **II**
dormir (ue) to sleep **I**
　　el **saco de dormir**
　　　　sleeping bag **II, 2.3**
dormirse (ue) to fall asleep **I**
dos two **I**
doscientos(as) two hundred **I, II**
el **drama** drama **5.1**
dramático(a) dramatic **5.3**
ducharse to take a shower **I, II**
dudar que… to doubt that… **II**
el **(la) dueño(a)** owner **II, 4.2**
dulce sweet **I**
la **durabilidad** durability **6.2**
durante during **I**

duro(a) hard, tough **I**

echar to throw out, away **II, 2.2;**
　　to send out
echar de menos to miss
el **ecosistema** ecosystem **2.2**
el **(la) ecoturista** ecotourist **2.3**
ecuatoriano(a) Ecuadorian **II**
la **edad** age **I**
la **edición** edition **II**
el **edificio** building **I**
el **(la) editor(a)** editor **I, II**
la **educación** education **II, 4.1**
　　la **educación física**
　　　　physical education **I**
educar al público
　　to educate the public **2.1**
el **efectivo** cash **I**
el **efecto** effect **II, 2.2**
el **ejército** army **3.3**
él he **I**
la **electricidad** electricity **II**
elegante elegant **II**
ella she **I**
ellos(as) they **I**
elogiar to praise **5.3**
embellecer to beautify **2.1**
emocionado(a) excited **I**
emocionante exciting **5.3**
empezar (ie) to begin **I, II**
el **(la) empleado(a)** employee **4.2**
el **empleo** employment, job **I, II, 4.2**
emprendedor(a) enterprising **4.1**
la **empresa**
　　business, company **II, 4.2**
empujar to push
en in **I**
　　en caso de que in case **3.1**
　　en cuanto as soon as **2.3**
　　en línea on-line **6.3**
　　en oferta on sale
　　en seguida at once **I**
　　en vivo y en directo
　　　　live (programming) **6.1**
enamorar(se) de to fall in love **II**
Encantado(a). Delighted/Pleased
　　to meet you. **I**
encantar to delight **II**
encargarse de to take charge of **4.1**

encender (ie) to turn on **1.3**
la **enchilada** enchilada **I**
enchufar to plug in **1.3**
encima (de) on top (of) **I, II**
encontrar (ue) to find, to meet **I**
el **encuentro** meeting **I**
la **energía** energy **II**
enero January **I**
la **enfermedad** sickness **II**
el **(la) enfermero(a)** nurse **II**
enfermizo(a) sickly
enfermo(a) sick **I**
los **enfermos** the sick **2.1**
enfrentar to confront **3.3**
enfrente (de) in front (of) **I, 6.2**
el **engaño** trick, deceit
englobar to encompass
enhorabuena congratulations **3.1**
el **enlace** link **6.3**
enojado(a) angry **I**
enojarse con to get angry with **II**
enorme huge, enormous **I, II**
la **ensalada** salad **I**
el **ensayo** essay **3.3**
enseñar to teach **I**
entender (ie) to understand **I**
entonces then, so **I**
entrar (a, en) to enter **I**
entre between **I**
el **entrenamiento** training **II, 4.2**
entrenarse to train **II**
entretenido(a) entertaining **6.1**
la **entrevista** interview **I, II**
el **(la) entrevistador(a)**
　　interviewer **II, 4.2**
entusiasmarse to get excited **1.3**
envidia
　　tener envidia to be envious **II**
el **episodio** episode **6.1**
el **equipaje** luggage **II**
el **equipo** team **I**
el **equipo estereofónico**
　　stereo equipment **6.2**
equivocarse to make a mistake **6.2**
esbelto(a) slender **1.1**
escalar montañas
　　to mountain climb **II, 1.2**
la **escalera** stairs **II**
la **escena** scene **II**
el **(la) esclavo(a)** slave **3.3**
escoger (escojo) to choose **II**
los **escombros** rubble, debris
esconder(se) to hide **II, 1.3**

escribir to write I, II
el (la) escritor(a) writer I, II
el escritorio desk I
escuchar to listen (to) I
la escuela school I, 5.1
el (la) escultor(a) sculptor II
la escultura sculpture II
ese(a) that I, 5.1
ése(a) that one I, 5.1
esencial
  Es esencial que…
    It's essential that… II
el esfuerzo
  hacer un esfuerzo
    to make the effort 2.1
eso that I, 5.1
el espacio de la televisión
  television program
la espada sword
el español Spanish (language) I
el (la) español(a) Spaniard II
especial special I
la especialidad de la casa
  specialty of the house II
especializarse to specialize 4.1
especialmente
  specially, especially I, II
las especies species 2.2
el espejo mirror I, II
esperar to wait for, to expect,
  to hope I, II
  Esperar que… to hope that… I
la esposa wife I
el esposo husband I
esquiar to ski I
  esquiar en el agua
    to water-ski 1.2
la esquina corner II
la estación de autobuses
  bus station I
el estacionamiento
  parking space II
las estaciones seasons I
el estadio stadium I
las estadísticas statistics 4.3
el estado civil civil status
  (married, divorced, or single)
  4.1
estadounidense of the U.S. II
estampado(a) embossed (fabric) 1.2
estar (estoy) to be I, II
  ¿A cuánto está(n)…?
    How much is (are)…? I

¿Cómo está usted?
  How are you? (formal) I
¿Cómo estás?
  How are you? (familiar) I
¿Está incluido(a)…?
  Is… included? I
estar a favor de to be for II, 2.1
estar bien informado(a)
  to be well informed II
estar de acuerdo to agree I, II
estar descompuesto(a)
  to be broken 6.2
estar dispuesto(a) a
  to be willing to 4.1
estar en contra de
  to be against 2.1
estar resfriado to have a cold II
estar roto(a) to be broken 6.2
no estar seguro(a) (de) que
  not to be sure that 3.2
la estatura height II, 4.1
el este east II
este(a) this I, 5.1
éste(a) this one I, 5.1
el estilo style 5.3
estirar(se) to stretch II
esto this I, 5.1
el estómago stomach I, II
estrecho(a) narrow I, II
la estrella star I
el estreno new release II
el estrés stress I, II
el (la) estudiante student II
estudiar to study I
  estudiar las artes marciales
    to study martial arts II
los estudios sociales
  social studies I
la estufa stove I, II
el examen test I
el exceso de equipaje
  excess luggage II
exclamar to exclaim II
exigir (exijo) to demand 3.1
el éxito success I, 3.1
  tener éxito to be successful II
explicar to explain II
la exportación exportation 4.3
exportar to export 4.3
la exposición exhibit II
expresivo(a) expressive 5.3
externo(a) external 6.3
el(la) extranjero(a) foreigner II

extraño(a) strange

la fábrica factory 4.3
fácil easy I
fácilmente easily I, II
la falda skirt I
faltar to lack II
la familia family II
la farmacia pharmacy, drugstore I
el faro lighthouse 3.3
fascinar to fascinate II
la fauna silvestre
  wild animal life 2.2
favorito(a) favorite I
el fax fax 6.3
  el fax multifuncional
    multifunctional
    fax machine 6.2
febrero February I
la fecha date I
  la fecha de nacimiento
    date of birth II, 4.1
  ¿Cuál es la fecha?
    What is the date? I
la felicidad happiness II
felicidades congratulations I
feliz happy I
felizmente happily I
feo(a) ugly I
feroz ferocious II
festejar to celebrate 3.2
la ficción fiction 5.1
la fiebre fever II
fiel faithful 1.1
la fiesta party I, II
  la fiesta continua
    party in stages 3.2
la figura figure 5.1
fijarse en to notice 6.2
el fin de semana weekend I
el final ending 5.3
el (la) financiero(a)
  financial expert 4.3
las finanzas finance 4.1
la firma signature II, 4.1
el flamenco
  flamenco-style dancing 5.1
el flan caramel custard I
el fleco fringe 1.2

el flequillo bangs **1.1**
flojo(a) loose **II**
la flor flower **I**
la flora silvestre wild plant life **2.2**
fluir to flow
la fogata campfire **II**
el fondo background **5.1**
la formación training, education **4.1**
formal formal **I, II**
formidable great **1.2**
formulista formulaic **5.3**
el fósforo match **II, 2.3**
la foto photo, picture
   sacar fotos
     to take pictures **I**
el (la) fotógrafo(a)
   photographer **I, II**
francés(esa) French **II**
frecuente frequent **I**
frecuentemente
   often, frequently **I, II**
frente a in front of, opposite **II**
la fresa strawberry **II**
   Hace fresco. It is cool. **I**
el frigorífico refrigerator **I**
los frijoles beans **II**
frío cold **I**
   Hace frío. It is cold. **I**
   tener frío to be cold **I**
la fruta fruit **I**
fue cuando it was when **II**
el fuego fire **I, 2.3**
la fuente source
fuera (de) outside (of) **I**
fuerte strong **I**
la fuerza motriz
   power, moving force
funcionar to work, to run **II, 6.2**
el fútbol soccer **I**
el fútbol americano football **I**

el gabinete cabinet **1.3**
las gafas de sol sunglasses **I**
la gala big, formal party **3.2**
la galería gallery **II**
la galleta cookie, cracker **I, II**
la gallina hen **I**
el gallo rooster **I**
la ganadería livestock industry **4.3**

el (la) ganadero(a) farmer **I**
el ganado livestock **4.3**
el (la) ganadora winner **I, II**
ganar to win **I**
ganarse la vida
   to earn a living **II, 4.2**
ganas
   tener ganas de… to feel like… **I**
la ganga bargain **II**
el garaje garage **II**
la garantía guarantee **6.2**
la garganta throat **II**
la gasolina gasoline **II**
gastar to spend **II**
los gastos expenses **II**
el (la) gato(a) cat **I**
la gaviota seagull
los (las) gemelos(as) twins **II**
el género genre **5.3**
la generosidad generosity **3.1**
generoso(a) generous **3.1**
genial wonderful **1.2**
el genio genius
la gente people **I**
   la gente sin hogar the homeless **2.1**
el (la) gerente manager **I, II, 4.2**
gestionar to arrange
el gimnasio gymnasium **I**
girar to turn **II**
los globos balloons **II**
el (la) gobernador(a) governor **3.3**
el gobierno government **3.3**
el gol goal **I**
gordo(a) fat **I**
la gorra baseball cap **I**
el gorro cap **I**
gozar de to enjoy, to have
la grabadora tape recorder **I, 6.2**
grabar to record **6.1**
Gracias. Thank you. **I, II**
   Gracias, pero no puedo.
     Thanks, but I can't. **I**
   Mil gracias. Many thanks. **3.1**
el grado degree **I**
el (la) graduando(a) graduate **3.1**
el gramo gram **I, II**
grande big, large **I**
la granja farm **I**
la gripe flu **II**
gritar to scream **II**
el grito shout
grueso(a) heavy **1.1**

el grupo group
   el grupo de conversación
     chat group **6.3**
   el grupo de noticias
     news group **6.3**
el guante glove **I**
guapo(a) good-looking **I**
el (la) guardaespaldas
   bodyguard **6.1**
guardar to hold, to keep **I, II**
guatemalteco(a) Guatemalan **II**
la guerra war
la guía telefónica
   phone directory **I**
los guineítos en escabeche
   small green bananas in garlic
   vinegar, red pepper and oil **3.2**
el guión script **5.3**
el (la) guionista scriptwriter **5.3**
la guitarra guitar **I**
   tocar la guitarra
     to play the guitar **I**
gustar to like **I, II**
   Le gusta… He/She likes… **I**
   Me gusta… I like… **I**
   Me gustaría… I would like… **I**
   Te gusta… You like… **I**
   ¿Te gustaría…?
     Would you like…? **I**
el gusto pleasure **I**
   a gusto comfortable, happy
   El gusto es mío.
     The pleasure is mine. **I**
   Mucho gusto.
     Nice to meet you. **I**
   Sí, con mucho gusto.
     Yes, gladly. **2.1**

la habanera
   habanera (Latin music) **5.2**
había there was, there were **II**
las habichuelas coloradas
   red beans **II**
las habilidades capabilities **II, 4.2**
la habitación bedroom, room **I, II**
hablar to talk, to speak **I, II**
   ¿Puedo hablar con…?
     May I speak with… **I**
hacer (hago) to make, to do **I, II**

¿Hace… que? How long has it been since…? **II**
Hace buen tiempo.
    It is nice outside. **I**
Hace calor. It is hot. **I**
Hace fresco. It is cool. **I**
Hace frío. It is cold. **I**
Hace mal tiempo.
    It is bad outside. **I**
Hace sol. It is sunny. **I**
Hace viento. It is windy. **I**
hacer alpinismo to go hiking **1.2**
hacer clic/doble clic
    to click/to double click **6.3**
hacer ejercicio to exercise **I**
hacer el papel to play the role **5.3**
hacer juego con…
    to match with…
hacer la cama to make the bed **I**
hacer la limpieza
    to do the cleaning
hacer montañismo
    to go mountaineering **1.2**
hacerle caso a uno
    to pay attention to **1.1**
hacer un esfuerzo
    to make the effort **2.1**
¿Qué tiempo hace?
    What is the weather like? **I**
hacerse pedazos to fall apart
hacia toward **II**
el halcón falcon **2.3**
el hambre (fem.) hunger
    tener hambre to be hungry **I**
la hamburguesa hamburger **I**
el hardware hardware **6.3**
la harina flour **I, II**
    la harina de trigo wheat flour
hasta until, as far as **I, II**
    Hasta luego. See you later. **I**
    Hasta mañana.
        See you tomorrow. **I**
hasta que until **2.3**
hay there is, there are **I**
    hay que one has to, one must **I**
    Hay sol. It's sunny. **I**
    Hay viento. It's windy. **I**
    No hay de qué. It's nothing. **3.2**
el hecho fact **II**
la heladería ice cream parlor **II**
el helado ice cream **I, II**
el (la) hermanastro(a)
    stepbrother/stepsister **II**

el(la) hermano(a) brother/sister **I**
los hermanos brother(s) and
    sister(s) **I**
el héroe hero **II**
la heroína heroine **II**
el hielo ice **I**
    sobre hielo on ice **I**
el hierro iron **4.3**
la hija daughter **I**
el hijo son **I**
los hijos son(s) and daughter(s),
    children **I**
hincharse to swell
la historia history, story **I, II**
el hockey hockey **I**
la hoja leaf **II**
la hoja de cálculo spreadsheet **6.3**
Hola. Hello. **I**
el hombre man **I**
    el (la) hombre/mujer de
        negocios businessman
        businesswoman **I, II**
el hombro shoulder **II**
hondureño(a) Honduran **II**
honrar to honor **3.3**
la hora hour
    ¿A qué hora es…?
        (At) What time is…? **I**
    ¿Qué hora es? What time is it? **I**
el horario schedule **I**
el hormigón concrete
el horno oven **I, II**
    el horno microondas
        microwave oven **II**
horrible horrible **1.2**
el horror horror **II**
hospedarse to stay at **II**
el hotel hotel **I**
hoy today **I**
    Hoy es… Today is… **I**
    ¿Qué día es hoy?
        What day is it? **I**
hubo there was, there were **II**
el (la) huésped(a) guest **II**
el huevo egg **I, II**
las humanidades humanities **4.1**
húmedo(a) humid **II, 2.3**
el huracán hurricane **II, 2.3**

el icono del programa
    program icon **6.3**
la ida y vuelta round trip **II**
la identificación identification **II**
el identificador de llamadas caller
    identification **6.2**
la ideología ideology **3.3**
la iglesia church **I**
Igualmente. Same here. **I**
la iguana iguana **2.3**
impaciente impatient **II**
el impermeable raincoat **I**
implicar to mean
imponer (impongo) to impose **3.1**
importar to be important **II**
    Es importante que…
        It's important that… **II**
imposible
    Me es imposible. It's just not
        possible for me. **2.1**
impresionante impressive **5.3**
la impresora printer **I**
incluido(a) included **I**
    ¿Está incluido(a)…?
        Is … included? **I**
incómodo(a) uncomfortable **1.2**
increíble incredible **II, 2.2**
los (las) indígenas indigenous,
    indigenous peoples
la industria industry **4.3**
la industria pesquera fishing
    industry **4.3**
la infección infection **II**
influir to influence **1.1**
informal informal **I**
la informática
    computer science **4.1**
la ingeniería engineering
    la ingeniería mecánica
        mechanical engineering **4.1**
el (la) ingeniero(a) engineer **II**
el inglés English (language) **I**
inglés(esa) English **II**
el ingreso entrance, admission
el inicio beginning
    los tambaleantes inicios
        shaky beginnings
inigualable unequalled **6.2**
la injusticia injustice **3.3**

inmediatamente immediately **II**
innovador(a) innovative **5.3**
inolvidable unforgettable **3.2**
la inquietud uncertainty
insistir en (que) to insist (on) **II, 3.1**
instituir to institute **2.2**
inteligente intelligent **I**
interesante interesting **I**
interesar to interest **II**
internacional international **II**
Internet
    navegar por Internet
        to surf the Internet **1.2**
interno(a) internal **6.3**
interpretar to interpret **5.1**
el (la) intérprete interpreter **4.3**
inútil useless **II, 2.2**
el invierno winter **I**
la invitación invitation **I, II**
invitar to invite
    Te invito.
        I'll treat you. I invite you. **I**
la inyección injection **II**
ir (voy) to go **I, II**
    ir a… to be going to… **I**
    ir al cine to go to the movies **I**
    ir al supermercado
        to go to the supermarket **I**
    ir de compras to go shopping **I**
    Vamos a… Let's… **I**
irónico(a) ironic **5.3**
irse (me voy) to leave, to go away **I**
italiano Italian **II**
la isla island **II**
la izquierda left
    a la izquierda (de)
        to the left (of) **I**

el jabalí wild boar
el jabón soap **I, II**
el jade jade **5.2**
el jaguar jaguar **II, 2.3**
jamás never
el jamón ham **I, II**
japonés(esa) Japanese **II**
el jarabe tapatío Mexican dance
    from Guadalajara **5.2**
el jardín garden **I, II**
el (la) jardinero(a) gardener **4.2**

la jarra pitcher **I**
los jeans jeans **I**
el (la) jefe(a) boss **I, II**
los jeroglíficos hieroglyphics **5.2**
la jota Aragonesa folk dance **5.1**
joven young **I**
los jóvenes young people **2.1**
las joyas jewelry **I**
la joyería jewelry store **I**
jubilarse to retire **4.2**
el juego interactivo
    interactive game **6.3**
jueves Thursday **I**
el (la) juez(a) judge **II, 4.2**
jugar (ue) to play **I, II**
    jugar al ajedrez to play chess **II**
el jugo juice **II**
el juguete toy **II**
la juguetería toy store **II**
julio July **I**
junio June **I**
juntar fondos to fundraise **2.1**
junto a next to **II**
juntos together **I**
justo(a) just, fair **3.3**
juzgar to judge

el kilo kilogram **I, II**

el laboratorio laboratory **4.3**
labrado(a) worked, cut **5.2**
lacio straight (hair) **II**
el lado side
    al lado (de) beside, next to **I, 6.2**
    por otro lado
        on the other hand **1.1**
    por un lado on the one hand **1.1**
el (la) ladrón(a) thief **II**
el lago lake **I**
la lámpara lamp **I, II**
la lana wool **I, 1.2**
el lápiz pencil **I**
largo(a) long **I**
    a lo largo de throughtout
la lástima

Es una lástima que…
    It's a shame that… **II**
¡Qué lástima! What a shame! **I**
lastimarse to hurt oneself **II**
la lata can **I, II, 2.2**
el lavabo washbowl **II**
el lavaplatos dishwasher **I, II**
lavar to wash **I**
    lavar los platos
        to wash the dishes **I, II**
lavarse to wash oneself **I, II**
    lavarse la cabeza
        to wash one's hair **I**
    lavarse los dientes
        to brush one's teeth **I, II**
la lección lesson **I**
la leche milk **I, II**
el lechón asado
    roast suckling pig **3.2**
la lechuga lettuce **I**
leer to read **I**
lejos (de) far (from) **I**
    ¿Queda lejos? Is it far? **I**
la lengua language **I**
lentamente slowly **I, II**
la lentejuela sequin **1.2**
los lentes de contacto
    contact lenses **1.1**
lento(a) slow **I**
la leña firewood **II, 2.3**
el león lion **II**
la letra lyrics **5.1**
el letrero sign **II**
levantar pesas to lift weights **I**
levantarse to get up **I, II**
la ley law **3.3**
liberal liberal **3.3**
la librería bookstore **I**
el libro book **I**
la licenciatura university degree **4.1**
el (la) líder leader **3.3**
la limonada lemonade **I**
limpiar el cuarto to clean the
    room **I, II**
la limpieza
    hacer la limpieza
        to do the cleaning **II**
limpio(a) clean **I, II**
    mantener (mantengo) limpio(a)
        to keep clean **II**
la línea line
    en línea on-line **6.3**
la linterna flashlight **II, 2.3**

listo(a) ready **I**
la literatura literature **I**
el litro liter **I, II**
la llama llama **I**
la llamada call **I**
llamar to call **I**
    ¿Cómo se llama?
      What is his/her name? **I**
    ¿Cómo te llamas?
      What is your name? **I**
    Dile/Dígale que me llame.
      Tell (familiar/formal) him or
      her to call me. **I**
    Me llamo… My name is… **I**
    Se llama…
      His/Her name is… **I**
la llave key **I, II**
el llavero keychain **1.2**
la llegada arrival **II**
llegar to arrive **I, II**
llenar to fill up **II**
lleno(a) full **II**
llevar
    to wear, to carry; to take along **I**
    llevar a cabo to accomplish **3.1**
    llevar(se) bien
      to get along with **II**
    llevarse bien/mal
      to get along well/badly with
      each other **1.3**
llorar to cry **II**
llover (ue) to rain **I**
la llovizna drizzle **II, 2.3**
la lluvia rain **I**
el lobo wolf **II**
local local **II**
el Localizador Unificador de
    Rescursos (LUR) URL **6.3**
la loción aftershave **II**
la loción protectora
    sunscreen **II**
loco(a) crazy **I**
lógico
    Es lógico que…
      It's logical that… **II**
lograr to achieve
el loro parrot **II, 2.3**
la lucha fight **3.3**
luchar contra to fight against **2.1**
lucir to appear
luego later **I**
    Hasta luego. See you later. **I**
    el lugar place **I**

lujoso(a) luxurious **I, II**
la lumbre light
el lunar beauty mark **1.1**
los lunares polka dots **1.2**
lunes Monday **I**
la luz light **I, II, 2.3**
    apagar la luz
      to turn off the light **I**

la madrastra stepmother **II**
la madre mother **I**
    el Día de las Madres
      Mother's Day **3.2**
la madrugada
    early morning, dawn **3.2**
la maestría master's degree **4.1**
el (la) maestro(a) teacher **I**
el maíz corn **4.3**
las malas hierbas weeds **1.3**
el malecón boardwalk
la maleta suitcase **II**
el (la) maletero(a) porter **II**
malo(a) bad **I**
    Es malo que… It's bad that… **II**
    Hace mal tiempo.
      It is bad outside. **I**
    lo malo the bad thing **1.1**
    Lo malo es que…
      The trouble is that… **II**
el mambo mambo
    (Latin music) **5.2**
mandar una carta
    to send a letter **I**
manejar to drive **I**
manipular to manipulate **6.1**
la mano hand **I, II**
la manta blanket **I, II, 2.3**
el mantel tablecloth **II**
mantener (mantengo)
    mantener limpio to keep clean **II**
mantenerse sano(a)
    to be healthy **II**
la mantequilla butter **I, II**
la mantequilla de cacahuate
    peanut butter **II**
la manzana apple **II**
mañana tomorrow **I**
la mañana morning **I**
    Mañana es… Tomorrow is… **I**

de la mañana in the morning **I**
Hasta mañana.
    See you tomorrow. **I**
por la mañana
    during the morning **I**
el mapa map **I**
el maquillaje makeup **II**
maquillarse to put on makeup **I, II**
la máquina contestadora
    answering machine **I**
el mar sea **I**
de maravilla marvelous **II**
la marca brand **6.2**
marcar to dial **I**
la marina navy
la marioneta marionette **II**
la mariposa butterfly **II, 2.3**
marrón brown **I**
martes Tuesday **I**
marzo March **I**
más more **I**
    más de more than **I**
    lo más the most **1.1**
    más… que more… than **I, 6.3**
la máscara mask
el mascarero maskmaker
las matemáticas mathematics **I**
la materia subject **I**
mayo May **I**
mayor older **I**
    mayor que older than **II**
la mayoría majority **II**
el (la) mecánico(a) mechanic **II, 4.2**
la medalla medallion **1.2**
la media hermana half-sister
la medianoche midnight **I, II**
mediante by means of
la medicina medicine **II**
medio(a) half **I, II**
el medio means, medium
el medio ambiente
    environment **II, 2.2**
el medio hermano half-brother **I**
el mediodía noon **I**
la mejilla cheek
mejor better **I**
    Es mejor que…
      It's better that… **II**
    lo mejor the best **1.1**
    mejor que better than **II**
la melodía melody **5.1**
el melón melon **II**
la memoria memory **6.3**

menor younger **I**
  menor que younger than **II**
menos to, before; less **I**
  menos de less than **I**
  lo menos the least **1.1**
  menos… que less…than **I**
el mensaje message **I**
  dejar un mensaje
    to leave a message **I**
  Deje un mensaje después del
    tono. Leave a message after
    the tone. **I**
  Quiero dejar un mensaje
    para… I want to leave a
    message for… **I**
mentir (ie) to lie
la mentira lie **II**
el menú menu **I**
el mercadeo marketing **4.1**
el mercado market **I**
merendar (ie) to have a snack **I**
el merengue
  merengue (Latin music) **5.2**
la merienda snack **I**
el mes month **I**
  el mes pasado last month **I**
la mesa table **I, II**
  poner la mesa to set the table **I**
  quitar la mesa to clear the table **I**
el (la) mesero(a) waiter (waitress) **I**
la meta goal **II**
meterse en to get into
el metro subway **I**
mexicano(a) Mexican **II**
mezclar to mix
la mezclilla denim **1.2**
mi my **I**
el micrófono microphone **6.3**
el microondas microwave **I**
el microprocesador
  microprocessor **6.3**
el miedo fear **I**
  tener miedo to be afraid **I, II**
mientras while **II**
miércoles Wednesday **I**
mil millones billion **4.3**
mil one thousand **I, II**
millón de millones trillion **4.3**
millón million **II**
mimado(a) spoiled **1.1**
la minería mining **4.3**
los minusválidos

  the physically challenged **2.1**
mío(a) mine
mirar to watch, to look at **I**
mismo(a) same **I**
la mitad de one half of **4.3**
la mochila backpack **I**
la moda fashion, style **1.2**
el módem modem **6.3**
moderno(a) modern **I, II**
modesto(a) modest **1.1**
mojado(a) wet **II**
molestar to bother **II**
el momento moment **I**
  Un momento. One moment. **I**
la monarquía monarchy **3.3**
el monedero change purse **1.2**
el monitor monitor **6.3**
la monja nun
el mono monkey **II**
  el mono araña
    spider monkey **2.3**
la montaña mountain **I**
  escalar montañas
    to mountain climb **II, 1.2**
el montañismo mountaineering **II**
  hacer montañismo
    to go mountaineering **1.2**
morado(a) purple **I**
moreno(a) dark hair and skin **I**
morir (ue) to die **II**
el mostrador counter **II**
mostrar (ue) to show **II**
el motivo purpose **3.2**
la moto(cicleta) motorcycle **I**
mover los muebles
  to move the furniture **I**
la muchacha girl **I**
el muchacho boy **I**
muchas veces often **4.2**
mucho often **I**
mucho(a) much, many **I**
los muebles furniture **I, II**
la mujer woman **I**
  la mujer de negocios
    businesswoman **I, II**
multinacional multinational **4.3**
el mundo world
la muñeca doll; wrist **II**
el muñeco de peluche
  stuffed animal **II**
el mural mural **5.2**
el (la) muralista muralist **5.2**

el museo museum **I**
la música music **I**
el musical musical **II**
el (la) músico(a) musician **II, 3.2**
muy very **I**

nacer to be born
el nacimiento
  la fecha de nacimiento
    date of birth **II, 4.1**
nada nothing **I**
  De nada. You're welcome.
nadar to swim **I**
  nadar con tubo de respiración
    to snorkel **2.3**
nadie no one **I**
la nariz nose **I, II**
la naturaleza nature **II, 2.2**
  la naturaleza muerta still life **5.1**
la navaja jackknife **II, 2.3**
navegar to navigate, sail
  navegar en tabla de vela
    to windsurf **1.2**
  navegar por Internet
    to surf the Internet **1.2**
  navegar por rápidos to go
    whitewater rafting **2.3**
la Navidad Christmas **3.2**
la neblina mist, fog **II, 2.3**
necesitar to need **I**
  Es necesario que…
    It's necessary that… **II**
negro(a) black **I**
nervioso(a) nervous **I**
  ponerse nervioso(a)
    to get nervous **1.3**
nevar (ie) to snow **I**
la neverita cooler **II**
ni nor, neither, not even **II**
ni… ni neither…nor **4.2**
nicaragüense Nicaraguan **II**
la nieta granddaughter
el nieto grandson
los nietos grandchildren
la nieve snow
la niña girl
el (la) niñero(a) babysitter **II, 4.2**
ninguno(a) none, not any **I**
el niño boy

la nitidez clarity, sharpness **6.2**
el nivel level
no no, not **I**
   ¡No digas eso! Don't say that! **I**
   ¡No me digas! Don't tell me! **I**
   ¡No te preocupes! Don't worry! **I**
la noche night, evening **I**
   Buenas noches. Good evening. **I**
   de la noche at night **I**
   por la noche
     during the evening **I**
el nombre name, first name **I**
normal normal **I**
normalmente normally **I, II**
el norte north **II**
nosotros(as) we **I**
la nota grade
   sacar una buena nota
     to get a good grade **I**
las noticias news **II**
   el grupo de noticias
     news group **6.3**
el noticiero news program **II**
novecientos nine hundred **I**
la novela novel **I**
el (la) novelista novelist **5.3**
noveno(a) ninth **I, II**
noventa ninety **I**
noviembre November **I**
el (la) novio(a)
   boyfriend/girlfriend **II**
la nube cloud **II, 2.3**
nublado cloudy **I**
   Está nublado. It is cloudy. **I**
nuestro(a) our **I**, ours **II**
nueve nine **I**
nuevo(a) new **I**
   el Año Nuevo New Year **3.2**
   el Nuevo Mundo New World **5.2**
el número number **I**; shoe size **II**
nunca never **I**
nutritivo(a) nutritious **II**

o or **I**
o…o either…or **4.2**
obediente obedient **II**
la obra work **I**; work of art **II**
   la obra de teatro
     theatrical production **II**
el (la) obrero(a) worker **II**
obtener (obtengo)
   to obtain, to get **II**
el océano ocean **II**
el ocelote ocelot **2.3**
ochenta eighty **I, II**
ocho eight **I, II**
ochocientos eight hundred **I, II**
octavo(a) eighth **I, II**
octubre October **I**
ocupado(a) busy **I**
ocurrir to occur **II**
   ocurrírsele (a uno)
     to dawn on, to occur to **6.2**
odiarse to hate each other **1.3**
el oeste west **II**
la oficina office **I**
ofrecer to offer **I, II**
   Le puedo ofrecer…
     I can offer you… **I**
   ¿Se le(s) ofrece algo más? May
     I offer you anything more? **I**
el oído inner ear **II**
oír (oigo) to hear **I**
ojalá que I hope that, hopefully **II**
el ojo eye **I, II**
la ola wave **I, II**
al óleo oil painting **5.1**
oler (huelo) (ue) a to smell
la olla pot **I**
el olmo elm
olvidar to forget **I**
   olvidársele (a uno) to forget **6.2**
once eleven **I**
la onda trendy thing to do; wave
ondulado(a) wavy **1.1**
el (la) operador(a) operator **I, II, 4.2**
opinar
   no opinar que
     not to be of the opinion that
oponerse a (me opongo)
   to oppose **1.3**
el (la) opresor(a) oppressor **3.3**
opuesto(a) opposite **1.1**
ordenar (las flores, los libros)
   to arrange (flowers, books) **I**
ordinario(a) ordinary **I**
la oreja ear **I, II**
el orfelinato orphanage
el orgullo pride **3.1**
orgulloso(a) proud **3.1**

original original **5.3**
la orilla edge; shore **II**
el oro gold **I**
la orquesta orchestra **3.2**
oscurecer to get dark **2.3**
la oscuridad darkness **2.3**
oscuro(a) dark **II**
   lo más oscuro
     the most incomprehensible,
     the least desirable
el oso hormiguero anteater **2.3**
el otoño fall **I**
otro(a) other, another **I**
ovalado(a) oval **1.1**

paciente patient **I**
el padrastro stepfather **II**
el padre father **I**
   el día de los Padres
     Father's Day **3.2**
los padres parents **I**
los padrinos godparents **3.1**
pagar to pay **I**
la página-web Web page **6.3**
el país country **I**
el paisaje landscape **5.1**
el pájaro bird **I**
la palma palm tree **II**
el palmar palm tree grove **II**
el palo stick
el pan bread **I, II**
   el pan dulce sweet roll **I**
la panadería bread bakery **I**
panameño(a) Panamanian **II**
la pandereta tambourine **5.1**
la pantalla screen **I**
los pantalones pants **I**
   los pantalones cortos shorts **I**
el pañuelo scarf **II**
la papa potato **II**
   las papas fritas French fries **I**
el papel paper **I**; role **II**
   hacer el papel to play the role **5.3**
la papelería stationery store **I**
el paquete package **I, 4.1**
el par
   un par de a pair of
para for, in order to **I**

para empezar to begin with I

para que so that 3.1

¿Para qué? For what purpose?

la parada stop, stand II

el paraguas umbrella I

paraguayo(a) Paraguayan II

parar to stop II

la pared wall I, II

el (la) pariente relative II

el parque park I

la partición

se harán particiones the
inheritance will be divided up

participar to participate 2.1

el partido game I

el (la) pasajero(a) passenger II

el pasaporte passport II

pasar to happen, to pass,
to pass by I

pasar la aspiradora
to vacuum I, II

pasar por debajo de la mesa
to go unnoticed

pasar un rato con los amigos
to spend time with friends I

pasarlo bien
to have a good time 3.2

las Pascuas Easter 3.2

pasear to go for a walk I

el pasillo aisle II

el paso the way 5.2

la pasta pasta I, II

la pasta de dientes toothpaste I

el pastel cake I; tamale-like mixture
of plantain, yuca and meat 3.2

la pastelería pastry shop I

la pastilla pill II

el pastor shepherd I

la patata potato I

patinar to skate I

los patines skates I

la patineta skateboard I

andar en patineta
to skateboard I

la patria mother country 3.3

el (la) patriota patriot 3.3

patriótico(a) patriotic 3.3

el patriotismo patriotism 3.3

patrocinar to sponsor 3.2

el pavo turkey 3.2

el payaso clown

la paz peace

el peatón (la peatona) pedestrian II

las pecas freckles 1.1

el pedazo piece I

pedir (i) to ask for, to order I, II

¿Me ayuda a pedir?
Could you help me order? I

el peinado hairdo

peinarse to comb one's hair I

el peine comb I

pelearse to fight II, 1.3

el pelícano pelican 2.3

la película movie I

peligroso(a) dangerous I, II, 2.3

Es peligroso que…
It's dangerous that… II

pelirrojo(a) redhead I

el pelo hair I

la pelota baseball I

el pelotón de fusilamiento firing
squad

el (la) peluquero(a)
hairstylist II, 4.2

los pendientes
dangling earrings 1.2

pensar (ie) to think, to plan I

la pensión pension; boarding
house II

peor worse I

lo peor the worst 1.1

peor que worse than II

pequeño(a) small I

la pera pear II

la percepción perception 6.1

perder (ie) to lose I

perdérsele (a uno)
to lose (something) 6.2

Perdona(e)… Pardon…

Perdona(e), ¿cómo llego a…?
Pardon, how do I get to…? I

perdonarse
to forgive each other 1.3

perezoso(a) lazy I

perfecto(a) perfect I

el perfil económico
economic profile 4.3

el perfume perfume II

el periódico newspaper I

el periodismo journalism II

el (la) periodista journalist I, II

el permiso permission II, 2.2

permitir to permit II, 2.1

pero but I

el (la) perro(a) dog I

caminar con el perro
to walk the dog I

el personaje character 5.3

la perspectiva perspective 5.1

pertenecer
to belong, to pertain II, 2.1

peruano(a) Peruvian II

pesado(a) boring, heavy 1.2

el pesado lecho heavy bed

el pescado fish I, II

el (la) pescador(a) fisherman II

pescar to fish II

pescar en alta mar
to go deep-sea fishing 1.2

pese a in spite of

el petróleo petroleum 4.3

el pez fish I, 2.3

el piano piano I

tocar el piano
to play the piano I

el picaflor hummingbird 2.3

picante spicy I

el pie foot I, II

a pie on foot I

la piedra stone, rock II, 2.2

la piel skin II

la pierna leg I, II

la pila battery 6.2

pilotar una avioneta to fly a
single-engine plane 1.2

el piloto pilot II

la pimienta pepper I, II

pintar to paint I

el (la) pintor(a) painter II

la pintura
painting, picture II; paint

la pirámide pyramid 5.2

la piscina swimming pool I

el piso floor, story II

el pizarrón chalkboard I

el placer pleasure

Es un placer. It's a pleasure. I

planchar (la ropa)
to iron (the clothes) I

el planeta planet II, 2.2

la planta plant I

la planta silvestre wild plant II

la planta baja ground floor II

el plástico plastic 2.2

la plata silver I

el plátano verde plantain II

el plato plate I
  lavar los platos
    to wash the dishes I, II
la playa beach I
la plaza town square I
la plena Afro-Caribbean dance 3.3
la pluma pen I
la población population II, 2.2
pobre poor II
la pobreza poverty II, 2.1
poco a little I
poder (ue) to be able, can I, II
  Gracias, pero no puedo.
    Thanks, but I can't. I
  Le puedo ofrecer…
    I can offer you… I
  No, de veras, no puedo.
    No, really, I can't. 2.1
  Si pudiera, lo haría.
    If I could, I would. 2.1
  ¿Podría(s) darme una mano?
    Could you give me a hand?
    2.1
  ¿Me puede atender? Can you
    help (wait on) me? II
  ¿Me puede(s) ayudar?
    Can you help me? 2.1
  ¿Me puede(s) hacer un favor?
    Can you do me a favor? 2.1
  ¿Puedes (Puede usted) decirme
    dónde queda…? Could you
    tell me where… is? I
  ¿Puedo hablar con…?
    May I speak with…? I
el poder power 3.3
el poema poem I
la poesía poetry I, 5.1
el (la) poeta poet 5.3
el (la) policía police officer II
el poliéster polyester 1.2
el pollo chicken I
  el pollo asado
    barbecued chicken II
el polvo dust I
  quitar el polvo to dust I, II
poner (pongo) to put I, II
  poner la mesa to set the table I
  ponerse (me pongo)
    to put on (clothes) I
  ponerse la ropa
    to get dressed I, II
  ponerse nervioso(a)
    to get nervous 1.3

el por ciento percent 4.3
por for, by, around I
  por favor please I
  por fin finally I
  por la mañana
    during the morning I
  por la noche
    during the evening I
  por la tarde
    during the afternoon I
  por otro lado
    on the other hand 1.1
  por todas partes
    everywhere I, 2.2
  por un lado on the one hand 1.1
¿Por qué? Why? I
  ¿Por qué no? Why not? 2.1
el porcentaje percentage 4.3
porque because I
portarse bien/mal
  to behave well/badly II
posible
  Es posible que…
    It's possible that… II
el postre dessert I
practicar to play, to practice I, II
  practicar deportes
    to play sports I
el precio price I
precioso(a) precious, valuable 5.2
precolombino(a)
  pre-Columbian 5.2
predecible predictable 5.3
preferir (ie) to prefer I, II
el prejuicio prejudice 2.1
el Premio Nóbel Nobel Prize 5.3
el prendedor pin 1.2
la prensa press
preocupado(a) worried I
preocuparse (por)
  to be worried about II
  ¡No te preocupes!
    Don't worry! I
preparar to prepare I
la preparatoria preparatory school
presentar to introduce
  Te/Le presento a…
    Let me introduce you
    (familiar/formal) to… I
preservar to preserve II, 2.1
el (la) presidente(a) president 3.3
el préstamo loan II
prestar to lend II

prevalecer to prevail
la primavera spring I
el primer plano foreground 5.1
primero first I
primero(a) first I, II
el (la) primo(a) cousin I
principal principal 4.3
prisa
  tener prisa to be in a hurry I
probable
  Es probable que…
    It's probable that… II
el problema problem I, II
la procesión procession 3.3
la producción production 5.1
producir to produce II
los productos forestales
  forestry products 4.3
la profesión profession I, II
el programa program II
  el ícono del programa
    program icon 6.3
  el programa anti-virus
    anti-virus program 6.3
  el programa de acción
    action program 6.1
  el programa de ciencia ficción
    science fiction program 6.1
  el programa de concurso
    game show 6.1
  el programa de entrevistas
    talk show 6.1
  el programa de horror
    horror program 6.1
  el programa de misterio
    mystery program 6.1
  el programa de reciclaje
    recycling program 2.2
prohibido(a) para menores
  R-rated (movie) 6.1
prohibir to prohibit 2.2
el promedio average 4.3
el pronóstico forecast II
pronto soon I, II
la propina tip I
  ¿Cuánto le doy de propina?
    How much do I tip? I
  dejar la propina
    to leave the tip I
el (la) proponente supporter 3.3
la prosa prose 5.3
¡Próspero Año Nuevo!
  Happy New Year! 3.2

el (la) protagonista protagonist 5.3
proteger to protect II, 2.2
   proteger las especies
      to protect the species I
la prueba quiz I
la publicidad publicity 4.1
el público audience 6.1
el pueblo town, village I
el puente bridge II
el puerco pork I
la puerta door I, II
puertorriqueño(a) Puerto Rican II
pues well I
el puesto position II, 4.2
la pulsera bracelet I
la puntualidad punctuality II, 4.2

qué what? I
   ¿A qué hora es…?
      (At) What time is… ? I
   ¿Qué desea(n)?
      What would you like? I
   ¿Qué día es hoy?
      What day is today? I
   ¡Qué (divertido)! What (fun)! I
   ¿Qué hora es? What time is it? I
   ¡Qué lástima! What a shame! I
   ¡Qué lío! What a mess! II, 2.2
   ¿Qué lleva? What is he/she
      wearing? I
   ¿Qué me (nos) recomienda?
      What do you recommend? II
   ¿Qué tal? How is it going? I
   ¿Qué tiempo hace?
      What is the weather like? I
quedar (en) to stay, to be (in a
   specific place), to agree on I
   ¿Puedes (Puede usted) decirme
      dónde queda…? Could you
      tell me where…is? I
   ¿Queda lejos? Is it far? I
quedársele (a uno)
   to leave something behind 6.2
los quehaceres chores I, II
quejarse to complain 1.3
la quemadura burn II
quemar to burn II
querer (ie) to want, to love I, II
   ¿Quieres beber…?
      Do you want to drink…? I

¿Quieres comer…?
   Do you want to eat…? I
Quiero beber…
   I want to drink… I
Quiero comer…
   I want to eat… I
Quiero dejar un mensaje
   para… I want to leave a
   message for… I
el queso cheese I
quién who I
   ¿De quién es…?
      Whose is…? I
   ¿Quién es? Who is it? I
   ¿Quiénes son? Who are they? I
el químico chemical II, 2.2
quince fifteen I
la quinceañera
   fifteenth birthday 3.2
quinientos(as) five hundred I
el quinto one fifth 4.3
quinto(a) fifth I, II, 4.3
el quiosco kiosk; newstand II
Quisiera… I would like… I
quitar
   quitar el polvo to dust I, II
   quitar la mesa
      to clear the table I
   quitarse la ropa
      to take off your clothes II
quizás perhaps II

el radio radio I
   el radio portátil
      portable radio 6.2
el radiocasete radio-tape player I
la radioemisora radio station 3.2
la radiografía X-ray II
la rana frog II
rápidamente quickly I, II
rápido(a) fast, quick I
la raqueta racket I
rara vez rarely I
raro(a) rare, strange II
   Es raro que… It's rare that… II
el rato
   un buen rato quite a while
el ratón mouse I
las rayas stripes II

el rayo thunderbolt, flash of
   lightning II, 2.3
la raza race I
la razón reason I
   Con razón. That's why. I
   tener razón to be right I
la reacción crítica
   critical response 6.1
real royal
el realismo mágico
   magical realism 5.3
la rebaja sale II
rebosante overflowing
la recepción
   reception/front desk II
el (la) recepcionista receptionist I
el receso break I
la receta prescription II
recibir to receive I
el reciclaje recycling II
reciclar to recycle II
reciente recent I
recientemente lately, recently I, II
el recital recital 5.1
recoger to collect; to pick up 2.1
las recomendaciones
   recommendations II
recomendar (ie) to recommend II
recordar (ue) to remember I
recuperar(se) to get better II
los recursos naturales
   natural resources II, 2.2
la red mundial World Wide Web 6.3
redondo(a) round 1.1
reducir to reduce II, 2.2
la refinería oil refinery 4.3
reflejar to reflect 5.2
el reflejo reflection
el refresco soft drink I
el refrigerador refrigerator II
el refugio de vida silvestre
   wildlife refuge 2.3
el regalo gift I
regar (ie) to water 1.3
regatear to bargain I
regresar to return, to go back I, II
   Regresa más tarde.
      He/She will return later. I
Regular. So-so. I
la reina queen 3.3
reír(se) (i) to laugh II
las relaciones públicas
   public relations 4.1

relajarse (i) to relax II
el relámpago lightning II, 2.3
el reloj clock, watch I
remar to row II
reparar to repair 1.3
el repertorio repertoire 5.1
repetir (i) to repeat II
el reportaje report II
el (la) reportero(a) reporter II
requerir (ie) to require II, 4.2
el requisito requirement II, 4.2
rescatar to rescue II
el rescate rescue II
la reserva reservation II
resfriado
    estar resfriado(a) to have a cold
      II
resistir to stand, to put up with 2.1
resolver (ue) to resolve II, 1.1
respaldado(a)
    supported by; backed by 6.2
respectivamente respectively
respetar to respect 1.1
respirar to breathe II
el restaurante restaurant I
el retrato portrait II
la reunión gathering II
reunir(se) to get together II
revisar to review, to check II
la revista magazine I
el rey king 3.3
rico(a) tasty I; rich II
ridículo
    Es ridículo que…
      It's ridiculous that… II
el riesgo
    correr riesgos to take risks 4.1
el río river I
riquísimo(a) very tasty I
la risa laugh, laughter II
el ritmo rhythm 5.1
rizado curly (hair) II
robar to steal II
el robo robbery II
rodeado(a) surrounded
la rodilla knee II
rogar (ue) to beg 3.1
rojizo(a) reddish 1.1
rojo(a) red I
el romanticismo romanticism 5.3
romántico(a) romantic II
romperromper la piñata
    to break the piñata II

rompérsele (a uno) to break 6.2
la ropa clothing
    ponerse la ropa
      to get dressed I, II
    quitar(se) la ropa
      to take off one's clothes II
rosado(a) pink I
roto(a)
    estar roto(a) to be broken 6.2
rubio(a) blond I
el ruido noise 3.2
las ruinas ruins 5.2

sábado Saturday I
la sábana sheet II
saber (sé) to know I, II
el sabor taste II
sabroso(a) tasty I, II
sacar to take II
    sacar fotos
      to take pictures I
    sacar la basura
      to take out the trash I
    sacar una buena nota
      to get a good grade I
el saco de dormir
    sleeping bag II, 2.3
la saeta Andalusian song 5.1
la sal salt I, II
la sala living room I, II
la sala de emergencia
    emergency room II
la salchicha sausage I; hot dog II
la salida departure II
salir (salgo) to go out, to leave I
saltar
    saltar la cuerda to jump rope II
¡Salud! Cheers! II
saludable healthy II
saludarse
    to greet (each other) II, 1.3
salvadoreño(a) Salvadoran II
salvaje wild II, 2.3
las sandalias sandals II
la sangre blood II
sano(a)
    mantenerse sano(a)
      to be healthy II
la sardana Catalan folk dance 5.1

la sátira satire 5.3
se recomienda discreción
    PG-13 rated (movie) 6.1
el secador de pelo hair dryer I, II
secarse to dry oneself I
seco(a) dry II
el (la) secretario(a)
    secretary I, II, 4.2
sed
    tener sed to be thirsty I
la seda silk 1.2
seguir (i) to follow, to continue II
segundo(a) second I, II
la seguridad security II
el seguro insurance II
    el seguro médico
      medical insurance 4.2
seguro(a) sure
    (No) es seguro que… It is (not)
    certain that… II
    (No) estoy seguro(a).
    I'm (not) sure. 3.3
    no estar seguro (de) que
    not to be sure that 3.2
seis six I, II
seiscientos six hundred I, II
la selva forest, jungle I, II, 2.2
el semáforo traffic light/signal II
la semana week I
    el fin de semana weekend I
    la semana pasada last week I
sembrar (ie) to plant 2.1
semejante a similar to 1.1
el semestre semester I
sencillo(a) simple, plain I, II
el sendero path, trail II, 2.3
sensacionalista sensationalized 6.1
sentar(se) (ie) to sit II
sentir (ie)
    sentir que to be sorry that II
    sentirse to feel II
    Lo siento mucho, pero…
    I'm sorry, but… 2.1
    Lo siento. I'm sorry. I
    sentirse frustrado(a)
    to feel frustrated 1.3
señalar to gesture
el señor Mr. I
la señora Mrs. I
la señorita Miss I
separar to separate II, 2.2
septiembre September I
séptimo(a) seventh I, II

la **sequía** drought **2.2**
ser (soy) to be **I, II**
    Es la…/Son las…
      It is… o'clock. **I**
    ser de… to be from… **I**
el **ser humano** human being **II, 2.1**
la **serie** series **II**
serio(a) serious **I**
la **serpiente** snake **II, 2.3**
el **servicio** service
    el **servicio de búsqueda**
      search engine **6.3**
    el **servicio social**
      social service **2.1**
los **servicios** bathrooms **II**
la **servilleta** napkin **II**
servir (i) to serve **I, II**
    servir de to be used as
sesenta sixty **I**
setecientos seven hundred **I, II**
setenta seventy **I**
sexto(a) sixth **I**
los **shorts** shorts **I**
si if **I, II**
sí yes **I**
    ¡Claro que sí! Of course! **I**
    Sí, con mucho gusto.
      Yes, gladly. **2.1**
    Sí, me encantaría.
      Yes, I would love to. **I**
siempre always **I**
siete seven **I**
el **siglo** century **5.1**
siguiente next **II**
la **silla** chair **I, II**
el **sillón** armchair **I, II**
simbólico(a) symbolic **5.3**
el **simbolismo** symbolism **5.3**
simpático(a) nice **I**
sin without **I**
sin embargo nevertheless **5.3**
el **sitio** site **6.3**
la **situación** situation **II**
el **smog** smog **II, 2.2**
sobrante leftover, surplus
sobre on **II**
    sobre hielo on ice **I**
el **sobre** envelope **4.1**
el (la) **sobrino(a)** nephew/niece **II**
sociable sociable **II**
la **sociedad anónima (S.A.)**
    corporation (Inc.) **4.3**
¡Socorro! Help! **II**

el **sofá** sofa, couch **I, II**
el **software** software **6.3**
el **sol** sun **I**
    las **gafas de sol** sunglasses **I**
    Hay sol./Hace sol. It's sunny. **I**
    tomar el sol to sunbathe **I**
soleado(a) sunny **II, 2.3**
solemne solemn **3.3**
solicitar to request, to apply for **II, 4.1**
la **solicitud** application **II, 4.1**
sólo only **I**
solo(a) alone **I**
la **solución** solution **2.1**
la **sombra** shade, shadow **II**
el **sombrero** hat **I**
la **sombrilla de playa**
    beach umbrella **II**
el **son** sound, rhythm **5.1**
sonar (ue) to sound, ring **3.3**
soñar (ue) to dream about
el **sonido** sound **I**
sonreír(se) (i) to smile **II**
la **sopa** soup **I**
sorprender to surprise **I, II**
la **sorpresa** surprise **I, II**
el **sótano** basement **1.3**
su your, his, her, its, their **I**
subir por to go up/to climb **II**
sucio(a) dirty **I, II**
las **sudaderas** sweats **1.2**
sudar to sweat **II**
el **sueldo** salary **II, 4.2**
suelen acercarse
    they often come up to
el **suelo** floor **I, II**
    barrer el suelo
      to sweep the floor **I**
suelto(a) loose **1.2**
el **sueño** sleep; dream
    tener sueño to be sleepy **I**
la **suerte** luck
    **La suerte viene a quien menos
      la aguarda.** Luck comes to he
      who least expects it.
    tener suerte to be lucky **I**
el **suéter** sweater **I**
suficiente enough **II**
sugerir (ie) to suggest **II**
sumar to add **4.3**
superarse to get ahead, to excel **4.1**
la **superficie** surface
el **supermercado** supermarket **I**

    ir al supermercado
      to go to the supermarket **I**
suplicar to ask, to plead **3.1**
el **sur** south **II**
el **surfing** surfing **I**
el **susto** shock, fright
suyo(a)
    yours (formal), his, hers, theirs

el **tablado** stage floor **5.1**
el **tablao** flamenco group **5.1**
tacaño(a) stingy **II**
el **taco** taco **II**
tal vez maybe **I**
    Tal vez otro día.
      Maybe another day. **I**
el **talento** talent **II**
la **talla** size (clothing) **II**
tallado(a) carved **5.2**
el **taller** workshop **I**
el **tamaño** size **6.3**
también also, too **I**
el **tambor** drum **5.1**
tampoco neither, either **I**
tan as **I**
    tan pronto como as soon as **2.3**
    tan… como as…as **I, II, 6.3**
el **tango** tango (Latin music) **5.2**
tanto as much **I**
    tanto como as much as **I**
    tantos… como as many…as **II, 6.3**
las **tapas** appetizers **I**
el **tapiz** tapestry **5.1**
la **taquería** taco restaurant **II**
la **taquilla** box office **II**
tarde late **I**
    Regresa más tarde.
      He/She will return later. **I**
la **tarde** afternoon **I**
    Buenas tardes. Good afternoon. **I**
    de la tarde in the afternoon **I**
    por la tarde during the
      afternoon **I**
la **tarea** homework **I**
la **tarjeta** card
    la **tarjeta de crédito** credit card **I**
    la **tarjeta de sonido**
      sound card **6.3**

la **tarjeta gráfica**
graphics card **6.3**
el **taxi** taxi, cab **I**
el (la) **taxista** taxi driver **I, II, 4.2**
la **taza** cup **I**
el **té** tea **I**
el **teatro** theater **I**
la **obra de teatro**
theatrical production **II**
el **techo** roof
la **tecla** key **6.3**
el **teclado** keyboard **I, 6.3**
la **técnica** technique **5.2**
el **técnico** technician **II, 4.2**
el **tejido** textile, weaving **5.2**
las **telecomunicaciones**
telecommunications **4.3**
el **teledrama** TV mini-series **6.1**
telefonearse
to telephone each other **1.3**
el **teléfono** telephone **I**
el **teléfono celular**
cellular telephone **6.2**
el **teléfono inalámbrico**
cordless telephone **6.2**
¿Cuál es tu teléfono? What is
your phone number? **I, II**
la **teleguía** television guide **6.1**
el **telemensaje** voice mail **6.2**
la **telenovela** soap opera **II**
la **teleserie** TV series **6.1**
el (la) **televidente** viewer **II**
la **televisión** television **I**
el **espacio de la televisión**
television program
la **televisión por cable**
cable television **6.1**
la **televisión por satélite**
satellite television **6.1**
ver la televisión
to watch television **I**
el **televisor** television set **I**
el **televisor portátil**
portable television **6.2**
el **tema** theme, subject **II**
el **tembleque**
coconut-milk custard **3.2**
la **temperatura** temperature **I**
el **templo** temple **5.2**
la **temporada**
season, period of time **1.2**
temprano early **I**
el **tenedor** fork **I**

tener (tengo) to have **I**
¿Cuántos años tiene…?
How old is…? **I**
tener calor to be hot **I**
tener cuidado to be careful **I, II**
tener en común
to have in common **1.1**
tener envidia to be envious **II**
tener éxito to be successful **II**
tener frío to be cold **I**
tener ganas de… to feel like… **I**
tener hambre to be hungry **I**
tener miedo to be afraid **I, II**
tener prisa to be in a hurry **I**
tener que to have to **I**
tener razón to be right **I**
tener sed to be thirsty **I**
tener sueño to be sleepy **I**
tener suerte to be lucky **I**
tener vergüenza
to be ashamed **II**
Tiene… años.
He/She is…years old. **I**
el **tenis** tennis **I**
teñido(a) dyed **1.1**
tercero(a) third **I, II**
el **tercio** third **4.3**
terminar to finish **I**
el **terreno** terrain, landscape
la **altura del terreno**
altitude of terrain
terrible terrible, awful **I**
los **textiles** textiles **4.3**
la **tía** aunt **I**
el **tiburón** shark **2.3**
el **tiempo** time; weather **I**
a tiempo on time **II**
Hace buen tiempo.
It is nice outside. **I**
Hace mal tiempo.
It is bad outside. **I**
¿Qué tiempo hace?
What is the weather like? **I**
tiempo completo full time
el **tiempo libre** free time **I**
la **tienda** store **I**
la **tienda de deportes**
sporting goods store **I**
la **tienda de música y videos**
music and video store **I**
la **tienda de campaña** tent **II, 2.3**
la **tierra** land **I, II, 2.2**
las **tijeras** scissors **I, II**

el **timbre** ring
tímido(a) shy **II**
la **tintorería** dry cleaner **II**
el **tío** uncle **I**
los **tíos** uncle(s) and aunt(s) **I**
típicamente typically **II**
típico(a) typical, regional **3.2**
la **tira cómica** comic strip **II**
el **titular** headline **II**
titularse to be called **5.3**
el **título** title **5.3**
la **tiza** chalk **I**
la **toalla** towel **I, II**
el **tobillo** ankle **II**
tocar to play (an instrument);
to touch **I, II**
¡A todos nos toca!
It's up to all of us! **2.2**
tocar el piano
to play the piano **I**
tocar la guitarra to play the
guitar **I**
todavía still, yet **I**
todo(a) all **I**
todo el mundo everyone **II**
todos los días every day **I**
la **toga** gown **3.1**
tomar to take, to eat or drink **I**
tomar decisiones
to make decisions **4.1**
tomar el sol to sunbathe **I**
tomar en cuenta
to take into account **6.2**
tomar un curso de natación
to take a swimming class **II**
el **tomate** tomato **I, II**
tonto(a) silly **I**
la **tormenta** storm **I**
el **toro** bull **I**
la **torta** sandwich **I, II**
la **tortilla española** potato omelet **I**
la **tortuga** turtle **II, 2.3**
la **tos** cough **II**
los **tostones** fried plantains **II**
trabajador(a) hard-working **I**
el (la) **trabajador(a) social**
social worker **4.3**
trabajar to work
trabajar de voluntario(a)
to volunteer **2.1**
el **trabajo** job, work
el **trabajo a tiempo completo**
full-time job **4.2**

el trabajo a tiempo parcial
part-time job **4.2**
la tradición tradition **5.2**
tradicional traditional **I, II**
traducir to translate **II**
el (la) traductor(a) translator **4.3**
traer (traigo) to bring **I**
¿Me trae…?
Could you bring me…? **I**
el tráfico traffic **I**
el traje suit **II**
el traje de baño bathing suit **I**
la trama plot **5.3**
tranquilamente calmly **I, II**
tranquilo(a) calm **I**
trasladarse to move **I**
tratar to treat **II**
tratar de to try to
tratarsede to be about **5.3**
trece thirteen **I**
treinta thirty **I**
el tren train **I**
trepar a un árbol to climb a tree **II**
tres three **I, II**
trescientos three hundred **I**
triangular triangular **1.1**
el trigo wheat **4.3**
triste sad **I**
Es triste que… It's sad that… **II**
la tristeza sadness **II**
la trompeta trumpet **5.1**
el tropiezo setback **3.1**
el trueno thunder **II, 2.3**
tu your (familiar) **I**
tú you (familiar singular) **I**
el tucán toucan **II, 2.3**
el turismo tourism **II**
tuyo(a) yours (familiar) **II**

ubicado(a) located
último(a) last **I**
único(a) unique, only **1.2**
la unidad monetaria currency **4.3**
unir to unite
la universidad university **II**
uno one **I, II**
Uruguay Uruguay **I**
uruguayo(a) Uruguayan **II**
usar to use, to wear, to take
a size **I, II**

usted you (formal singular) **I**
ustedes you (formal, plural) **I**
el (la) usuario(a) user **6.3**
útil useful **II**
la uva grape **I, II**

la vaca cow **I**
vaciar to empty **1.3**
vacío(a) empty **II**
valer (valgo)
to value, to be worth **II**
valer la pena
to be worthwhile **3.1**
el valle valley **II, 2.2**
valorar
to appreciate, to value **II, 2.1**
vanidoso(a) vain **1.1**
el vaso glass **I**
el vaso de glass of **I**
el vecindario neighborhood **II**
el (la) vecino(a) neighbor **I**
vegetariano(a) vegetarian **I**
veinte twenty **I**
veintiuno twenty-one **I**
las velas candles **II**
el venado deer **II, 2.3**
vencer (venzo)
to defeat, to overcome **3.1**
vender to sell **I**
venezolano(a) Venezuelan **II**
venir (vengo) to come **I, II**
la ventaja advantage **II, 4.2**
la ventana window **I, II**
la ventanilla window **II**
las ventas sales **4.1**
a la venta for sale **I**
ver (veo) to see **I, II**
¿Me deja ver…? May I see…? **I**
Nos vemos. See you later. **I**
ver la televisión
to watch television **I**
verse to look, to appear **1.1**
el verano summer **I**
la verdad truth **I**
Es verdad. It's true. **I**
No es verdad que…
It's not true that… **II**
verde green **I**
la verdura vegetable **I, II**
la vergüenza

tener vergüenza
to be ashamed **II**
el vestido dress **I**
vestir(se) (i) to dress oneself **II**
el vestuario wardrobe **1.2**
el (la) veterinario(a)
veterinarian **II, 4.2**
viajar to travel **I**
el viaje trip **I**
la victoria victory **3.3**
la vida life **I**
el video video **I**
alquilar un video
to rent a video **I**
la videocámara videocamera **6.2**
la videocasetera
videocassette recorder **6.1**
la videograbadora VCR **I**
el videojuego video game **I**
el vidrio glass **II, 2.2**
viejo(a) old **I**
el viento wind **I**
Hace viento. It's windy. **I**
Hay viento. It's windy. **I**
viernes Friday **I**
violento(a) violent **II**
el violín violin **5.1**
visitar to visit **I**
vivir to live **I**
Vive en… He/She lives in… **I**
Vivo en… I live in… **I**
volar (ue) to fly **II**
volar en planeador
to hang-glide **1.2**
el voleibol volleyball **I**
el (la) voluntario(a) volunteer **II**
volver (ue)
to return, to come back **I**
vosotros(as) you (familiar plural) **I**
votar to vote **2.1**
el vuelo flight **II**
vuestro(a) your (familiar plural) **I**,
yours **II**

y and **I**
y cuarto quarter past **I**
y media half past **I**
ya already, now **II**
¡Ya lo sé! I already know! **I**
ya no no longer **I**

el yeso cast **II**
yo I **I**
el yogur yogurt **I, II**

la zanahoria carrot **I, II**
el zapateado
   rhythmic heel tapping **5.1**
la zapatería shoe store **I**
el zapato shoe **I**
   el zapato de tacón
      high-heeled shoe **II**
las zonas de reserva ecológica
   conservation lands **2.2**
el zorrillo skunk **2.3**
el zumo juice **I**

GLOSARIO
inglés–español

This English–Spanish glossary contains all of the active vocabulary words that appear in the text as well as passive vocabulary from readings and culture sections. It contains all words listed in the Spanish–English Glossary.

abilities las habilidades **4.2**
to be able poder (ue) **I, II**
Abolition Day el día de la Abolición de Esclavitud **3.3**
to be about tratarse de **5.3**
academic el (la) académico(a) **4.3**
accessible accesible **6.2**
to accomplish llevar a cabo **3.1**
accountant el (la) contador(a) **I, II, 4.2**
accounting la contabilidad **4.1**
to achieve lograr
acquaintance la amistad **II**
actor/actress el (la) actor/actriz **II**
to adapt oneself adaptarse **4.1**
to add sumar **4.3**
address la dirección **I**
admiral el (la) almirante **3.3**
admission el ingreso
to advance avanzar **3.1**
advanced avanzado(a) **5.2**
advantage la ventaja **II, 4.2**
adventures las aventuras **II**
advice el consejo **II**
to advise aconsejar **II**
aerosol el aerosol **2.2**
afraid
to be afraid tener miedo **I, II**
I'm afraid that… Tengo miedo de que… **2.3**

after después (de) **I**
afternoon la tarde **I**
during the afternoon por la tarde **I**
Good afternoon. Buenas tardes. **I, II**
in the afternoon de la tarde **I**
after-shave lotion la loción **II**
afterward después **I**
against
to be against estar en contra de **II, 2.1**
age la edad **I**
ago hace…que **II**
to agree estar de acuerdo **I, II**
to agree (on) quedar (en) **I**
agriculture la agricultura **4.3**
agronomy la agronomía **4.1**
air el aire **I**
air conditioning el aire acondicionado **II**
air pollution la contaminación del aire **I, 2.2**
airline la aerolínea **II**
airplane el avión **I**
airport el aeropuerto **I**
aisle el pasillo **II**
alarm clock el despertador **I, II**
all todo(a) **I**
all around por todas partes **2.2**
It is up to all of us… A todos nos toca… **II, 2.2**
to allow dejar **3.1**

almost casi **II**
alone solo(a) **I**
alphabet el abecedario
already ya **II**
also también **I**
altitude la altura **II, 2.2**
altitude of terrain la altura del terreno
aluminum el aluminio **II**
always siempre **I**
ambulance la ambulancia **II**
ancestors los antepasados **3.3**
ancient antiguo(a) **I**
and y **I**
and so it was that así fue que **II**
anger la ira
angry enojado(a) **I**
to get angry with enojarse con **II**
animal el animal **I**
animated animado(a) **II**
ankle el tobillo **II**
anniversary el aniversario **II**
to announce avisar
another otro(a) **I**
to answer contestar **I**
answering machine la máquina contestadora **I**; la contestadora automática **6.2**
anteater el oso hormiguero **2.3**
apartment el apartamento **I**
to apologize disculparse **II**
to appear lucir; verse **1.1**
appetizers las tapas **I**
apple la manzana **II**

application la solicitud II, 4.1
to apply for solicitar II, 4.1
to apply oneself to dedicarse a 1.3
appointment la cita I
to appreciate
  valorar II; apreciar 3.1
April abril I
architect
  el (la) arquitecto(a) I, II, 4.2
architectural arquitectónico(a)
architecture la arquitectura I
Argentine argentino(a) II
to argue discutir 1.1
arm el brazo I, II
armchair el sillón I, II
army el ejército
around alrededorde II; por I
to arrange ordenar I; gestionar
arrangement el arreglo
arrival la llegada II
to arrive llegar I, II
art el arte I
article el artículo II
artisan el (la) artesano(a) I, II, 4.2
artist el (la) artista II
as como
  as…as tan…como I, II
  as contrasted with
    a diferencia de 1.1
  as far as hasta I
  as long as con tal de que 3.1
  as many…as tantos… como II,
    6.3
  as much as tanto como I
  as soon as en cuanto, tan
    pronto como 2.3
ashamed
  to be ashamed
    tener vergüenza II
to ask suplicar 3.1
to ask for pedir (i) I, II
aspirin la aspirina II
to assemble armar 1.3
assistant el (la) asistente II, 4.2
at a
  At…o'clock A la(s)… I
  at once en seguida II
athlete el (la) deportista II, 4.2
athletics el atletismo II
ATM machine
  el cajero automático II
to attend asistir a II; acudir a 3.3
attic el desván 1.3

audience el público 6.1
auditorium el auditorio I
August agosto I
aunt la tía I
author el (la) autor(a) II
autobiography la autobiografía 5.1
available disponible 6.3, accesible
  6.2
avenue la avenida I
average el promedio 4.3
award el premio I, II
awesome
  How awesome! ¡Qué chévere! I
awful terrible I

baby el bebé II
babysitter el (la) niñero(a) II, 4.2
backed by respaldado(a) 6.2
background el fondo 5.1
backpack la mochila I
bad malo(a) I
  the bad thing lo malo 1.1
  It is bad (weather) outside.
    Hace mal tiempo. I
  It's bad that… Es malo que… II
bag la bolsa I
balanced balanceado(a) II
bald calvo(a) 1.1
ball la bola, la pelota I
ballet el ballet 5.1
balloons los globos II
bananas
  small green bananas in garlic
    vinegar, red pepper and oil
    los guineítos en escabeche 3.2
band la banda 3.3
bangs el flequillo 1.1
bank el banco I
banker el (la) banquero(a) 4.3
barber
  el (la) peluquero(a) II
bargain la ganga II
to bargain regatear I
baseball el béisbol I
baseball cap la gorra I
basement el sótano 1.3
basketball el baloncesto I
bat el bate I
bath

  to take a bath bañarse I, II
bathing suit el traje de baño I
bathroom el baño, el servicio I, II
bathtub la bañera II
battery storage el acumulador; la
  batería, la pila 6.2
to be estar I, II; ser I, II
  to be enrolled in cursar
  to be against
    estar en contra de 2.1
  to be for estar a favor de 2.1
  to be raised criarse
  to be thrilled, touched
    emocionarse 3.1
  to be willing to
    estar dispuesto(a) 4.1
  to be worth valer II
  to be worthwhile valer la pena
    3.1
beach la playa I
beach umbrella
  la sombrilla de playa II
beans los frijoles II
beard la barba 1.1
beat el compás 5.1
to beautify embellecer 2.1
beauty la belleza II
beauty mark el lunar 1.1
because porque I
bed la cama I, II
  to go to bed acostarse (ue) I
  to make the bed hacer la cama I
bedroom la habitación I, II
beef la carne de res I, II
beeper el beeper 6.2
before antes (de) I
to beg rogar (ue) 3.1
to begin comenzar (ie) II;
  empezar (ie) I, II
to begin with para empezar II
beginning el inicio
  shaky beginnings
    los tambaleantes inicios
to behave well/badly
  portarse bien/mal II
behind detrás (de) I; atrás 6.2
belief la creencia 5.2
to believe creer I, 3.3
  not to believe that… no creer
    que… II
bell la campana 3.2
to belong pertenecer II, 2.1
below debajo de I, II

belt el cinturón **I**
benefits los beneficios **II, 4.2**
beside al lado (de) **I**
best lo mejor **1.1**
better mejor **I**
   better than mejor que **II**
   to get better recuperarse **II**
   It's better that…
      Es mejor que… **II**
between entre **I**
beverage la bebida **I**
big grande **I**
bike la bicicleta
   to ride a bike
      andar en bicicleta **I**
bill la cuenta **I, II**
billion mil millones **4.3**
biography la biografía **5.1**
bird el pájaro **I**; el ave (fem.)
birth
   date of birth la fecha de
      nacimiento **II, 4.1**
birthday el cumpleaños **I**
black negro(a) **I**
blanket la manta **I, II, 2.3**
blind ciego(a) **5.3**
blond rubio(a) **I**
blood la sangre **II**
blouse la blusa **I**
blue azul **I**
boa constrictor
   la boa constrictora **2.3**
to board (a plane) abordar **II**
boarding house la pensión **II**
boardwalk el malecón
boat el bote **II**
body el cuerpo **I, II**
bodyguard
   el (la) guardaespaldas **6.1**
Bolivian boliviano(a) **II**
book el libro **I**
bookstore la librería **I**
boots las botas **I**
bored
   to get bored aburrirse **II**
boring aburrido(a) **I**;
   pesado(a) **1.2**
born
   to be born nacer
boss el (la) jefe(a) **I, II**
to bother molestar **II**
bottle la botella **I, II, 2.2**
box office la taquilla **II**

boy el chico; el muchacho;
   el niño **I**
boyfriend
   el (la) novio(a) **II**
bracelet la pulsera **I**
brand la marca **6.2**
bread el pan **I, II**
bread bakery la panadería **I**
break el receso **I**
to break rompérsele (a uno) **6.2**
   to break down
      descomponérsele (a uno) **6.2**
   to break the piñata
      romper la piñata **II**
breakfast el desayuno **I**
   to have breakfast desayunar **I**
to breathe respirar **II**
bridge el puente **II**
to bring traer **I**
   Could you bring me …?
      ¿Me trae…? **I**
broken
   to be broken
      estar descompuesto(a),
      estar roto(a) **6.2**
brother el hermano **I**
brother(s) and sister(s)
   los hermanos **I**
brother-in-law el cuñado **II**
brown marrón **I**
brown (hair) castaño(a) **I**
brush el cepillo **I**
to brush one's hair cepillarse el
   pelo **II**
to brush one's teeth
   lavarse los dientes **I, II**
bucket el balde **II**
building el edificio **I**
bull el toro **I**
burn la quemadura **II**
to burn quemar **II**
bus el autobús **I**
bus station
   la estación de autobuses **I**
business el comercio **4.1**;
   la empresa **II, 4.2**
business administration la
   administración de empresas **4.1**
businessman
   el hombre de negocios **I, II**
businesswoman
   la mujer de negocios **I, II**
busy ocupado(a) **I**

but pero **I**
butcher's shop la carnicería
butter la mantequilla **I, II**
butterfly la mariposa **II, 2.3**
to buy comprar **I**
by por **I**
by means of mediante

cab el taxi **I**
cabinet el gabinete **1.3**
café el café **I**
cafeteria la cafetería **I**
cake el pastel **I**
calculator la calculadora **I**
call la llamada **I**
to call llamar **I**
   to be called titularse **5.3**
   Tell him or her to call me.
      Dile/Dígale que me llame. **I**
caller identification el
   identificador de llamadas **6.2**
calm tranquilo(a) **I**
calmly tranquilamente **I, II**
calorie la caloría **II**
camera la cámara **I, II**
camp el campamento **II, 2.3**
to camp acampar **1.2**
   to camp in the mountains
      acampar en las montañas **II**
campaign la campaña **2.1**
campfire la fogata **II**
can la lata **I, II, 2.2**
can poder (ue) **I, II**
   Can (May) I offer you anything
     more? ¿Se les ofrece algo
     más? **II**
   Can you do me a favor? ¿Me
     puede(s) hacer un favor? **2.1**
   Can you give me a hand?
     ¿Podría(s) darme una mano?
     2.1
   Can you help me? ¿Me
     puede(s) ayudar? **2.1**
   Can you help (wait on) me?
     ¿Me puede atender? **II**
   I can offer you… Le puedo
     ofrecer… **I**
   No, really, I can't. No, de veras,
     no puedo. **2.1**

Thanks, but I can't. Gracias, pero no puedo. I
can opener el abrelatas II, 2.3
Canadian canadiense II
candles las velas II
cap el gorro I; el birrete 3.1
   baseball cap la gorra I
capabilities las habilidades II
car el carro I, el coche 4.2
card la tarjeta
   graphics card la tarjeta gráfica 6.3
   sound card la tarjeta de sonido 6.3
cardboard, cardboard box el cartón II, 2.2
career la carrera II, 4.2
careful cuidadoso(a) I
   to be careful tener cuidado I, II
carefully cuidadosamente I
carrot la zanahoria I, II
to carry llevar I
to carry out a responsibility desempeñar un cargo 4.1
cartoons los dibujos animados 6.1
carved tallado(a) 5.2
cash el efectivo I
cash register la caja registradora II
cashier el (la) cajero(a) II
cassette el casete I
cast el yeso II
castanets las castañuelas 5.1
cat el (la) gato(a) I
cause la causa II
to celebrate celebrar, festejar 3.2
center el centro I
   community center el centro de la comunidad 2.1
   rehabilitation center el centro de rehabilitación 2.1
   shopping center el centro comercial I
centigrade el centígrado II, 2.3
century el siglo 5.1
ceramics la cerámica I
cereal el cereal I, II
certain
   It is not certain that… No es cierto que… II
   No es seguro que… II
chain la cadena 1.2
chair la silla I, II

chalk la tiza I
chalkboard el pizarrón I
change el cambio I
to change cambiar I
change purse el monedero 1.2
channel el canal II
   to change channels (TV) cambiar de canal 6.1
chapel la capilla
character el personaje 5.3
chat group el grupo de conversación 6.3
cheap barato(a) I
check la cuenta I
   Separate checks. Es aparte. I
   The check, please. La cuenta, por favor. I
to check revisar II
checked de cuadros I
checks los cheques II
cheek la mejilla
Cheers! ¡Salud! II
cheese el queso I
chemical el químico II, 2.2
cherry la cereza II
chess
   to play chess jugar al ajedrez II
chest el cofre
chicken el pollo I
   barbecued chicken el pollo asado II
Chilean chileno(a) II
Chinese chino(a) II
to choose escoger II
chores los quehaceres I, II
Christmas la Navidad 3.2
chronicler el (la) cronista 5.2
church la iglesia I
Cinderella Cenicienta
cinematographer el (la) cinematógrafo(a) 5.3
citizen el (la) ciudadano(a) 2.1
citizenship la ciudadanía II
city la ciudad I
city block la cuadra I
civil engineer el (la) ingeniero(a) civil 4.2
civil status el estado civil 4.1
civilization la civilización 5.2
to clap hands dar palmadas 5.1
clarity la nitidez 6.2
class la clase I
classroom la clase I

clean limpio(a) I, II
to clean hacer la limpieza II
to clean the room limpiar el cuarto I, II
to click hacer clic 6.3
   to double click hacer doble clic 6.3
climate el clima II, 2.2
climax el clímax 5.3
to climb a tree trepar a un árbol II
to climb mountains escalar montañas II
clip el broche
clock el reloj I
to close cerrar (ie) I
closed cerrado(a) I, II
closet el armario I, II
clothing la ropa II
cloud la nube II, 2.3
cloudy
   It is cloudy. Está nublado. I
clown el payaso
coat el abrigo I
coconut milk el agua de coco II
coconut-milk custard el tembleque 3.2
coffee el café I
coffee shop la cafetería I
cold
   to be cold tener frío I
   to have a cold estar resfriado(a) II
   It is cold. Hace frío. I
to collaborate with colaborar con 2.1
to collect coleccionar 1.2
Colombian colombiano(a) II
color el color I
   bright color el color brillante 1.2
   dark color el color oscuro 1.2
   solid color de un solo color 1.2
   What color…? ¿De qué color…? I
Columbus Day el día de la Raza 3.2
comb el peine I
to comb one's hair peinarse I
to come venir I, II
to come back volver (ue) I
comedy la comedia II
comedian, comedienne el (la) comediante II

comfortable
cómodo(a) II; a gusto **1.2**
comic strip la tira cómica II
comical cómico(a) I
to commemorate conmemorar **3.3**
commercial el anuncio II
common común II
to have in common
tener en común **1.1**
community la comunidad I
compact disc el disco compacto I
companion el (la) compañero(a) II
company (business) la empresa II
company la compañía I
to compare comparar **4.3**
to compete competir (i) II
competition la competencia **3.3**
to complain quejarse **1.3**
complicated complicado(a) II, **2.2**
computer la computadora I
computer science la computación
I; la informática **4.1**
concert el concierto I
concrete el hormigón
configuration la configuración **6.3**
to confront enfrentar **3.3**
to congratulate felicitar **3.1**
congratulations
felicidades I; enhorabuena **3.1**
to connect conectar **1.3**
to connect to conectarse a **6.3**
conqueror el(la) conquistador(a) **5.2**
conservation land las zonas de
reserva ecológica **2.2**
conservative conservador(a) **3.3**
to conserve conservar II, **2.1**
considerate considerado(a) **1.1**
constitution la constitución **3.3**
to construct construir II
consultation la consulta II
to consume less consumir menos
2.1
contact lenses
los lentes de contacto **1.1**
contemporary
contemporáneo(a) **5.3**
content contento(a) I
contest el concurso I, **3.2**
contract el contrato II, **4.2**
contrary
on the contrary al contrario II
to control controlar **6.1**
to convince convencer **6.2**

to cook cocinar I
cookie la galleta I, II
cool
It is cool. Hace fresco. I
cooler la neverita II
copper el cobre **4.3**
corn el maíz **4.3**
corner la esquina I
corporation (Inc.)
la sociedad anónima (S.A.) **4.3**
corral el corral I
correspondent
el (la) corresponsal **4.3**
cost costar (ue) I
Costa Rican costarricense II
costume el disfraz
cotton el algodón **1.2**
couch el sofá I
cough la tos II
to count contar (ue) I
counter el mostrador II
country el país, el campo I
mother country la patria **3.3**
countryside el campo I
court la cancha I
cousin el (la) primo(a) I
cow la vaca I
crab el cangrejo
cracker la galleta I
crazy loco(a) I
cream la crema I, II
to create crear II, **2.1**
creative creativo(a) **5.3**
credit card la tarjeta de crédito I
critic el (la) crítico(a) **5.3**
critical response
la reacción crítica **6.1**
criticism la crítica II
to cross cruzar I
crossing el cruce II
to cry llorar II
Cuban cubano(a) II
to culminate culminar **5.3**
cumbia la cumbia (Latin music) **5.2**
cup la taza I
curly (hair) rizado II
currency la unidad monetaria **4.3**
custom la costumbre **3.3**
customer el (la) cliente(a) II
customs la aduana II
cut labrado(a) **5.2**
to cut cortar
to cut oneself cortarse II

to cut the grass
cortar el césped II
cyberspace
relating to cyberspace
cibernético(a) **5.3**

daily diario(a) II
damaging dañino(a) **2.2**
dance la danza **5.2**
Afro-Caribbean dance
la bomba, la plena **3.3**
Aragonese folk dance
la jota **5.1**
Catalan folk dance
la sardana **5.1**
Mexican dance from
Guadalajara
el jarabe tapatío **5.2**
Mexican dance from Veracruz
la bamba **5.2**
to dance bailar I
dancer el (la) bailarín/bailarina II,
4.2; of flamenco el (la)
bailaor(a) **5.1**
dangerous peligroso(a) I, II, **2.3**
It's dangerous that…
Es peligroso que… II
daring atrevido(a) **1.1**
dark oscuro II
dark hair and skin moreno(a) I
to get dark anochecer,
oscurecer, atardecer **2.3**
darkness la oscuridad **2.3**
database la base de datos **6.3**
date la fecha I
date of birth
la fecha de nacimiento II, **4.1**
What is the date?
¿Cuál es la fecha? I
daughter la hija I
dawn el amanecer **2.3**;
la madrugada **3.2**
to dawn on ocurrírsele (a uno) **6.2**
day el día I
day before yesterday
anteayer I, II
What day is today?
¿Qué día es hoy?
dazzling deslumbrante **5.3**

debris los escombros
deceit el engaño
December diciembre **I**
to decide decidir **I**
to decipher descifrar **5.2**
decisions
   to make decisions
     tomar decisiones **4.1**
decorated decorado(a) **5.2**
decorations los adornos **II**
deer el venado **II, 2.3**
to defeat vencer **3.1**
degree el grado **I**
delicious delicioso(a) **I**
to delight encantar **II**
to demand exigir **3.1**
democracy la democracia **3.3**
democratic democrático(a) **3.3**
denim la mezclilla **1.2**
deodorant el desodorante **II**
departure la salida **II**
dependability la confiabilidad **6.2**
depressed deprimido(a) **I**
derivative derivado(a) **5.3**
descent la descendencia **5.2**
desert el desierto **I**
design el diseño **4.1**
designer el (la) diseñador(a) **1.2**
to desire desear **II**
desk el escritorio **I**
dessert el postre **I**
destruction la destrucción **II, 2.2**
detail el detalle **II**
to develop desarrollar **2.2**
development el desarrollo **II, 2.1**
to dial marcar **I**
dictionary el diccionario **I**
to die morir (ue) **II**
diet la dieta **II**
difficult difícil **I**
dining room el comedor **I, II**
dinner la cena **I, II**
   to eat dinner cenar **I, II**
diploma el diploma **3.1**
diplomat el (la) diplomático(a) **4.3**
to direct dirigir **5.3**
direction la dirección **I**
director el (la) director(a) **5.3**
dirty sucio(a) **I, II**
disadvantage la desventaja **II, 4.2**
to disconnect desconectar **1.3**
to disconnect from
   desconectarse de **6.3**

discount el descuento **6.2**
to discover descubrir **II, 2.2**
discovery el descubrimiento **3.3**
discrimination
   la discriminación **2.1**
to discuss discutir **1.1**
disk el disco **6.3**
dishwasher el lavaplatos **I, II**
to dislike caer mal **1.2**
disorganized desorganizado(a) **1.3**
distance la distancia **II**
to distinguish between
   distinguir entre **6.2**
diverse diverso(a) **II, 2.2**
to divide
   the inheritance will be divided
     up se harán particiones
to do hacer **I, II**
doctor el (la) doctor(a) **I**
doctorate el doctorado **4.1**
documentary el documental **6.1**
dog el (la) perro(a) **I**
   to walk the dog
     caminar con el perro **I**
doll la muñeca **II**
dollar el dólar **I**
Dominican dominicano(a) **II**
to donate donar **2.1**
door la puerta **I, II**
to doubt that dudar que **II**
   It's doubtful that…
     Es dudoso que…
down abajo **I, II**
downpour el aguacero **II, 2.3**
downtown el centro **I**
dozen la docena **I**
drama el drama **5.1**
dramatic dramático(a) **5.3**
to draw dibujar **II**
drawing
technical drawing
   el dibujo técnico **4.1**
dream el sueño
to dream about soñar (ue)
dress el vestido **I**
to dress oneself vestirse (i) **II**
   to get dressed
     ponerse la ropa **I, II**
   to get dressed up arreglarse **II**
drink la bebida **I**
to drink beber; tomar **I**
   Do you want to drink…?
     ¿Quieres beber…? **I**

   I want to drink…
     Quiero beber… **I**
to drive conducir **II**; manejar **I**
driver el (la) conductor(a) **II**
drizzle la llovizna **II, 2.3**
to drop caérsele (a uno) **6.2**
drought la sequía **2.2**
drugstore la farmacia **I**
drum el tambor **5.1**
dry seco(a) **II**
dry cleaner la tintorería **II**
to dry oneself secarse **I**
durability la durabilidad **6.2**
during durante **I**
to dust quitar el polvo **I, II**
dyed teñido(a) **1.1**

e-mail el correo electrónico **6.3**
   e-mail address
     la dirección electrónica **6.3**
each cada **I**
ear la oreja **I, II**
early temprano **I**
early morning la madrugada **3.2**
to earn a living
   ganarse la vida **II, 4.2**
earring el arete **I**
   earrings (dangling)
     los pendientes **1.2**
easily fácilmente **I, II**
east el este **II**
Easter las Pascuas **3.2**
easy fácil **I**
to eat comer **I, II**; tomar **I**
   Do you want to eat…?
     ¿Quieres comer…? **I**
   to eat breakfast desayunar **I**
   to eat dinner cenar **I, II**
   to eat lunch almorzar (ue) **I, II**
   I want to eat…
     Quiero comer… **I**
economic profile
   el perfil económico **4.3**
ecosystem el ecosistema **2.2**
ecotourist el (la) ecoturista **2.3**
Ecuadorian ecuatoriano(a) **II**
edge la orilla **II**
edition la edición **II**
editor el (la) editor(a) **I, II**

to educate the public
  educar al público **2.1**
education la educación **II**, **4.1**;
  la formación **4.1**
effects los efectos **II**, **2.2**
effort
  to make an effort
    hacer el esfuerzo **2.1**
egg el huevo **I**, **II**
eggnog el coquito **3.2**
eight ocho **I**, **II**
eight hundred ochocientos(as) **I**
eighteen dieciocho **I**
eighth octavo(a) **I**, **II**
eighty ochenta **I**
either…or o… o **4.2**
elbow el codo **II**
the elderly los ancianos **2.1**
electricity la electricidad **II**
electronic electrónico(a)
  electronic assistant
    el asistente electrónico **6.2**
  electronic mailbox
    el buzón electrónico **6.3**
elegant elegante **II**
elevator el ascensor **II**
eleven once **I**
elm el olmo
embroidery el bordado **5.2**
emergency room
  la sala de emergencia **II**
employee el (la) empleado(a) **4.2**
employment el empleo **II**
to empty vaciar **1.3**
empty vacío(a) **II**
enchilada la enchilada **I**
to encompass englobar
to end culminar **5.3**
ending el final **5.3**
energy la energía **II**
engineer el (la) ingeniero(a) **II**
engineering la ingeniería
  civil engineering
    la ingeniería civil **4.1**
  mechanical engineering
    la ingeniería mecánica **4.1**
English inglés(esa) **II**; (lang.) el
  inglés **I**
to enjoy gozar de; disfrutar de
  to enjoy oneself divertirse **II**
  to enjoy time with friends
    disfrutar con los amigos **II**

Enjoy! (your meal)
  ¡Buen provecho! **3.2**
enjoyable divertido(a) **I**
enormous enorme **I**, **II**
enough bastante; suficiente **II**
to enroll in cursar
to enter entrar (a, en) **I**
enterprising emprendedor(a) **4.1**
entertaining divertido(a) **II**;
  entretenido(a) **6.1**
entrance el ingreso
envelope el sobre **4.1**
environment
  el medio ambiente **II**, **2.2**
envious
  to be envious tener envidia **II**
episode el episodio **6.1**
eraser el borrador **I**
especially especialmente **I**, **II**
essay el ensayo **3.3**
even though aunque **II**
evening la noche **I**
  during the evening
    por la noche **I**
  Good evening. Buenas noches. **I**
every cada **I**
  every day todos los días **I**
everyone todo el mundo **II**
everywhere por todas partes **II**
to excel superarse **4.1**
excess luggage
  el exceso de equipaje **II**
to exchange cambiar **I**
excited emocionado(a) **I**
  to get excited entusiasmarse **1.3**
exciting emocionante **5.3**
to exclaim exclamar **II**
to exercise hacer ejercicio **I**
exhausted
  I'm exhausted.
    Estoy agotado(a). **2.1**
exhibit la exposición **II**
expandable ampliable **6.3**
to expect esperar **I**
expenses los gastos **II**
expensive caro(a) **I**
  It's very expensive!
    ¡Es muy caro(a)! **I**
to explain explicar **II**
to export exportar **4.3**
exportation la exportación **4.3**
expressive expresivo(a) **5.3**

external externo(a) **6.3**
eye el ojo **I**, **II**
eyeglasses los anteojos **1.1**

face la cara **I**, **II**
facing enfrente (de) **I**
fact el hecho **II**
factory la fábrica **4.3**
facts los datos **II**, **4.1**
fair justo(a) **3.3**
faithful fiel **1.1**
falcon el halcón **2.3**
fall el otoño **I**
  to fall apart hacerse pedazos
  to fall asleep dormirse (ue) **I**
  to fall down caerse **II**
  to fall in love with
    enamorarse de **II**
to be familiar with conocer **I**
  to be familiar with someone
    conocer a alguien **I**
family la familia **I**
far (from) lejos (de) **I**
  Is it far? ¿Queda lejos? **I**
farm la granja **I**
farmer el (la) ganadero(a) **I**; el (la)
  agricultor(a) **II**, **4.2**
to fascinate fascinar **II**
fashion la moda **1.2**
fast rápido(a) **I**
fastener el broche
fat gordo(a) **I**
father el padre **I**
Father's Day
  el día de los Padres **3.2**
favor
  to be in favor of
    estar a favor de **II**, **2.1**
favorite favorito(a) **I**
fax el fax **6.3**
  multifunctional fax machine
    el fax multifuncional **6.2**
February febrero **I**
to feed darle(s) de comer **I**
to feel sentirse (ie) **II**
  to feel frustrated
    sentirse frustrado(a) **1.3**
  to feel like tener ganas de **I**
fence la cerca **I**

ferocious feroz **II**
fever la fiebre **II**
fiction la ficción **5.1**
field el campo **I**
   field of study
     el campo de estudio **4.1**
fifteen quince **I**
fifteenth birthday
   la quinceañera **3.2**
fifth quinto(a) **I, II**
fifty cincuenta **I**
fight la lucha **3.3**
to fight pelearse **II, 1.3**
to fight against luchar contra **2.1**
figure la figura **5.1**
to fill llenar **II**
filmmaker el (la) cineasta **5.3**
finally por fin **I, II**
finance las finanzas **4.1**
financial expert
   el (la) financiero(a) **4.3**
to find encontrar (ue) **I**
fine arts las bellas artes **II**
fingers los dedos **II**
to finish terminar **I**
fire el fuego **II, 2.3**
firecracker el cohete **3.2**
firefighter el bombero **I, II, 4.2**
firewood la leña **II, 2.3**
firing squad el pelotón de
   fusilamiento
first primero(a) **I, II**
first name el nombre **I**
fish el pescado **I, II**; el pez **I, 2.3**
to fish pescar **II**
fisherman el (la) pescador(a) **II**
fishing industry la industria
   pesquera **4.3**
five cinco **I, II**
five hundred quinientos(as) **I**
flag la bandera **3.3**
flagpole el asta (fem.)
flamenco group el tablao **5.1**
   flamenco song el cante **5.1**
   tragic flamenco song
     el cante jondo **5.1**
   flamenco-style dancing
     el flamenco **5.1**
flash (of lightning) el rayo **II**
flashlight la linterna **II, 2.3**
flavor el sabor **I**
flight el vuelo **II**
flight attendant
   el (la) auxiliar de vuelo **II**

floor el suelo **1, II**; el piso **II**
flour la harina **I, II**
to flow fluir
flower la flor **I**
flu la gripe **II**
to fly volar (ue) **II**
   to fly a single-engine plane
     pilotar una avioneta **1.2**
fog la neblina **II, 2.3**
folk dance el baile folklórico **5.2**
to follow seguir (i) **II**
food el alimento **II**; la comida **I**
foot el pie **I, II**
   on foot a pie **I**
football el fútbol americano **I**
footprint la huella
for por; para **I**
   For what purpose? ¿Para qué?
   for sale a la venta
forecast el pronóstico
foreground el primer plano **5.1**
foreigner el (la) extranjero(a) **II**
forest el bosque **I, 2.2**;
   la selva **II, 2.2**
forestry products
   los productos forestales **4.3**
to forget olvidar **I, II**; olvidársele
   a uno, **6.2**
to forgive each other
   perdonarse **1.3**
fork el tenedor **I**
formal formal **I, II**
formulaic formulista **5.3**
forty cuarenta **I**
four cuatro **I, II**
four hundred cuatrocientos(as) **I**
fourteen catorce **I**
fourth cuarto(a) **I, II**
freckles las pecas **1.1**
free time el tiempo libre **I**
freezer el congelador **I, II**
French francés(esa) **II**
French fries las papas fritas **I**
frequent frecuente **I**
frequently frecuentemente **I, II**
fresco (art) el fresco **5.1**
fresh water el agua dulce **2.3**
Friday viernes **I**
friend el (la) amigo(a) **I**
   to spend time with friends
     pasar un rato con los amigos **I**
friendship la amistad **II**
fright el susto
to frighten asustar

fringe el fleco **1.2**
frog la rana **II**
from de; desde **I**
   from there desde allí **II**
front frente
   in front of delante de **II**; frente
    a **II**
front desk la recepción **II**
fruit la fruta **I**
frying pan la sartén **I**
fuel el combustible **II, 2.2**
full lleno(a) **II**
full time tiempo completo
full-time job
   el trabajo a tiempo completo **4.2**
fun divertido(a) **I**
to fundraise juntar fondos **2.1**
funny cómico(a) **I**
furniture los muebles **I, II**

G-rated (movie)
   apto(a) para toda la familia **6.1**
gallery la galería **II**
game el partido **I**
   interactive game
     el juego interactivo **6.3**
garage el garaje **II**
garbage la basura **2.1**
garden el jardín **I, II**
gardener el (la) jardinero(a) **4.2**
gathering la reunión **II**
generosity la generosidad **3.1**
generous generoso(a) **3.1**
genius el genio
genre el género **5.3**
German alemán(ana) **II**
to gesture señalar
to get conseguir **II**; obtener **II**
   to get ahead superarse **4.1**
   to get along convivir **2.1**
   to get along well/badly llevarse
    bien/mal **II, 1.3**
   to get better recuperarse **II**
   to get dark anochecer,
    oscurecer, atardecer **2.3**
   to get discouraged
    desanimarse **1.3**
   to get encouraged animarse **1.3**
   to get excited entusiasmarse **1.3**
   to get interested animarse **1.3**

to get into meter en
to get nervous
    ponerse nervioso(a) **1.3**
to get together reunirse **II**
to get up levantarse **I, II**
gift el regalo **I**
girl la chica; la muchacha; la niña **I**
girlfriend la novia **II**
to give dar **I**
   I'll give… to you for…
    Le dejo… en… **I**
glad
   to be glad that alegrarse de
   que **II**
glass el vaso **I**; el vidrio **II, 2.2**
   glass of el vaso de **I**
glasses los anteojos **1.1**
glove el guante **I**
to go ir **I, II**
   to be going to… ir a… **I**
   to go away irse **I**
   to go back regresar **II**
   to go deep-sea fishing
    pescar en alta mar **1.2**
   to go down bajar por **II**
   to go down a river by canoe
    bajar un río en canoa **II**
   to go for a walk pasear **I**
   to go hiking hacer alpinismo **1.2**
   to go mountaineering
    hacer montañismo **1.2**
   to go out salir **I**
   to go shopping ir de compras **I**
   to go to bed acostarse (ue) **I**
   to go to the movies ir al cine **I**
   to go to the supermarket
    ir al supermercado **I**
   to go unnoticed
    pasar por debajo de la mesa
   to go up subir por **II**
   to go whitewater rafting
    navegar por rápidos **2.3**
goal el gol **I**; la meta **II**
godparents los padrinos **3.1**
gold el oro **I**
good bueno(a) **I**
   Good afternoon.
    Buenas tardes. **I, II**
   Good evening.
    Buenas noches. **I, II**
   Good morning.
    Buenos días. **I, II**
   the good thing lo bueno **1.1**

It's good that…
   Es bueno que… **II**
Good-bye. Adiós.
good-looking guapo(a) **I**
gossipy chismoso(a) **4.2**
government el gobierno **3.3**
governor el (la) gobernador(a) **3.3**
gown la toga **3.1**
grade la nota
   to get a good grade
    sacar una buena nota **I**
graduate el (la) graduando(a) **3.1**
to graduate graduarse **3.1**
graduation ceremony
   la ceremonia de graduación **3.1**
grains los cereales **4.3**
gram el gramo **I**
grandchildren los nietos **I**
granddaughter la nieta **I**
grandfather el abuelo **I**
grandmother la abuela **I**
grandparents los abuelos **I**
grandson el nieto
grape la uva **II**
gray gris **I**
great grande **I**, formidable **1.2**
great grandfather el bisabuelo **II**
great grandmother la bisabuela **II**
green verde **I**
to greet saludar
to greet each other saludarse **1.3**
ground floor la planta baja **II**
group el grupo
   chat group
    el grupo de conversación **6.3**
   news group
    el grupo de noticias **6.3**
to grow crecer **II**
guarantee la garantía **6.2**
Guatemalan guatemalteco(a) **II**
guest el (la) huésped(a) **II**
guitar la guitarra **I**
   to play the guitar
    tocar la guitarra **I**
gymnasium el gimnasio **I**

H

habanera
   la habanera (Latin music) **5.2**
hair el pelo **I**; el cabello **1.1**

hair dryer el secador de pelo **I, II**
hairbrush el cepillo **II**
hairdo el peinado
hairstylist el (la) peluquero(a) **II,
4.2**
half medio(a) **I, II**
   half past y media **I**
half-brother el medio hermano **I**
half-sister la media hermana **I**
ham el jamón **I, II**
hamburger la hamburguesa **I**
hand la mano **I, II**
handbag la bolsa **I**
handicraft la artesanía **I**
to hang–glide
   volar en planeador **1.2**
to happen pasar **I**
happily felizmente **I**
happiness la felicidad **II**
happy alegre; contento(a); feliz **I**;
   a gusto
   to be happy that
    alegrarse de que **II**
   Happy New Year!
    ¡Próspero Año Nuevo! **3.2**
hard difícil; duro(a) **I**
   It's hard for me
    A mí me cuesta mucho
hard drive el disco duro **6.3**
hard-working trabajador(a) **I**
hardware el hardware **6.3**
harp el arpa (fem.) **5.1**
hat el sombrero **I**
to hate detestar **1.2**
   to hate each other odiarse **1.3**
to have tener **I, II**
   to have a cold estar resfriado(a)
   II
   to have a good time
    pasarlo bien **3.2**
   to have a snack merendar (ie) **I**
   to have breakfast desayunar **I**
   to have dinner, supper cenar **I**
   to have in common
    tener en común **1.1**
   to have just… acabar de… **I, II**
   to have to tener que **I**
he él **I**
head la cabeza **I, II**
headache el dolor de cabeza **II**
headline el titular **II**
headphones los audífonos **6.2**
healthy saludable **II**

to be healthy
mantenerse sano(a) II
to hear oír I
heart el corazón I
heat la calefacción II
heating la calefacción II
heavy grueso(a) **1.1**; pesado(a) **1.2**
heavy bed el lecho pesado
height la altura **II, 2.2;**
la estatura **4.1**
Hello. Hola. I
helmet el casco I
Help! ¡Socorro! II
to help ayudarse I, **1.3**
to help each other ayudarse **1.3**
Could you help me order?
¿Me ayuda a pedir? I
How can I help you?
¿Cómo puedo ayudarte? **2.1**
hen la gallina I
her su, ella I
here acá; aquí I
hero el héroe II
heroine la heroína II
hers suyo(a)
to hide esconderse II, **1.3**
hieroglyphics los jeroglíficos **5.2**
highway la carretera **4.1**
to hike
to go hiking hacer alpinismo **1.2**
hill la colina II, **2.2**
his su I; suyo(a) II
history la historia I
hockey el hockey I
to hold guardar II
homeless la gente sin hogar **2.1**
homework la tarea I
Honduran hondureño(a) II
to honor honrar **3.3**
to hope esperar I, II
I hope that ojalá que II
to hope that esperar que II
hopefully ojalá que II
horrible horrible **1.2**
horror el horror II
horse el caballo I
host(ess) el (la) anfitrión(a) **3.2**
hot caliente I; caluroso(a) II
to be hot tener calor I
It is hot. Hace calor. I
hot dog la salchicha II
hotel el hotel I
house la casa I

how cómo I
How are you? (familiar)
¿Cómo estás? I
How are you? (formal)
¿Cómo está usted? I
How awesome! ¡Qué chévere! I
How can I help you? ¿Cómo
puedo ayudarte(lo, la)? **2.1**
How do I look?
¿Cómo me veo? II
How does it look on you?
¿Cómo te queda? II
How is it going? II ¿Qué tal? I
How long has it been since…?
¿Cuánto tiempo hace que…?
II
How old is…?
¿Cuántos años tiene…? I
Pardon, how do I get to…?
Perdona(e), ¿cómo llego
a…? I
how much cuánto I
How much do I tip?
¿Cuánto le doy de propina? I
How much is (are)…?
¿Cuánto cuesta(n)…? I
¿A cuánto está(n)…? I
hug el abrazo II
huge enorme I
human being el ser humano II, **2.1**
human rights los derechos
humanos **2.1**
humanities las humanidades **4.1**
humid húmedo(a) II, **2.3**
hummingbird el picaflor **2.3**
humpback whale
la ballena jorobada **2.3**
hungry
to be hungry tener hambre I
hurricane el huracán II, **2.3**
to hurry
to be in a hurry tener prisa I
to hurt doler (ue) II
to get hurt lastimarse II
husband el esposo I

I

I yo I
ice el hielo I
on ice sobre hielo I

ice cream el helado I, II
ice cream store la heladería II
identification la identificación II
ideology la ideología **3.3**
if si I
iguana la iguana **2.3**
immediately inmediatamente II
impatient impaciente II
important
to be important importar II
It's important that…
Es importante que… II
to impose imponer **3.1**
impossible
It's impossible that…
Es imposible que…
impression
make a good (bad) impression
on someone
caerle bien (mal) a alguien II
impressive impresionante **5.3**
improbable
It's improbable that…
Es improbable que…
in en I
in a bad mood de mal humor II
in a good mood
de buen humor II
in back atrás **6.2**
in case en caso de que **3.1**
in front of
delante de I, II; frente a II
in order (to) para I
in spite of pese a
included incluido(a)
Is…included?
¿Está incluido(a)…? I
to increase aumentar **4.2**
incredible increíble II, **2.2**
Independence Day
el Día de la Independencia **3.2**
indigenous, indigenous peoples
los (las) indígenas
industry la industria **4.3**
inexpensive barato(a) I
infection la infección II
to influence influir **1.1**
informed
to be well informed
estar bien informado(a) II
informal informal I
information los datos II, **4.1**
injection la inyección II

injustice la injusticia **3.3**
inner ear el oído **II**
innovative innovador(a) **5.3**
inside (of) dentro (de) **I, II**
to insist insistir (en) **II**
to insist on insistir en que **3.1**
insolent descarado(a) **1.1**
to institute instituir **2.2**
insurance el seguro **II**
> **health insurance**
>> el seguro médico **4.2**

intelligent inteligente **I**
to interest interesar **II**
interested
> **to get interested** animarse **1.3**

interesting interesante **I**
internal interno(a) **6.3**
international internacional **II**
Internet
> **to surf the Internet**
>> navegar por Internet **1.2**

to interpret interpretar **5.1**
interpreter el (la) intérprete **4.3**
interview la entrevista **I, II**
interviewer
> el (la) entrevistador(a) **II, 4.2**

to introduce presentar
> **Let me introduce you**
> **(familiar/formal) to…**
>> Te/Le presento a… **I**

invitation la invitación **I, II**
to invite invitar
> **I invite you.** Te invito. **I**

iron el hierro **4.3**
to iron (the clothes)
> planchar (la ropa) **I**

ironic irónico(a) **5.3**
island la isla **II**
Italian italiano(a) **II**
its su **I**

jacket la chaqueta **I**
jackknife la navaja **II, 2.3**
jade el jade **5.2**
jaguar el jaguar **II, 2.3**
January enero **I**
Japanese japonés(esa) **II**
jeans los jeans **I**
jewelry las joyas **I**

jewelry store la joyería **I**
job el empleo **II, 4.2**
> **full-time job**
>> el trabajo a tiempo completo
>> **4.2**
> **part-time job**
>> el trabajo a tiempo parcial **4.2**

journalism el periodismo **II**
journalist el (la) periodista **I, II**
judge el (la) juez(a) **II, 4.2**
to judge juzgar
juice el jugo **II**; el zumo **I**
July julio **I**
to jump rope saltar la cuerda **II**
June junio **I**
jungle la selva **I, II, 2.2**
just justo(a) **3.3**

to keep guardar **II**
to keep clean mantener limpio(a) **II**
key la llave **I, II**; **(of an**
> **instrument)** la tecla **6.3**

keyboard el teclado **I, 6.3**
keychain el llavero **1.2**
kilogram el kilo **I**
kind
> **That's kind of you.**
>> Muy amable. **3.2**

king el rey **3.3**
kiosk el quiosco **II**
kiss el beso **II**
kitchen la cocina **I, II**
knee la rodilla **II**
knife el cuchillo **I**
to know conocer **I, II**; saber **I, II**
> **to know each other well/not**
> **very well** conocerse
>> bien/mal **1.3**
> **to know someone**
>> conocer a alguien **I**

knowledge el conocimiento **4.2**

laboratory el laboratorio **4.3**
to lack faltar **II**
lake el lago **I**
lamp la lámpara **I, II**

land la tierra **I, II, 2.2**
landscape
> el paisaje **I, 5.1**; el terreno

language la lengua **I, II**
laptop computer
> la computadora portátil **6.2**

large grande **I**
last último(a) **I**
> **last month** el mes pasado **I**
> **last name** el apellido
> **last night** anoche **I, II**
> **last week** la semana pasada **I**
> **last year** el año pasado **I, II**

late tarde **I**
> **late afternoon** el atardecer **2.3**

lately recientemente **I**
later luego **I**
> **See you later.**
>> Hasta luego; Nos vemos. **I**

laugh la risa **II**
to laugh reírse **II**
laughter la risa **II**
law el derecho; la ley **3.3**
law office el bufete **4.2**
lawnmower el cortacésped **1.3**
lawyer el (la) abogado(a) **II, 4.2**
lazy perezoso(a) **I**
leader el (la) líder **3.3**
leaf la hoja **II**
to learn aprender **I**
least
> **the least**
>> lo menos **1.1**
> **the least desirable**
>> lo más oscuro **1.1**

leather el cuero **1.2**
> **leather goods**
>> los artículos de cuero **I**

to leave salir; irse **I**
> **to leave a message**
>> dejar un mensaje **I**
> **Leave a message after the tone.**
>> Deje un mensaje después del
>> tono. **I**
> **to leave behind** dejar **I**;
>> quedársele (a uno) **6.2**
> **to leave the tip**
>> dejar la propina **II**

left la izquierda
> **to the left (of)**
>> a la izquierda (de) **I**

leftover sobrante
leg la pierna **I, II**

lemonade la limonada I
to lend prestar II
less menos
   less than menos de I
   less… than
     menos… que I, II
lesson la lección I
Let's… Vamos a…
letter la carta I
   to send a letter
     mandar una carta I
lettuce la lechuga I
level of high school curriculum
   el ciclo
liberal liberal 3.3
librarian el (la) bibliotecario(a) 4.3
library la biblioteca I
lie la mentira II
to lie mentir (ie)
to lie down acostarse (ue) II
life la vida I
to lift weights levantar pesas I
light la luz II, 2.3; lumbre
lightbulb la bombilla 1.3
lighthouse el faro 3.3
lightning el relámpago II, 2.3
   lightning flash el rayo 2.3
like como
to like gustar II; caerle bien 1.2
   He/She likes… Le gusta… I
   I like… Me gusta… I
   I would like… Me gustaría… I
   Would you like… ?
     ¿Te gustaría…? I
   You like… Te gusta… I
link el enlace 6.3
lion el león II
to listen (to) escuchar I
liter el litro I
literacy la alfabetización
literature la literatura I
to live vivir I
to live together convivir 2.1
lively animado(a) II
livestock el ganado 4.3
living room la sala I, II
llama la llama I
loan el préstamo II
local local II
located ubicado(a)
logical
   It's logical that…
     Es lógico que… II

long largo(a) I
to look verse 1.1
to look at mirar I
to look for buscar I
loose flojo(a) II; suelto(a) 1.2
to lose perder (ie) I
   to lose something
     perdérsele (a uno) 6.2
love el amor II
to love querer (ie) II
luck la suerte I
   Luck comes to him who least
     expects it. La suerte viene a
     quien menos la aguarda.
lucky
   to be lucky tener suerte I
luggage el equipaje II
lunch el almuerzo I
   to eat lunch almorzar (ue) I
luxurious lujoso(a) I, II
lyrics la letra 5.1

magazine la revista I
magical realism
   el realismo mágico 5.3
mail carrier
   el (la) cartero(a) I, II, 4.2
mailbox el buzón II
majority
   la mayoría II
to make hacer I, II
   to make a mistake
     equivocarse 6.2
   to make a toast brindar 3.1
   to make an effort
     hacer un esfuerzo 2.1
   to make decisions
     tomar decisiones 4.1
   to make the bed hacer la cama I
makeup el maquillaje II
   to put on makeup
     maquillarse I, II
to malfunction
   descomponérsele (a uno) 6.2
mambo (Latin music)
   el mambo 5.2
man el hombre I
manager el (la) gerente I, II, 4.2
to manipulate manipular 6.1

many mucho(a) I
map el mapa I
March marzo I
marionette la marioneta II
market el mercado I
marketing el mercadeo 4.1
to marry
   to get married (to) casarse (con)
     II
marvelous de maravilla II
mask la careta, la máscara
maskmaker el mascarero
Master's degree la maestría 4.1
masterpiece la obra maestra I
masthead el titular II
match el fósforo II, 2.3
to match with
   hacer juego con II, 1.2
mathematics las matemáticas I
to matter importar II
May mayo I
maybe tal vez I
   Maybe another day.
     Tal vez otro día. I
mayor el (la) alcalde(sa) 3.3
meal la comida I
to mean implicar
means el medio
meat la carne I
mechanic el (la) mecánico(a) II, 4.2
medallion la medalla 1.2
medicine la medicina II
medium medio
to meet encontrar (ue) I
melody la melodía 5.1
melon el melón II
memory la memoria 6.3
menu el menú I
merengue (Latin music)
   el merengue 5.2
message el mensaje
   I want to leave a message for…
     Quiero dejar un mensaje
     para… I
   to leave a message
     dejar un mensaje I
   Leave a message after the tone.
     Deje un mensaje después del
     tono. I
Mexican mexicano(a) II
microphone el micrófono 6.3
microprocessor
   el microprocesador 6.3

microwave el microondas I
    el horno microondas II
midnight la medianoche I
milk la leche I, II
milk shake el batido II
million millón I
mine mío(a) II
mining la minería 4.3
miniseries teledrama 6.1
mirror el espejo I, II
Miss (la) señorita I
to miss echar de menos
mist la neblina II, 2.3
mistake
    to make a mistake
        equivocarse **6.2**
to mix mezclar
modem el módem **6.3**
modern moderno(a) I, II
modest modesto(a) **1.1**
moment el momento I
One moment. Un momento. I
monarchy la monarquía **3.3**
Monday lunes I
money el dinero I
    money exchange el cambio I
monitor el monitor **6.3**
monkey el mono II
month el mes I
more más I
    more or less más o menos II
    more… than
        más… que I, II, **6.3**
    more than más de I
morning la mañana I
    during the morning por la
        mañana I
    Good morning. Buenos días. I
    in the morning de la mañana I
the most lo más **1.1**
mother la madre I
Mother's Day
    el Día de las Madres **3.2**
motorcycle la moto(cicleta) I
mountain la montaña I
    to mountain climb
        escalar montañas **1.2**
mountaineering el montañismo II
    to go mountaineering
        hacer montañismo **1.2**
mouse el ratón I
mouth la boca I, II
to move mover (ue) I; trasladarse

to move the furniture
    mover los muebles I
movie la película I
    to go to the movies
        ir al cine I
moving conmovedor(a)
moving force la fuerza motriz
Mr. (el) señor I
Mrs. (la) señora I
much mucho(a) I
    as much as tanto como I
multinational multinacional **4.3**
mural el mural **5.2**
muralist el (la) muralista **5.2**
museum el museo I
music la música I
    music and video store la tienda
        de música y videos I
musical el musical II
musical group el conjunto **3.3**
musician el (la) músico(a) II, **3.2**
must
    one must hay que I
mustache el bigote **1.1**
my mi I

name el nombre I
    His/Her name is…
        Se llama… I
    My name is… Me llamo… I
    What is his/her name?
        ¿Cómo se llama? I
    What is your name?
        ¿Cómo te llamas? I
napkin la servilleta II
narrow estrecho(a) I, II
natural resources
    los recursos naturales II, **2.2**
nature la naturaleza II, **2.2**
navy la marina
near (to) cerca (de) I
necessary
    It's necessary that…
        Es necesario que… II
neck el cuello II
necklace el collar I
to need necesitar I
neighbor el (la) vecino(a)
neighborhood el vecindario II

neither tampoco I; ni II
    neither… nor ni… ni **4.2**
nephew el sobrino II
nervous nervioso(a) I
    to get nervous
        ponerse nervioso(a) **1.3**
never nunca I
nevertheless sin embargo **5.3**
new nuevo(a) I
    new release el estreno II
    New World el Nuevo Mundo
        5.2
    New Year el Año Nuevo **3.2**
    New Year's Eve la despedida
        del año **3.2**
news las noticias II
    news group
        el grupo de noticias **6.3**
    news program el noticiero II
newspaper el periódico I
newstand el quiosco II
next siguiente II
    next to
        al lado de I, **6.2**; junto (a) II
Nicaraguan nicaragüense II
nice amable II; simpático(a) I
    It's nice outside.
        Hace buen tiempo. I
    Nice to meet you.
        Mucho gusto. I
niece la sobrina II
night la noche I
    at night de la noche I
nightfall el anochecer **2.3**
nine nueve I
nine hundred novecientos(as) I
nineteen diecinueve I
ninety noventa I
ninth noveno(a) I, II
no no I
no longer ya no I
no one nadie I
Nobel Prize el Premio Nóbel **5.3**
noise el ruido **3.2**
none ninguno(a) I
noon el mediodía I
nor ni II
normal normal I
normally normalmente I, II
north el norte II
nose la nariz I, II
not no I
not even ni II

notebook el cuaderno **I**
nothing nada **I**
>**It's nothing.** No hay de qué. **3.2**

to notice fijarse en **6.2**
nourishment la alimentación **II**
novel la novela **I**
novelist el (la) novelista **5.3**
November noviembre **I**
now ahora **I**
>**Right now!** ¡Ahora mismo! **I**

number, numeral la cifra **5.2**
>**What is your phone number?**
>¿Cuál es tu teléfono? **I**

nun la monja
nurse el (la) enfermero(a) **II**
nutritious nutritivo(a) **II**

obedient obediente **II**
to obtain obtener **II**
to occur ocurrir **II**
>**to occur to**
>ocurrírsele (a uno) **6.2**

ocean el océano **II**
ocelot el ocelote **2.3**
October octubre **I**
of de
>**Of course!**
>¡Claro que sí!; ¡Cómo no! **I**

to offer ofrecer **I, II**
office la oficina **I**
law office el bufete **4.2**
doctor's office el consultorio **II**
often mucho; frecuentemente **I;**
a menudo, muchas veces **4.2**
oil el aceite **I, II**
oil painting al óleo **5.1**
oil spill el derrame de petróleo **2.2**
old antiguo **I, II;** viejo(a) **I**
>**How old is…?**
>¿Cuántos años tiene…? **I**

older mayor **I**
>**older than** mayor que **II**

olives las aceitunas **I**
on en **I,** sobre **II**
>**on ice** sobre hielo **I**
>**on sale** en oferta
>**on the one hand**
>por un lado **1.1**
>**on the other hand**
>por otro lado **1.1**

on top (of) encima (de) **I, II**
once in a while
de vez en cuando **I**
one uno **I, II**
one fifth el quinto **4.3**
one half of la mitad de **4.3**
one hundred cien **I**
one third el tercio **4.3**
onion la cebolla **I, II**
on-line en línea **6.3**
only sólo **I,** único(a) **1.2**
open abierto(a) **I, II**
to open abrir **I**
>**to open the way**
>abrir el paso **5.2**

operator
el (la) operador(a) **I, II, 4.2**
opinion
>**not to be of the opinion that**
>no opinar que

to oppose oponerse a **1.3**
opposite frente a **II;** opuesto **1.1**
oppressor el (la) opresor(a) **3.3**
or o **I**
orange anaranjado(a) **I**
orchestra la orquesta **3.2**
to order pedir (i) **I, II**
>**Could you help me order?**
>¿Me ayuda a pedir? **I**

ordinary ordinario(a) **I**
organized organizado(a) **1.3**
original original **5.3**
orphanage el orfelinato
other otro(a) **I**
ought to deber **I, II**
our nuestro(a) **I**
ours nuestro(a) **II**
outdoors al aire libre **I**
outside afuera **II, 6.2**
>**outside (of)** fuera (de) **I, II**

oval ovalado(a) **1.1**
oven el horno **I, II**
to overcome vencer **3.1**
overflowing rebosante
owl el búho **2.3**
owner el (la) dueño(a) **II, 4.2**
ozone layer
la capa de ozono **II, 2.2**

package el paquete **I, 4.1**

pain el dolor
paint la pintura
to paint pintar **I**
painter el (la) pintor(a) **II**
painting el cuadro **5.1,** la pintura
II
>**historical painting**
>el cuadro histórico **5.1**

pair
>**a pair of** un par de **II**

palm tree la palma **II**
palm tree grove el palmar **II**
Panamanian panameño(a) **II**
pants los pantalones **I**
paper el papel **I**
parade el desfile **3.1**
Paraguayan paraguayo(a) **II**
Pardon, how do I get to…?
Perdona(e), ¿cómo llego
a…? **I**
parents los padres **I**
park el parque **I**
parking space
el estacionamiento **II**
parrot el loro **II, 2.3**
to participate participar **2.1**
party la fiesta **I, II**
>**big, formal party** la gala **3.2**
>**party in stages**
>la fiesta continua **3.2**

to pass (by) pasar **I**
passenger el (la) pasajero(a) **II**
passport el pasaporte **II**
password la contraseña **6.3**
pasta la pasta **I, II**
pastel el color claro **1.2**
pastry shop la pastelería **I**
path el sendero **II, 2.3;**
el camino **3.1**
patient paciente **I**
patriot el (la) patriota **3.3**
patriotic patriótico(a) **3.3**
patriotism el patriotismo **3.3**
paving stone la baldosa
to pay pagar **I**
>**to pay attention to**
>hacerle caso **1.1**

peace la paz
peanut butter
la mantequilla de cacahuate **II**
pear la pera **II**
pedestrian el peatón **II**
pelican el pelícano **2.3**

pen la pluma I; (animal) el corral I
pencil el lápiz I
pension la pensión II
people la gente I
pepper la pimienta I, II
percent el por ciento 4.3
percentage el porcentaje 4.3
perception la percepción 6.1
perfect perfecto(a) I
perfume el perfume II
perhaps quizás II
period la época I
period of time la temporada 1.2
permission el permiso II, 2.2
to permit permitir II, 2.1
perspective la perspectiva 5.1
Peruvian peruano(a) II
petroleum el petróleo 4.3
PG-13 rated (movie)
   se recomienda discreción 6.1
pharmacy la farmacia I
phone directory la guía telefónica I
   to phone each other
      telefonearse 1.3
photographer el (la)
   fotógrafo(a) I, II
physical education
   la educación física I
physically challenged
   los (las) minusválidos 2.1
piano el piano I
   to play the piano
      tocar el piano I
to pick up recoger 2.1
picture
   to take pictures
      sacar fotos I
piece el pedazo I, II
pig el cerdo I
pill la pastilla II
pillow la almohada II, 2.3
pilot el piloto II
pin el prendedor 1.2
pink rosado(a) I
pitcher la jarra I
pity
   It's a pity that…
      Es una lástima que… II
place el lugar I
plaid de cuadros I
plain sencillo(a) I
to plan pensar (ie) en + *infinitive* I
planet el planeta II, 2.2

plant la planta I
to plant sembrar (ie) 2.1
plantain el plátano verde II
   fried plantains los tostones II
plastic el plástico 2.2
plate el plato I
play la obra de teatro 5.1
   to play jugar (ue) I, II; practicar
      I; (an instrument) tocar I, II
   to play chess jugar al ajedrez II
   to play sports
      practicar deportes I
   to play the guitar
      tocar la guitarra I
   to play the piano
      tocar el piano I
   to play the role
      hacer el papel 5.3
to plead suplicar 3.1
please por favor I
pleased contento(a) I
   Pleased to meet you.
      Encantado(a). I
   It's a pleasure. Es un placer. I
   The pleasure is mine.
      El gusto es mío. I
plot la trama 5.3
to plug in enchufar 1.3
poem el poema I
poet el (la) poeta 5.3
poetry la poesía I, 5.1
police officer el (la) policía I
polka dots los lunares 1.2
pollutant el contaminante 2.2
to pollute contaminar II
pollution la contaminación II
polyester el poliéster 1.2
ponytail la cola de caballo 1.1
poor pobre II
population la población II, 2.2
pork el puerco I
   pork rinds los chicharrones I
   to eat pork rinds
      comer chicharrones I
porter el (la) maletero(a) II
portrait el retrato II
position el puesto II, 4.2
possible
   It's just not possible for me
      Me es imposible 2.1
   It's possible that…
      Es posible que… II
post office el correo I

pot la olla I
potato la patata I; la papa II
poverty la pobreza II, 2.1
power
   la fuerza motriz; el poder 3.3
to practice practicar I, II
to praise elogiar 5.3
pre-Columbian
   precolombino(a) 5.2
precious precioso(a) 5.2
predictable predecible 5.3
to prefer preferir (ie) I, II
prejudice el prejuicio 2.1
preparatory school la preparatoria
to prepare preparar I
prescription la receta II
to preserve preservar II, 2.1
president el (la) presidente(a) 3.3
press la prensa
pretty bonito(a) I
to prevail prevalecer
price el precio I
pride el orgullo 3.1
principal principal 4.3
print estampado(a) 1.2
printer la impresora I
prize el premio
probable
   It's probable that…
      Es probable que… II
problem el problema I, II
procession
   el desfile 3.1; la procesión 3.3
to produce producir II
production la producción 5.1
profession la profesión I, II
program el programa I, II
   action program
      el programa de acción 6.1
   anti-virus program
      el programa anti-virus 6.3
   horror program
      el programa de horror 6.1
   mystery program
      el programa de misterio 6.1
   program icon
      el icono del programa 6.3
   science fiction program
      el programa de ciencia
      ficción 6.1
to prohibit prohibir 2.2
prose la prosa 5.3
protagonist el (la) protagonista 5.3

to protect proteger II, 2.2
   to protect the species
      proteger las especies II
proud orgulloso(a) 3.1
provided that con tal de que 3.1
public relations
   las relaciones públicas 4.1
publicity la publicidad 4.1
pudding el budín 3.2
Puerto Rican puertorriqueño(a) II
punctuality la puntualidad II, 4.2
purple morado(a) I
purpose el motivo 3.2
to push empujar
to put poner I, II
   to put on (clothes) ponerse I
   to put on makeup maquillarse I
   to put up with resistir
pyramid la pirámide 5.2

qualified capacitado(a) II, 4.1
quality la calidad I
quarter cuarto(a) I, II
   quarter past y cuarto I
queen la reina 3.3
quick rápido(a) I
quickly rápidamente I, II
quiet
   Be quiet! ¡Cállate! I
quiz la prueba I

R-rated (movie)
   prohibido(a) para menores 6.1
racket la raqueta I
radio el radio I
   portable radio
      el radio portátil 6.2
   radio station la radioemisora
      3.2
   radio-tape player
      el radiocasete I
rain la lluvia I
to rain llover (ue) I

raincoat el impermeable I
to raise criar
rare raro(a) II
   It's rare that… Es raro que… II
rarely rara vez I
reach
   within reach
      dentro del alcance 5.3
to read leer I
ready listo(a) I
   to get ready (dressed)
      arreglarse II
to realize darse cuenta de II
Really? ¿De veras? II
reason la razón I
to receive recibir I
recent reciente I
recently recientemente I, II
reception desk la recepción II
receptionist el (la) recepcionista I
recital el recital 5.1
to recommend recomendar (ie) II
recommendations
   las recomendaciones II
to record grabar 6.1
to recycle reciclar II
recycling el reciclaje II
   recycling program
      el programa de reciclaje 2.2
red rojo(a) I
red beans
   las habichuelas coloradas II
reddish rojizo(a) 1.1
redhead pelirrojo(a) I
to reduce reducir II, 2.2
to reflect reflejar 5.2
reflection el reflejo
refrigerator
   el frigorífico I; el refrigerador II
regional típico(a) 3.2
rehabilitation center
   el centro de rehabilitación 2.1
relative el (la) pariente II
to relax relajarse II
reliability la confiabilidad 6.2
to remember recordar (ue) I;
   acordarse (ue) de
remote control el control remoto
   6.1
to rent a video
   alquilar un video I
to repair reparar 1.3
to repeat repetir (i) II

repertoire el repertorio 5.1
report el reportaje II
reporter el (la) reportero(a) II
to request solicitar II, 4.1
to require requerir (ie) II, 4.2
requirement el requisito II, 4.2
rescue el rescate II
to rescue rescatar II
reservation la reserva II
to resolve resolver (ue) II, 1.1
to respect respetar 1.1
respectively respectivamente
to rest descansar I
the rest of the people los demás II
restaurant el restaurante I
résumé el currículum II;
   el currículum vitae 4.1
to retell contar (ue) I
to retire jubilarse 4.2
to return regresar II; volver (ue) I
   He/She will return later.
      Regresa más tarde. I
   to return (an item)
      devolver (ue) I, 6.2
review la crítica II
to review revisar II
rhythm el ritmo 5.1; el compás 5.1
rhythmic heel tapping
   el zapateado 5.1
rice el arroz I
   dessert of rice, cinnamon,
      and coconut milk
      el arroz con dulce 3.2
   dessert of sweet rice and milk
      el arroz con leche 3.2
   rice and pigeon peas
      el arroz con gandules 3.2
rich rico(a) II
ridiculous
   It's ridiculous that…
      Es ridículo que… II
right el derecho 3.3
   to the right (of)
      a la derecha (de) I
   (human) rights
      los derechos (humanos) 2.1
   right, correct acertado(a)
   to be right tener razón I
ring el anillo I; (of telephone) el
   timbre
to ring sonar (ue) 3.3
risk
   to take risks correr riesgos 4.1

river el río **I**

road el camino **I, 3.1;** la carretera **4.1**

roast suckling pig el lechón asado **3.2**

robbery el robo **II**

rock la piedra **2.2**

role el papel **II**
  to play the role hacer el papel **5.3**

romantic romántico(a) **II**

romanticism el romanticismo **5.3**

roof el techo

room el cuarto **I**; la habitación **II**

rooster el gallo **I**

round redondo(a) **1.1**

to row remar **II**

royal real

rubble los escombros

ruins las ruinas **5.2**

rule la regla **I**

to run correr **I**

to run out of acabársele (a uno) **6.2**

sad triste **I**
  It's sad that... Es triste que... **II**

sadness la tristeza **II**

salad la ensalada **I**

salary el sueldo **II, 4.2**

sale la rebaja **II**

sales las ventas **4.1**

sales agent el (la) agente de ventas **4.3**

salesperson el (la) dependiente **II**

salsa la salsa **I**

salt la sal **I, II**

Salvadoran salvadoreño(a) **II**

same mismo(a) **I**

sand la arena **II**

sandals las sandalias **II**

sandwich (sub) la torta **I, II**

satellite dish la antena parabólica **6.1**

satire la sátira **5.3**

Saturday sábado **I**

sausage el chorizo; la salchicha **I, II**

to save ahorrar **II**

savings account la cuenta de ahorros **II**

saxophone el saxofón **I**

to say decir **I**
  Don't say that! ¡No digas eso! **I**
  to say goodbye despedirse **I, II**
  to say hello to each other saludarse **1.3**

scarcely apenas **II**

scared
  to be scared of asustarse de **II**

scarf la bufanda **I**; el pañuelo **II**

scene la escena **II**

schedule el horario **I**

scholarship la beca

school la escuela **I, 5.1;** el colegio

science las ciencias **I**
  science fiction la ciencia ficción **II**

scissors las tijeras **I, II**

to scream gritar **II**

screen la pantalla **I**

script el guión **5.3**

scriptwriter el (la) guionista **5.3**

to scuba-dive bucear **2.3**

sculptor el (la) escultor(a) **II**

sculpture la escultura **II**

sea el mar **I**

seagull la gaviota

search la búsqueda
  search engine el servicio de búsqueda **6.3**

to search buscar **I**

season la temporada **1.2**

seasons las estaciones **I**

seat el asiento **II**

second segundo(a) **I, II**

secretary el (la) secretario(a) **I, II, 4.2**

security la seguridad **II**

to see ver **I, II**
  May I see...? ¿Me deja ver...? **I**

self-portrait el autorretrato **5.1**

to sell vender **I**

semester el semestre **I**

to send mandar
  to send a letter mandar una carta **I**
  to send out echar

sensationalized sensacionalista **6.1**

to separate separar **II, 2.2**

September septiembre **I**

sequin la lentejuela **1.2**

series la serie **II**

serious serio(a) **I**

to serve servir (i) **I, II**

to set the table poner la mesa **I**

setback el tropiezo **3.1**

seven siete **I, II**

seven hundred setecientos(as) **I**

seventeen diecisiete **I**

seventh séptimo(a) **I, II**

seventy setenta **I**

shade la sombra **II**

shadow la sombra **II**

shame la lástima
  What a shame! ¡Qué lástima! **I**

shameless descarado(a) **1.1**

shampoo el champú **I, II**

to share compartir **I, 1.1**

shark el tiburón **2.3**

sharpness la nitidez **6.2**

to shave afeitarse **I**

she ella **I**

sheet la sábana **II**

shell el caracol **II**

shepherd(ess) el (la) pastor(a) **I**

ship el barco **I**

shirt la camisa **I**

shock el susto

shoe el zapato **I**
  high-heeled shoe el zapato de tacón **II**
  shoe size el número **II**
  shoe store la zapatería **I**

shopping
  to go shopping ir de compras **I**
  shopping center el centro comercial **I**

shore la orilla, **II**

short (height) bajo(a) **I; (length)** corto(a) **I**

shorts los shorts; los pantalones cortos

should deber **I**

shoulder el hombro **II**

shoulder bag el bolso **1.2**

shout el grito

show
  game show el programa de concurso **6.1**
  talk show el programa de entrevistas **6.1**

to show mostrar (ue) **II**

shower

to take a shower ducharse I, II
sick enfermo(a) I
   the sick los enfermos 2.1
sickly enfermizo(a)
sickness la enfermedad II
sidewalk la acera II
sign el letrero II
signature la firma II, 4.1
silk la seda 1.2
silly tonto(a)
silver la plata I
silverware los cubiertos II
similar to semejante a 1.1
simple sencillo(a) I, II
to sing cantar I
   to sing in the chorus
      cantar en el coro II
singer el (la) cantante II;
   (of flamenco) el (la) cantaor(a)
   5.1
sink (bathroom) el lavabo II
sister la hermana I
   sister-in-law la cuñada II
to sit down sentarse(ie) II
site el sitio 6.3
situation la situación II
six seis I
six hundred seiscientos(as) I
sixteen dieciséis I
sixth sexto(a) I, II
sixty sesenta I
size (clothing) la talla II; **(shoe)** el
   número II
el tamaño 6.3
to skate patinar I
skateboard la patineta I
to skateboard andar en patineta I
skates los patines I
to ski esquiar I
skin la piel II
skirt la falda I
skunk el zorrillo 2.3
sky el cielo 2.2
slave el (la) esclavo(a) 3.3
to sleep dormir (ue) I
sleeping bag
   el saco de dormir II, 2.3
sleepy
   to be sleepy tener sueño I
slender esbelto(a) 1.1
slow lento(a) I
slowly lentamente I, II
small pequeño(a) I

to smell of oler (ue) a
to smile sonreírse (i) II
smog el smog II, 2.2
snack la merienda I
to snack merendar (ie) I
snake la serpiente II, 2.3
to snorkel nadar con tubo de
   respiración 2.3
snow la nieve I
to snow nevar (ie) I
so entonces I
So-so. Regular. I
so that para que 3.1
soap el jabón I
soap opera la telenovela II
soccer el fútbol I
   soccer ball el balón 1.1
sociable sociable II
social sciences
   las ciencias sociales 2.1
social service
   el servicio social 2.1
social studies
   los estudios sociales I
social worker
   el (la) trabajadora social 4.3
sock el calcetín I
sofa el sofá I, II
soft drink el refresco I
software el software 6.3
solemn solemne 3.3
solution la solución 2.1
some alguno(a) I
someone alguien I
   to know (be familiar with)
      someone conocer a alguien I
something algo I
sometimes a veces I, II, 4.2
son el hijo I
son(s) and daughter(s) los hijos I
soon pronto I
sorry
   to be sorry that sentir (ie) que II
   I'm sorry. Lo siento. I
   I'm sorry, but…
      Lo siento mucho, pero… 2.1
soul el alma (fem.)
sound el sonido I; el son 5.1
to sound sonar (ue) 3.3
soup la sopa I
soup kitchen
   el comedor de beneficencia 2.1
source la fuente

south el sur II
Spaniard el (la) español(a) II
Spanish el español I
to speak hablar I, II
   May I speak with…?
      ¿Puedo hablar con…? I
speaker el altoparlante 6.2
special especial I
to specialize especializarse 4.1
specially especialmente I
specialty of the house la
   especialidad de la casa II
species las especies 2.2
speech el discurso 3.1
to spend gastar II
spicy picante I
spider monkey el mono araña 2.3
spoiled mimado(a) 1.1
to sponsor patrocinar 3.2
spoon la cuchara I
sport el deporte I
   to play sports
      practicar deportes I
sporting goods store
   la tienda de deportes I
spreadsheet la hoja de cálculo 6.3
spring la primavera I
square cuadrado(a) 1.1
squid los calamares I
stadium el estadio I
stage floor el tablado 5.1
staircase las escaleras II
stairs las escaleras II
stand la parada II
to stand (endure) resistir
to stand out destacarse 1.2
star la estrella I
to start comenzar (ie) II
   to start the day amanecer 2.3
station el canal II
stationery store la papelería I
statistics las estadísticas 4.3
to stay (at) hospedarse (en) II
steak el bistec I
to steal robar II
stepbrother el hermanastro II
stepfather el padrastro I
stepmother la madrastra II
stepsister la hermanastra II
stereo equipment
   el equipo estereofónico 6.2
stick el palo
still todavía I

still life la naturaleza muerta 5.1
stingy tacaño(a) II
stock exchange
    la bolsa de valores 4.3
stomach el estómago I, II
stone la piedra II
stop la parada II
to stop parar II
    to stop doing something
        dejar de
storage battery el acumulador I
store la tienda I
storm la tormenta I
story la historia II; (of building) el
    piso II
    short story el cuento 5.1
    short story writer
        el (la) cuentista 5.3
stove la estufa I, II
straight (hair) lacio II
straight ahead derecho I
strange raro(a) II; extraño(a)
strangers los desconocidos
strawberry la fresa II
street la calle I
stress el estrés II
to stretch estirarse II
striped con rayas I
stripes las rayas II
strong fuerte I
student el (la) estudiante I
to study estudiar I
    to study martial arts
        estudiar las artes marciales II
stuffed animal
    el muñeco de peluche II
style la moda 1.2; el estilo 5.3
subject
    la materia I; el tema II
subscriber el (la) abonado(a)
subway el metro I
success el éxito 3.1
successful
    to be successful tener éxito II
suddenly de repente II
sugar el azúcar I
to suggest sugerir (ie) II
suit el traje II
suitcase la maleta II
summer el verano I
sun el sol I
to sunbathe tomar el sol I
Sunday domingo I, II

sunglasses las gafas de sol I
sunny soleado(a) II, 2.3
    It is sunny. Hace sol.; Hay sol. I
sunscreen la loción protectora II
suntan lotion el bronceador I
supermarket el supermercado I
    to go to the supermarket
        ir al supermercado I
supper la cena I
    to have supper cenar I
support el apoyo 3.1
    to support each other
        apoyarse 1.3
supported (by) respaldado(a) 6.2
supporter el (la) proponente 3.3
sure seguro(a)
    I'm (not) sure.
        (No) Estoy seguro(a). 3.3
    not to be sure that
        no estar seguro (de) que 3.2
to surf the Internet
    navegar por Internet 1.2
surface la superficie
surfing el surfing I
surname el apellido I
surplus sobrante
surprise la sorpresa I, II
to surprise sorprender I, II
surrounded rodeado(a)
surrounding area el contorno
to sweat sudar II
sweater el suéter I
sweats las sudaderas 1.2
to sweep the floor barrer el suelo
    I; barrer el piso II
sweet dulce I
    sweet roll el pan dulce I
to swell hincharse
to swim nadar I
swimming pool la piscina I
sword espada
symbolic simbólico(a) 5.3
symbolism el simbolismo 5.3

T

T-shirt la camiseta I
table la mesa I, II
    to clear the table quitar la mesa I
    to set the table poner la mesa I
tablecloth el mantel II
taco el taco II

taco restaurant la taquería II
to take tomar I; sacar II; cursar
    to take a bath bañarse I
    to take a shower ducharse I
    to take a swimming class
        tomar un curso de natación II
    to take a walk, stroll, or ride
        dar una vuelta II
    to take advantage of aprovechar
    to take along llevar I
    to take apart desarmar 1.3
    to take care of cuidar de I, 2.1
    to take charge of
        encargarse de 4.1
    to take into account
        tomar en cuenta 6.2
    to take off one's clothes
        quitarse la ropa II
    to take out the trash
        sacar la basura I
    to take pictures
        sacar fotos I
    to take risks correr riesgos 4.1
talent el talento II
to talk hablar I, II
talk show el programa de
    entrevistas
tall alto(a) I
tamale-like mixture of plantain,
    yuca and meat el pastel 3.2
tambourine la pandereta 5.1
tango (Latin music) el tango 5.2
tape recorder la grabadora I, 6.2
tapestry el tapiz 5.1
taste el sabor II
tasty rico(a); sabroso(a) I, II
taxi el taxi I
taxi driver el (la) taxista I, II 4.2
tea el té I
to teach enseñar I
teacher el (la) maestro(a) I
team el equipo I
technician el (la) técnico II, 4.2
technique la técnica 5.2
telecommunications
    las telecomunicaciones 4.3
telephone el teléfono I
    cellular telephone
        el teléfono celular 6.2
    cordless telephone
        el teléfono inalámbrico 6.2
    to telephone each other
        telefonearse 1.3

television
cable television
la televisión por cable **6.1**
portable television
el televisor portátil **6.2**
satellite television
la televisión por satélite **6.1**
television guide la teleguía **6.1**
television mini-series
el teledrama **6.1**
television program
el espacio de la televisión
television series la teleserie **6.1**
television set el televisor **I**
to watch television
ver la televisión **I**
to tell decir (i) **I, II**; contar (ue) **I**
Don't tell me! ¡No me digas! **II**
to tell each other gossip
contarse (ue) chismes **1.3**
to tell each other secrets
contarse (ue) secretos **1.3**
Tell him or her to call me.
Dile/Dígale que me llame. **I**
to tell jokes
contar (ue) chistes **II**
temperature la temperatura **I**
temple el templo **5.2**
ten diez **I**
tennis el tenis **I**
tent la tienda de campaña **II, 2.3**
tenth décimo(a) **I, II**
terrain el terreno
altitude of terrain
la altura del terrano
terrible terrible **I**
test el examen **I**
textiles los textiles **4.3**
to thank agradecer **3.1**
Many thanks. Mil gracias. **3.1**
Thank you. Gracias. **I**
Thanksgiving el Día de Acción, de
Gracias **3.2**
that aquello, eso **I, 5.1**
that ese(a) **I, 5.1**
that (over there)
aquel(la); aquello **I, 5.1**
that one ése(a) **I, 5.1**
that one (over there) aquél(la) **I, 5.1**
theater el teatro **I**
theatrical production
la obra de teatro **II**
their su **I**
theirs suyo(a) **II**

theme el tema **II**
then entonces **I**
there allá/allí **I**
there is/are hay **I**
there was/were había **II**; hubo **II**
they ellos(as) **I**
thief el ladrón (la ladrona) **II**
thin delgado(a) **I**
thing la cosa **I**
to think pensar (ie) **I**; creer **I**
Do you think so? ¿Tú crees? **II**
I don't think that…
No creo que… **2.3**
I think so. / I don't think so.
Creo que sí/no. **I**
third tercero(a) **I, II**; tercio **4.3**
thirsty
to be thirsty tener sed **I**
thirteen trece **I**
thirty treinta **I**
this este(a); esto **I, 5.1**
this one éste(a) **I, 5.1**
thousand mil **I**
threatening amenazador(a) **5.3**
three tres **I**
three hundred trescientos(as) **I**
throat la garganta **II**
throughout a lo largo de
to throw out, away echar **II, 2.2**
thunder el trueno **II, 2.3**
thunderbolt el rayo **II, 2.3**
Thursday jueves **I, II**
ticket el boleto **II**
tight apretado(a) **II**
tile la baldosa
time el tiempo **I**
(At) What time is…?
¿A qué hora es…? **I**
free time el tiempo libre **I**
on time a tiempo **II**
to spend time with friends
pasar un rato con los amigos **I**
What time is it? ¿Qué hora es? **I**
tip la propina **I**
How much do I tip?
¿Cuánto le doy de propina? **I**
to leave the tip
dejar la propina **I**
tired cansado(a) **I**
to get tired cansarse de **II**
title el título **5.3**
to a
to the left (of)
a la izquierda (de) **I**

to the right (of)
a la derecha (de) **I**
toast el brindis **3.1**
to make a toast brindar **3.1**
today hoy **I, II**
Today is… Hoy es… **I**
What day is today?
¿Qué día es hoy? **I, II**
together juntos **I**
to get together reunirse **II**
tolling of the bell
la campanada **3.2**
tomato el tomate **I**
tomorrow mañana **I**
See you tomorrow.
Hasta mañana. **I**
Tomorrow is… Mañana es… **I**
too también **I**
too much demasiado(a) **I, II**
tooth el diente **I, II**
toothbrush
el cepillo de dientes **I, II**
toothpaste la pasta de dientes **I, II**
toucan el tucán **II, 2.3**
to touch tocar
tough duro(a) **I**
tourism el turismo **II**
toward hacia **II**
towel la toalla **I**
town el pueblo **I**
town square la plaza **I**
toy el juguete **I, II**
toy store la juguetería **II**
tradition la tradición **5.2**
traditional tradicional **I, II**
traffic el tráfico **I**
traffic light/signal el semáforo **II**
trail el sendero **II**
train el tren **I**
to train entrenarse **II**
training la capacitación **II**, la
formación **4.1**; el entrenamiento
4.2
to translate traducir **II**
translator el (la) traductor(a) **4.3**
trash la basura **I**
trash can el basurero **II, 1.3**
to travel viajar **I**
travel agent
el (la) agente de viajes **II**
traveler's checks
los cheques de viajero **II**
to treat tratar **II**
I'll treat you. Te invito. **I**

tree el árbol **I, 2.1**
trendy thing to do la onda
triangular triangular **1.1**
trick el engaño
trillion billón **4.3**
trip el viaje **I**
trouble
> **The trouble is that…**
>> Lo malo es que… **II**
trumpet la trompeta **5.1**
truth la verdad **I**
> **It's not true that…**
>> No es verdad que… **II**
> **It's true.** Es verdad. **I**
to try to tratar de
Tuesday martes **I**
tuna el atún **II**
turkey el pavo **3.2**
to turn doblar **I;** girar **II**
> **to turn off** desconectar **1.3**
> **to turn off the light**
>> apagar la luz **I**
> **to turn on** encender (ie) **1.3**
turtle la tortuga **I, 2.3**
twelve doce **I**
twenty-one veintiuno **I**
twenty veinte **I**
twins los (las) gemelos(as) **II**
two dos **I**
two hundred doscientos(as) **I**
typical típico(a) **3.2**
typically típicamente **II**

ugly feo(a) **I**
umbrella el paraguas **I**
> **beach umbrella**
>> la sombrilla de playa **II**
uncertainty unquietud
uncle el tío **I**
uncle(s) and aunt(s) los tíos **I**
uncomfortable incómodo(a) **1.2**
under(neath) debajo (de) **I, II**
to understand
> comprender; entender (ie) **I**
understanding comprensivo(a) **1.1**
unequalled inigualable **6.2**
unforgettable inolvidable **3.2**
unfortunately
> desafortunadamente **II**

unique único(a) **1.2**
to unite unir
university la universidad **II**
university degree
> la licenciatura **4.1**
unless a menos que **3.1**
unoriginal derivado(a) **5.3**
unpleasant desagradable **1.1**
to unplug desenchufar **1.3**
until hasta (que) **I, II, 2.3**
up arriba **I, II**
URL el Localizador Unificador
> de Recursos (LUR) **6.3**
Uruguayan uruguayo(a) **II**
to use usar **I, II**
> **to be used as** servir (i) de
useful útil **II**
useless inútil **II, 2.2**
user el (la) usuario(a) **6.3**

to vacuum pasar la aspiradora **I, II**
vacuum cleaner la aspiradora **I**
vain vanidoso(a) **1.1**
Valentine's Day
> el día de la Amistad **3.2**
valley el valle **II, 2.2**
valuable precioso(a) **5.2**
to value valorar **2.1**
VCR la videograbadora **I**
vegetable la verdura **I, II**
vegetarian vegetariano(a) **I**
Venezuelan venezolano(a) **II**
very muy **I**
vest el chaleco **II**
veterinarian
> el (la) veterinario(a) **II, 4.2**
victory la victoria **3.3**
video el video **I**
> **video game** el videojuego **I**
> **to rent a video**
>> alquilar un video **I**
> **videocamera** la videocámara **6.2**
> **videocassette recorder**
>> la videocasetera **6.1**
viewer el (la) televidente **II**
village el pueblo **I**
violent violento(a) **II**
violin el violín **5.1**
to visit visitar **I**

voice mail el telemensaje **6.2**
volleyball el voleibol **I**
volunteer el (la) voluntario(a) **II**
to volunteer
> trabajar de voluntario(a) **2.1**
to vote votar **2.1**

to wait for esperar **I**
waiter el mesero **I**
waitress la mesera **I**
to wake up despertarse (ie) **I**
to walk andar **II**
> **to walk the dog**
>> caminar con el perro **I**
wall la pared **I, II**
wallet la cartera **I,** la billetera **1.2**
to want querer (ie) **I, II, 3.1**
> **Do you want to drink…?**
>> ¿Quieres beber…? **I**
> **Do you want to eat…?**
>> ¿Quieres comer…? **I**
> **I want to drink…**
>> Quiero beber… **I**
> **I want to eat…**
>> Quiero comer… **I**
> **I want to leave a message for…**
>> Quiero dejar un mensaje
>> para… **I**
war la guerra
wardrobe el vestuario **1.2**
warm caliente **I**
to wash lavar **I**
> **to wash one's hair**
>> lavarse la cabeza **I**
> **to wash oneself** lavarse **I, II**
> **to wash the dishes**
>> lavar los platos **I, II**
waste el desperdicio **2.2**
wastebasket el basurero **1.3**
watch el reloj **I**
to watch mirar **I**
> **to watch television**
>> ver la televisión **I**
water el agua (fem.) **I**
to water regar (ie) **1.3**
waterfall la catarata
to water-ski esquiar en el agua **1.2**
wave la ola **I, II;** la onda
wavy ondulado(a) **1.1**

way el paso **5.2**
   to open the way abrir el paso
we nosotros(as) **I**
to wear llevar **I**
   What is he/she wearing?
      ¿Qué lleva? **I**
weather el tiempo **I**
   What is the weather like?
      ¿Qué tiempo hace? **I**
weaving el tejido **5.2**
Web page la página-web **6.3**
wedding la boda **II**
Wednesday miércoles **I, II**
to weed desyerbar **1.3**
weeds las malas hierbas **1.3**
week la semana **I**
weekend el fin de semana
weights
   to lift weights levantar pesas **I**
welcome bienvenido(a) **I**
   You're welcome. De nada. **I**
well bien; pues **I**
well-being el bienestar **II**
west el oeste **II**
wet mojado(a) **II**, húmedo(a) **2.3**
what cuál(es); qué **I**
   What (fun)! ¡Qué (divertido)! **I**
   What a mess! ¡Qué lío! **II, 2.2**
   What a shame! ¡Qué lástima! **I**
   What day is today?
      ¿Qué día es hoy? **I**
   What do you recommend?
      ¿Qué me recomienda? **II**
   What is he/she like?
      ¿Cómo es? **I**
   What is your phone number?
      ¿Cuál es tu teléfono? **I**
   What would you like?
      ¿Qué desea(n)? **II**
wheat el trigo **4.3**
wheat flour la harina de trigo
when cuando; cuándo **I, 2.3**
   when I/he/she was young
      cuando era niño(a) **II**
whenever cuando **I**
where dónde; adónde **I**
   Could you tell me where… is?
      ¿Puedes (Puede usted)
      decirme dónde queda…? **I**
   Where are you from?
      ¿De dónde eres? **I**
   Where is he/she from?
      ¿De dónde es?

   Where will it take place?
      ¿Dónde tiene lugar?
which (ones) cuál(es) **I**
while mientras **II**
   quite a while un buen rato
white blanco(a) **I**
who quién(es) **I**
   Who are they? ¿Quiénes son? **I**
   Who is it? ¿Quién es? **I**
   Whose is…? ¿De quién es…? **I**
why por qué **I**
   That's why. Con razón. **I**
   Why not? ¿Por qué no? **2.1**
wicker rocking chair
   el columpio de mimbre
wide ancho(a) **I, II**
wife la esposa **I**
wild salvaje **II, 2.3**
   wild animal life
      la fauna silvestre **2.2**
   wild boar el jabalí
   wild plant la planta silvestre **II**
   wild plant life
      la flora silvestre **2.2**
   wildlife refuge el refugio de
      vida silvestre **2.3**
willing
   to be willing to
      estar dispuesto(a) a **4.1**
to win ganar **I**
wind el viento **I**
window
   la ventana **I, II**; la ventanilla **II**
to windsurf
   navegar en tabla de vela **1.2**
windy
   It is windy.
      Hace viento; Hay viento. **I**
wing el ala (fem.)
winner el (la) ganador(a) **I, II**
winter el invierno **I**
with con **I**
   with me conmigo **I**
   with you contigo **I**
without sin **I**
wolf el lobo **II**
woman la mujer **I**
wonderful genial **1.2**
wonderfully a las mil maravillas
wool la lana **I, 1.2**
work of art la obra **II**
to work trabajar **I**; funcionar **II, 6.2**
worked labrado(a) **5.2**

worker el (la) obrero(a) **II**
workshop el taller **I**
world el mundo **I**
World Wide Web
   la red mundial **6.3**
worried preocupado(a) **I**
   to be worried about
      preocuparse por **II**
to worry preocuparse **II**
   Don't worry!
      ¡No te preocupes! **I**
worse peor **I**
   worse than peor que **II, 6.3**
the worst lo peor **1.1**
worthwhile
   to be worthwhile
      valer la pena **3.1**
wrist la muñeca **II**
to write escribir **I**
writer el (la) escritor(a) **I, II**
   short story writer el (la)
   cuentista

x-ray la radiografía **II**

year el año **I**
   He/She is… years old.
      Tiene… años. **I**
   school year el año escolar **II**
yellow amarillo(a) **I**
yes sí **I**
   Yes, gladly.
      Sí, con mucho gusto. **2.1**
   Yes, I would love to.
      Sí, me encantaría. **I**
yesterday ayer **I, II**
yet todavía **I**
yogurt el yogur **I, II**
you tú **(familiar singular)**,
   usted **(formal singular)**,
   ustedes **(plural)**,
   vosotros(as) **(familiar plural) I**
young joven **I**
young people los jóvenes **2.1**

younger menor **I**
    younger than menor que **II**
your su **(formal)**, tu **(familiar)**,
    vuestro(a) **(plural familiar) I**
yours tuyo **(familiar)**, suyo(a)
    (formal), vuestro(a) **(plural**
      familiar) II

zero cero **I, II**

Índice

a, 457, 459
+ *person*, 306, 360
abrir, past participle, 46
abilities, personal, 265, 275
accents
in commands, 116
in imperfect subjunctive, 192
in present progressive, 264
accessories. *See* clothing and accessories
el Aconcagua, 15
actions, past, 12, 42, 43, 46, 238
address, forms of, 120, 267
adjectives
demonstrative, 336
descriptive, 34, 35, 36, 38, 39, 53
possessive, 308
adverbs
of frequency, 284
of location, 433, 445
advertisements, 426–427, 428–429, 431
affirmative expressions, 284, 297
agreement
of demonstrative adjectives and pronouns with nouns, 336
of reflexive pronoun and verb, 84
el Alcázar de Colón, 1
algún/alguna, 284
Allende, Isabel, 314–315, 384
almorzar, preterite tense, 16
andar
preterite tense, 24
progressive with, 266
Andes region, 14, 15, 22–23
animals, 7, 68, 152, 153, 154, 155, 156, 168–169, 173
antonyms, 34
apologizing, 438
aquel (-la/-los/-las), 336
aquello, 336
-ar verbs

command forms, 114, 116
conditional tense, 119
imperfect tense, 42
past participles, 46
present subjunctive, 136, 137
present participle, 264
present tense, 4
preterite tense, 12
architecture, 366–367
Argentina, 14, 15, 230, 231, 305, 325
Arias, Óscar, 103
Arreola, Juan José, 164
Arroyo, Fian, 78
arts and artists, 335, 349, 361, 371
Latin American, 102, 103, 250, 324
preColumbian, 103, 353, 354
Spain, 325, 330–333, 344–345
asking
for assistance, 113
questions, 262
Atahualpa, 353
aunque, 412, 432
Aztecs, 317, 35

«Baby H.P.», 166–167
Baja California, 7
el balboa, 140
Barragán, Luis, 366–367
barrios, in U.S., 3, 40
Bemberg, María Luisa, 391
Berni, Antonio, 250
Bolívar, Simón, 399
Bolivia, 22, 23, 399, 464–465
Borges, Jorge Luis, 270–271
Botero, Fernando, 324
bueno, 454
buscar
command forms, 114
present subjunctive, 136
preterite tense, 16

... que, 210
business, types of, 311, 321

c –> qu verbs, 16, 114, 136, 210
caer
command forms, 114
present subjunctive, 138
present tense, 8
camping, 163, 173
-car verbs
command forms, 114
present subjunctive, 136
careers. *See* professions
Caribbean countries, 10–11, 176–177
La casa de Bernarda Alba, 388, 389
celebrations
graduation, 182–183, 184–185, 188, 190, 193
New Year's, 204–205
patriotic, 226, 227, 242–243, 247
cellular telephones, 399, 440–441
Central America, 102, 103, 154–155, 161
currencies, 140, 300
ecotourism, 152–153
history, 6–7
-cer verbs, subjunctive, 163, 233
cerrar, present subjunctive, 139
certainty, expressing, 160, 238
characteristics, physical, 34, 35, 36, 53
Chayanne, 215
Chile, 14, 15, 292, 301
chores, household, 79–80, 99
Cisneros, Sandra, 92
citizenship, 108–109, 110–111, 113, 117
clothing and accessories, 34, 35, 56–57, 58–59, 61, 70–71, 75
cognates, 108
Colombia, 1, 23, 354

el colón, 140
colors, 34, 35, 36, 61
Columbus, Christopher, 227,
    228–229, 234, 242
comer
    future tense, 65
    imperfect tense, 42
    past participle, 46
    present participle, 264
    present subjunctive, 136
    present tense, 4
    preterite tense, 12
¿cómo?, 254
Como agua para chocolate, 387
commands
    formation of, 114, 116
    pronouns with, 115, 358, 360,
        380, 382
community participation, 108, 109,
    117
el compadrazgo, 87
comparisons, 53, 452
computers, 446–447, 448–449, 451,
    467
con, 457
conditional
    perfect tense, 289
    sentences, 214
    tense, uses of, 119, 414
conjunctions, 201, 432
    of time, 162, 173, 234
    + *subjunctive,* 190, 234, 432
connecting cultures
    architecture, 366
    disguises, 218–219
    ecotourism, 168–169
    fashion, 70
    Internet, 464
    language and culture, 316
    literacy, 144
    movies, 390
    music, 94
    national celebrations, 242
    planning for the future, 292
    technology, 440
conocer
    present subjunctive, 138
    present tense, 8
    preterite vs. imperfect, 410
    vs. **saber,** 83
¿Conoces a alguien que… ?, 210
el Cono Sur, 14–15, 250–251, 361
contrasting, 422, 445

contribuir, present subjunctive,
    137
Cortés, Hernán, 352
Costa Rica, 7, 105, 152, 154, 157,
    168–169
coursework, 256–257, 275
¿cuál?, 262, 338
el/la cual, 362
cuando, 162, 412
Cuba, 10–11, 49, 176
cubrir, past participle, 46
currencies, 140, 300, 312
cuyo, 362
cyberspace, 448, 450, 469

D

Dalí, Salvador, 325
dances, regional, 331, 349, 359, 371
dar
    commands, 116
    present subjunctive, 138
    present tense, 8
    preterite tense, 24
dates, giving, 40
de, 457, 459
de la Hoya, Oscar, 28
de la Renta, Oscar, 70–71
decir
    commands, 116
    conditional tense, 119
    future, 65
    past participle, 46
    present subjunctive, 138
    preterite tense, 24
demonstrative adjectives and
    pronouns, 336
des-, 78
describing people, 34, 35, 36, 38,
    39, 40, 53
después (de) que, 412
direct object pronouns, 358
disagreement/denial, expressing,
    212, 232
Dominican Republic, 1, 10–11, 190,
    191, 227, 234, 237, 242–243
dormir
    present subjunctive, 139
    present tense, 4
    preterite tense, 20
double

negative, 284
    object pronouns, 380
doubt, expressing, 160, 212, 223,
    232, 238
Duarte, Juan Pablo, 237, 243

E

e –> i/ie stem-changing verbs
    present subjunctive, 139, 210
    present tense, 4
    preterite tense, 20
«Ébano real», 196, 197
ecology. *See* environment,
    pollution
economics, of South America, 300,
    301, 302
ecotourism, 152–153, 168–169
Ecuador, 23
educar, subjunctive, 210
education in Hispanic countries,
    277, 285
El Salvador, 7, 103, 152, 154
elections, 108, 110–111
electronics, 426–427, 431, 445,
    448–449. *See also* computers
emotions, expressing, 84, 158, 232.
    See also feelings
en, 457, 459
    cuanto, 412
environment, 108, 131, 132, 135,
    142, 143, 149, 161. *See also*
    ecotourism
-er verbs
    command forms, 114, 116
    conditional tense, 119
    imperfect tense, 42
    past participles, 46, 264
    present participles, 264
    present subjunctive, 136, 139
    present tense, 4
    preterite tense, 12
escribir
    past participle, 46
    present participle, 264
    present subjunctive, 136
ese (a/os/as), 336
eso, 336
Esquivel, Laura, 387
estar
    commands, 116

present subjunctive, 138
preterite tense, 24
vs. **ser,** 40
este (a/os/as), 336
esto, 336
expressing
certainty, 238, 412
contrasts, 432, 445
disagreement/denial, 212, 232
distance, 457
doubt, 160, 173, 212, 223, 232, 238, 412
emotions, feelings, 84, 99, 158, 173, 232
frequency, 284
likes/dislikes, 62, 63, 75
location, 40, 63, 433
motion, 457, 459
opinions, 63, 108, 109, 136, 137, 412
origin, 40
possession, 40, 308
speculation, 68, 119
state or condition, 40
thanks, 209, 223
uncertainty, 160, 162, 210, 238
what might happen, 68
what might have been, 289
what will (may) have happened, 310
wishes, 188, 232, 286

fabrics, 61, 75
fashions. *See* clothing and accessories
feelings, 75, 84, 99
Ferré, Rosario, 177
film making, 384, 395
directors, 390–391
fine arts, 330–333, 335, 337, 339–341, 343
flamenco, 331, 335
food, 304, 324, 399
fruit, 177
Mexican, 29, 103
Puerto Rican, 204, 217, 223
frequency, expressing, 284
future perfect tense, 310
future tense, 65, 68, 414

g –> gu verbs, preterite tense, 16
g –> j, 210
-gar verbs
command forms, 114
present subjunctive, 136, 233
García, Cristina, 48
García Lorca, Federico, 388
García Márquez, Gabriel, 462–463
geography, of South America, 300, 301
-ger verbs, subjunctive, 189, 210, 233
gestures, 265
godparents, 87
Goya, Francisco de, 330
graduation, 182–183, 184–185, 188, 190, 193, 201
Guarga, Rafael, 250
Guatemala, 103, 152, 154, 354
guayabera, 3
Guerra, Juan Luis, 177
Guillén, Nicolás, 196
gustar, verbs like, 62

haber
conditional tense, 119, 289
future tense, 65, 310
past subjunctive, 286
in present and past perfect tenses, 46, 141, 286
present subjunctive, 138
hablar
imperfect subjunctive, 192
imperfect tense, 42
past participle, 46
present subjunctive, 136
present tense, 4
preterite tense, 12
hacer
command forms, 114
conditional tense, 119
future tense, 65
past participle, 46
present subjunctive, 138
present tense, 8

preterite tense, 24
handcrafts, 103
hasta que, 412
hay. *See* **haber**
¿Hay algo/alguien que… ?, 210
help, offering and accepting, 113
historic events, 226, 227
holidays, 213, 223
patriotic, 226, 227, 242–243, 247
Honduras, 7, 103, 153, 154
housework, 78–79, 99
humor devices, 166–167

identifying, 40
imperfect
subjunctive, 192, 232
tense, 42
vs. preterite tense, 43, 410
impersonal
expressions, 137, 149, 188
constructions with **se,** 89
Incas, 353, 360, 398
indefinite clauses, 210, 234
indicative vs. subjunctive, 238, 412, 432
indirect object pronouns, 62, 360
interactions, 44, 53, 86, 99
Internet, 448, 456, 460, 464–465, 469
interrogative words, 262, 275
invitations, 416
ir
a + *infinitive,* 65, 116, 414
imperfect subjunctive, 192
imperfect tense, 42
progressive with, 266
present subjunctive, 138
preterite tense, 24
-ir verbs
command forms, 114, 116
conditional tense, 119
imperfect tense, 42
past participles, 46
present participles, 264
present subjunctive, 136, 139
present tense, 4
preterite tense, 12
irregular verbs
command forms, 114, 116

conditional tense, 119
future tense, 65
imperfect subjunctive, 192, 232
imperfect tense, 42
past participles, 46
present subjunctive, 136, 137, 138, 139
present tense, 8
preterite tense, 16, 20, 24
la Isla de Ometepe, Nicaragua, 158
Izquierdo, María, 102

jeroglíficos, 370, 371
job hunting, 263, 280–281, 283, 308
jota, 331

La casa en Mango Street, 92, 93
el lago Titicaca, 23
Latin America
    artists, 102, 324, 399
    authors, 166, 177, 196, 240, 270, 314, 384, 386, 387, 388, 462–463
    dances, 359, 371
    native languages, 317, 356
    preColumbian civilizations, 352–353, 354
Legorreta, Ricardo, 366–367
leisure activities, 67, 75
likes and dislikes, 62, 63, 75
listening strategies
    analyzing the appeal in radio, 428
    using advance knowledge of the topic, 332
    improving your auditory memory, 354
    anticipating, comparing, and contrasting issues, 110, 132
    identifying important computer vocabulary, 450
    recognizing descriptions, 36
    distinguishing admiring and

critical remarks, 58
    evaluating discussions, 376
    identifying key information for careers, 280
    observing interview techniques, 206
    listening and taking notes, 228
    determining your purpose for listening, 154
    keeping up with what is said, 406
    summarizing, 258
    transitions, 184
literacy, 144–145
literature
    describing, 374–375, 377, 379, 395
    genres, 342, 349
llegar, preterite tense, 16
location, 40, 63, 433, 445
lo que, 364
Los Angeles, Greater Eastside, 91

Machu Picchu, 353, 354
Madrid, **el Prado,** 330, 332, 333, 341
malo, 454
maps
    Caribbean, 176–177
    Central America, 102–103, 154, 316–317, 324
    Christopher Columbus's voyages, 228–229
    Costa Rica, national parks, 157
    Hispaniola, 242
    South America, 251, 317, 324–325, 399
    Spain, 316, 325
    United States, 28–29
Martí, José, 240
más... que, 454
masks, 218–219
mate, 251
material. *See* fabric
Matute, Ana María, 345
Mayans, 353, 370
Menchú, Rigoberta, 122–123
menos... que, 454
mentir, present subjunctive, 139

Mexican cuisine, 29
Mexico, 102, 103, 354
    environmental and social service, groups, 118, 142
    history, 6
Miami, 3
Miró, Joan, 1
Mission District, San Francisco, 3
Moctezuma, 352
monetary units. *See* currencies
Montevideo, Uruguay, 1
morir, past participle, 46
movies, 404–405. *See also* film making
el Museo del Prado. *See* **el Prado**
music, 94–95, 107, 176, 177, 211, 215, 325, 339, 349
musical instruments, 94, 176, 251, 337, 349

names, use of, 50, 120, 267, 269
native languages, 316, 317, 356
necesitar... que, 210
negative
    commands, 114, 116, 136, 382
    expressions, 284, 297
neuter demonstratives, 336
New Year's, 204–205
New York, Hispanics in, 3
nicknames, 44
Nicaragua, 103, 153, 154
    la Isla de Ometepe, 158
    literacy program, 144–145
ningún/ninguna, 284
no hay... que + *subjunctive,* 210
nominalization, 382, 385
nonexistent, 210
nosotros commands, 116
numbers, 305

o -> u/ue stem-changing verbs
    present subjunctive, 139
    present tense, 4
    preterite tense, 20
object pronouns, 306, 358, 360, 380

placement of, 46, 264, 382
Ochoa, Ellen, 29
ofrecer
  command forms, 114
  present subjunctive, 138
oír
  command forms, 114
  present subjunctive, 138
  ojalá que, 286
opinions, expressing, 63, 108, 109, 136, 137, 412

Panama, xxxiv, 105, 153, 154, 354
para vs. **por,** 63
Paraguay, 14, 293, 301
participles
  past, 46
  present, 264
past
  participles, 46
  perfect subjunctive, 286
  perfect tense, 46
  progressive tense, 268
pastimes, 67, 75
patriotism, 226, 227, 231, 237, 242–243, 247
Paula, 314, 315
pedir
  present subjunctive, 139
  present tense, 4
  preterite tense, 20
pensar, present tense, 4
people, describing, 34, 35, 36, 38, 39, 53
pero vs. **sino,** 435
personality traits, 39, 53
Peru, 23, 378. *See also* Machu Picchu
pets, 68
physical characteristics, 34, 35, 36, 39, 53
Pizarro, Francisco, 353
poder
  conditional tense, 119
  future tense, 65
  preterite tense, 24, 410
  preterite vs. imperfect tense, 410
la poesía negra, 196, 197
poetry, 196, 197, 240–241
politics, 236, 247

pollution, 131, 135, 149. *See also* environment
Ponce de León, Juan, 11
poner
  command forms, 114
  conditional tense, 119
  future tense, 65
  past participle, 46
  present subjunctive, 138
  present tense, 8
  preterite tense, 24
por vs. **para,** 63
possessive pronouns, 308, 321
el Prado, 330, 332, 333, 341
preColumbian civilizations, 352–353, 354
prefixes, 78
La Prensa, 29
prepositions, 457, 459
  of location, 433, 445
present
  participles, 264
  perfect subjunctive, 141, 158, 160, 232
  perfect tense, 46, 141
  progressive tense, 40, 264, 266
  subjunctive, 136, 138, 160, 210, 232
  tense, 4, 65
preterite tense, 12, 16, 20
  irregular, 24
  vs. imperfect, 43, 410
producir, preterite tense, 24
professions, 254–255, 269, 276, 278–279, 280, 281, 283, 288, 291, 297, 307, 321
pronouns
  demonstrative, 336
  object, 62, 306, 350, 360, 380, 382
  placement of, 46, 84, 115, 264, 358, 360, 380
  possessive, 308, 321
  reflexive, 84
  relative, 341, 362, 364
  subject, 306
Prud'homme, Emilio, 237
Puente, Tito, 94–95
Puerto Rico
  authors, 177, 386
  food, 204, 217
  history, 10–11, 177, 216, 226
  holidays, 212, 226
  masks, 218–219

  music, 211
  New Year's in, 204–205

quantities, 305, 321
que, 341, 362
¿qué?, 262, 338
querer
  conditional tense, 119
  future tense, 63
  preterite tense, 24
  preterite vs. imperfect, 410
  … que, 210
el quetzal, 140
quien(es), 341

reading strategies
  comparing famous authors, 344
  complex sentences, 122
  contrasts, 92
  drama, 388
  fact vs. interpretation, 418
  humor, 166
  identity and fantasy, 270
  metaphors, 196, 240
  paraphrase, 462
  poetry, 240
  speculate about author, 314
  verb tenses, 48
reciprocal verbs, 86
recoger, subjunctive, 210
recommendations, making, 238
reflexive pronouns, 84, 264
reflexive verbs, 84, 86
  commands, 116
regional expressions. *See* Spanish
relative pronouns, 341, 362, 364
Repertorio español, 29
reported speech, 414
requests, making and responding to, 113, 127
resolver, past participle, 46
Reverón, Armando, 399
Reyes, José, 237
romper, past participle, 46
Ruinas de Copán, 103

saber
  commands, 116
  conditional tense, 119
  future tense, 65
  present subjunctive, 138
  present tense, 8
  preterite tense, 24
  vs. **conocer**, 83
salir
  command forms, 114
  conditional tense, 119
  future tense, 65
  present subjunctive, 138
  present tense, 8
  preterite vs. imperfect, 410
salsa music, 211
sardana 331
Saura, Carlos, 390
se
  impersonal, 89
  for unplanned occurences, 437
Segarra, Araceli, 67
seguir
  command forms, 114
  progressive with, 266
la Selva de Darién, xxxiv
sembrar, subjunctive, 210
sentir, preterite tense, 20
sequence of tenses, 415
ser
  commands, 116
  + **de,** 40
  imperfect tense, 42
  present subjunctive, 138
  preterite tense, 24
  vs. **estar,** 40
shopping for electronics, 426–427,
  428, 431, 445
si clause, 214
sino vs. **pero,** 435
social concerns, 108–109, 117, 127,
  144–145
Soñar en Cubano, 48–49
Sosa, Sammy, 88
South America, 14–15, 22–23,
  250–251
  economics, 300–301, 302
Spain, 316, 325
  art, 1, 325, 330, 333, 340, 343

authors, 344–345
cities of, 19
dancing, 331, 339
history, 18–19, 325, 341
Madrid, 330, 332, 348
Spanish
  dialects, 316
  regional expressions, 38, 69, 83,
    112, 122, 136, 216, 266, 290,
    311, 356, 408, 435, 452
speaking strategies
  adding details to descriptions,
    39
  brainstorming, 74
  comparing and evaluating
    computer configurations, 468
  comparing and evaluating films
    in a chat group, 456
  considering factors, 444
  describing celebrations, 231
  describing people and actions,
    39, 52, 222, 231
  discussing books and dancing,
    370, 379
  discussing Latin American
    dance, 370
  encouraging participation, 222
  expressing feelings, 98, 246, 263,
    274
  expressing (lack of) support,
    148
  expressing yourself, 246
  giving advice and best wishes,
    200
  guessing cognates, 305
  identifying general ideas, 126
  interviewing, 206, 283, 296
  making excuses, 438
  naming social problems and
    solutions, 121
  negotiating, 414
  organizing ideas, 348
  reassuring others, 172
  restating, 159
  retelling memories, 422
  reviewing a film, 394
  socializing, 209
  speculating about the past, 320
  using familiar vocabulary, 61
speculation, 68, 119
sports, 67, 75, 152, 153, 194. *See also*
  camping

statistics, 305
stem-changing verbs
  present subjunctive, 139
  present tense, 4
  preterite tense, 20
strategies
  connecting cultures, 70, 94, 144,
    242, 292, 316, 366, 440
  listening, 36, 58, 80, 110, 132,
    154, 206, 228, 258, 280, 332,
    406, 428
  reading, 48, 92, 123, 166, 270,
    314, 344, 354, 388, 418, 444,
    462
  speaking, 39, 52, 61, 74, 98, 121,
    126, 137, 148, 159, 172, 200,
    206, 209, 231, 246, 263, 274,
    283, 296, 305, 320, 348, 370,
    379, 413, 422, 438, 468
  writing, 100, 174, 248, 322, 396,
    470
subject pronouns, 306
subjunctive
  for expressing wishes, 188
  with conjunctions, 190
  imperfect of, 192
  with non-existent and
    indefinite, 210
  for disagreement and denial, 212
  summaries, 230, 232
  past perfect, 286
subjunctive vs. indicative, 238, 430
suggestions, making, 238
summarizing, 414
superlatives, 454
synonyms, 34

los taínos, 177, 216, 317
tan(to)… como, 454
tan pronto como, 412
technology. *See* electronics
tejidos guatemaltecos, 103
el teleférico de Monserrate, 1
telenovelas, 412, 418–419
telephones. *See* cellular telephones
television, 408, 416, 418, 437
  equipment, 413, 423
  schedules, 404–405, 406, 409,
    423

tener
command forms, 114
conditional tense, 119
future tense, 65
present subjunctive, 138
preterite tense, 24, 410
que, 459
Tenochtitlán, 352
textiles, 304, 353, 354
thanks, expressing, 209, 223
¿Tienes algo que...?, 210
Tikal, 364
time
conjunctions of, 162, 173, 234
extent of, 63
of day, 163
telling, 40, 42
titles, professional, 269
Torres, Fina, 392
traer
command forms, 114
present subjunctive, 138
present tense, 8
preterite tense, 24
tú commands, 114

Ud(s). command forms, 114
Unamuno, Miguel de, 344
uncertainty, expressing, 160, 238
United States, xxxiv, 28–29
barrios, 3, 40
Hispanic immigrants, 2, 48

unplanned events, 437, 445
-uir verbs, present subjunctive, 137
Uruguay, 1, 14, 15, 293, 301

Velázquez, Diego de, 330
Venezuela, 23, 399, 418
venir
conditional tense, 119
future tense, 65
present subjunctive, 138
preterite tense, 24
ver
imperfect tense, 42
past participle, 46
present tense, 8
preterite tense, 24
verb tenses
conditional, 119, 214, 414
conditional perfect, 289
future, 65
future perfect, 310
imperfect, 42
past perfect, 46
past progressive, 268
present, 4, 8
present perfect, 46, 141
present progressive, 40, 264
preterite, 12, 16, 20, 24, 43
preterite vs. imperfect, 43, 410
sequence of, 415
subjunctive

imperfect, 192, 232
past perfect, 286
present perfect, 141, 232
present, 136, 138, 139, 232
Versos sencillos 240, 241
vivir
imperfect tense, 42
past participle, 46
present tense, 4
preterite tense, 12
volver
past participle, 46
present subjunctive, 139

weather, 164, 173
wishes, expressing, 188, 232, 286
words within words, 34, 56, 108
workplace
expectations, 287, 290
qualifications, 261, 263, 275
writing, 342, 374–375

z –> c 16, 114, 136
-zar verbs
command forms, 114
present subjunctive, 136

Créditos

Acknowledgments

91 Excerpt from *La casa en Mango Street* by Sandra Cisneros. Copyright © 1984 by Sandra Cisneros. Published by Vintage Español, a division of Random House, Inc. Translation copyright © 1994 by Elena Poniatowska. Reprinted by permission of Susan Bergholz Literary Services, New York. All rights reserved. **94–95** Excerpts and adapted material from "El Legendario Rey del Mambo" by Mark Holston, from *Américas* magazine, volume 42, no. 6, 1990-91. Courtesy of *Américas* magazine, a bimonthly magazine published in English and Spanish by the General Secretariat of the Organization of American States. **95** Cover of *The Best of Tito Puente, el Rey del Timbal* CD, by Tito Puente. Copyright © 1997. Reprinted by courtesy of RCA, Beverly Hills, U.S.A. **167** Excerpts from "Baby H.P." by Juan José Arreola, from *Confabulario*. Copyright © 1952 by Fondo de Cultura Económica. Reprinted by permission of Fondo de Cultura Económica. **176** Cover of *Live in New York* CD, by Los Muñequitos de Matanzas. Copyright © 1992. Reprinted by courtesy of QBADISC, New York, U.S.A. **197** "Ébano real" by Nicolás Guillén, from *Antología Mayor*. Copyright © 1972. Reprinted by permission of Editorial Letras Cubanas, Havana, Cuba. **207** Cover of *Gilberto Santa Rosa... de corazón* CD, by Gilberto Santa Rosa. Copyright © 1997. Reprinted by courtesy of Sony Discos, Inc, Miami, U.S.A. **211** Cover of *Hecho en Puerto Rico* CD, by Willie Colón. Copyright © 1993. Reprinted by courtesy of Sony Discos, Inc, Miami, U.S.A. **300** Cover for *Américas*. Copyright © 1998. Reprinted by courtesy of OAS, Washington, U.S.A. **313** Material for statistics on Paraguay from *Almanaque Mundial 1998*. Used with the permission of Andrés Jorge González Ortega, Editorial Director of *Almanaque Mundial*. **325** Cover of *Chant* CD, by The Benedictine Monks of Santo Domingo de Silos. Copyright © 1994. Reprinted by courtesy of EMI Odeon, Madrid, Spain. **329** *Flamenco*. Copyright © Teatro de Madrid, Spain. **332** Map of Museo del Prado. Reprinted by courtesy of Museo del Prado, Madrid, Spain. **342** *Pajarico*. Copyright © P.C. Filmart, Madrid, Spain. *Mensaka*. Copyright © Tornasol Films, Madrid, Spain. Reprinted by courtesy of Filmart and Tornasol Films. **355** *Arqueología Mexicana*. Copyright © 1998. Reprinted by courtesy of Editorial Raíces, Mexico City, Mexico. **363** Excerpt from *Guía Oficial, Templo Mayor*. Copyright © 1996. Reprinted by courtesy of INAH-JGH Editores, Mexico City. **373** *Libros de Madrid*. Copyright © Ediciones La Librería, Madrid, Spain. **384** *La Nación*. Copyright © 1998. Reprinted by courtesy of La Nación, Santiago, Chile. **389** Excerpt from *La casa de Bernarda Alba* by Federico García Lorca. Copyright © Herederos de Federico García Lorca. Reprinted by permission of Mercedes Casanovas Agencia Literaria, Barcelona. **403** Cover of *TV y Novelas* magazine. Copyright © 1998. Reprinted by courtesy of Editorial América, S.A., Virginia Gardens, U.S.A. **407** *El Comercio*. Copyright © 1996. Reprinted by permission of C.A. *El Comercio*, Quito, Ecuador. **419** Excerpts from "Amor mío: Brillo afuera, oscuridad en casa..." by Friné Sánchez Brandt, from *Venezuela Farándula* No. 1039. Copyright © Revista Ronda C.A. **425** *La Nación*. Copyright © 1998. Reprinted by courtesy of La Nación, Santiago, Chile. **447** *Monkey On Line*. Copyright © 1998. Reprinted by permission of Monkey On Line, Quito, Ecuador. **461** *Virus informático, Diccionario de Microinformática, Disco Duro*. Copyright © 1998. Reprinted by courtesy of Hobby Post, Madrid, Spain. **464** *Tiempos Digital*. Copyright © 1998. Reprinted by permission of *Los Tiempos Digital*, Cochabamba, Bolivia. **465** *Bolivia Net*. Copyright © 1998. Reprinted by permission of Bolivia Net, Bolivia.

Photography

1 *top* Jean-Leo Dugast/ Panos Pictures; *bottom right* Tony Morrison/South American Pictures; *center right* Getty Images; *center left* Robert Frerck/ Odyssey/Chicago; 3 *top left* Sophie Dauwe/Robert Fried Photography; *top right, top center* Richard B. Levine; *center* Courtesy of Kiwanis Club of Little Havana, Miami; *bottom left* Randy Taylor/ Getty Images; *bottom right* RMIP/Richard Haynes; 7 *top left* Tropix/C. Moulds; *top right* Corbis Sygma; *bottom right* Barrett & MacKay Photo; *bottom left* Gerald & Buff Corsi; 11 *top left, top right* Suzanne Murphy-Larronde; *bottom left* Jan Buchovsky / Dave Houser; *bottom right* David Simson / Stock Boston; 15 *top left* DDB Stock Photo; *top right* Gordon Gahanngs / National Geographic Image Collection; *center* George Mobley/ National Geographic Image Collection; *bottom left* Robert Frerck / Odyssey Productions; *bottom right* Image State; *bottom center* Ned Gillette / Image State; 16 Bob Daemmrich / Stock Boston/ PictureQuest; 19 *top left* Walter Bibikow / Viesti Associates, Inc.; *top right* Joe Viesti/ Viesti Associates, Inc.; *bottom right* Paul A. Hein / Unicorn Stock Photo; *bottom left* Eric Vandeville / Getty Images; 23 *top right* R. E. Barber / Visuals Unlimited; *top left* David R. Frazier Photolibrary / Photo Researchers, Inc.; *center right* David Matherly/ Visuals Unlimited; *bottom left* Wolfgang Kaehler; *bottom right* Tony Morrison/ South American Pictures; 26 *top* David Young-Wolff/PhotoEdit; *bottom* Joe Viesti/ Viesti Associates, Inc.; 28 *right* Courtesy of La Prensa; 29 *top left* Kevork Djansezian/AP Wide World Photos, *top right* Ken O'Donoghue, *bottom right* Courtesy of Repertorio Español, NYC, *bottom left* Courtesy of NASA; 30 *top* Spencer Grant/PhotoEdit, *center* Frank Siteman/Stock Boston, *bottom* Nova Online/Getty Images; 31 *top right* Bob Daemmrich/The Image Works; 32-33 Tom Stack & Associates; 33 *inset* Courtesy of Miami Mensual; 34-35 RMIP/Richard Haynes; 37 School Division, Houghton Mifflin Co.; 38 RMIP/Richard Haynes; 43 Bob Daemmrich/Stock Boston; 51 School Division, Houghton Mifflin Co.; 52 Bob Daemmrich/The Image Works; 54-55 Bob Daemmrich Photography; 56-57 RMIP/Richard Haynes; 58 *left* School Division, Houghton Mifflin Co., *right* Robert Burke/Getty Images; 59 *top left* Hughes Martin/Sharpshooters, *top right* Claude Poulet/Getty Images; *bottom right* Telegraph Colour Library/Getty Images, *center* Paul Howell/Getty Images; 60 RMIP/Richard Haynes; 65 Frank Siteman/Stock Boston; 67 *bottom* NOVA Online/Getty Images; 68 Myrleen Ferguson/PhotoEdit; 70 *left* Gerardo Somoza/Corbis Outline, 70 *right*, 71 *top left* Courtesy of Oscar de la Renta; 71 *center* Bill Davila/Retna Ltd., *bottom right* Ted Mehieu/Corbis; 76-77 RMIP/Richard Haynes; 78-79 *background* Ken O'Donoghue; 80-81 RMIP/Richard Haynes; 87 Nancy Sheehan; 88 Elsa Hasch/Allsport; 89 W.B. Spunburg/PhotoEdit/PictureQuest; 94 *top* Dan Dion/Getty Images, *bottom* Spencer Grant/PhotoEdit; 95 *top* PPI Entertainment Group; 102 *right* "Family Portrait," María Izquierdo. Museo de Arte Moderno, Mexico City/Schalkwijk/Art Resource, NY; 103 *top left* Bill Cardoni/Getty Images, *top center* Suzanne Murphy-Larronde/Getty Images, *top right* Mes Beveridge/Visuals Unlimited, *center* J. Becker, *bottom right* Ken O'Donoghue, *bottom left* Gianni Vechiatto; 104 *top right* Gianni Vecchiatto, *center left* RMIP Richard Haynes, *bottom left* Alyx Kellington/DDB Stock Photo, *bottom right* Unicorn Stock Photos; 105 *center left* "Family Portrait," María Izquierdo. Museo de Arte Moderno, Mexico City, Schalkwijk/Art Resource, New York, *center* bradleyireland.com, *center right* Gary Payne/Getty Images, *bottom* Patricia A. Eynon; 106-107 Daniel Aguilar/Hulton Archive/Getty Images; 108 RMIP/Richard Haynes; 109 *left* Michael Newman/PhotoEdit, *right* PhotoEdit; 110-111 *balloons* PhotoDisc, *all others* RMIP/Richard Haynes; 115 PhotoEdit; 118 Jeff Greenberg/Photo Researchers; 120 Bob Daemmrich; 122 *top inset* Sipa Press, 122-123 *background* RMIP/Richard Haynes; 123 *top inset* Gary Payne/Getty Images; 125 Unicorn Stock Photos; 126 *bottom right* Bob Daemmrich/Stock Boston; 127 *top left* Skjold/The Image Works, *top right* Patricia A. Eynon; 128-129 Ulrike Welsch; 130 *top* Courtesy of NASA, *center* Steve Winter/Black Star/PictureQuest; 131 *top left* Robert Frerck/Woodfin Camp, *center left* Rob Crandall/Stock Boston/PictureQuest, *bottom left* School Division, Houghton Mifflin Co., *top right* Visuals Unlimited, *center right & bottom right* Jay Ireland & Georgienne E. Bradley/Earth Images; 140 School Division, Houghton Mifflin Co.; 141 Patricia A. Eynon; 144 Patricia Ramsy/eStock; 145 *pen* School Division, Houghton Mifflin Co., *background* Alyx Kellington/DDB Stock Photo, *bottom left inset* Michelle Bridwell/PhotoEdit; 147 Chris R. Sharp/DDB Stock Photo; 152 *toucan* Gail Shumway/Getty Images, *butterfly* Leroy Simon/Visuals Unlimited, *anteater* Claus Meyer/Black Star/PictureQuest, *skunk* Renee Lynn, *falcon* John S. Dunning/Animals Animals, *deer* Robert A. Luback/Animals Animals, *monkey* Richard K. LaVal/Animals Animals, *others* bradleyireland.com/Earth Images; 153 *snorklers & tropical fish* bradleyireland.com, *whale* W. Ober/Visuals Unlimited, *jaguar* Mes Beveridge/Visuals Unlimited, *snake* Merli/Visuals Unlimited, *pelican* Lynn M. Stone/Animals Animals, *shark* Stuart Westmorland; 155 *top* School Division, Houghton Mifflin Co.; 157 *monkey* Richard K. LaVal, *whale* W. Ober/Visuals Unlimited, *hummingbird & tortoise* Bradleyireland.com/Earth Images, *skunk* Renee Lynn, *snake* Jim Merli/Visuals Unlimited, *pelican* Lynn M. Stone/Animals Animals; 158 Kim Exterberg; 165 *left* Fian Arroyo, *right* La Finca de Mariposa, Frente Club entre Los Reyes, Costa Rica; 168 *top right* PictureQuest, *bottom left* Richard K. LaVal, *bottom right* Gail Shumway/Getty Images; *background* bradleyireland.com; 169 *bottom left* Leroy Simon/Visuals Unlimited, *bottom right* Roy Morsch/Corbis; 171 Bob Mun; 172 Patricia A. Eynon; 177 *top left* Richardo Figueroa/AP Wide World Photos, *top right* School Division, Houghton Hough Gerardo Somoza/Corbis Outline, *bottom right* Robert Frerck/Odyssey Productions, *bottom left* Tom & Therisa Stack; Division, Frerck/Odyssey Productions, *center left* Joe Viesti/The Viesti Collection, *center* School Division, Houghton Mifflin Co.; 184 Tom Tom & Therisa Stack; 179 *top left* Andre Jenny/Stock South/Atlanta/PictureQuest, *center left* Suzanne ffi Trelles, *center right* Tom Bean/DRK Photo; 180-181 Tom & Therisa Stack; 181 *inset* School Division, left School Division, Houghton Mifflin Co., *others* Tom & Therisa Stack; 182-183 *background* School top left & center left* Tom & Therisa Stack, *top right & bottom left* School Division, Houghton Mifflin ol Division, Houghton Mifflin Co.; 188 Tom & Therisa Stack; 190 Tom Bean/Corbis; 191 Bill

R9

editos

Associates, Inc., *bottom left* Barry Barker/Odyssey Productions; **366** Marc Liberman/The Salk Institute; **367** *top* Courtesy of Ricardo Legorreta, *bottom* Courtesy of Camino Real, Cancun; **372-373** Erica Lananan/Black Star/PictureQuest; **374, 376** Michael Newman/PhotoEdit; **377** School Division, Houghton Mifflin Co.; **380** *center* Photofest; **384** credit unavailable; **386** Emece Editores S.A. Argentina; **390** Photofest; **391** *top left* Photofest, *top center* Robin Holland/Corbis Outline, *center left* Jane Brown/Camera Press/Retna, Ltd. USA, *center right* Carrion/Corbis Sygma; **394** *bottom right* The Granger Collection, New York; **398** *bottom* PhotoDisc, Inc.; **399** *top left* Courtesy, Maloka Park, Columbia; *center* Bettmann/Corbis; *bottom left* Robert Frerck/Odyssey Productions; *bottom right* Courtesy, Colección Museo Armando Reverón; **400** *top right* Courtesy of Maloka Park, Colombia, *center left* Myrleen Ferguson/PhotoEdit; **401** *top right* Eduardo Gil/Black Star; **402-403** Eduardo Gil/Black Star; **404** *top left* Wolfgang Kaehler; *center* Owen Franken/Stock Boston; *bottom* Getty Images; *top right* Bud Gray/Motion Picture & Television Archive; **405** *top left* Photofest; *top center* Getty Images; *top right* © 1993 Warner Bros. Inc/Photofest; *center left* Guido A. Rossi/Getty Images; *center* © 1994 Warner Bros. Animation/Motion Picture & Television Archive; *center right* Nubar Alexanian/Stock Boston; *bottom* David Young-Wolff/PhotoEdit; **406** *From left to right*: Guido A. Rossi/Getty Images; Getty Images; Wolfgang Kaehler; Photofest; **407** School Division, Houghton Mifflin Company; **409** Al Bello/Allsport/Getty Images; **416** José Peláez Photography; **417** Randy Green/Getty Images; **418** Courtesy Venevision; **422** *left* Merleen Ferguson/PhotoEdit; **424-425** RMIP/Richard Haynes; **426** *top left, center left* School Division, Houghton Mifflin Company; *bottom left, top right, center right, bottom right* PhotoDisc, Inc.; **427** *top left, bottom left to bottom right* PhotoDisc, Inc.; *top center, top right, center left to center right* School Division, Houghton Mifflin Company; **428** *clockwise from top left:* School Division, Houghton Mifflin Company; PhotoDisc, Inc.; Spencer Jones; School Division, Houghton Mifflin Company; **429** School Division, Houghton Mifflin Company; **430** *top left, top center left, center right, bottom right* PhotoDisc, Inc.; *top right, bottom left* School Division, Houghton Mifflin Company; **435** Jeff Greenberg/Index Stock; **439** PhotoDisc, Inc.; **440-441** *background* Telegraph Color Library/Getty Images; **440** *bottom left* Bob Daemmrich/The Image Works; **441** *bottom left* Mark E. Gibson/Visuals Unlimited; *top center* Degas-Parra/Ask Images/The Viesti Collection, Inc.; **444** *all* PhotoDisc, Inc. *except top right*; **447** Courtesy, Monkey Online; **448** Tony Anderson/Getty Images; **449** Mark M. Lawrence/Corbis; **451** School Division, Houghton Mifflin Company; **453** Jeff Greenberg/Index Stock; **455** *answering machine, white telephone, portable personal stereo* School Division, Houghton Mifflin Company; *all others* PhotoDisc, Inc. ; **462** Francoise de Mulder/Corbis; **463** UPI/Bettman/Corbis; **464-465** *background* Vera Lentz/Black Star/PictureQuest; **464** *bottom right* Editorial Canas S.A. Bolivia; **465** *left* HighINFO, creators of Bolivianet; **466** School Division, Houghton Mifflin Company; **468** *top left, top center* PhotoDisc, Inc.; *top right* Mark M. Lawrence/Corbis; *bottom right* Jeff Greenberg/Index Stock; **470** David Young-Wolff/PhotoEdit.

All other photography by Martha Granger/EDGE Productions

Illustration Credits

5, 13, 17 Donna Ruff; **21** Nikki Middendorf; **36** Donna Ruff; **47** Bryan Leister; **48** *top* Kent A. Barton; **48-49** *bottom spread* Rubén de Anda; **50** Rick Powell; **53** Fian Arroyo; **69** Mike Deitz; **72** Rick Powell; **78-79** Fian Arroyo; **82** Jim Nuttle; **83** Matthew Pippin; **85** Fian Arroyo; **91** Randy Verougstraete; **92** *top* Kent A. Barton, *bottom* Neverne Covington; **93** Neverne Covington; **99, 112** Jim Nuttle; **121** Fian Arroyo; **135** Randy Verougstraete; **143** Frian Arroyo; **154, 156, 157** Gary Antonetti; **166** *top* Kent A. Barton, **166** *bottom*, **167** Catherine Leary; **173** Fian Arroyo; **186, 187, 194** Donna Ruff; **195** Nikki Middendorf; **196** *top* Kent A. Barton; **196-197** Fabricio Vanden Broeck; **201** Mike Deitz; **204** Mike Reed; **209** Rick Powell; **216** Susan Blubaugh; **217** Mike Deitz; **223** Fian Arroyo; **228-229** Neverne Covington; **235** Susan Blubaugh; **238** Neverne Covington; **239** Susan Blubaugh; **240** Kent A. Barton; **240-241** Enrique O. Sánchez; **247** Jim Nuttle; **255** Fian Arroyo; **265** Matthew Pippin; **269** Susan Blubaugh; **275** Fian Arroyo; **291** *left two columns* Fian Arroyo, *bottom right* Mike Reed; **297** Fian Arroyo; **304** Jim Nuttle; **307** Rick Powell; **313** Catherine Leary; **314** *top* Kent A. Barton; **314-315** Mike Reagan; **319** Stacey Schuett; **321** Catherine Leary; **337, 338** Stacey Schuett; **343** Fian Arroyo; **357** Patrick O'Brien; **365, 387** Fian Arroyo; **388** *top* Kent A. Barton, *bottom* Fundacion Federico García Lorca; **395** Fian Arroyo; **408** Patrick O'Brien; **417** Neverne Covington, Inc.; **421** Rick Powell; **423** Catherine Leary; **434** Stacey Schuett; **439, 442** Nikki Middendor; **443** Bryan Leister; **449** Fian Arroyo; **452** Patrick O'Brien; **458, 469** Neverne Covington, Inc.